Beam Effects, Surface Topography, and Depth Profiling in Surface Analysis

METHODS OF SURFACE CHARACTERIZATION

Volume 1 VIBRATIONAL SPECTROSCOPIES OF MOLECULES ON SURFACES
Edited by John T. Yates, Jr., and Theodore E. Madey

Volume 2 ION SPECTROSCOPIES FOR SURFACE ANALYSIS
Edited by A. W. Czanderna and David M. Hercules

Volume 3 SURFACE INFRARED AND RAMAN SPECTROSCOPY
Methods and Applications
W. Suëtaka with the assistance of John T. Yates, Jr.

Volume 4 SPECIMEN HANDLING, PREPARATION, AND TREATMENTS IN SURFACE CHARACTERIZATION
Edited by Alvin W. Czanderna, Cedric J. Powell, and Theodore E.Madey

Volume 5 BEAM EFFECTS, SURFACE TOPOGRAPHY, AND DEPTH PROFILING IN SURFACE ANALYSIS
Edited by Alvin W. Czanderna, Theodore E. Madey, and Cedric J. Powell

---

Beam Effects, Surface Topography, and Depth Profiling in Surface Analysis

Edited by

Alvin W. Czanderna
National Renewable Energy Laboratory
Golden, Colorado

Theodore E. Madey
Rutgers, The State University of New Jersey
Piscataway, New Jersey

and

Cedric J. Powell
National Institute of Standards and Technology
Gaithersburg, Maryland

PLENUM PRESS • NEW YORK AND LONDON

Library of Congress Cataloging-in-Publication Data

Beam effects, surface topography, and depth profiling in surface analysis / edited by Alvin W. Czanderna, Theodore E. Madey and Cedric J. Powell.
p. cm. -- (Methods of surface characterization ; v. 5)
Includes bibliographical references and index.
ISBN 0-306-45896-9
1. Surfaces (Technology)--Analysis. 2. Materials--Effect of radiation on. I. Czanderna, Alvin Warren, 1930- . II. Madey, Theodore E. III. Powell, C. J. (Cedric John) IV. Series.
TA418.7.B43 1998
620'.44--dc21 98-41250
CIP

ISBN 0-306-45896-9

A Division of Plenum Publishing Corporation
233 Spring Street, New York, N.Y. 10013

http://www.plenum.com

10 9 8 7 6 5 4 3 2 1

Printed in the United States of America

Contributors

Peter R. Boudewijn, Philips Research Laboratories, 5656 AA Eindoven, The Netherlands
Alvin W. Czanderna, National Renewable Energy Laboratory, Golden, CO 80401
Linda S. Dake, National Renewable Energy Laboratory, Golden, CO 80401
John A. Dagata, National Institute of Standards and Technology, Gaithersburg, MD 20899
Andrew S. D'Souza, Pennsylvania State University, University Park, PA 16802
Kiyoshi Iizuka, Nikon Corporation, Tsukuba, Ibaraki, 300-26, Japan
David E. King, National Renewable Energy Laboratory, Golden, CO 80401
Carlo G. Pantano, Pennsylvania State University, University Park, PA 16802
J. Roland Pitts, National Renewable Energy Laboratory, Golden, CO 80401
Alan M. Then, Pennsylvania State University, University Park, PA 16802
John H. Thomas III, 3-M Corporate Research Laboratory, 3-M Co., St. Paul, MN 55144
Theodore V. Vorburger, National Institute of Standards and Technology, Gaithersburg, MD 20899
Helmut W. Werner, Philips Research Laboratories, 5656 AA, Einhoven, The Netherlands and Technical University, Vienna Austria
Günter Wilkening, Physikalisch-Technische Bundestalt, 38023 Braunschweig, Germany

Foreword

Many books are available that detail the basic principles of the different methods of surface characterization. On the other hand, the scientific literature provides a resource of how individual pieces of research are conducted by particular laboratories. Between these two extremes the literature is thin but it is here that the present volume comfortably sits. Both the newcomer and the more mature scientist will find in these chapters a wealth of detail as well as advice and general guidance of the principal phenomena relevant to the study of real samples.

In the analysis of samples, practical analysts have fairly simple models of how everything works. Superimposed on this ideal world is an understanding of how the parameters of the measurement method, the instrumentation, and the characteristics of the sample distort this ideal world into something less precise, less controlled, and less understood. The guidance given in these chapters allows the scientist to understand how to obtain the most precise and understood measurements that are currently possible and, where there are inevitable problems, to have clear guidance as the extent of the problem and its likely behavior.

In many instances, when first considering a sample we wish to know something of the surface topography. Chapter 4 provides a full review of the possibilities from the finest level used in metrological laboratories to the practical requirements of engineers. Modern, functional scanned probes are discussed together with their limitations. Traceability and calibration are an important interest of the authors and here clear, practical advice is given. More traditional and optical methods are also covered.

Following a study of the physical characterization of the surface, the surface may be analyzed by Auger electron spectroscopy (AES), X-ray photoelectron spectroscopy (XPS), or secondary ion mass spectrometry (SIMS). Both AES and XPS may damage the surface being studied. AES is more damaging than XPS but since XPS is often used to study more delicate materials, the problems require understanding for both techniques. Chapters 1 and 2 cover these aspects clearly and in depth, with many excellent examples for different classes of materials. Electron stimulated adsorption and desorption are covered for AES and in sufficient detail to alert users of other techniques, where an electron beam is added for charge neutralization or stabilization, to the importance of the problem.

The greater part of surface analysis deals with composition-depth profile measurements. Either by using an addition ion gun or by using the probe ion gun, the surface is removed layer by layer by sputtering while surface analysis measurements are conducted. This provides the composition-depth profile. Unfortunately, the measured profile is rarely a perfect representation of the original sample. As the sputtering proceeds, compositional changes occur, elements are redistributed and recombined, and the surface topography changes. By understanding these phenomena the correct instrumental parameters may be chosen to minimize these unwanted effects. Thus, today, we see an increasing amount of work showing depth resolutions at the level of better than 1% of the depth sputtered for depths in the range 100 nm to 1000 nm. The sputtering condition has many degrees of freedom, from ion species and energy to current density and angle of incidence. Varying these parameters leads to many interesting effects detailed in Chapters 3 and 5 with many examples and much good advice on what should and should not be done to obtain meaningful results. A major problem for many practical analysts is the development of surface topography during sputtering. An understanding of this problem has been greatly assisted by the development of scanned probes already discussed in Chapter 4.

In all chapters, the reader will find a clear exposition of the phenomena needed to understand the material response with equations kept to a minimum. All chapters have a very full set of figures so that the importance and behavior of the item being discussed can be properly appreciated. As is well known, a picture is worth ten thousand words[1] but a diagram may be assimilated in the time taken to read only twenty words. For the reader to obtain more details, very extensive bibliographies are provided. For many of the phenomena, there is a broad understanding of what is going on but there is not always sufficient knowledge of the required parameters to provide simple rules or predictive behavior. In these cases, one must resort to descriptions of observed behavior, classified into whatever groupings are possible. Where this is appropriate, extensive examples of behavior are presented so that the analyst can appreciate what has happened or is likely to happen even if he or she cannot obtain the ideal result!

(1) F R Barnard, Printer's Ink, 10 March 1927

Martin P. Seah
National Physical Laboratory
Teddington, United Kingdom

About the Series

Many techniques are now being used to characterize the chemical and physical properties of surfaces with an emphasis on surface composition, and improvements in these techniques are reported each year. While most of the techniques are generally relatively simple in concept, their successful application involves the use of complex instrumentation, adequate understanding of the physical principles, avoidance of many problems, identification of possible artifacts, and careful analysis of the data.

We have designed this series to assist newcomers to the field of surface characterization although we hope that the series will also be of value to more experienced workers. The approach is pedagogical or tutorial. Our main objective is to describe the principles, techniques, and methods that are considered important for surface characterization, with emphasis on how important surface characterization measurements are made and how to ensure that the measurements and interpretations are satisfactory to the greatest extent possible. At this time, we have planned six volumes, but others may follow.

The first volume provides a description of methods for vibrational spectroscopy of molecules on surfaces. Most of these techniques are still under active development; commercial instrumentation is not yet available for some techniques, but this situation could change in the next few years. The current state-of-the-art of each technique is described as are the relative capabilities. An important component of the first volume is a summary of the relevant theory.

The second volume contains descriptions of ion spectroscopies for surface analysis and is the first of two volumes on the techniques and methods of electron and ion spectroscopies that are in widespread use for surface analysis. These two volumes address techniques for which commercial instrumentation is available. The books are intended to fill the gap between a manufacturer's handbook and review articles that highlight the latest scientific developments.

The third volume is concerned with infrared (IR) and Raman spectroscopies that provide detailed molecular-level information about the species present on surfaces in ultrahigh vacuum as well as in various absorbing media. This book is written for those who are not spectroscopists but who are beginning to use IR or Raman spectroscopy to investigate surfaces of practical materials.

The fourth volume is about techniques for specimen handling, preparation, and treatments in surface analysis and other surface studies. It provides a compilation of methods that have proven useful for those engaged in surface science, applied surface science, thin film deposition, surface analysis, and related areas.

This fifth volume provides descriptions of common artifacts and problems associated with the bombardment of solid surfaces by photons, electrons, and ions. Descriptions are also given of methods for characterizing the surface topography and of sputter depth profiling for measuring the near-surface composition of a solid.

Surface-characterization measurements are being used increasingly in diverse areas of science and technology. We are confident that this series will be useful in ensuring that these measurements can be made as efficiently and reliably as possible. Comments on the series are welcomed, as are suggestions for volumes on additional topics.

C.J. Powell
Gaithersburg, Maryland
A.W. Czanderna
Golden, Colorado
D.M. Hercules
Nashville, Tennessee
T.E. Madey
Piscataway, New Jersey
J.T. Yates, Jr.
Pittsburgh, Pennsylvania

Preface

The determination of surface composition is essential in many areas of science and technology and many approaches may be used for this purpose. Each method has particular strengths and limitations that often are directly connected to the physical processes involved. Typically, atoms on the surface and in the near-surface region may be excited by photons, electrons, ions, or neutral species and the detected particles are usually ions, electrons, or photons that have been emitted, ejected, or scattered. A major part of any surface analysis involves using photon, electron, and ion beams that can damage the sample being studied but the resulting artifacts are rarely discussed in detail in the literature. The purpose of this book is to present a discussion of the damage and artifacts resulting from the beams used in surface compositional analysis or for sputter depth profiling. Characterization of the resulting surface topography is usually necessary for assessing the extent of beam damage, but is also important for reasons given below.

Each of the five chapters in this volume contains extensive references to prior work and sufficient figures to illustrate the concepts being described. Tables are used to summarize information about a concept, technique, method, or process.

The first chapter deals with photon beam damage in the surface and near-surface of solids, and especially the damage from X-rays used in X-ray photoelectron spectroscopy. Electrostatic charging of samples in photoemission experiments, calibration of the binding energy scale, use of the Auger parameter, and X-ray absorption and damage are emphasized.

In the second chapter, the fundamentals of electronic-excitation processes including electron-beam interactions with solid surfaces, electron stimulated desorption or adsorption, decomposition or oxidation of surface layers, and the thresholds for electron-beam damage are discussed. Charging and electromigration in insulators, electron-beam-induced heating, and electron-beam effects during Auger electron spectroscopy surface analysis of several classes of solids are also considered.

The third chapter is concerned with ion-beam-bombardment effects of solid surfaces at energies used for sputter depth profiling. The topics discussed include the fundamentals of ion-beam/solid interactions such as penetration, implantation, atomic mixing, and sputtering; structural changes that range from bond stretching to amorphizing the solid and the effect of the changes on the properties of the solid; ion-beam-induced topography from microscopic to macroscopic dimensions; compositional changes from differential sputtering and other chemical effects in organic, alloy, semiconductor, metal oxide and other compound materials; and the effects of the sample, ion beam, and instrument used on depth resolution.

The fourth chapter is an overview of profiling methods used for the characterization of surface topography. The results obtainable with profiling instruments, a description of stylus-type instruments, methods for optical profiling techniques, types of scanned probe microscopy instruments, applications of the instruments, and intercomparisons among them are presented. A major focus in this chapter is on scanned probe microscopy (SPM) and includes discussions of instrument calibration, characterization of displacement-type instruments, force-based SPM, methods based on probe-sample interactions, and applications of SPM in industries ranging from data storage to electrochemical science.

The fifth chapter provides an overview of sputter depth profiling (SDP) for near-surface compositional analysis. Sputter depth profiling is compared with other destructive and non-destructive methods of depth profiling. The physical basis of sputtering, data for sputtering yields, determination of information depth, and experimental aspects of SDP are presented and followed by a discussion of the analysis of SDP, artifacts, and limitations on the obtainable depth resolution. Finally, application of SDP to a variety of materials and thin-film multilayers and the interpretation of the results obtained are discussed.

The editors are deeply grateful to the authors whose work made this book possible, and for taking the time from their active research programs to prepare their contributions.

Alvin W. Czanderna
Golden, Colorado
Theodore E. Madey
Piscataway, New Jersey
Cedric J. Powell
Gaithersburg, Maryland

Contents

1

Photon Beam Damage and Charging at Solid Surfaces

John H. Thomas III

1. Introduction

Particle-excited surface spectroscopies such as Rutherford backscattering (RBS), secondary-ion mass spectrometry (SIMS), and Auger electron spectroscopy (AES) produce data that may be strongly affected by the interaction of a particle (ion, electron, etc.) with the solid surface of the sample. As described in other chapters in this book, some of these effects can be large and totally obscure the information. Others can be handled with proper mathematical algorithms or instrumentally.[1] Photon-excited spectroscopies [X-ray photoelectron spectroscopy (XPS) and ultraviolet photoelectron spectroscopy (UPS)] were naturally assumed to be the least "damaging" or perturbing of the available surface methods to the spectroscopist.[2–7] This assumption is mostly true in applications to inorganic material systems. However, when X-ray excited spectroscopy is applied to organics, and to some inorganic polymeric materials, photon absorption can result in materials modification and, in some cases, decomposition. X-ray- or ultraviolet-induced changes are likely to be more common than the literature leads one to believe. Photons used in the production of photoelectrons in insulators or semiconductors cause other effects, such as surface charging and differential charging.[3,8,9] These phenomena can ultimately cause considerable difficulty in data interpretation. This chapter is concerned with these problems, namely how they affect data acquisition, and which

---

John H. Thomas III • 3M Corporate Research Laboratory, St. Paul, MN 55144.

Beam Effects, Surface Topography, and Depth Profiling in Surface Analysis, edited by Czanderna *et al.* Plenum Press, New York, 1998

methods reduce their effect. Because photons are generally encountered only in XPS, UPS, and synchrotron-excited photoemission, this section is oriented toward understanding problems that can occur in the surface or near-surface regions.

2. Electrostatic Charging of Samples in Photoemission Experiments

2.1. Electrostatic Surface Charging

Figure 1 is a block diagram of a typical photoemission experiment. Soft X-ray photons of energy $h\nu$ irradiate a material surface, with fluxes ranging from 10^{12} to 10^{14} photons/cm^2-s, and produce a flux of photoemitted electrons that, given enough energy, will leave the material surface. Photoelectrons, emitted into the analysis vacuum chamber, are ultimately collected and their kinetic energy analyzed to determine the chemical elements of a surface of unknown composition. Most photoemission experiments are concerned with accurately measuring the kinetic energy of the emitted photoelectrons that escape without energy loss. Kinetic energy, E_K, is commonly measured to an accuracy of 0.1 eV in 1000 eV or 1 part in 10^4 and is converted to a relative electron binding energy, E_B, by using accurate elemental standards of known binding energy to determine the energy scale. Modern methods also include using a field emission electron source powered from an accurate voltage source that can be used to flood the entrance slit of the electron spectrometer. The electron energy is accurately adjusted and used to adjust the spectrometer power supplies for calibration purposes. Binding energy is referenced to the Fermi level of a solid conductive material and is corrected for the spectrometer

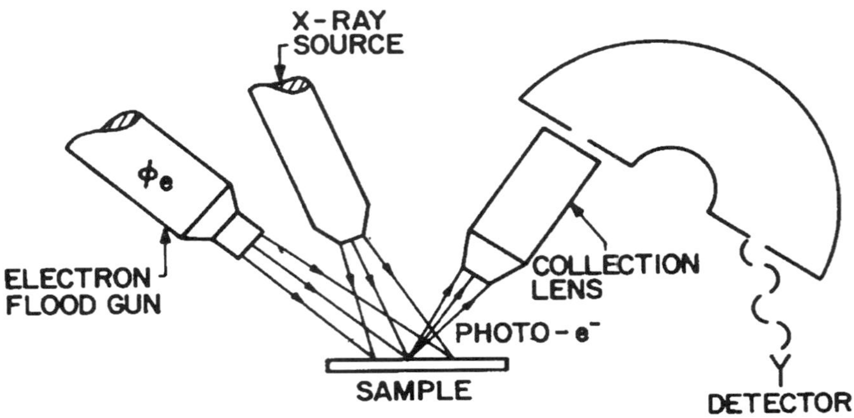

Figure 1. Schematic of XPS–UPS apparatus showing the flood gun and X-ray source as typically found in modern instrumentation.

contact potential (the difference in work function of the sample and spectrometer detector material). Numerous methods to mechanize binding-energy corrections have been developed and described in detail.[(6,9,10)] An understanding of the effects of photon irradiation on core-level and valence-level binding energies is very important in the interpretation of photoemission spectra, especially of poorly conducting materials.

When photons create a photoelectron current from an insulating surface, a net positive charge is left behind. The positive charge accumulates because the rate of electron current flow into the insulator is insufficient to compensate the positive charge. Hnatowich *et al.*[(11)] were the first to investigate this phenomenon systematically as to how it relates to photoelectron emission spectral interpretation.

Although it is not our intent to present a detailed account of the theory of photoelectron production, a brief review is necessary to appreciate how the surface potential depends on this physical process. In general, the photoelectron current, I, can be represented by[(1,4, 12–15)]

$$I_A = n_A \, \sigma_i(A) \, F_i(\theta) \, \lambda_i \, M\vartheta \tag{1}$$

where n_A is the number of A atoms, i is the core-level designation (e.g., $2p$), $\sigma_i(A)$ is the photoionization cross section of the i core level, $F_i(\theta)$ is the angular dependence of the photoelectron yield, λ_i is the electron escape depth, M is the instrumental property correction, and ϑ is the X-ray photon flux.

Because all the parameters affect the photoelectron current produced from an insulating (or any) surface, the magnitude of the surface charging potential, ϕ_s, under a given set of X-ray irradiation conditions depends on the physics of photoelectron excitation in and from the sample surface. In addition to these parameters, the surface and bulk conductivity affect ϕ_s. This effect is readily demonstrated by comparing the photoelectron spectrum from a conductive gold foil with one evaporated onto an insulator surface such as glass, as shown in Fig. 2. In the spectrum, the grounded conductive gold foil $4f_{7/2}$ peak is observed at 84.0 eV.[(16)] Under identical irradiation conditions, the $4f_{7/2}$ peak of the insulated gold film is shifted to a lower kinetic energy by ~1.7 eV. The shift in kinetic (or binding) energy indicates the average value of ϕ_s. The value of ϕ_s varies in magnitude with the total X-ray flux which generates the photoelectron current.

Surface charge buildup and subsequent peaks shift are larger when a monochromated X-ray source of photons is employed, such as those used in many modern analytical systems. Ebel and Ebel[(17)] showed that when a standard nonmonochromated X-ray source is used, this shift is partially compensated. Most nonmonochromated X-ray sources use an aluminum window to contain high-energy backscattered electrons generated at the anode and to act as an X-ray photon low-pass filter. A typical source is shown in Fig. 3. Varying the X-ray tube parameters (operating voltage, current, and window material) separates the effect

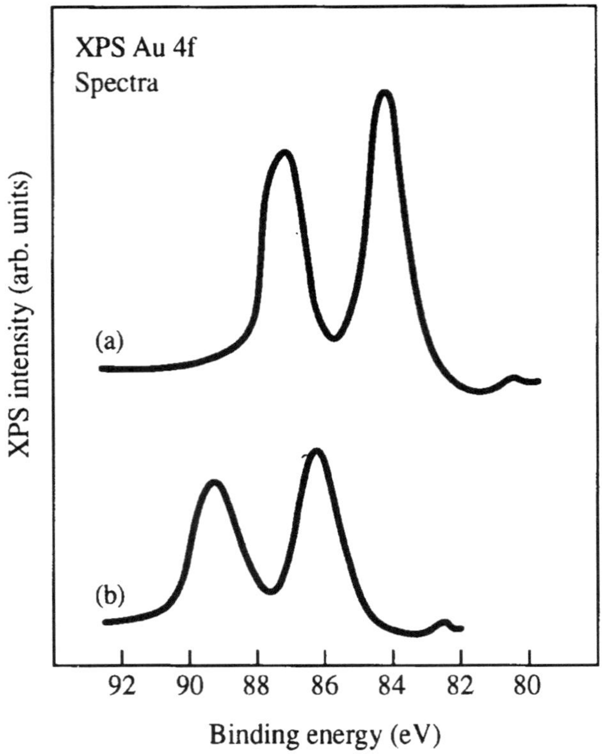

Figure 2. XPS Au 4*f* spectra for (a) conductive Au (grounded) and (b) an isolated Au film on a glass substrate (not grounded) (adapted from Ref. 16).

of the window on surface charge compensation from other operating parameters. Backscattered electrons from the X-ray tube anode excite low-energy secondary electrons from the window material (aluminum), and these electrons can flood the emitting sample surface. In addition, X-ray absorption in the window material also produces secondary electrons and primary fluorescent X-rays from the window. Figure 4 shows a typical spectrum of secondary electrons emitted from an aluminum window.[17] Ebel and Ebel[17] also demonstrated that the secondary electron emission current is governed by the X-ray tube current and not the operating voltage. By applying Kirchhoff's current law to the irradiated sample surface, one can readily determine ϕ_s if all currents are accurately known. Refer to Fig. 5. The current sum at the node is

$$\phi_s/R_S + \phi_s/R_B + i_{\text{ph}} + i_{\text{sec}} = 0 \tag{2}$$

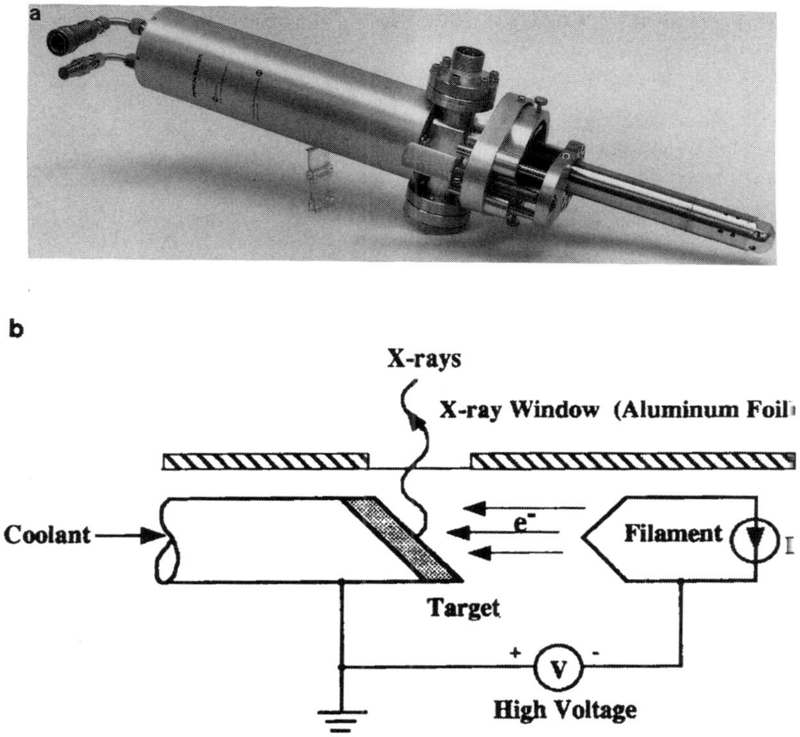

Figure 3. (a) Typical X-ray source and (b) diagram (courtesy of Physical Electronics Inc.). The diagram is typical of a nonmonochromatic X-ray source and shows the aluminum window, filament, and anode target of the system.

where R_S and R_B are the surface and bulk resistances of the irradiated sample, i_{ph} is the total photoelectron current leaving the surface, and i_{sec} is the total incident secondary-electron current from all sources, including the X-ray tube window area. The surface current, ϕ_s/R_B, varies linearly with X-ray flux, as expected, and is material dependent. If the current leaving the sample surface is large enough, the surface charges positively; if the electron current from the X-ray tube window or a separate source of electrons such as a flood gun (low energy) is large enough, the surface can charge negatively. This produces peaks at a higher kinetic energy than expected. Thus, the photoelectron peak energy depends on the surface potential, ϕ_s, and the correct peak energy is obtained when ϕ_s is zero.

Generally, the presence of the X-ray tube window and its low-energy secondary-electron emission is advantageous. When monochromatic X-rays are employed,

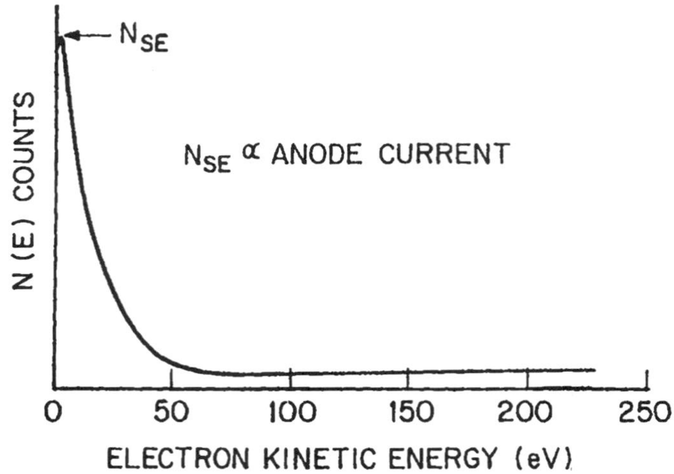

Figure 4. Kinetic energy spectrum of secondary electrons emitted from a nonmonochromated X-ray source with an aluminum window (adapted from Ref. 17).

this form of surface charge compensation of insulating surfaces does not occur, and in cases of high-resistance insulators (e.g., Teflon) the surface potential can be ~100 V or more, and a spectrum cannot be obtained without using an external source of low-energy electrons. This situation was typical of the Hewlett–Packard 5950 series of instruments where a well-designed electron flood source was used to compensate for surface charging. Positive surface charge can be compensated by using a low-energy (typically <10 eV) electron source as shown in Fig. 6. A fundamental

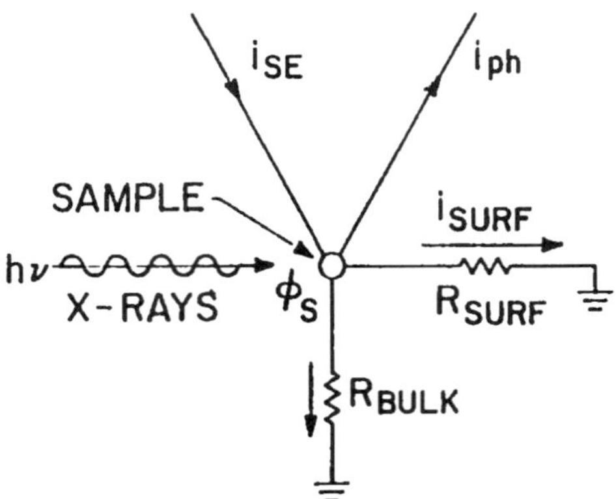

Figure 5. Equivalent circuit of an insulating sample surface showing surface resistance, bulk resistance, exterior electron sources, and the photoexcited electrons.

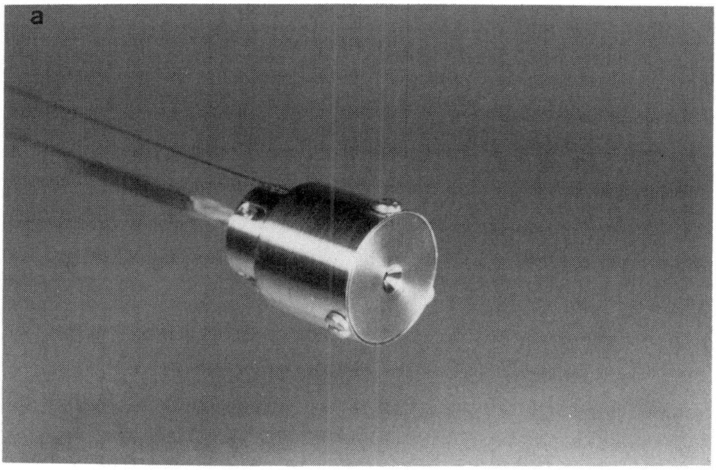

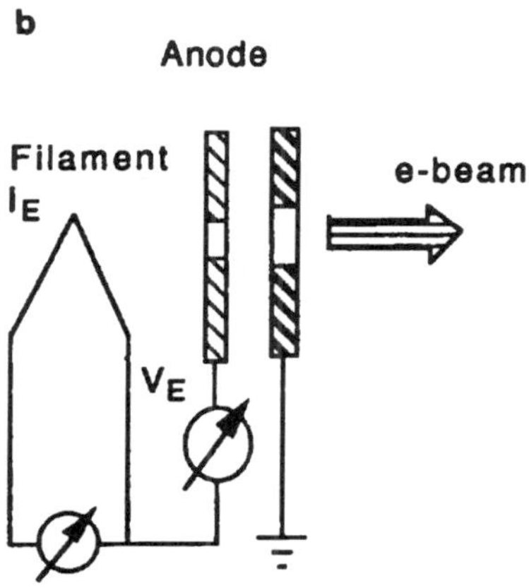

Figure 6. (a) Electron flood gun used in charge neutralization (courtesy of Kimball Physics, Inc.). (b) The source consists of a filament and anode (a diode system). The apertures are designed to provide good electron optics to the sample surface.

requirement is that the electron flux from the gun be laterally uniform and cover the area of X-ray irradiation. It is necessary to use spectral references to adjust the energy scale, which is typically the gold $4f_{7/2}$ peak at 84.0 eV binding energy.

The uniformity of the neutralization can be improved by placing a grounded window over the insulating surface. Generally, there is some lateral conductivity on the surface, and this mechanical fixturing can help reduce charging with the use of the flood gun. The window should be approximately the same size as the image of the entrance slit of the electron monochromator and lens system. The HP 5950 series instruments employed this system. Charging in Auger electron spectroscopy can be reduced with this type of fixturing.[1] Refer to Chapter 2 of this volume for further information.

Electron damage of materials is covered in detail in Chapter 2, so it is not discussed in detail here. However, when using an electron flood gun on sensitive materials such as organics, some polymers, and some inorganic compounds such as SiO_2, the investigator should be aware of the possible problems that may be encountered. Most materials survive low-energy electron bombardment intact. There are always exceptions. Some hydrocarbons and molecular layers can dissociate via dissociative attachment (negative ion intermediate) with thresholds of <10 eV.

2.2. Differential Surface Charging

Differential surface charging is a special case of simple surface charging where the surface potential varies laterally and can be a source of spectral misinterpretation[18] when studying, for example, powders or heterogeneous catalysts on insulating or conducting surfaces. When XPS is used to study functional groups on insulating surfaces, a multipeaked spectral response may be due to differential surface charging rather than to the presence of different chemical states. Differential surface charging occurs when several or many regions or phases within the analysis area are charged at a different surface potential as a result of being spatially isolated (they are not electrically continuous; that is, they have different local values of R_S and R_B). Differential charging appears as a "chemically shifted" peak or peaks, or peak broadening in the core-level spectrum and may be interpreted in this way unless the interpretation reduces to an absurdity for the particular problem. Careful study of the spectra as a function of X-ray source operating conditions or as a function of electron flood gun conditions will generally allow a distinction to be made between chemically shifted peaks and differential charging peaks. Chemically shifted peaks will remain at a constant separation from the unshifted peak position, but the charge peaks will, under proper conditions, merge with the unshifted peaks. It is also possible to minimize the full width at half-maximum (FWHM) of the peaks as a function of flood gun conditions.

Early in the history of the application of photoemission spectroscopy to surface analysis, studies were performed to demonstrate and understand differential charging. Dickinson *et al.*[19] describe an experiment using the ionic conductivity change of Ag_2I powder as a function of temperature to demonstrate differential charging. The Ag_2I powder was pressed into a conductive mesh screen made of silver and was mounted on a hot–cold stage in the analysis chamber. The Ag 3*d* XPS spectrum was studied at a temperature low enough (–50°C) that Ag_2I was in an insulating state (low ionic conductivity). A similar measurement was performed at 100°C where the ionic conductivity is high. At the low temperature, the Ag $3d_{5/2}$ peak is at 368.5 eV, as shown in Fig. 7. At 100°C, the peak shifts to 367.4 eV. At room temperature, the Ag $3d_{5/2}$ spectrum consists of two components, one at 368.5 eV and the other at 367.4 eV. At –50°C, the powder is insulated from the silver mesh and a surface charge is accumulated on the particle surfaces not in direct contact with the mesh. This situation produced a surface potential, ϕ_s, of 1.1 V. Conversely, at 100°C, the powder is a good ionic conductor, and it makes excellent electrical contact with the wire mesh, where $\phi_s = 0$ V. At room temperature, the nonzero ionic conductivity allows the powder in a region near the mesh to make electrical contact, while the powder in the open region of the mesh are insulated. Hence, two Ag $3d_{5/2}$ peaks are observed. Since the surface potential varies from particle to particle, the charging has been termed differential charging. Although Dickinson *et al.*[(19)] varied other parameters in their experiment, these data clearly demonstrate the phenomenological aspects of differential surface charging.

An example of differential surface charging was also presented by Brinen[(18)] in a study of surface catalysts. In this case, an electron flood gun was used to

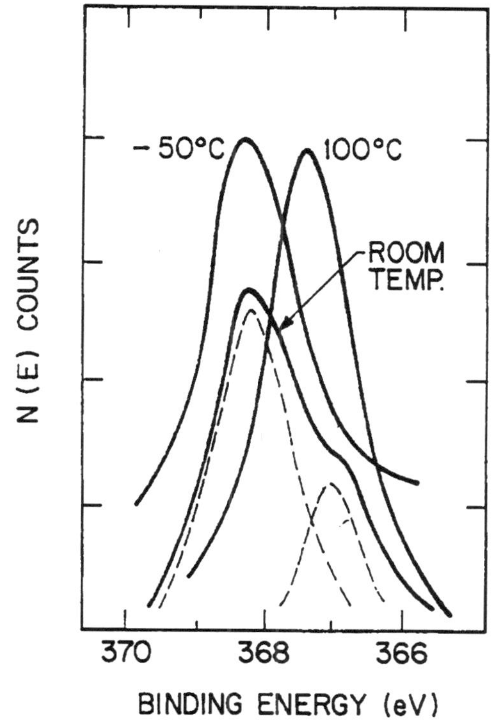

Figure 7. XPS Ag $3d_{5/2}$ spectrum for silver iodide at –50°C, room temperature, and 100°C (adapted from Ref. 18).

demonstrate the nature of the charging and to compensate for it. Brinen also observed that the charge-shifted peak was broadened as a result of the variation in the surface potential from particle to particle on the catalyst surface.

Like surface charging, differential charging is more prominent when monochromatic X-ray sources are employed. Careful investigation of the effect of the electron flood gun parameters (beam potential and current) on the spectral peaks in question generally resolves the peaks resulting from differential charging and acceptably compensates the surface. If the sample consists of a heterogeneous mixture of insulating particles containing common elements such as SiO_2 in Al_2SiO_5, differential charging would be expected. The magnitude of the charge would depend on the parameters that govern the physical process of electron photoemission, particle size, surface contamination, etc.(20,21)

For practical applications of XPS or UPS, differential charging or simple surface charging is no longer a serious problem to the analyst. Many commercial instrument manufacturers include sample charge neutralization systems typically in the form of a low-electron energy flood gun oriented toward the analysis area (Figs. 1 and 6). A typical gun consists of a filament for the electron source and an accelerating electrode that includes a rough focusing aperture. The accelerating voltage and filament current (controlling the emission current) are externally adjustable. With the gun pointed toward the sample, the investigator simply needs to adjust the emission current and voltage conditions to compensate for the observed charging effect.

Adjusting the flood gun for optimal conditions is usually simple. The voltage and current are increased until the peak kinetic energy increases to a stationary value. This condition usually corresponds to a minimum in the peak width. Unneutralized spectra of insulating surfaces tend to produce broad photoelectron peaks that cannot be adequately fitted with conventional functions. In other cases, the spectra may be fitted adequately but may contain peaks that are an artifact of fitting. In apparatus equipped with real-time imaging of the electron flux at the dispersion plane of the electron analyzer, the adjustment of the neutralizing parameters can be conveniently observed in real time.

2.3. Lateral and In-depth Charge Effects

Improved methods have been developed to neutralize surface charging in a uniform fashion. In the 1980s, Surface Science Instruments, Inc. developed a charge neutralization method(22,23) based on placing a grounded wire-mesh screen a few millimeters above the surface being measured with a standard low-energy flood gun. The wire mesh was nickel and had a transmissivity of 90%.(24) Figure 8 shows the C 1*s* FWHM of polypropylene as a function of flood gun voltage,(24) first using only the electron flood gun and then using a screen with the flood gun. Functionally, Bryson(24) describes the effect as redistributing the neutralized surface uniformly

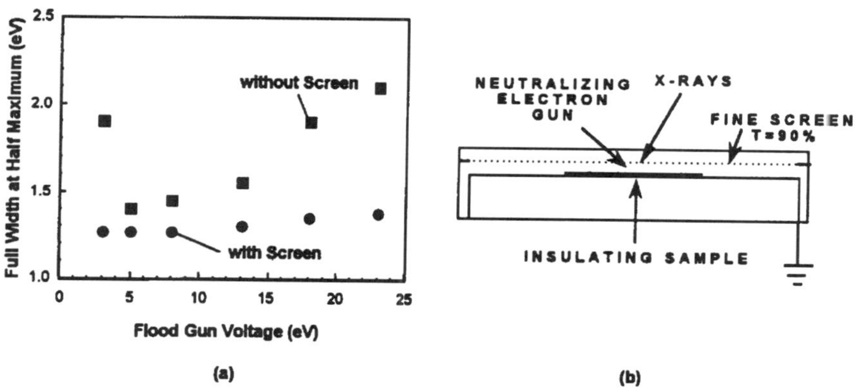

Figure 8. (a) XPS C 1*s* FWHM of polypropylene as a function of flood gun voltage position with and without the use of a conducting screen above the specimen (adapted from Ref. 24).

over the screened area. Thus, the photoelectron flux generated by a monochromated source is uniform and produces a uniform surface charge distribution. It is not clear why this method works, but it appears to do better than a simple flood gun and achieves smaller values of the FWHM.[25] Their results demonstrated that polymeric surfaces could be more accurately characterized by this method.

Another method, developed by Service Physics Inc., employs a "high"-voltage flood gun, that is, one that will operate up to around 200 V and supply a microampere unfocused beam to the sample surface. This gun has been installed in our SSX-100 system. Using polyethyleneterephthalate (PET) as a test sample, and operating the gun at successively higher voltages, the smallest FWHM was found at around 150 V and a low beam current. Under various conditions, charging is minimized when the photoelectron peak is narrowest. Like the low-energy version of electron flood gun charge neutralization, a sharp region (V_E and I_E, the accelerating voltage and emission current, respectively) of operating conditions is experimentally observed. In addition to producing high-resolution spectra of charged surface species, the beam penetrates the surface layer to provide in-depth charge neutralization. Figure 9 shows the PET C 1*s* spectrum, using the normal low-voltage SSX-100 neutralizer, and is compared to the higher-voltage neutralizer. The high-energy gun produces a spectrum in which the C 1*s* components of PET are clearly observable as compared with the poorly resolved spectrum produced by the standard low-energy flood gun system (Fig. 9b). Experience in our laboratory has been that this instrument provides good neutralization and does not damage most common polymer surfaces.

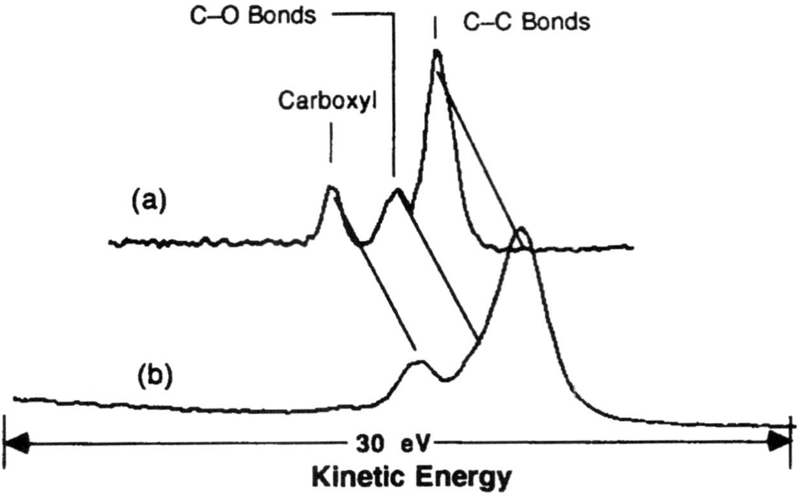

Figure 9. XPS C 1*s* spectrum of polyethylene terephthalate (PET) using (a) a higher-voltage neutralizer (Service/Physics, Inc.) and (b) compared to the normal low-voltage Surface Science Instruments SSX-100 neutralizer. The C–O bonds in PET are not resolved using the normal low-voltage neutralizer.

Samples covered with thin layers should be treated differently because of the possibility of differential charging as a function of depth. In-depth charging can occur, for example, when measuring thin SiO_2 layers on silicon. Lau[26] has shown that the induced field at the Si surface due to photoemission can result in band bending. When the take-off angle is varied from a normal to a grazing angle with the surface, numerical analysis shows that the Si 2*p* peak shifts to higher binding energy and broadens with the sampled depth, assuming that depletion at the surface is complete (that is, band bending is equal to the band gap). The depletion region is a function of the silicon properties (e.g., doping). In another study Lau[26] used a thin oxide film (on aluminum) that has a large potential gradient as a result of photoemission from the surface. By varying the take-off angle, he showed that peak broadening results from the convolution of shifted O 1*s* peaks at various depths; that is, the shift of the O 1*s* peak depends on the sampled depth. In a recent review, Barr[27] discussed a similar phenomenon.

Special procedures are needed to minimize charging effects on samples with thin insulating layers. When the substrate is grounded with the flood gun "on," the spectra appear more distorted than if no flood gun were used or if the sample were electrically isolated from the spectrometer ground.[28] Yu and Hantsche[28] suggest that the effects of vertical differential charging can be minimized by isolating the sample electrically and analyzing it as an insulator.

2.4. Small-Spot Analysis Charging Effects

Small-spot X-ray sources (~50 to 150 μm in diameter) used to reduce the surface analysis area achieve high photon flux at the analysis region. In comparison with flood-type X-ray sources, the flux can be as much as a factor of 1000 times higher for the same irradiated area.[29–31] When small-spot instruments are used to study insulating materials, there can be a local charge gradient in the vicinity of the X-ray spot. This situation results in differential charging effects if the analyzed area (the image of the entrance slits of the electron analyzer on the sample surface) is larger than the X-ray irradiation spot.[29] For optimal efficiency in producing and collecting spectra, the analyzed and X-ray spot-size area should be equal. Nevertheless, the variation in flux distribution can produce a lateral variation in surface charging. An electron flood gun can readily compensate for this effect. It has been found that the higher-energy electron guns work better than standard low-energy flood guns. The low-energy flood gun combined with a screen does, however, provide adequate correction.[22] For thin films, where the electron escape depth is in the range of the film thickness, techniques such as insulating the film, as described earlier, should be tried.[28]

Although this section is oriented toward photoelectron spectroscopies, the same phenomena have been observed in electron-excited AES (Chapter 2). The surface potential buildup occurs in the same fashion as that observed in UPS or XPS and can reach very high voltages, in some cases nearly as high as the primary electron beam potential. When an electron beam of energy E_0 impacts a surface, secondary electrons, primary backscattered electrons, and, of course, Auger electrons are produced. A large fraction of the primary-electron beam is backscattered from the surface at energies equal to or less than E_0. If the total electron flux leaving the surface is larger than the incident flux, charging occurs. The surface potential depends on the electron excitation process, the backscattering factor,[32] and the total secondary-electron emission coefficient,[33] and is highly material dependent. Both simple and differential charging can occur in heterogeneous semiconducting or insulating material systems. Flood gun neutralization has been employed in AES where the surface potential is not so large. The angle of incidence, primary-electron beam energy, and current can be varied along with neutralization parameters to produce adequate spectra in many cases.

3. Energy Scale Calibration

In view of the problems encountered in surface charging, a summary of methods used to correct and calibrate the binding-energy scale is presented. For a more complete discussion, the reader is referred to the literature.[1,9,10, 33–36]

Surface charging of insulating materials causes the photoelectron kinetic energy to be shifted by an amount equal to the surface potential, ϕ_s. Photoelectrons excited from the sample leave with a kinetic energy of roughly $E_K \approx h\nu - E_B$. The measured kinetic energy of a photoelectron peak, $E_M = E_K - (\phi_s + \mathrm{wf}_{\mathrm{sp}})$, differs from that of a conductor by ϕ_s; in this relation, $\mathrm{wf}_{\mathrm{sp}}$ is the spectrometer work function.

As pointed out for commercial instruments not equipped with an electron flood gun, the measured kinetic energy of electrons emitted from insulating substrates is not easily predicted because of the effect of ϕ_s. To calibrate the kinetic energy scale, reference spectra should be obtained at the same surface potential conditions. This has been done by using the C 1*s* peak energy of adventitious carbon contamination from air exposure or residual gases in the analysis chamber. Another possibility is to use a thin film of a conductor with well-known photoelectron transitions such as gold, which floats electrically on the insulator, to measure the surface potential. By measuring the C 1*s* binding energy or the Au $4f_{7/2}$ binding energy as a standard, the energy scale of the unknown can be roughly corrected. Both of these methods have been widely used to obtain "absolute" binding energies, but they suffer from inaccuracies. One of these includes a possible nonlinearity in the power supply;[(37)] that is, a single binding-energy reference cannot be used with confidence. Also, the standard's binding energy varies with surface coverage. Finally, possible interactions of the film with the surface of the insulator may cause perturbations in the reference binding energy or surface potential.

Evans[(9)] presents a good review of the problems encountered in using the various calibration methods. Adventitious carbon contamination is the least accurate method for determining the surface potential and binding-energy calibration. The C 1*s* binding energy varies with coverage, substrate material, and instrument. It also depends on the chemical nature of the unknown surface (ionic, covalent, or metallic), as shown in Fig. 10 for SiO_2, Al_2O_3, and Pd. Kohiki and Oki[(38)] and others[(39,40)] have shown that the C 1*s* binding energy has a nearly constant binding energy at 284.9 eV if measured after precleaning the unknown surface by, for example, argon-ion sputtering. Because of the high temperature required to evaporate gold, the surface in the vicinity of the deposited gold could be melted, resulting in a change in the correction for the surface potential (where Au $4f_{7/2} = 83.95$ eV). In the early 1980s, it was demonstrated that the Au $4f_{7/2}$ binding energy can vary with surface coverage and can be material dependent. Kohiki suggests,[(41)] for gold in electrical equilibrium with an insulating surface, that the shift in binding energy can also result from extra-atomic relaxation. In using either gold or adventitious carbon as a reference, the deposited thickness must be large enough to effectively screen the interfacial interactions. Also, the carbonaceous deposit must be thin enough to be at the same potential as the unknown surface when charged to a steady state.

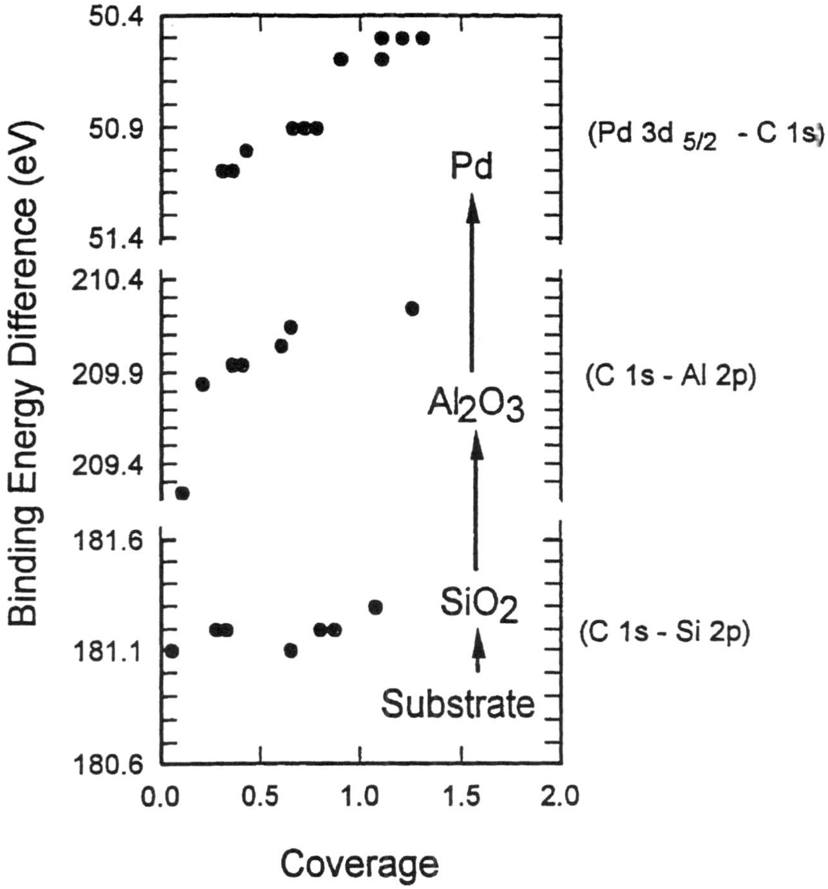

Figure 10. Binding-energy differences (C 1*s* – Si 2*p*, C 1*s* – Al 2*p*, and Pd $3d_{5/2}$ – C 1*s*) for adventitious carbon on SiO_2, Al_2O_3, and Pd substrates as a function of carbon contamination coverage in monolayers (ML).[38]

Lewis and Kelly used an electron flood gun to produce an accurately calibrated energy scale.[16] If a surface is entirely insulated from contact with the spectrometer ground and isolated from any possibility of random electron irradiation by using a monochromated X-ray source, the measured kinetic energy is determined primarily by the electron flood conditions. As an experimental example, they placed strips of gold and aluminum on a clean Teflon mount with the capability of grounding the strips. At ground potential, E_K (and hence the relative binding energy) of the Au 4*f* and Al 2*p* peaks is fixed. The aluminum surface is oxidized with its native oxide. The chemically shifted peak due to Al in the surface oxide shifts with electron flood gun conditions as shown in Fig. 11a. When the metal strips are ungrounded, not only are the elemental peaks fixed in kinetic energy but the surface potential of the native oxide on Al remains fixed under varying charge conditions as shown in Fig. 11b. As Lewis and Kelly show, the electron flood gun current does not affect the binding energy above 0.2 mA emission current. This method has been generally applied to insulators (polymers, etc.) to fix the kinetic energy of the photoemitted electrons. It is necessary to have a system that is equipped with a monochromatic X-ray source to ensure that the sample is not exposed to other electron sources. Nonmonochromatic X-ray sources, typically equipped with a window in front of the source, emit a shower of low-energy secondary electrons from the window material, as mentioned earlier.

Finally, a more sophisticated method of calibration has been proposed and employed by Anderson *et al.*,[42] which they named FRESCA (Fermi-level referenced electron spectroscopy for chemical analysis). This technique uses a highly

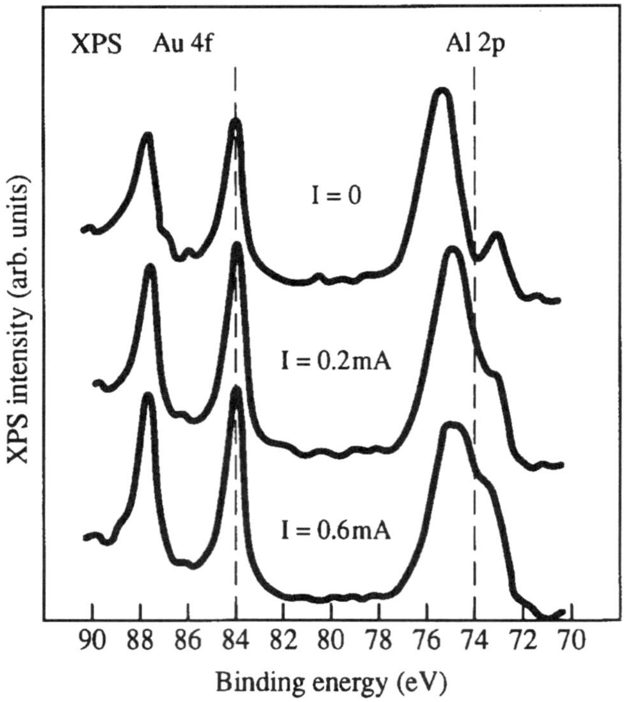

Figure 11. Relative XPS intensities for (a) Au 4*f* and Al 2*p* from grounded foils at various electron flood gun settings, and (b) the same samples ungrounded at various flood gun emission settings.[16] The vertical dashed lines are guides to the eye in (a) and (b).

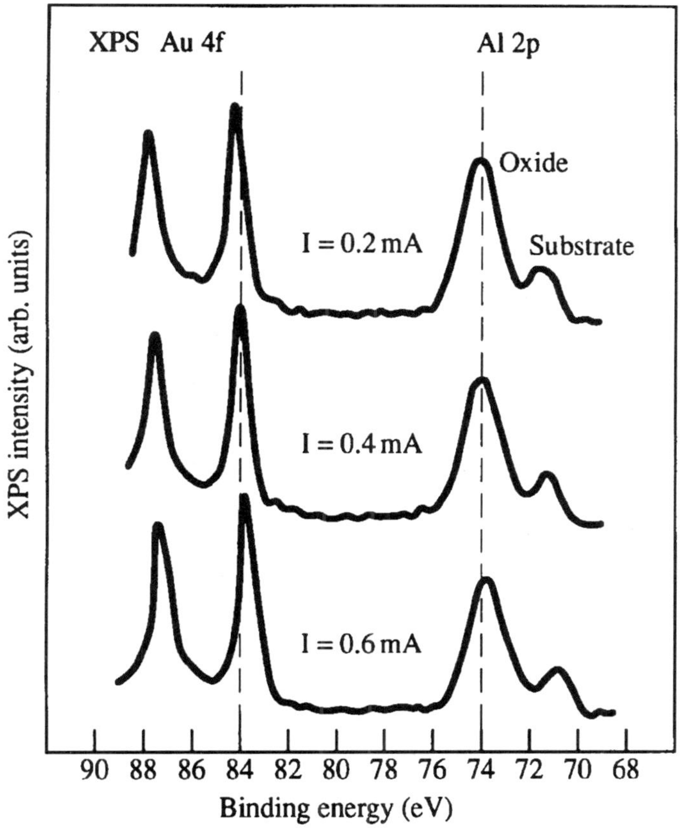

Figure 11. Continued.

stable and accurately calibrated electron gun with a field emitter. It is implicitly assumed that the gun voltages are known to better accuracy than the analyzer voltages. By examining the elastically scattered electron signal from a conductive sample surface with the electron energy analyzer of the analytical instrument, one may compare the kinetic energy of the electron source with the spectrometer power supplies. For example, the absolute binding energy of the gold $4f_{7/2}$ peak was determined to be about 84.3 eV, or within the FWHM of the field emission source. The discrepancy between this binding energy for the Au $4f_{7/2}$ peak and what is generally recognized as the $4f_{7/2}$ binding energy has not been explained.[35] In a further study[37] it was found that some power supplies used for supplying bias to the electron energy analyzer are nonlinear. Consequently, the use of a single energy reference standard is only useful near the electron peak in question.

Using absolute binding energies to predict chemical information is difficult and time consuming. Peak structure, peak binding energy differences, and, as discussed subsequently, the Auger parameter, are generally used to characterize materials. The Au $4f_{7/2}$ peak at 83.98 eV and the Cu $2p_{3/2}$ peak at 932.67 eV have been used as primary calibration energies for spectrometer calibration.(34,35,46) Powell has shown that other photoelectron lines can be used as secondary standards.(35)

Surface charging of insulators, including most polymers, encountered in photoelectron spectroscopy, is well understood, and with proper calibration standards and flood guns available the energy scale can be determined to within ±0.1 eV. With the use of an electron flood gun on materials not readily damaged by low-energy electrons, the surface charging can be compensated instrumentally. Readers interested in the details of binding-energy calibration are referred to the literature.(6,34,35,42)

4. The Auger Parameter: Charge-Independent Chemical Identification

In Sec. 3 it was demonstrated that the binding-energy scale can be corrected for surface charge effects by using thin-film overlayers on the sample (for example, adventitious carbon or gold layers). At best, the absolute binding energies depend on the accuracy of the referencing method. Many materials do not exhibit large chemical shifts in core binding energies, and the chemical shift in the presence of surface charging becomes of the same order of magnitude as the accuracy of the binding-energy determination.

Early studies of the chemical shift phenomena showed that the chemical shift in X-ray-excited Auger peaks was significantly larger than the corresponding core-level shifts.(43) Using this information, Wagner proposed the use of the extra-atomic relaxation energy, ΔE_{ea}, to characterize surfaces chemically, where $\Delta E_{\text{ea}} \cong \Delta E_K(\text{XYY}) - \Delta E_K(\text{X})$, where $E_K(\text{XYY})$ is the Auger kinetic energy for the XYY Auger peak and $E_K(\text{X})$ is the kinetic energy of the photoelectron peak for the X core level.(43) The Auger parameter, α, is defined by Wagner as $\alpha = E_K(\text{XYY}) - E_K(\text{X})$.(44) It is useful to consider the chemical shifts in the photoelectron kinetic energy $E_K(\text{X})$ or binding energy $E_B(\text{X})$ as

$$\Delta E_B(\text{X}) = \Delta\varepsilon(\text{X}) - \Delta E_{\text{ea}}(\text{X}^+) \tag{3}$$

and the chemical shift in Auger kinetic energy as

$$\Delta E_K(\text{XY}^+\text{Y}^+) = \Delta\varepsilon(\text{X}) - 2\Delta\varepsilon(\text{Y}) + \Delta E_{\text{ea}}(\text{Y}^+\text{Y}^+) - \Delta E_{\text{ea}}(\text{X}^+) \tag{4}$$

where $\Delta\varepsilon(\text{X})$ is the change in energy of the ionized core level. The difference defines

$$\Delta\alpha = \Delta E_{ea}(XYY) - \Delta E_B(X) = \Delta E_{ea}(Y^+Y^+) - 2\,\Delta E_{ea}(X^+) \quad (5)$$

where $\Delta\varepsilon$ is assumed to be zero. This expression is independent of ϕ_s. This result makes the Auger parameter of considerable importance when studying insulators. A modified Auger parameter[(45)] was later introduced and defined as

$$\alpha' = \alpha + h\nu = E_K(\text{Auger}) + E_B(\text{photoelectron}) \quad (6)$$

which is independent of the photon energy, $h\nu$. The parameter α' is also independent of the surface potential since it depends only on an energy sum.

Wagner *et al.*[(46)] have compiled a detailed listing of the Auger parameter values which can be used to generate a "chemical state plot." A chemical-state plot is obtained by plotting the kinetic energy of the most intense Auger peak versus the binding energy of the most intense core-level peak for a given element. The modified Auger parameter appears as a line at 45° with a positive slope. Scatter plots of many compounds for a particular element show groupings around compound types. Figure 12 shows a chemical-state plot for fluorine compounds where

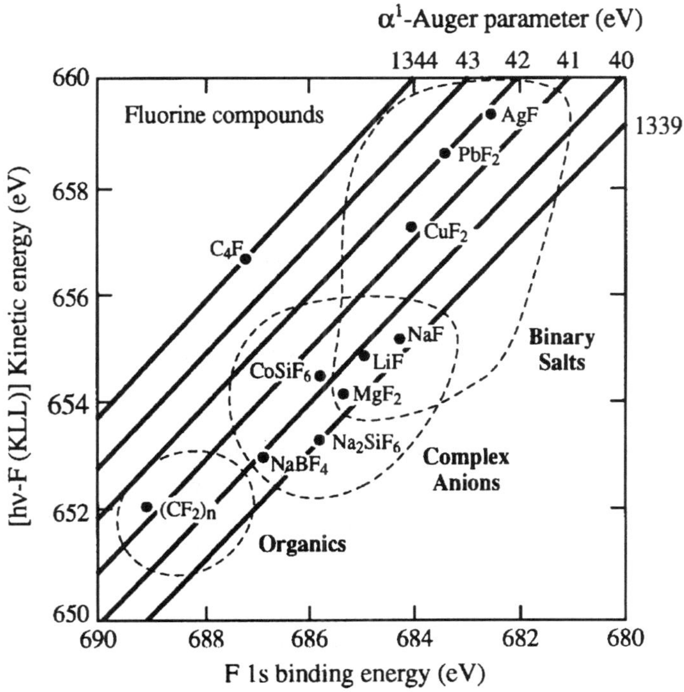

Figure 12. The modified Auger parameter, per Eq. (6), α', for fluorine compounds as compiled by C. D. Wagner *et al.*[(45)]

α′ ranges from 1338.6 to 1343.6 eV, a spread of 5 eV. Plotting an unknown compound on the chemical-state plot often allows the compound type to be identified.

The use of the Auger parameter approach depends on the availability of Auger and photoelectron peaks in a single spectrum. Many manufacturers of XPS apparatus have available dual-anode X-ray sources of various materials that permit excitation of Auger lines not readily excited by the Mg or Al K_α X-ray lines normally employed. The use of Al K_α monochromated X-rays limits the number of elements that can be examined with the Auger parameter method. Auger peaks of energy greater than the primary-photon energy of the source can be excited from the bremsstrahlung component of the radiation of nonmonochromated sources.[47] The bremsstrahlung radiation intensity can be increased by using a thin beryllium window in place of the normal aluminum window. Further details and applications of the Auger parameter method are found in the literature[45,46] and references therein.[46,48]

5. Photon Damage

5.1. Photon Absorption Processes

In XPS or UPS data acquisition, photon irradiation of an unknown surface typically does not result in any substantial material damage,[2–4] and the spectra acquired are representative of an undamaged surface. However, materials can degrade through photon absorption and subsequent chemical reaction within the depth of penetration of the incident radiation. Photon interaction with a material produces photoexcited electrons, which themselves produce excitations (e.g., plasmons and phonons). These interactions may result in spectral features that change with irradiation time or that occur as a result of a specific irradiation (photochemically induced peaks). Inorganic, organic, and polymeric materials have been observed to exhibit photosensitivity and form, for example, the basis of the photographic process.[49,50] To understand the possible materials modifications that may occur and to predict photosensitivity, it is necessary to discuss briefly the process of radiation absorption in materials. A few examples will be presented to demonstrate possible analytical problems caused by material photosensitivity.

Small-spot X-ray sources and synchrotron radiation sources are used to decrease the analysis area in XPS and UPS studies.[29–31,51,52] As the analysis area is decreased by using focused X-rays, the possible damage to materials being studied increases dramatically as a result of the increased X-ray dose. Cazaux[53] has shown that the minimum analysis area is related to the detectability requirements of the specific experiment and a critical X-ray dose, which is the maximum dose a sample can take without observable degradation. The results of his theory (Fig. 13 for XPS)

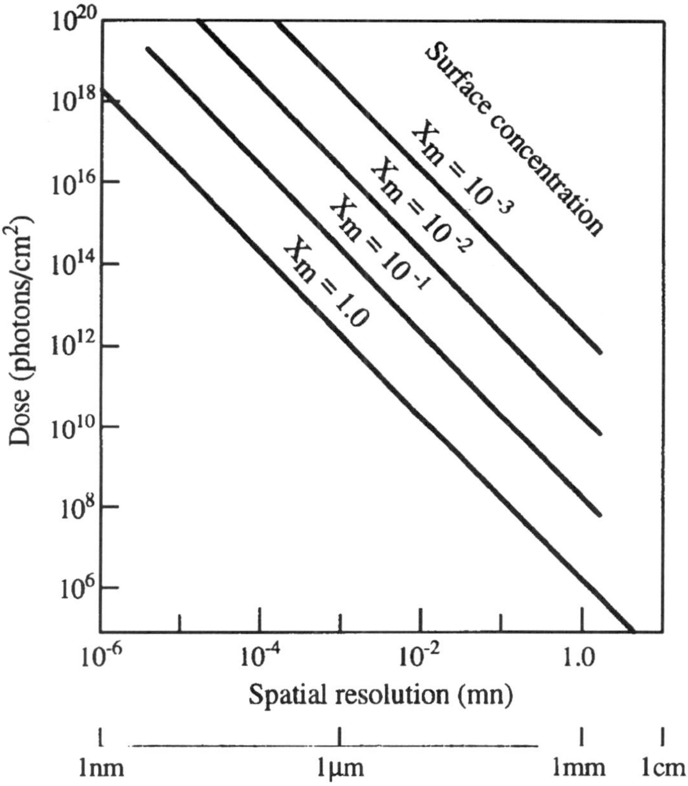

Figure 13. A theoretical prediction by Cazaux[53] shows how the spot size for small-spot XPS analysis, X-ray dose, and minimum detectable surface concentration (χ in monolayers) are related. As the spot size is decreased or the dose is increased, the damage is increased.

show that the critical dose is inversely proportional to the spatial resolution for a fixed detection sensitivity of an element, X_m. It is important when studying materials that exhibit photosensitivity at high spatial resolution that the possible occurrence of damage should be investigated prior to performing the XPS or UPS measurements.

Photons incident on a material are absorbed within the volume of penetration by internal photoemission of energetic electrons, which may then produce ions, free electrons, and excited states.[33,54,55] The energetic electrons interact with the surrounding atoms or molecules to produce other excitations, resulting in a cascade of events that terminate with electronic thermalization (Fig. 14). These events may render the structure unstable and result in thermal decomposition or induce chemical changes. At low photon energies, in the ultraviolet (UV) region, for example, a

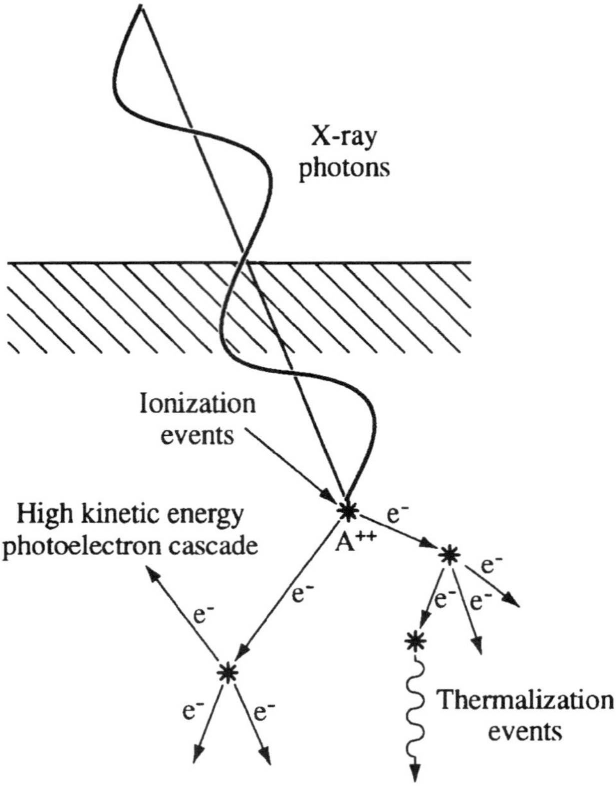

Figure 14. Electronic events leading to thermalization as a result of X-ray absorption in a typical material.

resonance in photon absorption can occur when the photoionization cross section for a particular excitation is a maximum as a function of photon energy.[54–56] At high photon energies (~1 keV), photoionization cross sections are smaller and resonances are not as likely to occur.[54–57] Damage is a function of dose or time, hence even at higher photon energies damage can occur through the interaction of secondary electrons excited in the material.

Surface damage can arise from photon-stimulated desorption (PSD) processes, a direct photon absorption process. Madey[55] divided the fundamental mechanisms for PSD into three physical steps. First, a fast electronic excitation occurs (typically from a core level in XPS) through photon absorption. This results in a surface electronic rearrangement that results in the formation of a repulsive state that will repel an ion, neutral atom, or molecule from the surface. An ionic species leaving the surface will change its final state to conserve energy and momentum. Fundamental excitations from single-crystal surfaces have been employed to study and categorize these processes.

The cross section for radiation damage can be estimated theoretically for fixed experimental requirements, such as detection limit, spatial resolution, instrumental parameters, etc.[54–58] Various electronic excitations are responsible for sample damage and contribute to the overall damage cross section. First, ion desorption can occur by Auger-stimulated desorption (ASD).[55] A "Coulomb explosion" event may also occur. This is a deep core ionization followed by an Auger cascade process that has a final state consisting of many holes. The positive ion(s) in a multiply charged state "explode" from the surface.[54] Second, bond cleavage can occur by direct core-level photoionization or excitation of valence levels.[33,54,55] Finally, secondary electrons produced through photon absorption by the aforementioned, or other, related mechanisms can also produce bond cleavage. Secondary electrons produce damage in the same fashion as expected by electron excitation in AES or scanning electron microscopy.[33]

Contributions to the damage cross section can be examined directly with XPS or UPS. In the Coulomb explosion, XPS can be used to monitor the core levels involved in this process. When XPS is used to study photoinduced damage by valence-electron excitation, core levels indirectly yield useful information through the "chemical shift." Generally, the aforementioned phenomena result in the production of energetic internal electrons and secondary electrons that can damage the material.

Electron spectroscopy can be used to study directly the phenomena of radiation damage in some instances, preferably *in situ* under ultrahigh vacuum conditions. The production of defects and chemical changes, such as color centers[49] and polymerization or depolymerization,[56,58] are readily monitored. Because XPS can be used to obtain quantitative surface composition, defect generation can be monitored as a function of time, yielding useful kinetic data that can be used to infer the basic generation mechanisms. Since many of these mechanisms are encountered in technologically important materials (photoresists, scintillation materials, curing of polymers, etc.), degradation and damage can be viewed in a positive way as permitting defect examination of photosensitive materials and their photoinduced time-dependent chemistry.

Direct UV-induced photodegradation or curing of polymers is common and represents a simple case of direct photon-induced material damage or change (see the literature on photoresists). To describe this phenomenon, valence states of the molecule are shown by an idealized energy diagram in Fig. 15. If a photon of energy $h\nu$ is absorbed by the ground state S_0, an electron from this state makes a transition to the excited state S_1 by the indicated vertical transition. Deexcitation can occur by various processes (phonon production, reradiation, etc.) and is indicated in Fig. 15 by the down arrow from S_1 to S_0. This process may not be damaging. However,

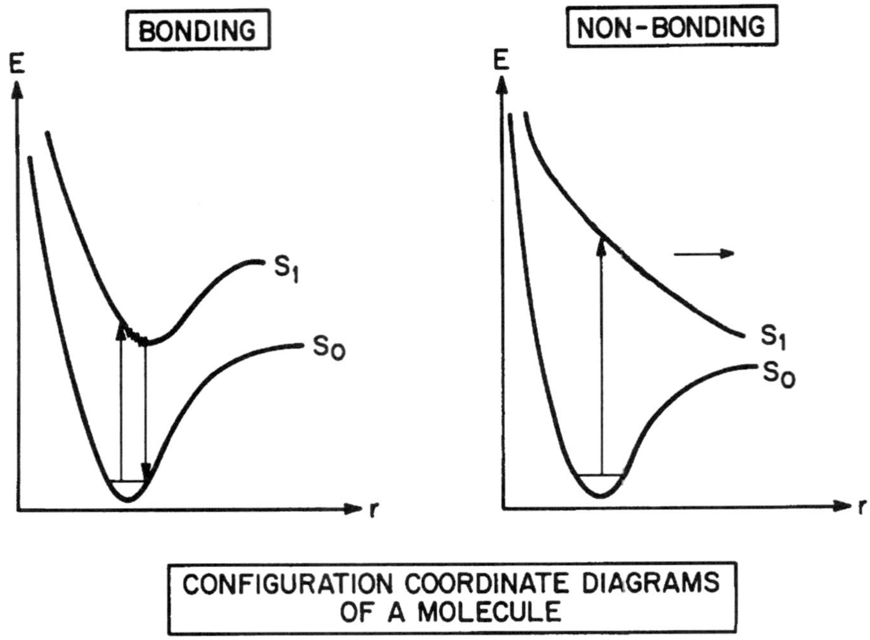

Figure 15. Idealized diagram showing energy versus the configuration coordinate, Γ, for bonding and nonbonding excitations in a molecule for states S_0 and S_1.

if the excitation decays to a nonbonding state, bond cleavage occurs and the material may dissociate. This is similar to the surface-photon-stimulated desorption process.

The energy required to dissociate a material depends on the bond dissociation energy and is proportional to the enthalpy of bond formation.[54–56,58,59] Photosensitivity depends on materials properties and may exhibit a resonance condition (a maximum sensitivity at a specific photon energy). The resonance peak photon energy for dissociation may not scale directly with the total bond energy, in general. Similar effects occur at higher photon energies when resonance occurs with available core levels. In XPS or UPS, resonance events occur in materials containing elements that exhibit a maximum in the photoionization cross section[57] for a particular incident photon energy. For example, the theoretical photoionization cross sections are near maximum for the Ga $2p_{1/2}$ level at a binding energy of 1143 eV or the La $3p_{1/2}$ level at a binding energy of 1205 eV with a Mg X-ray excitation source with the K_α characteristic energy equal to 1253.6 eV.

Radiation absorption in inorganic, organic, and elemental materials depends on the nature of the material (i.e., metal, semiconductor or insulator). In metals, radiation has little or no deleterious effect since electronic excitations simply contribute to the delocalized conduction–electron population, and both electronic excitations and heat are readily conducted away from the absorption site. This is also true in narrow-band-gap semiconducting materials. In ionic or covalent materials, however, radiation can produce ions in the solid that may migrate in the presence of internal fields. This situation results in a change in material composition. For example, metallic azides can be reduced by UV irradiation to produce the elemental metal and nitrogen gas.[60] Reduction of this type depends on the formation and mobility of the metal ions within the material.[60] Eland discusses the process of photoinduced dissociation in detail and reports the results of several experimental methods that can be used to study the details of radiation absorption and product formation.[61]

In relationship to surface analysis methods (XPS, UPS, etc.) that use X-rays or UV radiation as an excitation source, radiation absorption depends on the mass absorption coefficient. The inverse of the corresponding linear absorption coefficient is large relative to the sampled depth. Therefore, the damaged region in a photosensitive material can be considered compositionally uniform in depth. When photoreactions produce mobile species such as metallic ions or other atoms, it is possible that the near-surface region of the material may differ from the irradiated bulk as a result of reactions near the surface. In addition, the vacuum–surface interface may represent an active surface for readsorption of effluent species.

5.2. Radiation Damage to Inorganic Materials

Inorganic materials tend to be stable during UV or X-ray irradiation. However, it has been noted that some compounds were degraded[2] or, in some cases, decomposed during X-ray irradiation.[62] For maximal valency oxides (e.g., Ta_2O_5, Nb_2O_5, Y_2O_3, WO_3, etc.), theory[63] predicts that electron- and photon-stimulated desorption can occur at the surface. Both have been studied in great detail.[54,55,61]

In copper compounds, Klein *et al.*[64] showed that decomposition is more likely to be due to thermal effects from the X-ray source. They suggest that anhydrous $CuSO_4$ be used as a detector to determine whether heating is sufficient to degrade a sample. Figure 16 plots the Cu^+ to Cu^{2+} $2p$ peak intensity ratio versus irradiation time for a $CuSO_4$ sample. Cu^{2+} species appear to be continuously reduced to Cu^+ as a result of heating from the X-ray source. However, it is still possible that X-ray absorption can produce decomposition at a low rate. In the following, these processes and some experimental results will be discussed.

Radiation damage to inorganic materials has been studied by solid-state physicists for many decades. Perhaps the most prominent forms of "damage" are the photographic process and F-center formation. Both are observed in ionic crystals, such as alkali halides, and proceed by similar mechanisms. The formation of mobile ionic species and lattice vacancies by direct photon absorption is responsible for both the photographic process and the creation of F-centers[49] in ionic crystals. Other electronic features, such as excitonic states or bands, can also be produced

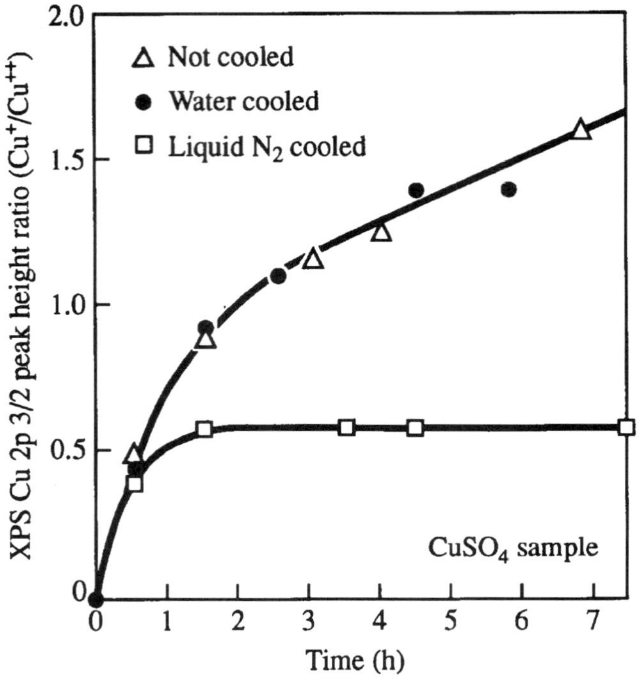

Figure 16. The reduction of Cu^{2+} to Cu^{+} in $CuSO_4$ as a result of heat generation by X-ray absorption. Three conditions are shown: not cooled, water cooled, and liquid-nitrogen cooled.[64]

through UV or X-ray absorption and are observable in the photoemission spectra. Therefore, XPS and UPS can be used to directly probe these important materials properties.

The production of F-center defects (an anion vacancy) can be used as a model for X-ray absorption in ionic crystals (Fig. 17). F-center production requires that an anion be displaced from its lattice position through photon absorption. Energetically, this is not directly possible using UV or X-rays, although internally emitted photoelectrons can have energies in excess of 1 keV. If an anion surrounded by cations in a lattice is multiply ionized, however, the anion can be electrostatically displaced from its lattice position by unequal electrostatic forces acting on it, producing a Schottky defect. In the photographic process, highly mobile cations are produced that form aggregates and thereby change the optical density of the material.[49] The production of mobile cations is the result of a purely electronic process.[65] Because of the relatively large ionic mobility in some ionic crystals (for example, silver sulfide), silver ions can readily move in the presence of internal or external fields. The observed result may be time-varying spectra both in amplitude and peak position, which likely result from surface charging, as discussed earlier.

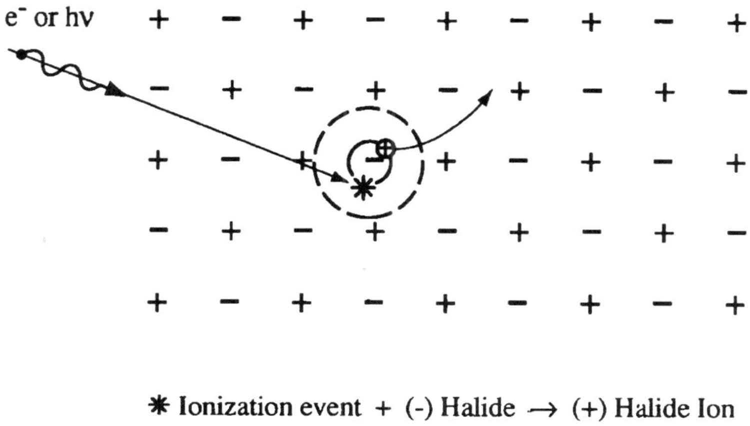

Figure 17. F-center generation in an ionic crystal by X-ray or electron bombardment.

In 1975, Burroughs *et al.*[66] studied the X-ray damage produced by Al K_α X-rays in a series of platinum(IV) complexes that included platinum ethylene diamine chloride ($Pt(C_2H_5NH_2)_4Cl_4$). X-ray absorption resulted in a reduction of the compound from Pt^{4+} to Pt^{2+} and a subsequent evolution of chlorine. The change in the Pt 4*f* spectrum from a 4^+ state to a 2^+ state is shown in Fig. 18 and occurs over a relatively long time (more than 17 h). Kinetic studies showed the reaction is first order, which best fits a model involving exciton capture. Exciton capture results in the production of interstitial chlorine, thus leaving an anion vacancy. Therefore, radiation damage does not occur through a direct photon absorption process, but rather the photon absorption process produces a shower of high-energy photoelectrons that ultimately produce excitons or electron–hole excitations that cause platinum reduction. In fact, it is likely that low-energy electron damage is similar to soft X-ray absorption. This example shows a time-varying spectrum resulting from X-ray damage. Various materials exhibit similar properties.[66–75]

In their studies of inorganic salts, Copperthwaite and LLoyd[69] showed that the decomposition of $NaClO_4$ and $NaClO_3$ results from the absorption of Al K_α X-rays. Decomposition occurs below room temperature, producing Cl^- and releasing oxygen. The product of the reduction is NaCl. However, the product is damaged and contains a large density of F-centers. Figure 19 shows the Cl 2*p* and O 1*s* spectra as a function of irradiation time. It is interesting that ClO_4^- decays in steps by the sequential release of oxygen. Irradiation by small X-ray doses produces mostly ClO_3^-, and the process terminates in the production of Cl^-. Copperthwaite[65] suggested that the damage to ionic crystals produced by X-rays extends to depths comparable to the X-ray penetration depth and appears to be purely electronic in origin. Sasaki *et al.*[72] demonstrated that electron bombardment of chlorates and

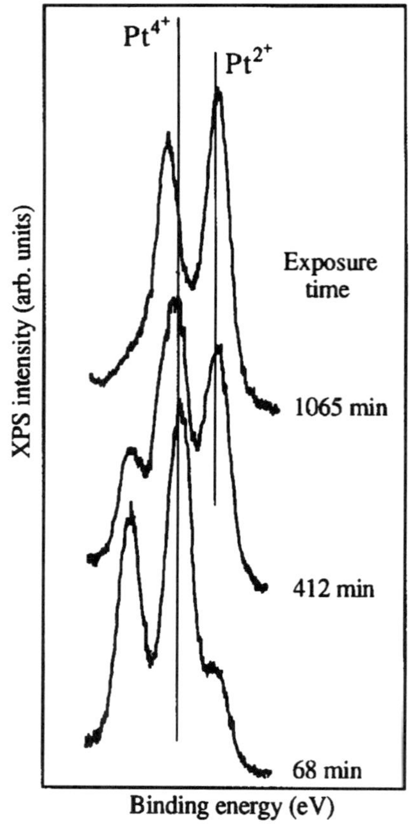

Figure 18. Relative intensity of XPS Pt 4*f* spectra as a function of X-ray exposure time.[66] As the exposure time is increased, Pt^{4+} of platinum ethylene diamine chloride is reduced to Pt^{2+}.

halides produces the reduced halide. Thus, Cl^- is produced by a process analogous to X-ray-induced damage in the work of Copperthwaite and Lloyd.[69] This result confirms the similarity of the damage produced by low-energy and X-ray-induced damage to crystals. Both XPS and UPS provide a direct means of studying the electronic properties and defect structure of ionic crystals. It is thought that extensive irradiation of many materials, particularly by X-rays at ~1 keV, will result in some form of damage. At energies of ~1 keV and flux levels of ~10^{14}/cm^2-s, often employed in photoelectron spectroscopy, damage proceeds very slowly and, during the course of a typical experimental study, is simply not observed. More careful investigations of damage to inorganic materials are needed to determine whether, for the high photon fluxes encountered in small-spot XPS analysis,

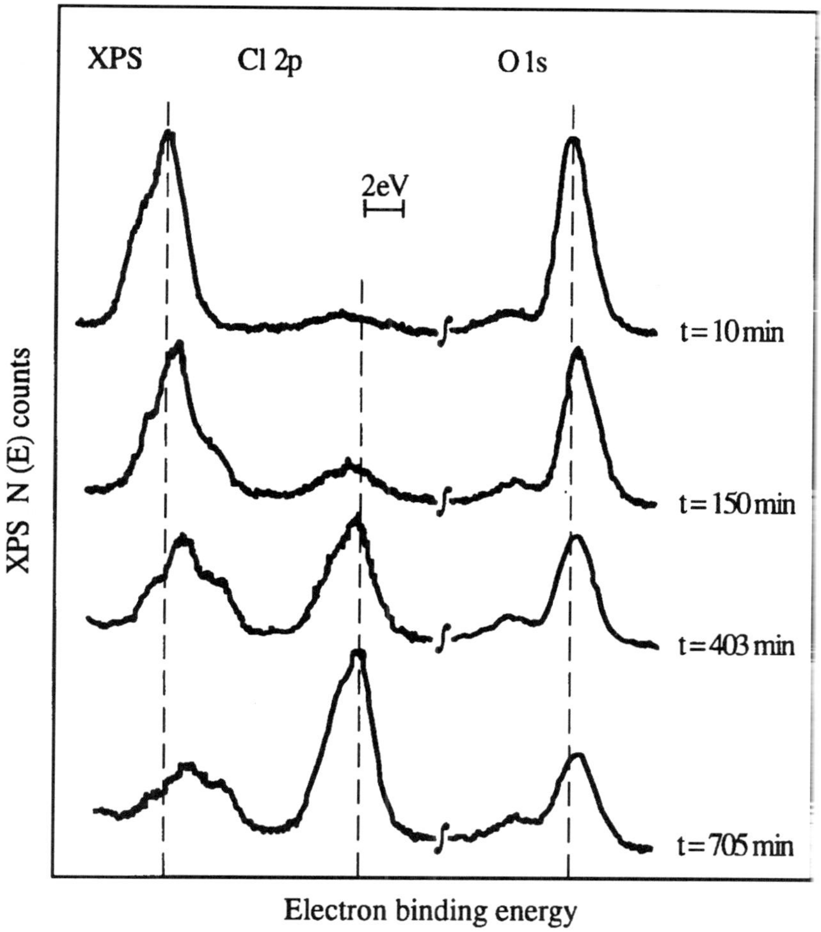

Figure 19. XPS Cl $2p$ and O $1s$ spectra as functions of X-ray exposure of $NaClO_4$.[69] The surface is reduced to NaCl through, in part, the formation of F-centers. The vertical lines are guides to the eye.

materials are prone to decomposition. Cazaux[53] presents a theoretical formulation that addresses this problem in some detail.

5.3. Photon Damage to Polymers

Polymers, as noted earlier, are less stable than the inorganic crystals discussed in Section 5.2 and decompose by electronic and thermal processes. As for damage to ionic crystals, XPS and UPS provide a means of dynamically studying photon-induced processes, such as photodecomposition or photooxidation.[76–78] Degradation products of polymers produced by direct photon-induced processes are difficult to predict. In polymers, degradation can occur by photon-induced cross-linking, chain scission, or structural modification to the backbone or side groups. Inductive effects also occur. For example, photon absorption in a side group may produce ionization or scission at a site that is physically removed from the absorption site. In a study aimed at optimizing the photosensitivity of X-ray photoresists based on poly-(ω-chloro-olefin sulfones),[79] resonant X-ray absorption by chlorine on a side group was intended to produce a scission in the backbone and release SO_2. Experimentally, it was found that resonant X-ray absorption by the Cl did not occur and the side group actually stabilized the polymer.[79]

Benzene groups or ring structures are even more stable and absorb energy by virtue of the delocalized nature of the ring bond.[54,58] In fact, it has been observed that benzene groups can act as an "energy sponge" absorbing energy from remote sites. Polymers containing benzene ring structures tend to be stable under X-ray damage. It is clear from these examples that radiation damage and photon absorption are very complex and that any photon-induced reaction may not be predictable from simple arguments. Here is a "weak" rule of thumb for polymer degradation or reaction:[58] (1) Polymers such as $-(CH_2-CAB)_n-$ tend to cross-link, and (2) other forms of backbone structures tend to fragment or depolymerize at weak links, for example, at the SO_2 link in a polychloro-olefin sulfone.[78]

A few examples of polymers that degrade under UV irradiation are listed in Table 1 and show the relationship between the enthalpy of formation and the resonant photon energy of maximum sensitivity.[58,60] Since 1 eV/bond is equal to 23.05 kcal/mole, it is clear that bond dissociation occurs for excitations from near the maximum in the valence-band density of states because the photon energy for maximum sensitivity is in reasonable agreement with the enthalpy of formation.[59]

One of the most widely studied polymers is polyvinyl chloride because of its wide application as an insulator in electronic materials. It is unstable in a radiation environment, is thermally unstable, and readily degrades at low temperatures.[58] Stabilizers are added to the polymer to absorb energy and reduce the amount of

Table 1. UV Photodegradation of Selected Polymers

Polymer	Photon energy for maximum sensitivity (eV)	Dissociation energy kcal/mole
Polypropylene	3.35	77
Polyethylene Terephthalate	3.82	88
Polyvinyl chloride	3.87	89
Polystyrene	3.90	90
Polyethylene	4.13	96
Polycarbonate	4.20/3.59	97/83

evolved species. Numerous papers relating to polymer surfaces have ben published, so only a few examples are presented here.

Polyvinyl chloride (PVC) degrades by photoionization, resulting in the production of HCl through H and Cl bond cleavage.[81] Alkyl radicals form through photoionization of Cl and C core and valence levels by X-rays during XPS analysis. It is thought that a double bond spontaneously forms at two adjacent vacated carbon sites; that is,

$$-(CH_2-CHCl)- + h\nu \rightarrow -(CH{=}CH)- + HCl\ (gas) \tag{7}$$

Figure 20 shows the C 1*s* spectrum as a function of X-ray irradiation time. The component peak due to Cl bonding decreases in amplitude with time, and the reaction proceeds by first-order kinetics.[81] A shift in binding energy results from two effects: a decrease in the total photoionization cross section per repeat unit, which changes the positive charge accumulation on the surface, and the formation of delocalized bonds, which tend to increase the local conductivity of the surface.

Irradiation of PVC by X-rays produces a brown coloration of the material that is similar to that observed during thermal degradation.[82,83] This result indicates the formation of polyenes ($-(HC{=}CH)_n-$ groups). In the study of thermal degradation, following the crude rule of thumb, PVC degrades by cross-linking and appears to degrade by a mechanism similar to that shown before. Here, the photoelectron spectrum varies in time both in peak amplitude and position as a result of irradiation.

Poly-ω-chloro-olefin sulfones,[79] mentioned earlier, are candidates for X-ray photoresist materials; here X-rays interact with the polymer to release SO_2 from the backbone. Films placed in an XPS UHV analysis chamber depolymerize so rapidly under Mg K_α radiation that the evolution of SO_2 and polymer fragments produce a pressure increase large enough to trip pressure safety interlocks.[84] Decreasing the X-ray source flux to a small fraction of its normal operating condition allows stable operating conditions to be obtained. Caution should always be exercised when studying known X-ray- or UV-sensitive polymers in a UHV analysis chamber.

As a final example of polymer photosensitivity, the photoaging of bisphenol-A-polycarbonate[85,86] using XPS is discussed. Although polycarbonate does not appear to degrade during routine surface analysis, the material is photosensitive to UV radiation and will oxidize in air when exposed to natural sunlight. Clark and Munro[85] studied this material under various illumination conditions to determine the basic surface oxidation mechanism. Ultraviolet light spectrally peaked at ~350 nm produces a change in both oxygen and carbon chemistry with irradiation time. The C 1*s* spectrum is shown in Fig. 21 as a function of exposure. High-binding-energy peaks of the spectrum are assigned to C–O functionalities, and these change in intensity as a function of exposure. The unexposed C 1*s* spectrum contains peaks

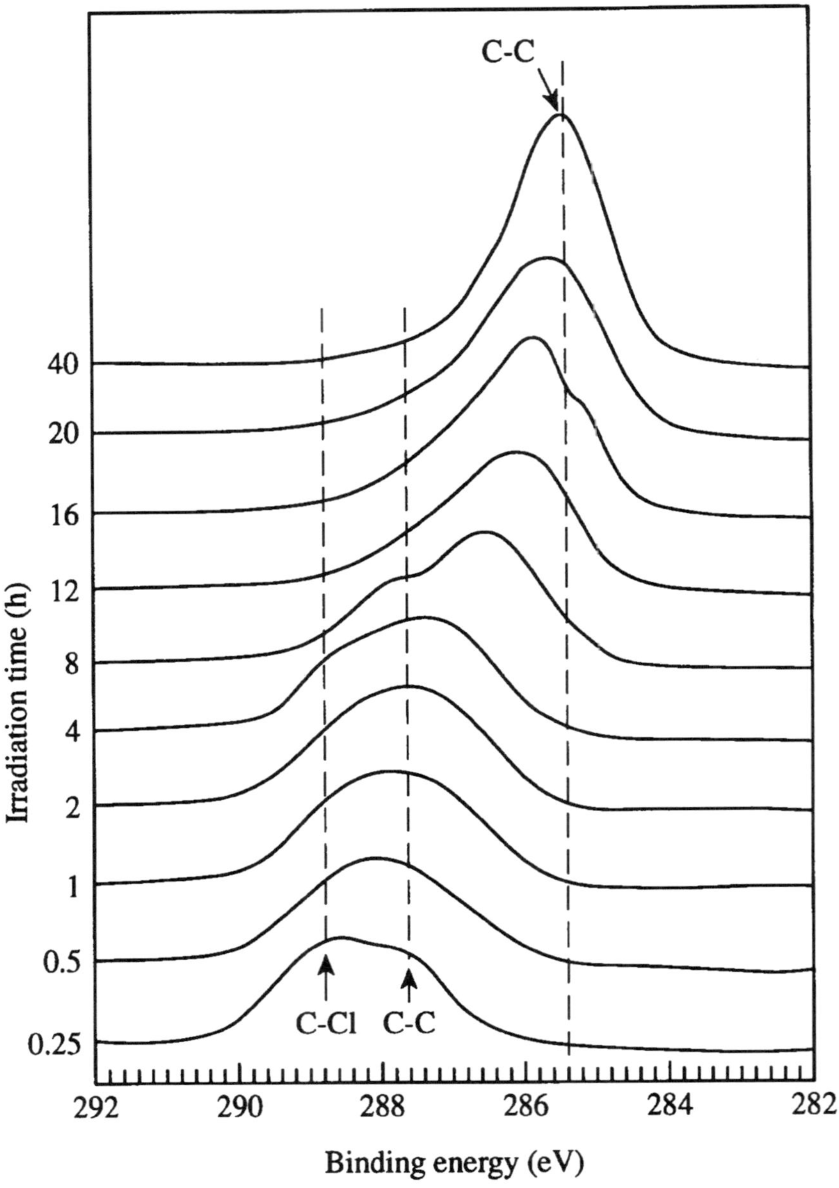

Figure 20. XPS C 1*s* spectra of polyvinyl chloride as a function of X-ray exposure. The C–Cl component is observed to decrease in amplitude as a function of exposure, and the peaks shift to a lower binding energy as a result of the effects discussed in the text.[81]

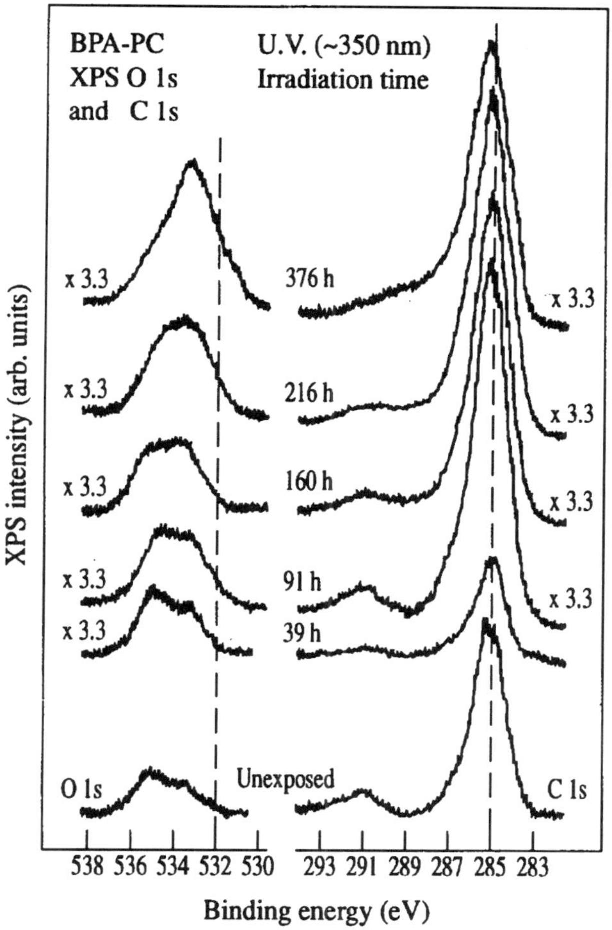

Figure 21. XPS C 1*s* and O 1*s* spectra of bisphenol-*a*-polycarbonate (BPA-PC) as a result of UV exposure at ~350 nm.[85]

resulting from C–H, C–O, and O=COO. As the exposure time is increased, these peaks change in intensity and new peaks appear that result from –C=O and O–C=O functionalities. In addition, the O 1*s* peak intensity increases in amplitude. A $\pi \rightarrow \pi^*$ shakeup peak resulting from the benzene ring structure is also observed and decreases along with the O=COO functionality with exposure.

Photooxidation of bisphenol-A-polycarbonate is very complex, as shown by the carbon chemistry in Fig. 21. Surface photooxidation involves the benzene ring and O=COO functionalities. The O–C=O and C=O functionalities increases in amplitude with a decrease in benzene ring functionality and gem dimethyl groups.

Clark and Munro[85] demonstrated that with proper UV radiation sunlight-induced degradation of the polycarbonate can be approximated in a laboratory experiment. There are many more recent photoinduced degradation results described in the *Journal of Polymer Degradation and Stability*.

Photon-stimulated desorption has also been used to study surface electronic structure of fluorine on silicon (111), as described by Yarmoff and Joyce.[87] The reader is referred to these references and others cited in similar articles.

6. Closing Comments

In this brief presentation of photon-induced effects, X-ray and UV photoelectron spectroscopies were shown to be capable of inducing surface chemical changes and surface charging artifacts. Contrary to popular belief that photon-excited surface analysis techniques are innocuous, surface modification by UV and X-ray absorption is more prevalent than many expect. For typical XPS analysis, degradation phenomena in most materials occur on a time scale much longer than the scale of a measurement, as a result of the low X-ray flux employed and the low photoelectron cross section. UV excitation can be significantly more damaging to photosensitive materials because of a typically higher radiated flux and the larger photoionization cross sections encountered.

Surface charging and artifacts induced by UV and X-ray irradiation, which occur primarily in the studies of insulators, are readily identified and corrected by using electron flood gun techniques to replace the electrons on the electron-deficient surface. Charging artifacts are difficult to correct when the material under investigation is photosensitive. Using an electron flood gun will simply accelerate the photochemical degradation process. In some instances, finding a set of low-damage, low-charging parameters for the flood gun may not be possible. Generally, other methods, such as infrared spectroscopy, can be employed to complete the investigation.

In studying photon-induced chemical changes, the use of mass spectrometry and optical emission spectroscopy in conjunction with electron, ion, or photon excitation provide powerful experimental methods in addition to photoelectron spectroscopy. Eland[61] describes these combinations and others as they are applied to fragmentation. Additional fundamental principles have been studied by the surface science community from photon- or electron-stimulated surface desorption.[84]

The references in this chapter are intended as guides to the vast amount of literature about photon-induced damage phenomena. The reader is encouraged to investigate references included in papers and books cited and other references in the archived literature. In the specific case of polymer degradation, the *Journal of*

Polymer Degradation and Stability and the *Journal of Polymer Science* contain numerous relevant papers.

Acknowledgments

The author would like to thank his friends and colleagues at RCA Laboratories (now David Sarnoff Research Center, Inc.) for their encouragement in writing this manuscript. He also thanks Mr. G. A. Korba at the 3M Corporate Research Laboratory for obtaining some of the data.

References

1. This volume, Chapters 2, 3, and 5.
2. K. Siegbahn, C. N. Nordling, A. Fahlman, R. Nordberg, K. Hamrin, J. Hedman, G. Johansson, T. Bermark, S. E. Karlsson, I. Lindgren, and B. Lindberg. *ESCA: Atomic, Molecular and Solid State Structure Studied by Means of Electron Spectroscopy,* Almqvist and Wiksells, Uppsala (1967).
3. W. M. Riggs, in: *Methods of Surface Analysis* (A. W. Czanderna, ed.), Elsevier, Amsterdam (1975), pp. 103–158.
4. C. J. Powell in: *Applied Electron Spectroscopy for Chemical Analysis* (H. Windawi and F-L. Ho, eds.), Wiley-Interscience, New York (1982) Chap. 2.
5. C. J. Powell, D. M. Hercules, and A. W. Czanderna, in: *Ion Spectroscopies for Surface Analysis* (A. W. Czanderna and D. M. Hercules, eds.), Plenum, New York (1991), pp. 417–437.
6. D. Briggs and M. P. Seah, eds., *Practical Surface Analysis*, Wiley, New York (1990).
7. S. H. Hercules and D. M. Hercules, in: *Characterization of Solid Surfaces*, (P. F. Kane and G. R. Larabee, eds., Plenum Press, New York (1974).
8. R. E. Honig, *Thin Solid Films* **31**, 89 (1976).
9. S. Evans, in: *Handbook of X-ray and Ultraviolet Photoelectron Spectroscopy*, (D. Briggs ed.), Heyden, London (1977), Chap. 3.
10. P. Swift, E. Shuttleworth, and M. P. Seah, in: *Practical Surface Analysis*, (D. Briggs and M. P. Seah, eds.), Wiley, New York (1983), pp. 437–444.
11. D. J. Hnatowich, J. Hudis, M. L. Perlman, and R. C. Ragaini, *J. Appl. Phys.* **42**, 4883 (1971).
12. C. D. Wagner, in: *Quantitative Surface Analysis of Materials*, ASTM STP-643 (N. S. McIntyre, ed.), ASTM, Philadelphia (1978), pp. 31–45.
13. C. J. Powell, in: *Quantitative Surface Analysis of Materials*, ASTM STP-643 (N. S. McIntyre, ed.), ASTM, Philadelphia (1978), pp. 5–29.
14. M. P. Seah, in: *Practical Surface Analysis*, D. Briggs and M. P. Seah, eds.), Wiley, New York (1983), pp. 247–283.
15. D. Briggs, in: *Handbook of X-ray and Ultraviolet Photoelectron Spectroscopy*, (D. Briggs, ed.), Heyden, London (1977), Chap. 4.
16. R. T. Lewis and M. A. Kelly, *J. Electron Spectrosc. Rel. Phenom.* **20**, 105 (1980).
17. M. F. Ebel and H. Ebel, *J. Electron Spectrosc. Rel. Phenom.* **3**, 169 (1974).
18. T. Dickinson, A. F. Povey, and P. M. A. Sherwood, *J. Electron Spectrosc. Rel. Phenom.* **2**, 441 (1973).
19. J. S. Brinen, *J. Electron Spectrosc. Rel. Phenom.* **5**, 377 (1974); in: *Applied Surface Analysis*, ASTM STP-699 (T. L. Barr and L. E. Davis, eds.), ASTM, Philadelphia (1978), pp. 24–41.
20. C. D. Wagner, *J. Electron Spectrosc. Rel. Phenom.* **18**, 345 (1980).

21. G. Grimvall, *Physica Scripta* **9**, 43 (1974).
22. C. E. Bryson *Surf. Sci.* **189/190**, 50–58 (1987).
23. C. E. Bryson III and D. L. Jones, *Electron Spectroscopy System for Chemical Analysis of Electronically Isolated Specimens*, U. S. Patent # 4,680,467.
24. G. Barth, R. Linder, and C. E. Bryson, *Surf. Interface Anal.* **11**, 307–11 (1988).
25. Comments made by C. E. Bryson, Surface Interface Inc., 1993.
26. W. M. Lau, *J. Appl. Phys.* **65**, 2047 (1989).
27. T. L. Barr, *J. Vac. Sci. Technol.* **A 7**, 1677 (1989).
28. X.R. Yu and H. Hantsche, *Surf. Interface Anal.* **20**, 555 (1993); see also, *J. Electron Spectrosc. Rel. Phenom.* **59** (1992), special issue.
29. R. Chaney, *Surf. Interface Anal.* **10**, 36–47 (1987).
30. M. A. Kelly, *Research and Development*, Jan. 1984, p. 80.
31. C. D. Wagner and A. Joshi, *Surf. Interface Anal.* **6**, 215 (1984).
32. P. M. Hall and J. M. Morabito, *Surf. Sci.* **83**, 391 (1979).
33. Leon Sanche, in: *Excess Electrons in Dielectric Media*, (J. P. Jay-Gerin and C. Ferradini, eds.), CRC Press, Boca Rotan, FL (1991), p. 1.
34. M. P. Seah and M. T. Anthony, *Surf. Interface Anal.* **6**, 95 (1984).
35. C. J. Powell, *App. Surf. Sci.* **89**, 141 (1995).
36. R. J. Bird and P. Swift, *J. Electron Spectrosc. Rel. Phenom.* **21**, 227 (1980).
37. R. N. Lee, *J. Electron Spectrosc. Rel. Phenom.* **28**, 195 (1982).
38. S. Kohiki and K. Oki, *J. Electron Spectrosc. Rel. Phenom.* **33**, 375 (1984).
39. N. I. Nefedov, Y. V. Salyn, G. Leonhardt, and R. Scheibe, *J. Electron Spectrosc. Rel. Phenom.* **10**, 121 (1977).
40. D. T. Clark, A. Dilks, and H. R. Thomas, *J. Polym. Sci. Polym. Chem. Ed.* **16**, 1461 (1978).
41. S. Kohiki, *Appl. Surf. Sci.* **17**, 497 (1984).
42. C. R. Anderson and R. N. Lee, *J. Electron Spectrosc. Rel. Phenom.* **34**, 173 (1984); C. R. Anderson, R. N. Lee, J. F. Morar, and R. L. Park, *J. Vac. Sci. Technol.* **20**, 617 (1982).
43. C. D. Wagner and P. Biloen, *Surf. Sci.* **35**, 82 (1973).
44. C. D. Wagner, *Faraday Discuss. Chem. Soc.* **60**, 291 (1975).
45. C. D. Wagner, D. E. Passoja, H. F. Hillery, T. F. Kinisky, H. A. Six, W. I. Jansen, and J. A. Taylor, *J. Vac. Sci. Technol.* **21**, 933 (1982); C. D. Wagner, in: *Practical Surface Analysis*, (D. Briggs and M. P. Seah, eds.), J. Wiley, New York (1990), pp. 595–634.
46. C. D. Wagner, W. M. Riggs, L. E. Davis, J. F. Moulder, and G. E. Muilenberg, *Handbook of X-ray Photoelectron Spectroscopy*, Perkin-Elmer, Eden Prairie MN (1979); J. F. Moulder, W. F. Stickle, P. E. Sobol, and K. D. Bomben, *Handbook of X-ray Photoelectron Spectroscopy*, Perkin-Elmer, Eden Prairie MN (1992).
47. J. E. Castle and R. H. West, *J. Electron Spectrosc. Rel. Phenom.* **16**, 195 (1979).
48. C. D. Wagner and J. A. Taylor, *J. Electron Spectrosc. Rel. Phenom.* **28**, 211 (1982); *J. Vac. Sci. Technol. A* **1** (1983).
49. N. F. Mott and R. W. Gurney, *Electronic Processes in Ionic Crystals*, Oxford University Press (1948).
50. W. E. Garner and I. Maggs, *Proc. Roy. Soc. A.* **172**, 299 (1939).
51. M. Schaible, H. Petersen, W. Braun, and E.-E. Koch, *Rev. Sci. Instrum.* **60**, 2172 (1989).
52. F. Lodders, A. Goldman, D. Rudolph, G. Schmahl, and W. Braun, *J. Electron Spectrosc. Rel. Phenom.* **60**, 1 (1992).
53. J. Cazaux, *Ultramicroscopy* **17**, 43 (1984); *Appl. Surf. Sci.* **20**, 457 (1985).
54. See, for example, A. Charlesby, *Atomic Radiation and Polymers*, Pergamon Press, Oxford (1960); T. A. Carlson, in: *Desorption Induced by Electronic Transitions*, (N. H. Tolk, M. M. Traum, J. C. Tully, and T. E. Madey, eds.), Springer-Verlag, Berlin (1983).

55. T. E. Madey, in: *Analytical Electron Microscopy*, (D. C. Joy, ed.), San Francisco Press, San Francisco (1987), p. 345.
56. J. A. Brydson, *Plastic Materials, 4th ed.*, Butterworth Scientific, London (1981) Chap. 5.
57. J. H. Scofield, *J. Electron Spectrosc. Rel. Phenom.* **8**, 129 (1976).
58. T. Kelen, *Polymer Degradation*, Van Nostrand-Reinhold, New York (1983), Chap. 7.
59. L. Pauling, *The Nature of the Chemical Bond, 3rd ed.* Cornell University Press, Ithaca, NY (1960).
60. C. I. Chang, J. Neuberger and R. Fischer, *Phys. Stat. Solidi* **B 93**, 255 (1979).
61. J. H. D. Eland, in: *Electron Spectroscopy*, Vol. 3 (C. R. Brundle and A. D. Baker, eds.), Academic Press, New York (1979), Chap. 5; T. E. Madey and J. T. Yates, Jr., *J. Vac. Sci. Technol.* **8**, 525 (1972).
62. H. P. Fritzer, D. T. Clark, and I. S. Woolsey, *Inorg. Nucl. Chem. Lett.* **10**, 247 (1974).
63. T. E. Madey, private communication, 1995.
64. J. C. Klein, C. P. Li, D. M. Hercules, and J. F. Black, *Appl. Spectrosc.* **38**, 729 (1984).
65. R. G. Copperthwaite, *Surf. Int. Anal.* **2**, 17 (1980).
66. P. Burroughs, A. Hamnett, J. F. McGilp, and A. F. Orchard, *J. Chem. Soc. Faraday Trans. II* 177 (1975).
67. P. Burroughs, A. Hamnett, and A. F. Orchard, *J. Chem. Soc. Dalton Trans.*, 565 (1974).
68. B. Wallbank, C. E. Johnson, and I. G. Main, *J. Electron Spectrosc. Rel. Phenom.* **4**, 263 (1974).
69. R. G. Copperthwaite and J. Lloyd, *J. Chem. Soc. Dalton Trans.* 1117 (1977); *J. Chem. Soc. Faraday Trans. I* **74**, 2252 (1978).
70. J. H. Burgess, L. T. Taylor, and J. G. Dillard, *J. Electron Spectrosc. Rel. Phenom.* **8**, 483 (1976).
71. B. A. DeAngelis, *J. Electron Spectrosc. Rel. Phenom.* **9**, 81 (1976).
72. T. Sasaki, R. S. Williams, J. S. Wong, and D. A. Shirley, *J. Chem. Phys.* **68**, 2718 (1978); **69**, 4374 (1978); **71**, 4601 (1979).
73. A. F. Povey and P. M. A. Sherwood, *J. Chem. Soc. Faraday Trans. II*, 1240 (1974).
74. D. Briggs, V. A. Gibson, and J. K. Becconsall, *J. Electron Spectrosc. Rel. Phenom.* **11**, 343 (1977).
75. M. Batista-Leal, J. E. Lester, and C. A. Lucchesi, *J. Electron Spectrosc. Rel. Phenom.* **11**, 333 (1977).
76. A. Dilks, *Degrad. Stab. Polym.* **1**, 600 (1983).
77. D. T. Clark, *Physicochem. Aspects Polym. Surf.* **1**, 3 (1983); *Pure Appl. Chem.* **54**, 415 (1982); *ACS Symp. Ser.*, Vol. 162, American Chemical Society, Washington, DC (1981) p. 247; *Pure Appl. Chem.* **57**, 941 (1985).
78. D. T. Clark, A. Dilks, and H. R. Thomas, *Dev. Polym. Degrad.* **1**, 87 (1977).
79. M. Kaplan, *Polym. Eng. Sci.* **23**, 957 (1983).
80. G. Scott, in: *Ultraviolet Light Induced Reactions in Polymers*, (S. S. Labana, ed.), *Am. Chem. Soc. Symp. Series*; Vol. 25, American Chemical Society, Washington DC (1976), Chap. 24.
81. H. P. Chang and J. H. Thomas III, *J. Electron Spectrosc. Rel. Phenom.* **26**, 203 (1982).
82. Z. Mayer, *J. Macromol. Sci. Rev. Macromol. Chem.* **10**, 263 (1974).
83. J. F. Rabek, B. Ranby, B. Ostensson, and P. Flodin, *J. Appl. Polym. Sci.* **24**, 2407 (1979).
84. J. H. Thomas III, unpublished results, 1982.
85. D. T. Clark and H. S. Munro, *Polym. Degrad. Stab.* **8**, 195 (1984).
86. A. Factor and M. L. Chu, *Polym. Degrad. Stab.* **2**, 203 (1980).
87. J. A. Yarmoff and S. A. Joyce, *J. Vac. Sci. Technol. A* **7**, 2445 (1989).

2

Electron Beam Damage at Solid Surfaces

Carlo G. Pantano, Andrew S. D'Souza and Alan M. Then

1. Introduction

It has been recognized for some time that the use of electron beams to analyze the chemical and structural characteristics of solid materials can simultaneously modify, or otherwise damage, the specimen. These effects were probably first observed during electron microprobe analyses, and especially in transmission electron microscopy of biological materials. It has become increasingly evident, though, that these effects occur not only in biological and other organic materials (1) but also in metallic, ionic, and covalent inorganic materials (2). The damage in electron microanalysis and microscopy usually occurs in bulk because these techniques use electron beams with high energy and high current density.

In Auger electron spectroscopy (AES) and other surface-sensitive techniques, including electron energy loss spectroscopy (EELS) and low-energy electron diffraction (LEED), the physical effects of electron beam damage may be less severe, but nevertheless they often impose a greater limitation upon the practical application of these methods. The reasons are fairly clear. First, the damage is often localized at the outermost surface of a solid, and because of the surface sensitivity of AES the consequences are immediately evident. Second, the surface, in contrast

---

Carlo G. Pantano, Andrew S. D'Souza and Alan M. Then • Department of Materials Science and Engineering, Pennsylvania State University, University Park, PA 16802.

Beam Effects, Surface Topography, and Depth Profiling in Surface Analysis, edited by Czanderna *et al.* Plenum Press, New York, 1998

to the bulk, is most susceptible to many of the mechanisms associated with electron beam radiation damage. Finally, a primary effect of electron beam irradiation is composition changes in the surface region, and, clearly, this limits the primary objective of most AES analyses, which is elemental composition analysis at surfaces.

Because AES is so extremely sensitive to electron beam irradiation, it is a powerful technique for the study of electron and other irradiation effects at surfaces. For this reason the literature is rich in information concerning the behavior of solid surfaces under electron irradiation. These studies have revealed that the mechanisms and extent of the damage depend in a complex way upon the instrumental parameters used in the analysis, the specific material, and the condition of the surface. The dependence of the electron beam damage upon the original condition of the surface is especially significant because AES is often used to examine and characterize unknowns. Thus, the philosophy of this chapter is that one must be familiar with the mechanisms of electron beam damage, with their dependence upon instrumental parameters and material type, and with the reported occurrences of electron beam damage in practical applications. It is only in this way that the surface scientist can anticipate and account for the effects of electron beam damage in an analysis.

Electron beam damage and related effects influence the specimens during analysis in a variety of ways. Often, these effects occur in combination, but for the purposes of this presentation the following list provides a convenient framework:

(i) Changes in composition
(ii) Changes in chemical structure
(iii) Lattice or other structural modification
(iv) Generation or annihilation of electronic defects and color centers
(v) Electrical charging
(vi) Heating or melting

These effects manifest themselves in the spectra in the form of peak shape changes, peak shifts, changes in peak intensity, or instability; if the damage extends to the subsurface, the composition depth profile may be distorted. In some cases, visual deterioration, coloration, or etching of the specimen surface may be observed.

The objectives of this chapter are to describe the fundamental mechanisms associated with each of these effects, to show examples of how they manifest themselves in AES and related analyses, and to offer some recommendations for minimizing or recognizing the extent of the damage, or both. The materials most susceptible to electron beam damage are identified and, wherever possible, quantitative estimates of the damage rates have been provided. It will be evident, though, that insulating, ionic, or organic materials are especially susceptible to these phenomena, and where high-current-density beams are used for microanalyses, the

Table 1. Bibliography on Electron Beam Damage at Solid Surfaces

References	Topic
1–7	Reviews of beam damage in electron microscopy
8–16	General discussions of electron beam damage in surface analysis
17–33	Fundamentals of electron-stimulated desorption
34–136	Studies of electron beam interactions at solid surfaces
137–160	Charging of insulators due to electron beam irradiation
161–180	Electron beam effects at glass surfaces
181–226	Electron beam damage during surface analyses

damage can occur within seconds! A rather extensive set of references was compiled to meet these objectives; these are summarized in Table 1.

2. Fundamentals

In general, the incident electron beam interacts with atoms and molecules in both the surface and bulk of the material. Most of the consequent effects result from electronic excitation processes, but elastic collisions with atoms, charge trapping, and secondary-electron emission may also contribute. These microscopic phenomena then lead to more macroscopic events, which include adsorption, desorption, decomposition, etching, oxidation, diffusion, electromigration, amorphization, heating, and melting.

In this section, brief descriptions of the most important phenomena, and the mechanisms through which they occur, are presented. More detailed descriptions can be found in the references cited.

2.1. Electronic Excitation Processes

2.1.1. Electron-Stimulated Desorption

Electron-stimulated desorption (ESD) is largely a monolayer effect wherein atomic or molecular species are ejected from the surface during electron irradiation and, thereby, the surface composition or chemical structure of surface molecules is altered. It can be viewed as a multiple-step process involving electronic excitation of surface species followed either by escape of an excited fragment or by deexcitation and recapture at the surface. Desorption occurs whenever the excitation process leads to repulsive interactions between the excited species and surface. The desorbed species include ions and excited neutrals, but in most cases the cross sections for neutral desorption are larger.

Electronic excitation is usually assumed to be localized and may involve valence-level or core-level states. The fact that valence-electron excitations whose threshold energies may be as low as 5 eV can lead to ESD means that not only the primary electrons but also the secondary electrons can contribute to the excitation. This is probably a contributing factor to the large cross sections for ESD in insulating materials, since they generally exhibit higher secondary-electron emission coefficients than metals. The secondary-electron effect may also explain the dependence of ESD-induced sample damage upon the condition of the surface; i.e., the presence of contaminants or oxide layers that influence the work function (and thereby the secondary yield) may control the extent of ESD. It is also important to recognize that the core-level excitations, which trigger Auger or other analytical signals, may also contribute to ESD. Fortunately, the cross sections for ESD are

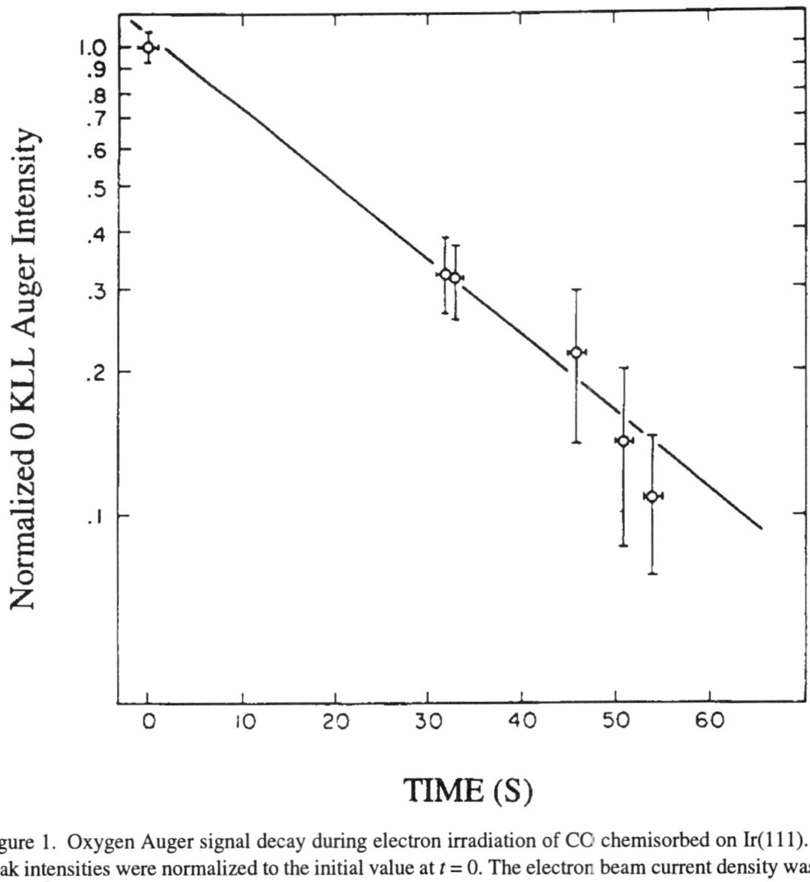

Figure 1. Oxygen Auger signal decay during electron irradiation of CO chemisorbed on Ir(111). The peak intensities were normalized to the initial value at $t = 0$. The electron beam current density was 0.6 mA/cm^2 (From Ref. 52).

seldom as high as those for core-level ionization. In metals, at least, this probably results from screening of the electronic excitations by conduction electrons.

When the electronic excitation is localized within the states that bond molecules to one another or to a surface, the molecule will desorb intact. This situation is easy to recognize and interpret in most surface analyses. In many instances, however, electronic excitation is intramolecular. This type of excitation usually leads to dissociation of the surface species and to more complex effects in the analysis. Figures 1 and 2 show the time-dependent change in the Auger intensity of oxygen and carbon, respectively, during electron beam irradiation of CO ad-

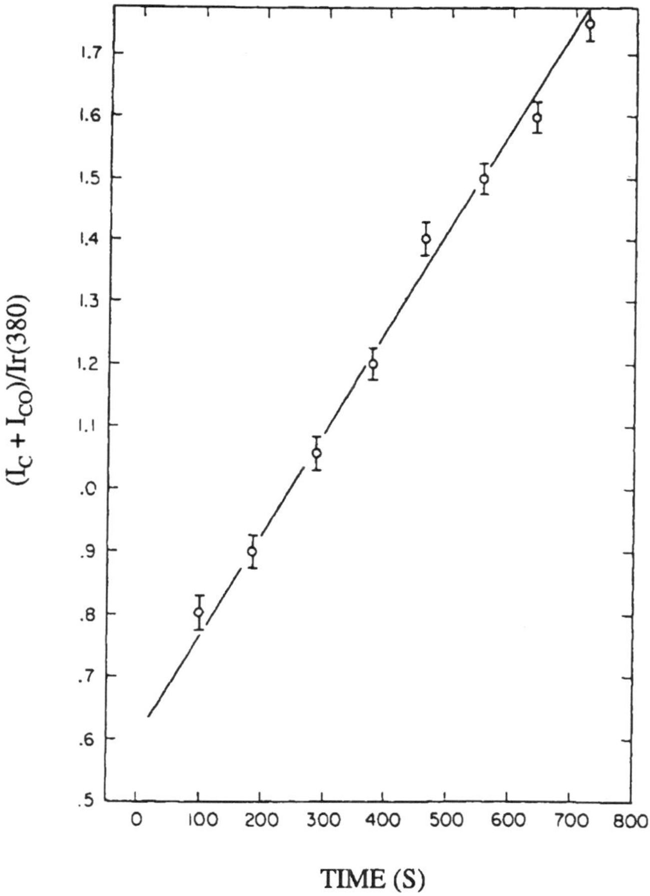

Figure 2. Time-dependent increase in the carbon Auger signal during electron irradiation of CO chemisorbed on Ir(111); the carbon accumulation is associated with the oxygen desorption in Fig. 1. The electron beam current density was 0.6 mA/cm^2 (From Ref. 52).

sorbed on Ir(111). Note that the oxygen signal decreases while the relative carbon signal increases. Here, the net effect of electron irradiation is the dissociation of CO, which leads to desorption of oxygen and the accumulation of carbon on the surface. In addition, it was shown by Hooker and Grant[(192)] that this electron beam effect was apparent not only in the *peak intensity changes* due to desorption but also in *peak shape changes* associated with the dissociation. Figure 3 shows, for example, the Auger spectra of CO on Ni(110) before and after electron irradiation. Note the decrease in O KLL intensity, but, more dramatic, the change in peak shape

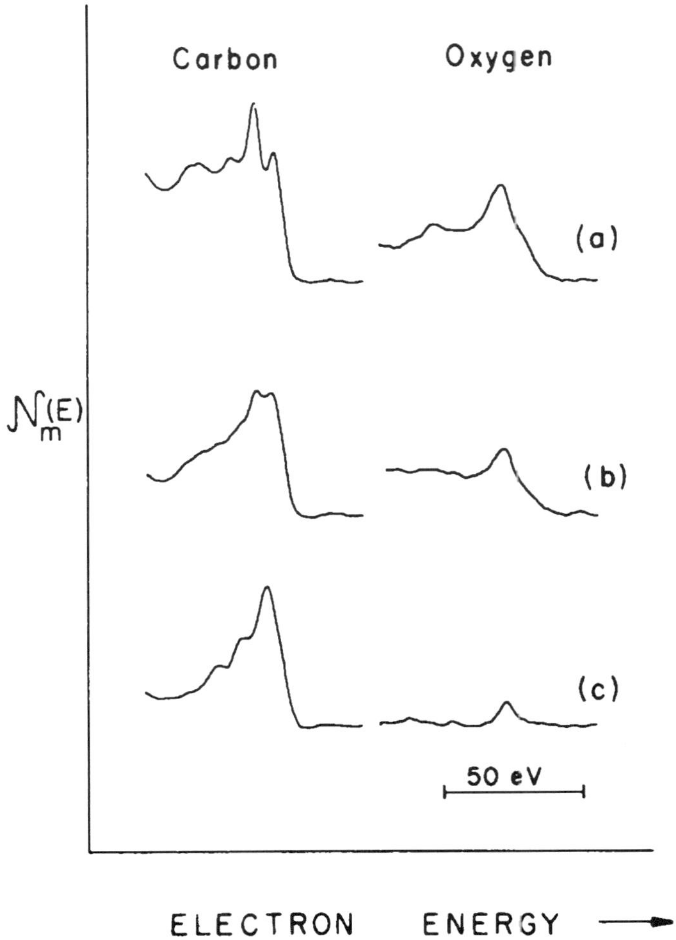

Figure 3. Changes in shape of the C KLL and O KLL Auger peaks during electron irradiation of CO chemisorbed on N(110). The peaks in (a) were recorded with minimal exposure to the electron beam, (b) after 10 min exposure, and (c) after 40 min exposure to the electron beam (From Ref. 192).

of the C KLL. Initially, peak shape is characteristic of carbon in CO; after irradiation it transforms to the shape associated with elemental carbon.

Although ESD is often a problem in Auger and other surface analyses, it can, if properly executed, be used for the characterization of surfaces and adsorbed species on surfaces. Increased understanding of desorption mechanisms, and the development of methods for the detection of desorbing species, has led to the establishment of powerful new techniques for surface science.[(34–136)] In addition to ESD, these include electron-stimulated desorption-ion angular distribution (ESDIAD) and Auger-stimulated desorption (ASD); photon-stimulated desorption (PSD) and angle-resolved PSD are related methods. In turn, these methods have contributed to the interpretation of desorption effects in practical surface analyses.

2.1.2. Electron-Stimulated Adsorption

Electron-stimulated adsorption (ESA) is also a monolayer-level effect where gas-phase species adsorb or react with the surface under electron irradiation. In the absence of electron beam irradiation, the adsorption processes are considerably slower or nonexistent. Figure 4 shows, for example, the Auger spectra for CdS surfaces—in the presence of 2×10^{-8} Torr H_2O—with and without electron irradiation. Note that in the absence of electron irradiation, the adsorption of oxygen is not observed, whereas in the presence of a 4-keV electron beam at 4 $\mu A/mm^2$ an oxygen peak appears. These authors reported ESA of oxygen at water vapor partial pressures as low as 10^{-10} Torr!

The generally accepted model considers electronic excitation or dissociation of a molecule in the gas phase above the surface, or of a molecule weakly associated with the surface, to be fundamental to the ESA process. That is, electron beam irradiation results in the formation of excited molecules or molecular fragments that adsorb or react with the surface at a much higher rate than observed in the absence of excitation. Many examples have been cited in the literature where this model is sufficient to explain the observations.[(39–44)] In more recent years, ESA processes in the absence of electronic excitation of the gas-phase species, per se, have been reported. In these cases, it is proposed that excited species or defect sites within the substrate surface result from electron irradiation, and these enhance the adsorption or reaction of ground-state gas-phase species. This mechanism has been especially evident IN compound semiconductors.[(70,80)] Figure 5 shows how the rate of oxygen adsorption, during electron irradiation at 4 keV and 0.2 mA/cm_2, varies among a variety of semiconductors. The rates correlate with the band gap of semiconductors. This is taken as evidence that electron irradiation excites the surface, rather than the gaseous adsorbate, and thereby stimulates adsorption.

Usually, one can verify that ESA is responsible for sample damage by observing the change in surface composition with the partial pressure of the adsorbate. Figure 6 shows the Auger signals measured during exposure of a Si(111) surface

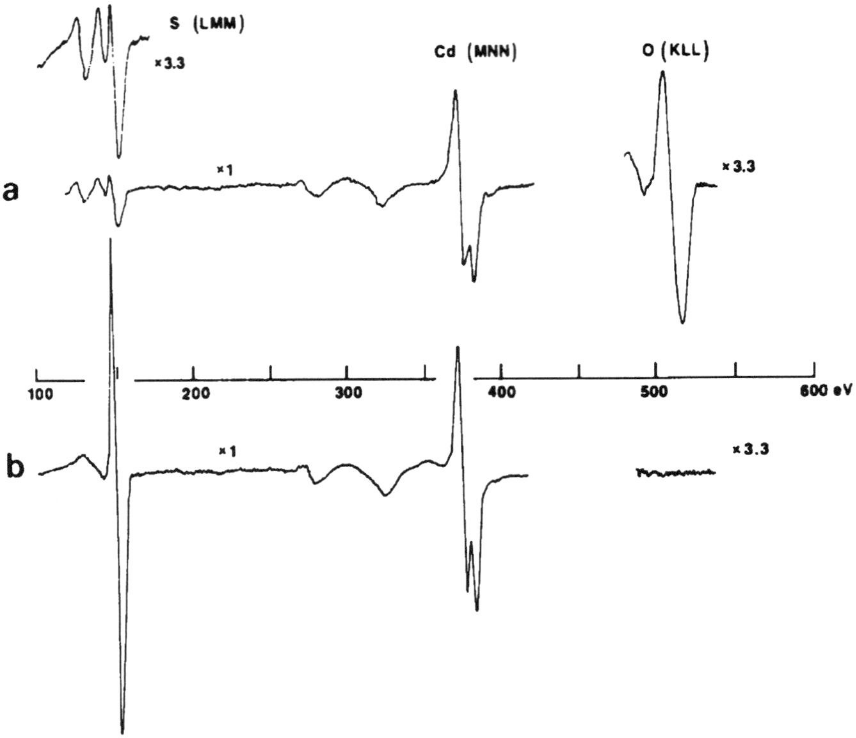

Figure 4. Auger spectra of a CdS(0001) surface (a) after 60 min of electron irradiation at 0.4 mA/cm^2 in the presence of 2×10^{-8} Torr H_2O, and (b) after 15 min exposure to several Torr of H_2O but without electron irradiation (From Ref. 62).

to oxygen at 20°C, with and without continuous irradiation by a 2500-eV beam at 0.25 mA/cm^2. A strong and rapid increase in the oxygen Auger signal occurs at low oxygen exposure. This increase is independent of the electron beam irradiation and, thus, has been attributed to the formation of an oxygen monolayer on the surface. But, at higher oxygen pressures, electron beam exposure enhances the continued oxidation of the surface. This result is due to dissociation of O_2 by the electron beam and the diffusion of these dissociated species into the bulk. The Auger signal tends toward saturation at a high pressure of oxidation in the "beam-on" condition because the oxide layer thickness approaches the oxygen Auger electron escape depth. (This in-depth effect of ESA is discussed later.) These investigators[41] noted that at a lower beam energy of 500 eV, the effect was less prevalent, and it was

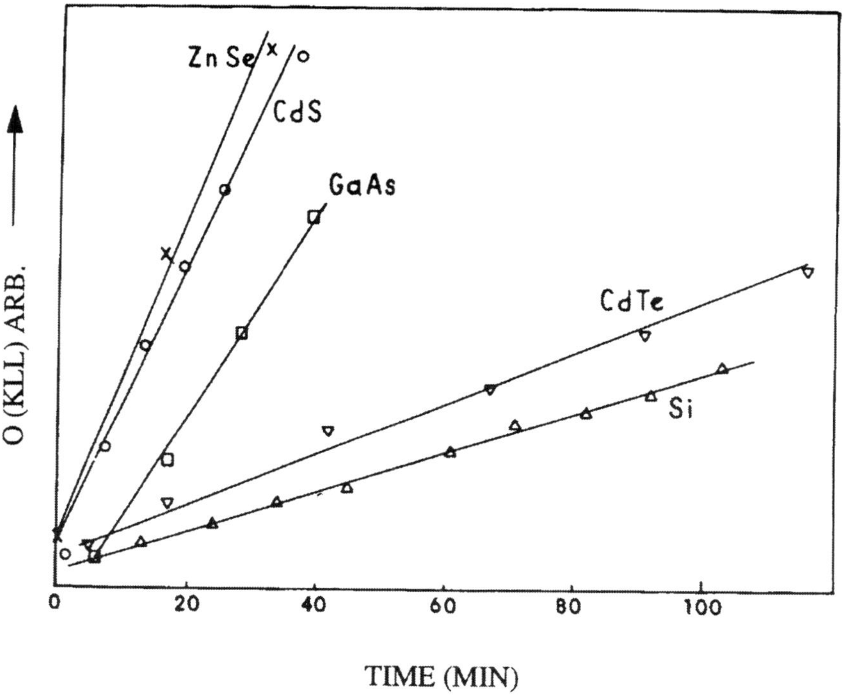

Figure 5. The time-dependent oxidation (represented by the normalized oxygen Auger signal intensity) of semiconductor surfaces during electron irradiation in the presence of 2×10^{-3} Torr H_2O; the electron beam current density was 0.2 mA/cm^2. The band gaps for these semiconductors are 1.11 eV (Si), 1.58 eV (CdTe), 1.43 eV (GaAs), 2.42 eV (CdS), and 2.7 eV (ZnSe) (From Ref. 70).

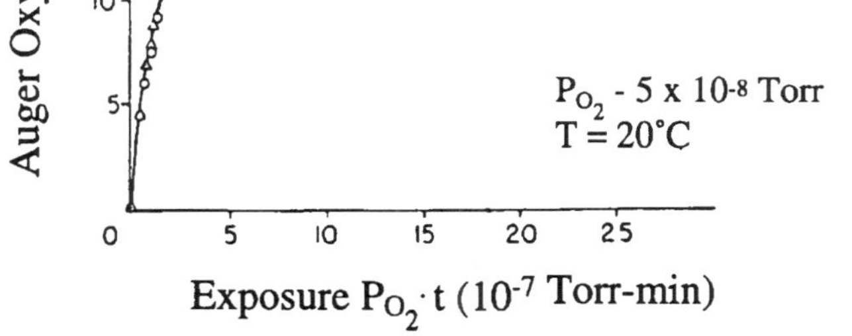

Figure 6. Auger O KLL signal showing the enhanced oxidation of a Si(111) surface resulting from ESA of O_2; the beam current density was 0.25 mA/cm^2 (From Ref. 41).

suggested that this resulted from a decrease in the number and energy of backscattered and secondary electrons that also dissociate O_2 at the surface.

In some instances, it has been observed that the ESA effect is independent of the partial pressure of the adsorbing species. This observation was made by Fontaine *et al.*[67] in their study of electron beam effects on oxygen-exposed aluminum surfaces. They concluded that the electron beam stimulated the chemisorption of oxygen, which was diffusing laterally into the irradiated area. This effect was not observed on clean aluminum surfaces, but only on surfaces contaminated with adsorbed oxygen and oxygen-bearing species. Thus, the effects of ESA on real surfaces—where oxide and other contamination layers may already exist—can be quite complex.

Naturally, ESA is most prevalent on clean metal surfaces, although the effect is also observed on elemental semiconductor, compound semiconductor, and even oxide surfaces. The gas-phase species, which are found to be most often involved

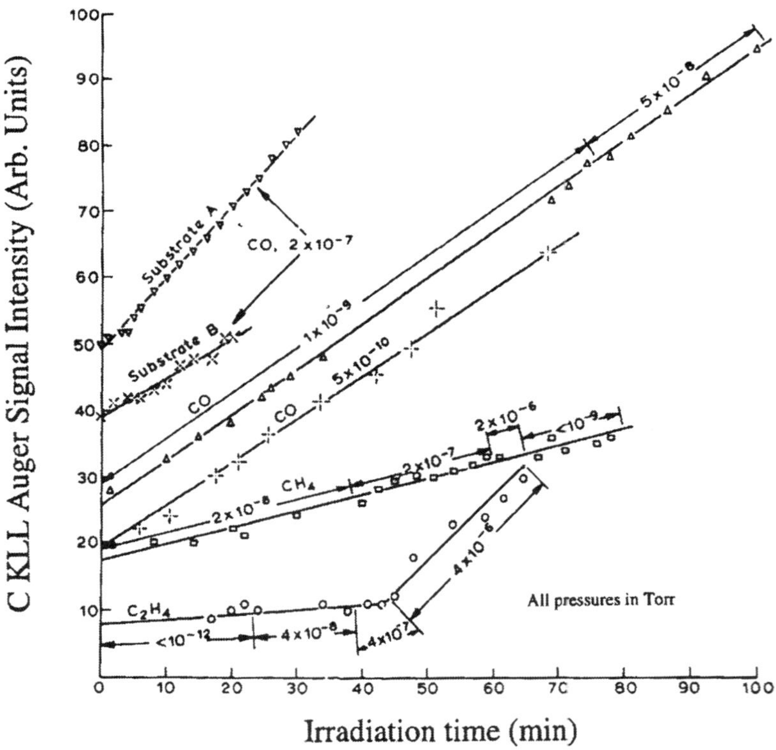

Figure 7. Time-dependent increase in the carbon KLL Auger signal surface resulting from ESA on Si(111); the composition and partial pressure of the carbonaceous adsorbates was varied. Beam current density was not reported (From Ref. 40).

in ESA phenomena, are H_2O, O_2, CO, and CO_2. Not surprisingly, the ESA process is also often accompanied by an ESD process. It has been pointed out that the ESA of H_2O at a metal surface often results in oxygen or hydroxyl chemisorption and desorption of hydrogen. Conversely, the ESA of CO or CO_2 may result in carburization of a metal oxide surface (to metal carbide) and consequent desorption of oxygen.

Because of the prevalent use of AES for the analysis of carbon on surfaces, it is especially important to recognize the ESA of carbon. In a classic study by Joyce and Neave,[40] it was shown that the ESA of carbon on silicon can be observed in the presence of a wide variety of carbonaceous gas-phase species. Figure 7 shows that exposure to CO, CH_4, and C_2H_4 results in the accumulation of carbon on the surface. The rate and extent of adsorption depend upon the chemistry of the gas-phase species, whereas the quantity of carbon accumulated is a function of partial pressure. CO is especially susceptible to ESA. Again, the surface condition also affects the interaction; even though substrate A and substrate B were prepared in an identical fashion, the rate and extent of carbon accumulation are measurably different.

2.1.3. Decomposition of Surface Layers and Thin Films

The ESD process, although fundamentally a surface effect, can result in in depth decomposition of the surface. The electronic excitations that initiate the ESD process occur in depth, and once desorption has occurred within the outermost monolayer of the substrate, an activity gradient exists to drive the out-diffusion of species liberated by bond-breaking events in the subsurface. Here, we refer exclusively to ionic or covalent compounds that may comprise the substrate, a surface layer, or a deposited thin film.

These in-depth effects have been observed in alkali halides and alkaline-earth halides, but are most prevalent in oxides. According to Knotek and Feibelman,[17 18] the materials most likely to decompose under electron bombardment are maximal valency compounds in which the cation and anion have Pauling electronegativity differences >1.7; these materials include SiO_2, Al_2O_3, B_2O_3, TiO_2, V_2O_5, WO_3, and MoO_3. In this case, the decomposition is essentially a reduction of the oxide because of O^+ desorption. Figure 8a shows how the progressive decomposition of silica glass manifests itself in the Si LVV spectra. The growth of a peak at 92 eV results from elemental Si, and, although not shown here, a corresponding decrease in the oxygen peak intensity is observed because of oxygen desorption. Figure 8b shows that this decomposition can be significantly reduced by decreasing the beam current density. This observation verifies that the effect is associated with electronic excitation, and, as will be shown, the rate can be quantified in terms of beam current density and a material-dependent cross section for beam damage.

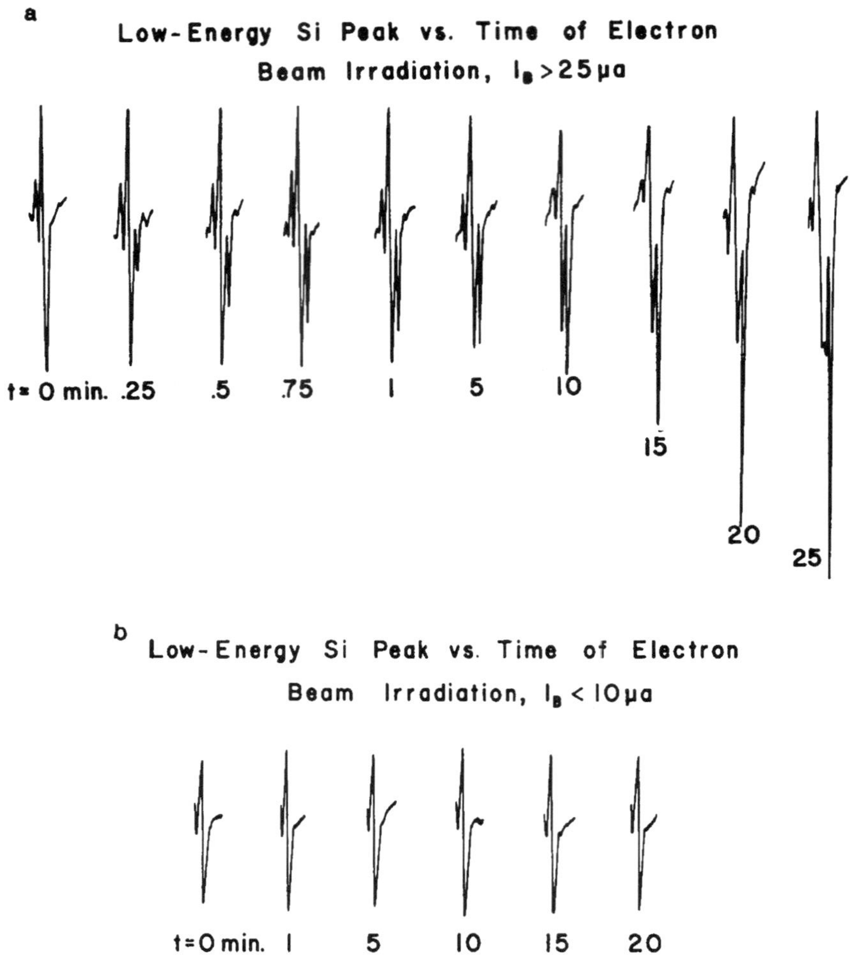

Figure 8. (a) Si LVV Auger peak shape changes because of electron-beam-stimulated decomposition of a SiO_2 glass surface at >2.5 mA/cm^2, and (b) the stability exhibited after reducing the beam current density to ~0.1 mA/cm^2.

The most important and prevalent examples of these oxide decomposition effects in AES are summarized in Fig. 9. These spectra provide clear indications of electron-beam-induced decomposition. In this case the changes in peak shape verify modification of the chemical structure and composition of the oxide surface. The spectra on the far left were recorded prior to any appreciable irradiation, the center spectra after an intermediate dose, and those on the far right after extensive irradiation. These spectra probably correspond to the progressive growth of a

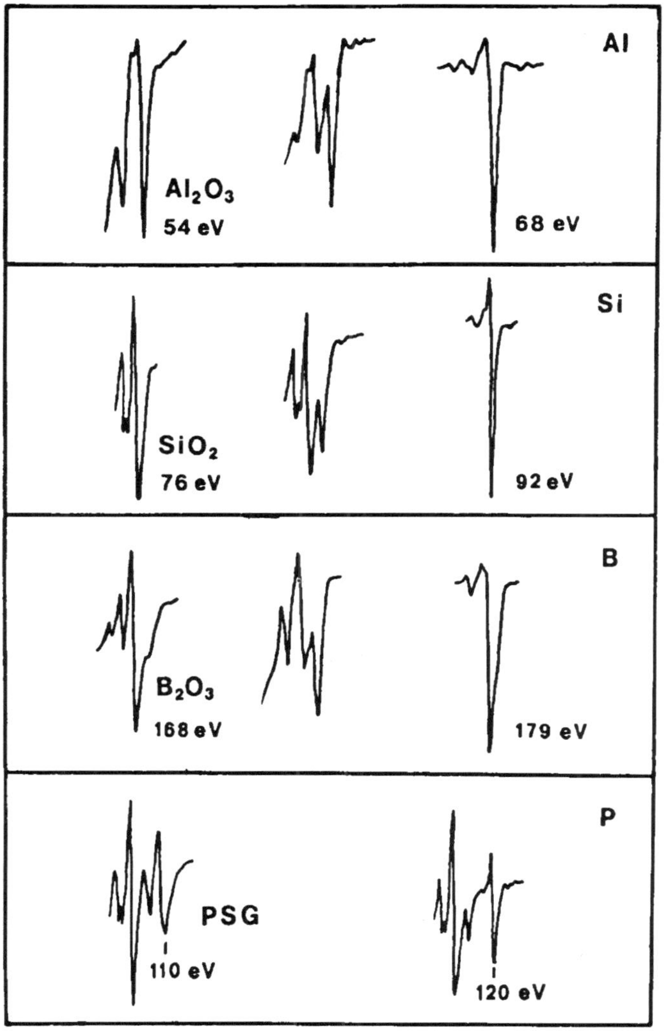

Figure 9. Effects of electron-stimulated decomposition upon the spectra of Al_2O_3, SiO_2, B_2O_3 and phosphosilicate glass. The spectra on the left were obtained at a beam current density less of than 1 mA/cm^2. At higher beam current density, a progressive reduction of the oxides is verified with growth of the "elemental" Auger peaks for Al, Si, B, and P (From Ref. 12).

metallic surface layer, which results from decomposition, until its thickness exceeds the escape depth of the Auger electrons. A similar effect has been reported during the decomposition of CaF_2.[65]

The electron-beam-induced decomposition of oxide surfaces does not always result in the formation of a metallic layer on the surface, i.e., complete reduction of the oxide. Sasaki *et al.*[54] used XPS to monitor the electron-beam-induced decomposition of chromates and tungstates. Figure 10 shows changes in the XPS Cr 2*p* spectra, with increasing electron dose, that correspond to the reduction of Cr(VI) to Cr(III). Figure 11 reveals that after extensive irradiation the decomposed

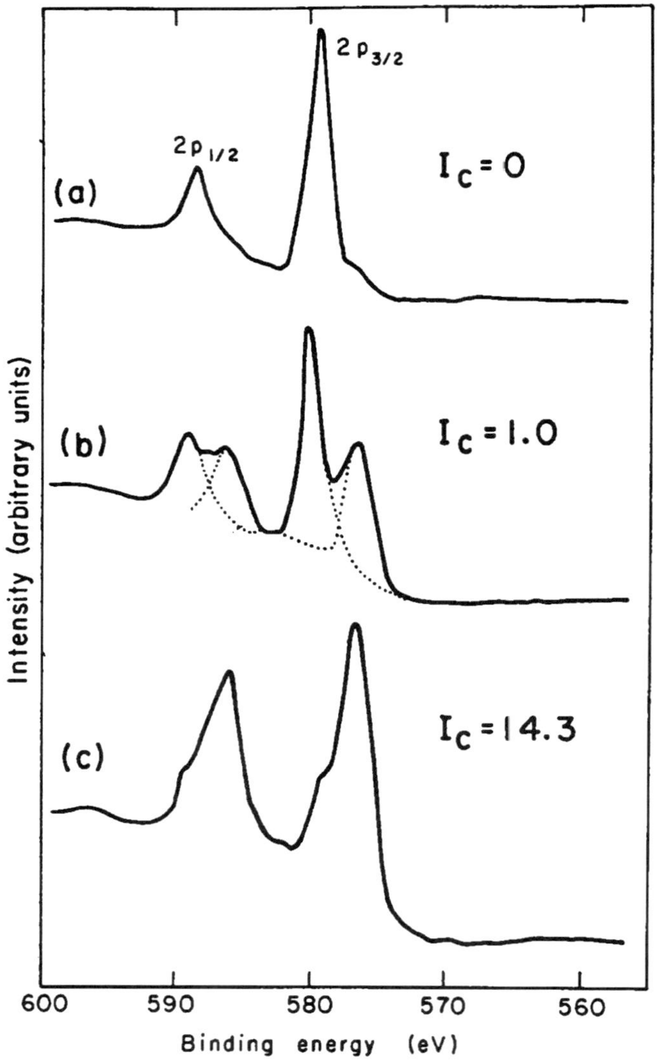

Figure 10. XPS Cr 2*p* spectra of Li_2CrO_4 (a) before and (b,c) after exposure to electron irradiation (I_c = total dose in Coulombs/cm^2); the new peaks (b) and (c) represent Cr^{3+} that resulted from the electron-stimulated decomposition of Li_2CrO_4. (From Ref. 54).

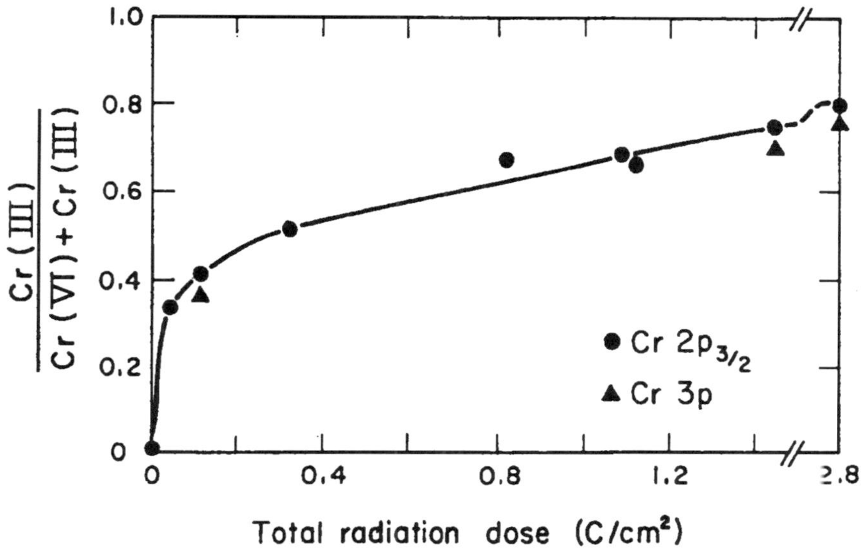

Figure 11. Electron-stimulated decomposition of Li_2CrO_4 ($Cr^{6+} \rightarrow Cr^{3+}$) measured by XPS as a function of electron dose. A reduction to elemental Cr was not observed (From Ref. 54).

surface layer is approximately 80% Cr(III). For bombardment by 0.5–1.5-keV electrons, at least, the formation of metallic Cr was not observed.

2.1.4. Oxidation of Surface Layers and Thin Films

In a manner analogous to ESD-induced decomposition of surface layers, the ESA of oxygen can result in the in-depth oxidation of metallic and elemental semiconductor surfaces. Here, the higher concentration of adsorbed oxygen under the electron beam results in diffusion of oxygen into the bulk. Figure 12 shows the oxygen depth profiles observed at a Ni surface after 10 and 100 min of electron irradiation in the presence of ~10^{-7} Torr H_2O. It was estimated that this oxidized surface layer was many nanometers thick, i.e., greater than the escape depth of the O KLL Auger electrons. Thus, it is not surprising that the oxygen Auger signal at the surface saturates at a comparable value after 10 or 100 min of electron bombardment. This observation means that even though changes in the Auger spectra associated with ESA may cease, e.g., the increase in oxygen signal intensity, the sample modifications may be continuing in depth. This in-depth oxidation effect has been reported for Si and Al in the presence of O_2 and for Ni in the presence of H_2O. Also note that, although this effect has been studied and reported only for oxygen, it seems likely that ESA of carbon or hydrogen can result in in-depth carburization or hydrogenation.

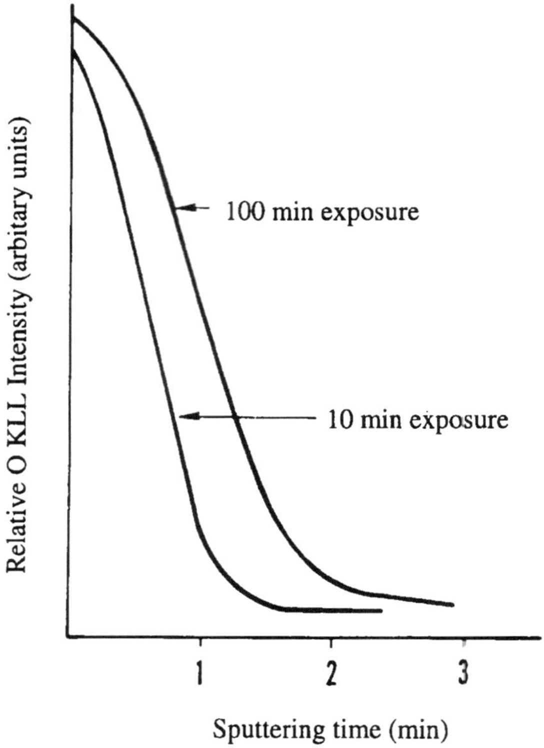

Figure 12. Auger O KLL relative intensity measured at a Ni surface after 10 and 100 min exposure to an ~50 mA/cm^2 electron beam in the presence of 1.5×10^{-7} Torr H_2O; the in-depth effect of ESA is evident even though the surface coverage with oxygen is saturated after 2–3 min (From Ref. 48).

On the other hand, some materials show a saturation of the adsorbed species at a thickness of less than one monolayer. Although a well-established model for this effect does not exist, it seems likely that a steady-state surface composition is reached in response to the relative rates of ESA and ESD. That is, partial oxidation of the surface resulting from ESA may turn off the ESA process or result in a kinetic equilibrium between ESA and ESD.

2.1.5. Thresholds for Sample Damage Resulting from Electronic Excitation

Electron-stimulated desorption, adsorption, decomposition, oxidation, etc., of specimens under electron beam irradiation is fundamentally dependent on an initial electronic excitation step. Figure 13 illustrates this relationship schematically. It reveals that the rate of sample damage, resulting from any of these phenomena, can

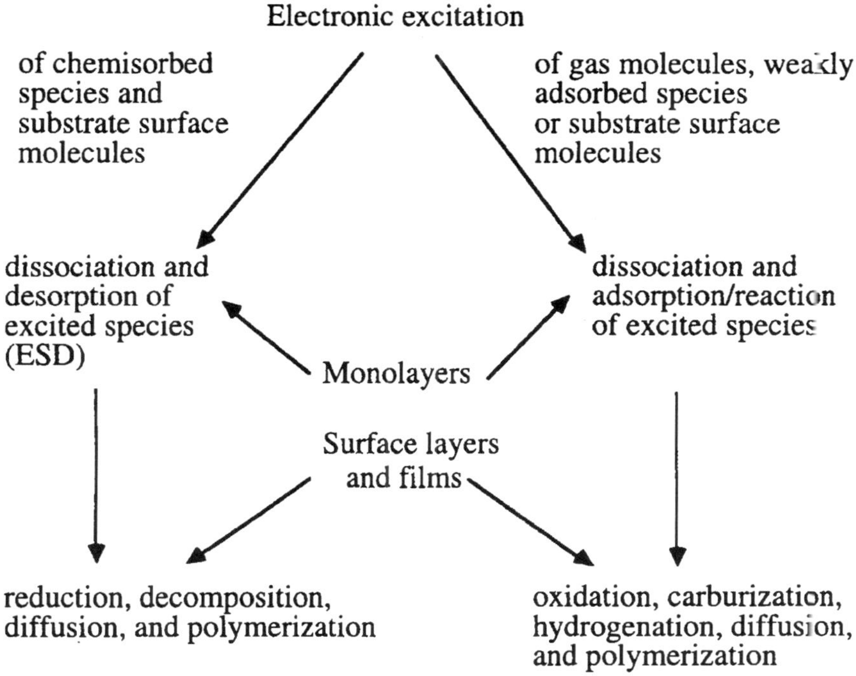

Figure 13. Potential consequences of electron beam irradiation at solid surfaces; it is implied that electronic excitation or ionization of surface species is fundamental to most beam damage mechanisms.

be defined on the basis of the initial electronic excitation. That is, a threshold or cross section can be defined to characterize material dependence, and this can be used to calculate the rate of sample damage in terms of the incident electron flux density. The cross section may depend on the electron energy, but this dependence is usually weak in the energy range of interest for surface analyses; it will be ignored in this treatment. The influence of secondary-electron emission will not be treated explicitly, SO the calculated damage thresholds may be underestimated by a factor of 2 or 3.

The concentration of "undamaged" surface species or surface sites, $N(t)$, is assumed to obey the first-order relation

$$N(t) = N(0)\exp(-I_e Q/A\varepsilon) \qquad (1)$$

where $N(0)$ is the initial undamaged concentration of surface species or surface sites per square centimeter within the analysis region, I_e is the total beam current in amperes, A is the irradiated area in cm^2, ε is the electronic charge (1.6×10^{-19} C), and Q is an effective cross section (cm^2) for the electron-stimulated change in the ratio $N(t)/N(0)$. The use of this first-order rate equation is based upon the simple ESD of monolayers.[19,20,23,24]

The electron dose, D, in Coulombs/cm^2 is given by the product of current density and time; i.e.,

$$D\ (\mathrm{C/cm^2}) = I_e t/A \tag{2}$$

In ESD or ESA, it is reasonable to assume that a 10% change in the surface coverage, $\Delta N/N$, is easily detected. Thus, a damage threshold or critical dose, D_d, can be calculated, assuming that "detectable damage" has occurred when $\Delta N/N =$ 0.1 (i.e., $N(t)/N(0) = 0.9$). Thus, from Eqs. (1) and (2), the electron dose required to cause detectable damage to a monolayer, D_d, is

$$D_d = \{\ln N(0)/N(t)\}\varepsilon/Q \tag{3}$$

A similar criterion can be used to approximate the electron dose required to oxidize or decompose surface layers in depth. In this case, the cross section, Q, must be normalized to the number of atomic or molecular layers, N, affected. In AES the analysis depth is approximately three times the inelastic mean free path, λ_e. The number of layers within the range of detection is $N = 3\lambda_e/a$, where a is the associated atomic or molecular dimension, or typically 10–20 atomic layers. Thus, the electron dose, D'_d, necessary to damage N layers, assuming a detectable level of concentration change $\Delta C/C = 0.1$, where C is the concentration of atoms or molecules per cubic centimeter, is

$$D'_d \sim 0.11 N\varepsilon/Q \tag{4}$$

A typical cross section for dissociative ionization of adsorbed or gaseous molecules by 100 eV to >1000 eV electrons is ~10^{-16} to 10^{-17} cm^2. One can assume that each of these excitations leads to an ESD or ESA event, and in this case, the damage threshold, D_d is about 10^{-4} to 10^{-3} C/cm^2. Because many of the excited species undergo deexcitation and/or neutralization, the doses calculated with Eq. (4) overestimate the threshold for damage. In fact, the cross sections for ESD, at least, have been measured and are more typically 10^{-18} cm^2. Thus, damage thresholds of the order 10^{-2} C/cm^2 for 10 to 20 monolayers are more reasonable.

These estimates of threshold dose can be related to the instrumental electron beam current density or damage rate through Eqs. (2), (3), and (4), i.e.:

$$D\ (\mathrm{C/cm^2}) = J\ (\mathrm{amps/cm^2}) \cdot t\ (\mathrm{s}) \tag{5}$$

so the time in seconds required for 10% damage to a monolayer is

$$t = 0.11\varepsilon/QJ = D'_d/J \tag{6}$$

and for N molecular layers

$$t = 0.11\varepsilon N/QJ = ND'_d/J \tag{7}$$

A 1-μA electron current in a 0.3-mm full-width half-maximum (FWHM)-diameter beam is typical in many surface analyses and yields a current density of ~1 mA/cm^2. Thus, the damage to a monolayer would be observed in about 10 s, and the damage to 10 molecular layers in about 100 s. Of course, the electron beam current density used in Auger microanalysis can be as high as 1 A/cm^2, so the damage rate may be 100–1000 times greater.

These estimates, and the first-order rate Eq. (1), are consistent with real data concerning electron beam damage. The data in Fig. 1, for example, are a semilog plot that corresponds to Eq. (1). The "damage" in this case is the desorption of oxygen resulting from the dissociation of CO chemisorbed on Ir(111). The damage cross section, Q, calculated from these data, is ~10^{-17} cm^2; this result corresponds to a damage threshold, D_d, of 10^{-3} C/cm^2. The cross section for ESD of oxygen on Ir(111) has been independently determined to be 7×10^{-19} cm^2. This difference reflects the more complex nature of CO dissociation or desorption or both. It suggests, at least, that the desorption of oxygen occurs through some excited state of CO, e.g., $CO_{ads} \rightarrow C_{ads} + O\uparrow$, and not through an adsorbed state of oxygen created through the dissociation of CO, e.g., $CO_{ads} \rightarrow C_{ads} + O_{ads} \rightarrow C_{ads} + O\uparrow$.[51]

The simple excited-state model can be used to estimate the damage thresholds of other surfaces and materials where detectable electron beam effects have already

Table 2. Critical Doses (D_c) and Times (t) for Electron Beam Damage in Various Materials

Material	Energy (keV)	D_c (C/cm^2)	t @1 A/cm^2	Ref.
Si_3N_4	2	Stable	—	193
Al_2O_3	5	10	3 h	9
Cu, Fe pthalocyanines	1	1	15 min	196
SiO_2	2	0.6	10 min	196
Li_2WO_4	~1	0.05	8 min	54
NaF, LiF	0.1	0.06	60 s	38
$LiNO_3$, $LiSO_4$	1	0.05	50 s	54
KCl	1.5	0.03	30 s	38
TeO_2	2	0.02	20 s	202
H_2O (film)	1.5	0.01	10 s	59
Native oxides	5	0.002	2 s	8
Formvar	75	0.001	2 s	225
C_6H_{12} (film)	0.1	0.0003	0.3 s	203
CH_3OH (film)	1.5	0.0003	0.3 s	59
Na_3AlF_6	3	0.0001	0.1 s	51

been observed. The critical dose, D_c, is listed in Table 2 for electron beam damage to various materials; D_c was calculated for each on the basis of some reported change in the Auger spectra. The values of D_c may or may not correspond exactly to D'_d, i.e., a 10% change in composition over an analysis depth of 10 atomic layers. But the fact that so many organic and inorganic solids exhibit critical doses in the predicted range of 10^{-4} to 10^{-2} C/cm^2 indicates that the initial electronic excitation step is rate limiting to the beam damage mechanism. The relative stability of the various material types can also be observed in these data. The time required for detectable damage in each of these materials was calculated, using the reported critical doses, assuming a current density of 1 mA/cm^2. Not surprisingly, the bulk oxides and nitrides are much more stable than halides and organics.

2.2. Charging Insulators

Electrical charging of insulator surfaces is a severe limitation in AES and other surface analytical techniques where electron beams are used. The effects are complex because they depend on the bulk electrical properties of the material, the condition of the surface, and the instrumental beam parameters.

The electrostatics associated with charging insulators under electron bombardment have been described in detail by Cazaux.[146] In AES and other surface analytical techniques, he defines the surface electric field in terms of a thin positively charged surface layer and negative subsurface space charge, as shown in Fig. 14.

The positively charged surface layer develops because of the emission of secondary electrons, while the negative space charge in the subsurface results from the penetration and trapping of primary electrons. The parameters that influence the sign and magnitude of the field are the stored or steady-state charges per unit area, q_+ and q_-, and the charge densities, ρ_+ and ρ_-:

$$q_+ = J_0\delta t, \qquad \rho_+ = \frac{J_0\delta t}{d_s}, \qquad q_- = J_0(1-\eta t), \qquad \rho_- = \frac{J_0(1-\eta t)}{d_p} \tag{8}$$

where J_0 is the incident primary-beam current, δ is the secondary-electron emission coefficient, η is the backscatter coefficient, t is the time, d_s is the thickness of the positively charged surface layer, and d_p is the penetration depth of the incident electrons.

When $1 - \eta > \delta$, the total negative charge q_- eventually exceeds q_+ and the surface potential can become negative. This situation is usually unmanageable, especially in good insulators that can trap high-energy primary electrons in the band gap. The negative surface potential increases linearly with time until the electric field strength builds to its breakdown value or until it deflects or retards the primary-electron beam. The net effect is an unstable surface potential. This charging

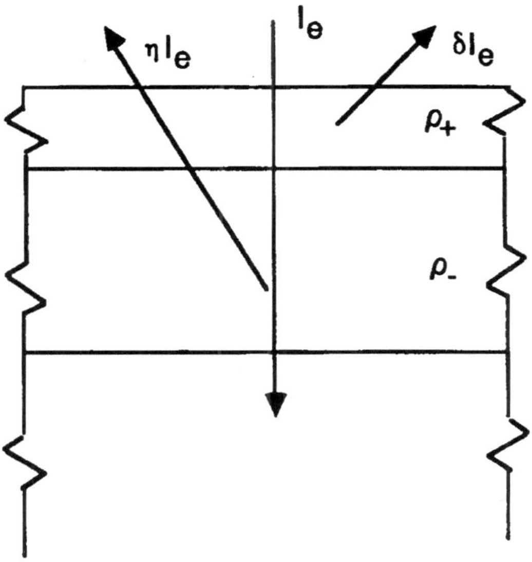

Figure 14. Schematic representation of the surface potentials developed in an insulator during electron beam bombardment.

and discharging of the surface shifts the energies of the ejected Auger and other secondary electrons in and out of the correct energy for a particular Auger peak and, thereby, precludes the acquisition of meaningful AES spectra. For a poor insulator, the charge distributions δ_+ and δ_- may reach a steady state and the surface potential, although negative, should remain nearly constant. This situation shifts the Auger spectra to higher energy. Unfortunately, the shift can be sizable, can limit the maximum beam current that can be used, and can often continue to increase slowly with time. The surface potential can also distort the shape of the measured Auger peaks.

The ideal, and more common, situation occurs when $\delta > 1 - \eta$. Here, the total positive charge q_+ exceeds q_-, and the surface potential builds positively. However, the positive increase in surface potential is limited because the low-energy secondary electrons cannot overcome the increasing surface potential barrier. Thus, the surface potential is inherently stabilized at an equilibrium positive value and results in a uniform shift of the Auger spectrum to lower energy. These shifts are typically 2–10 eV and are exceedingly stable. The situation $\delta > 1 - \eta$ can be achieved by (i) adjusting the primary-beam energy to a value between the so-called crossover energies, E_{p1} and E_{p2} where $\delta > 1$ (as seen in Fig. 15), (ii) increasing the angle of incidence between the primary beam and the sample normal to 45–70° (that increases δ), or (iii) increasing the sample temperature (that increases the conductivity). In many samples, $\delta > 1 - \eta$ is achieved by serendipity because the surface

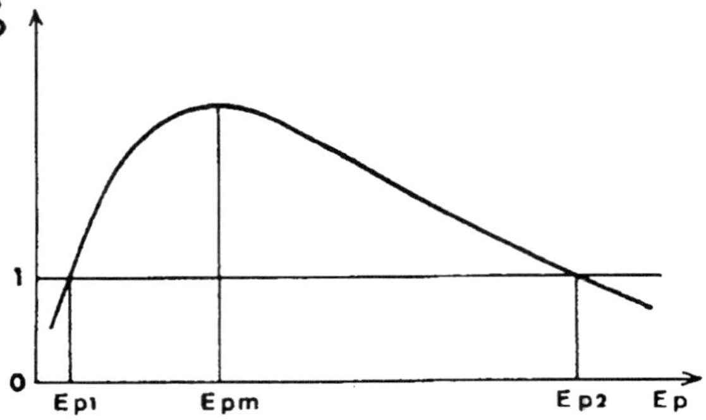

Figure 15. Variation in the secondary-electron coefficient (δ) as a function of the primary-electron beam energy E_p.

is conductive, because mobile ions are present in the sample, or because the irradiation enhances the bulk conductivity. In general, though, the conditions where $\delta > 1 - \eta$ vary with the material and are obtained by trial and error.

2.3. Electromigration in Insulators

Although the attainment of a stable surface potential is necessary for the acquisition of meaningful spectra for insulators, electron bombardment may still result in other forms of sample damage. Specifically, electron bombardment is found to cause electromigration of mobile ions, and, in many cases, it enhances the ESD of ions. Ironically, it is probably this ion migration that provides the conductivity necessary to attain a stable surface potential and, thereby, the measurement of the Auger spectra of many insulators. This situation is especially common in glasses. Of course, the net effect of this electromigration is a change in the composition as a function of depth in the surface layer being analyzed. It is seldom possible to eliminate this effect, so one is forced to recognize the impact of the effect and deal with it in the data analysis.

The Cazaux model[146] of an insulator surface charged by an incident electron beam can also be used to explain these electromigration effects. If the surface charges negatively because $\delta < 1 - \eta$, positive ions will migrate toward the surface. In Fig. 16, an example of this situation is shown by the Auger spectra that were measured at the cleavage surface of Na-βAl_2O_3 before and after electron bombardment. The electron beam voltage was 5 keV, which yielded a large negative charging of the surface and resulted in sodium migration toward the surface. An increase in the Na KLL Auger signal, after only 3 min of bombardment at $1 A/cm^2$, is clearly evident. Figure 17 shows that the increase in the Na KLL Auger signal depends upon the beam current density and the beam energy, but, most importantly, verifies

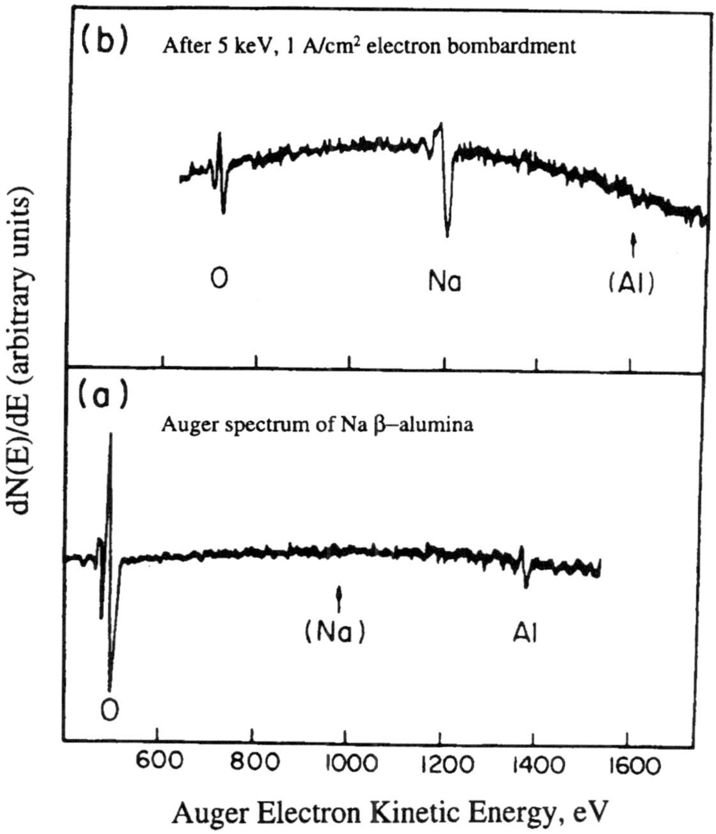

Figure 16. Auger spectra for Na-βAl_2O_3 (a) initially at 0.17 A/cm^2, and (b) after 3 min irradiation with a 1-A/cm^2 electron beam. Shift in the (b) spectrum to higher energy indicates negative charging of the surface (From Ref. 69).

that the electromigration of sodium is associated with negative charging of the surface. The surface charge was measured by the shift in the Auger spectra.

In the more common situation where the insulator surface charges positively because $\delta > 1 - \eta$, Cazaux's model[146] reveals that the electric field within the positively charged surface layer is nonuniform. This situation results in an inversion point within the positively charged surface layer that, in most cases, is probably within the escape depth of Auger electrons. The nonuniform field results in the migration of, for example, positive ions into the bulk if they reside at a depth greater than the inversion point and to a simultaneous migration of positive ions toward the surface if they reside at a depth less than the inversion point.

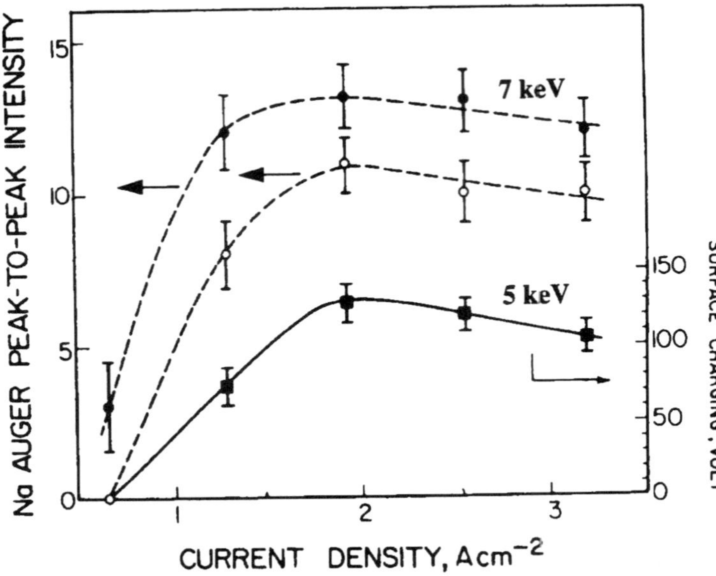

Figure 17. Auger signal changes resulting from electromigration of sodium to the surface of Na-βAl_2O_3 with increasing current density of electron bombardment. Note that the outward migration of sodium correlates with negative charging of the surface (From Ref. 69).

Figure 18 shows Na depth profiles measured by Rutherford backscattering after electron beam irradiation of a soda–silica glass. The presence of a Na-depleted region near the surface is clearly evident after the irradiation, and, correspondingly, a region of Na enrichment in the subsurface can be seen. There is no question that migration of Na has occurred into the bulk, although most of the Na depletion at the surface may result from ESD (see the sequel). The amount of Na migration increases with the beam current density and beam energy, while the depth at which the subsurface sodium accumulates increases with beam energy.

Figure 19 provides additional evidence that electron bombardment can result in ion migration into the bulk. Here, Auger analysis during sputter depth profiling was performed for a 100-nm-thick soda–silica glass film on a steel substrate. It can be seen that sodium has migrated to the glass–substrate interface. Sodium accumulates at this interface because the field terminates at the metal substrate or because sodium ions have a negligible diffusivity in the steel. The profile obtained "before irradiation" shows the migration that occurred resulted from ion sputtering and electron irradiation used to obtain the Auger spectra. The other profile was obtained after an electron irradiation of the film that was sufficient to reduce the Na KLL Auger signal at the surface to 25% of its initial value. Thus, the difference in the

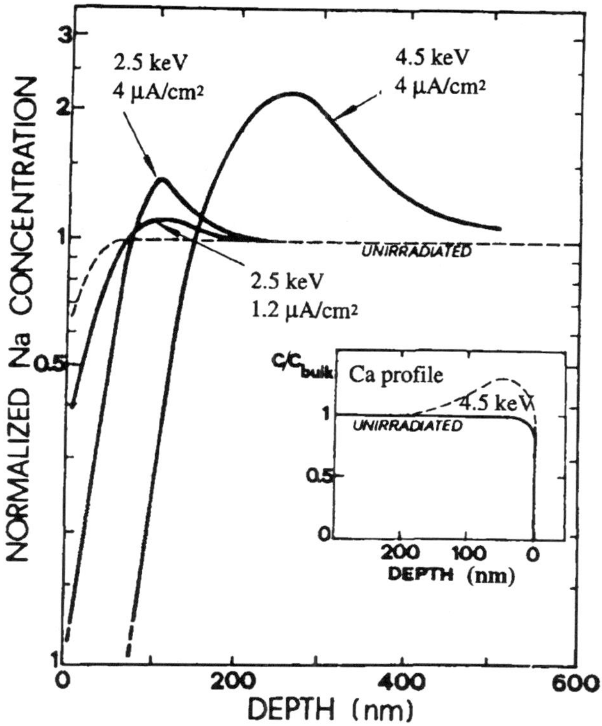

Figure 18. Sodium and calcium concentration depth profiles (obtained by Rutherford backscattering) in glass after electron irradiation under the designated conditions (From Ref. 176).

sodium profiles is the result of electron irradiation only. In another study [167] it was shown that cooling the sample to liquid-nitrogen temperature significantly reduced the sodium migration during electron irradiation.

In general, the Auger signal depends on many factors, including beam current density, beam energy, sample temperature, depth of the inversion point, the diffusion coefficient of the migrating ion, and the escape depth of the detected Auger electrons. It is important to recognize that the change in the measured signal intensity results from changes in *both* the quantity and the distribution of the detected species in the volume defined by the escape depth. If the migration of an ionic species to the surface occurs faster than migration into the bulk or the depth of the inversion point, d_s, is greater than the escape depth ($d_s > \lambda_e$), an (initial) increase in signal will be observed; later, this signal may decrease. Alternatively, the inversion point may be very close to the surface ($d_s < \lambda_e$) or migration into the bulk may be most rapid. In this case, a continuous decrease in signal may be observed.

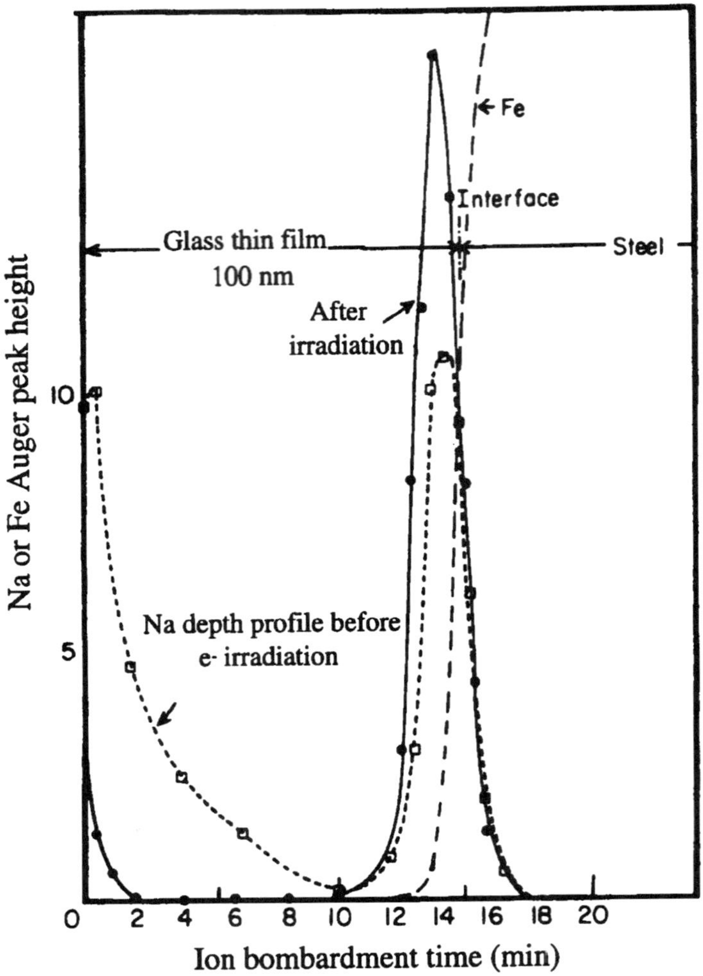

Figure 19. Auger peak to peak intensities during sputter depth profiling obtained before and after "pre-irradiation" of a glass thin film on steel with a 5-mA/cm^2 electron beam (From Ref. 169).

2.4. Electron-Beam-Induced Heating

The irradiation of solids with energetic electrons leads to power dissipation in the surface and, thereby, can cause heating of the specimen. Pittaway[224] has modeled the surface temperature rise in a semi-infinite solid under a stationary electron beam. The beam is considered to have a stationary Gaussian profile in this treatment. His expression for the temperature, T, at the center of the beam spot is

$$T = E_0 t \tan^{-1} 2(\kappa_t/d^2)^{0.5}/\pi^{1.5} K d \quad (9)$$

where E_0 is the beam energy deposited per second (in watts), K is the thermal conductivity of the semi-infinite solid, d is the width parameter for the Gaussian beam profile, κ is the thermal diffusivity, and t is the time. Thus, the steady-state temperature rise (at $t = \infty$) is directly proportional to the beam power and inversely proportional to the thermal conductivity. Obviously, beam heating will be most prevalent for insulating materials and for Auger microanalyses where a large beam energy and a small spot size are used.

Gossink *et al.*[170] used Pittaway's expression to calculate the beam heating effect at the surface of a multicomponent silicate glass. A 50-μm-diameter beam at 3 keV and 10 μA was assumed (~500 mA/cm^2), and this gave an equilibrium temperature rise of 184°C. They further reported that this temperature rise would occur within 0.5–0.6 ms. We deposited thin-film thermocouples on glass microscope slides and measured local temperature rises of about 150–250°C at 1 mA/cm^2. Naturally, the size and mounting of the specimen also influences the beam heating effect, but these data give an order-of-magnitude estimate of the associated temperature rise for massive specimens. In the case of thin glass fibers and thin glass membranes, visible melting or softening has been observed. Since the glass transition temperatures were approximately 500–600°C, it is clear that, in some instances, beam heating can be rapid and result in high temperatures in the beam impact region.

The temperature rises in the sample surface resulting from beam heating do not, *a priori*, influence the surface analysis or constitute beam damage. However, the temperature change can assist or influence other beam damage mechanisms. This situation can arise in beam-induced electromigration and segregation because the diffusion process is thermally activated.

3. Electron Beam Effects in Auger Surface Analyses

In most practical surface analyses, one or more of the fundamental phenomena described in the last section may play a role in sample damage. Often a combination of effects are observed simultaneously. Also, one beam damage mechanism can modify the surface and, thereby, create a susceptibility to another effect. For example, the beam-induced decomposition of metal halides or sulfides can result in subsequent electron-stimulated adsorption and oxidation. In this section a few of the more practical aspects of sample damage will be exemplified and related to the fundamental mechanisms of beam damage.

3.1. Physical Effects

In some instances, the most significant effect of electron beam damage does not appear in the surface-sensitive spectra but in the form of visible damage to the specimen, such as discoloration, etching, blistering, crystallization, cracking, or melting. The formation of spots and discoloration, directly at the irradiated spot, are a fairly common phenomenon. A common source of blackening and discoloration resulting from electron beam irradiation is carbon deposition. This is observed on "real" surfaces that have been exposed to environments where hydrocarbons, in particular, may be adsorbed on an oxide surface or an oxidized metal surface. Generally, the hydrocarbons undergo a beam-induced decomposition to leave a

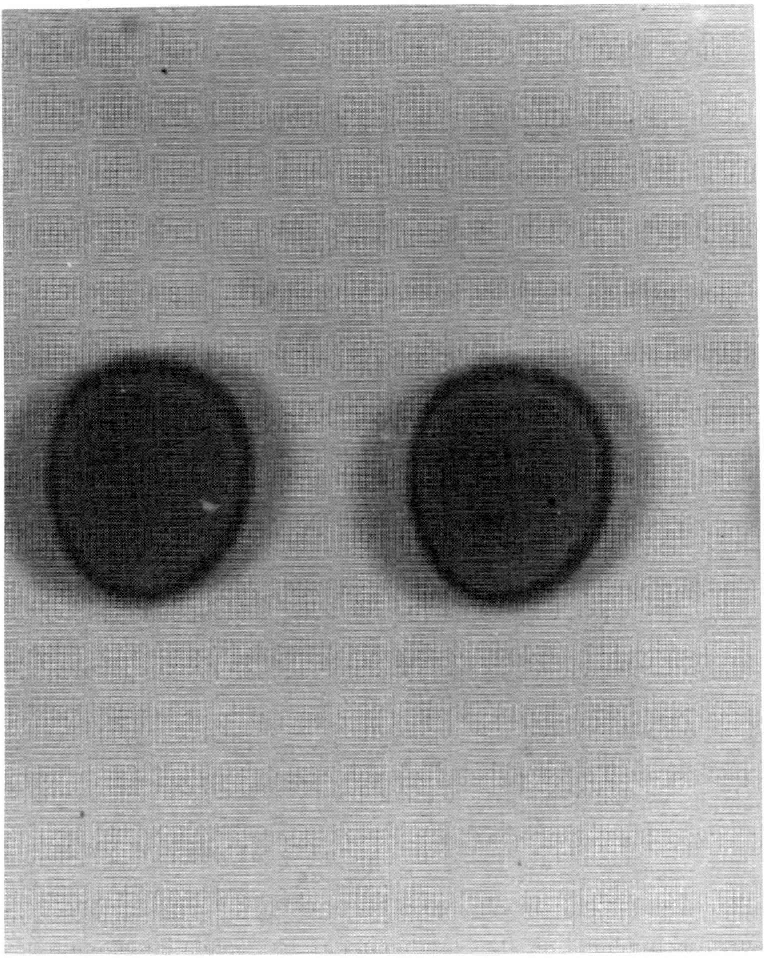

Figure 20. Optical microphotographs showing the physical effects of electron beam damage in a phosphosilicate glass film. These imprints replicate the ~80 μm FWHM electron beam diameter (From Ref. 12).

visible film of elemental carbon. For clean surfaces, especially metals, the electron-stimulated adsorption and dissociation of CO may leave a layer of carbon on the surface. With organic polymers or inorganic materials with organic coatings, localized discoloration resulting from decomposition and consequent carbon-layer formation is more the rule than the exception.

The localized discoloration of specimens during electron beam irradiation is not limited to carbon film formation. Figure 20 shows the electron beam imprints left on a phosphosilicate glass thin film on a silicon substrate. There was evidence that desorption of phosphorus and decomposition of the oxide were responsible for the damage in this case. The formation of spots under the electron beam is also common with other bulk or thin-film glasses. Such damage often occurs when the glass contains an alkali oxide because this species is easily reduced by the electron beam, and its mobility readily results in the precipitation of microscopic alkali-metal particles. A dark, blackened spot, usually within the subsurface is often produced by the beam. The only consequence of this effect in the Auger spectra is a decrease in alkali signal.

Figure 21 shows the "tracks" left in the surface of an Nb_2O_5 ceramic specimen resulting from electron beam irradiation. The tracks were formed by moving the beam from one spot to another, and it thus seems clear that the damage is nearly instantaneous in this case. It was verified by the investigators that ESD of O^+, and subsequent decomposition of the Nb_2O_5 to NbO_x and then NbO^+Nb, results in the

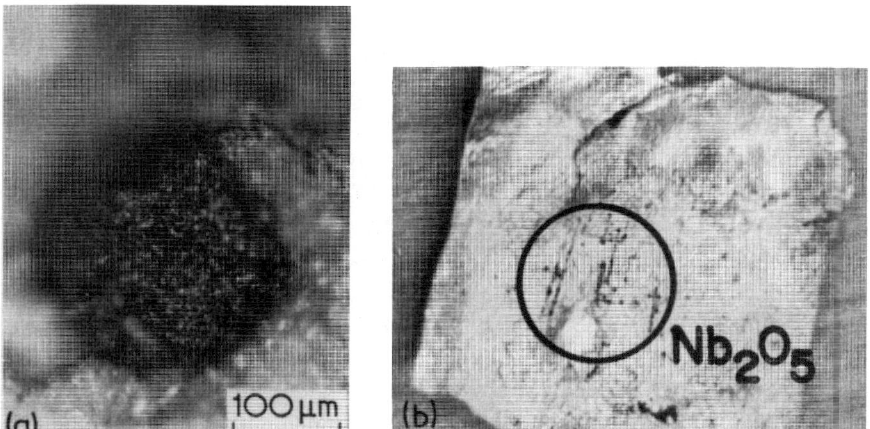

Figure 21. Physical effects, which also include a color change, resulting from electron beam irradiation of a Nb_2O_5 powder compact; the microphotograph in (b) shows the tracks and spots formed by moving a 2-keV beam at 1 mA/cm^2 fromregion to region; (a) is a magnified image of a single spot after 1 h irradiation (From Ref. 57).

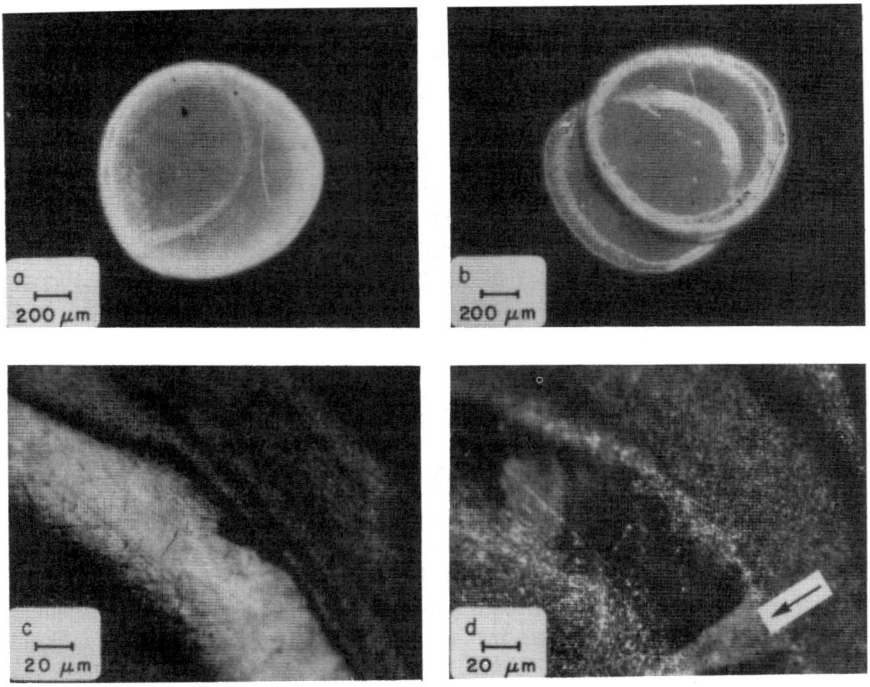

Figure 22. Optical photomicrographs showing the physical effects of electron beam damage in a UF4 thin film: (a) after 5 min exposure; (b,c,d) after 30 min exposure to a 2-mA/cm^2 electron beam. (a)–(c) were taken under bright-field illumination, (d) under crossed polarizers. The corresponding Auger spectra showed that the decomposition left a metallic uranium residue, and the polarized light micrograph in (d) verified its recrystallization (From Ref. 200).

deposition of metal and suboxide precipitates on the surface.(57) The color change from white to black is stable in air at room temperature for many months.

In alkali and other metal halides, the rate of ESD can be sufficient to "etch" the sample. This effect has been reported for bulk specimens, e.g., KCl and CsI, as well as for thin films and powder compacts.(36,66) Ellis reported on the decomposition and etching of UF_4 thin films on UO_2 during AES analysis.(200) Figure 22 shows these spots and the crystallized metal residues within them. The damage is clearly visible, and the Auger spectrum is characteristic of uranium metal. More recently, Bouquet *et al.*(211) reported that craters were produced under the electron beam in powder compacts of UF_4. Again, the physical damage was clearly evident and, in fact, has been used as a convenient method for measuring the electron beam diameter.

3.2. Contaminated, Oxidized, or Coated Surfaces

A frequent objective in many AES surface analyses is the routine characterization of surfaces. In some cases, an analysis is sought for contamination or oxidation, whereas in others it may be in response to an intentional surface treatment or coating process. However, in all cases, it is the composition of a thin surface layer or film, which necessarily contains gases and other species adsorbed from the ambient atmosphere, that is of primary interest. Unfortunately, these surface layers can be especially susceptible to electron beam damage.

The analysis of surface layers on corroded or oxidized metal surfaces can be susceptible to a variety of beam damage mechanisms. These mechanisms include desorption of water, decomposition of oxides, carbide formation, and color center formation. Figure 23 shows beam-induced changes in the Auger spectra of thin anodic films on Ni and a Ni alloy (Hastalloy B-2). In both cases, there is desorption of carbon, but, in the Ni alloy, the carbon signal does not decrease to zero, even after a long time. This result occurs because of electron-stimulated segregation of

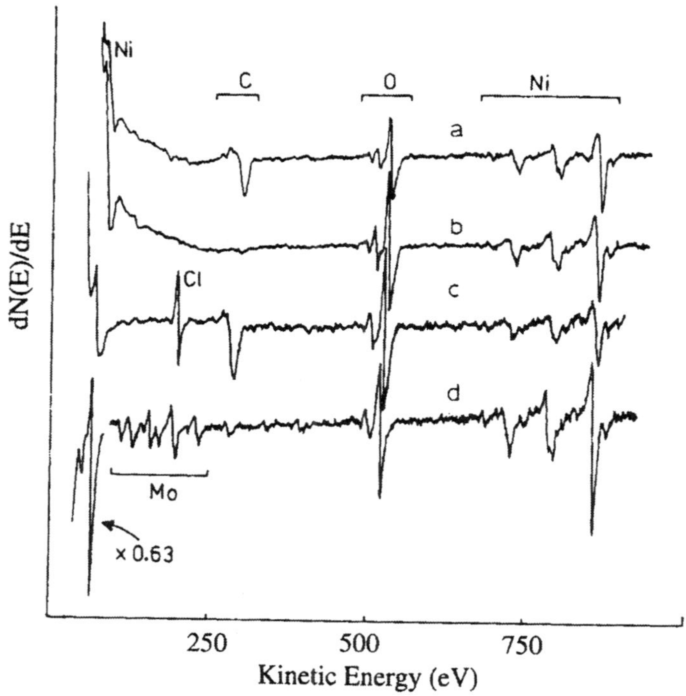

Figure 23. Auger spectra of anodized surface layers on (a) Ni after 22 min of irradiation, (b) Ni after 5 h irradiation, (c) Hastelloy B-2 initially, and (d) Hastelloy B-2 after 80 h irradiation. Beam current density was 3 mA/cm^2 (From Ref. 206).

Mo to the surface of the alloy where Mo–carbide forms. This conclusion is obtained from the shape of the carbon peak after electron irradiation and a corresponding change in shape and intensity of the oxygen peak. The situation is complex here, but, clearly, electron-stimulated desorption, segregation, oxide reduction, and carbide formation have occurred. In thick anodic films on the Ni alloy, these effects were also observed. Figure 24 shows the time-dependent changes in the various Auger signals where Cl and C desorption are observed initially, followed by the decomposition of oxide and corresponding formation of a Mo carbide. Burstein[206] actually used these beam-induced changes in the surface composition to characterize the nature of thick multilayer oxide films on Hastalloy B-2. The point is that caution must be exercised in the AES analysis of surface reaction-product layers,

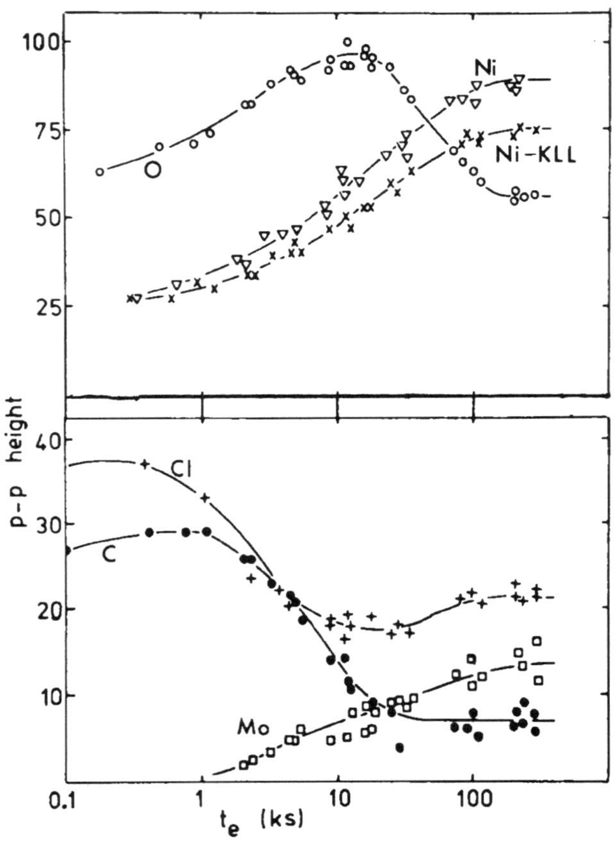

Figure 24. Time-dependent changes in Auger intensities during electron-beam-induced decomposition of a thick anodic film on Hastelloy B-2; the beam current density was 3 mA/cm^2 (From Ref. 206).

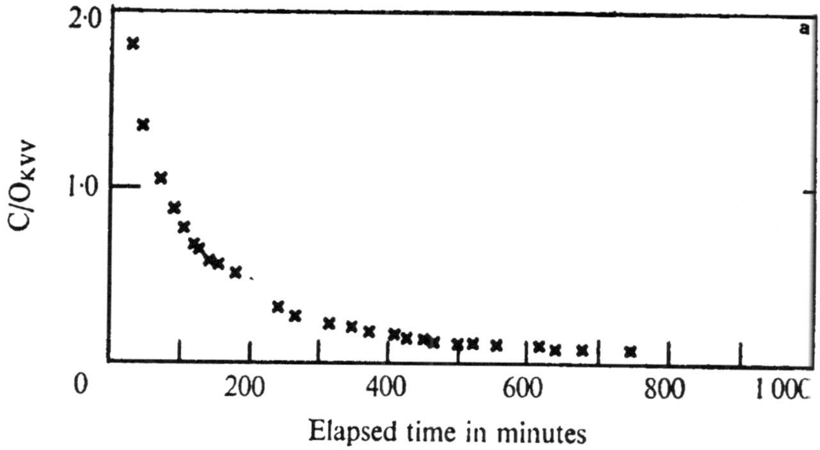

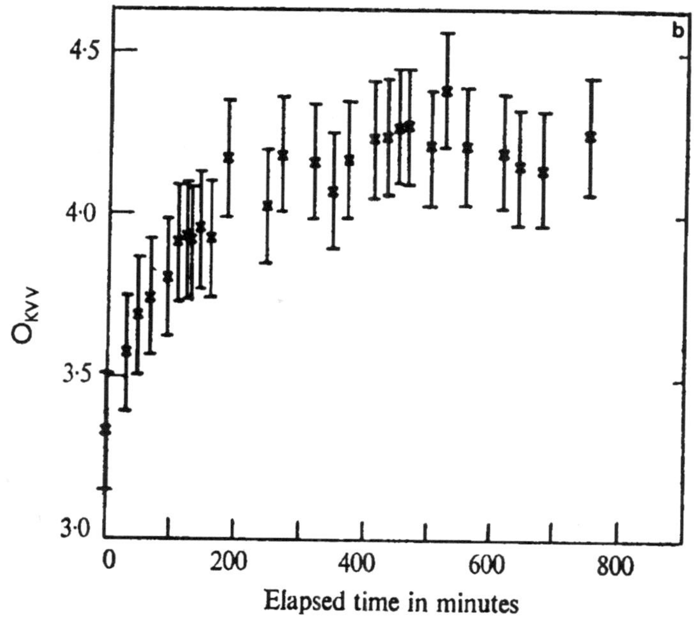

Figure 25. Electron-beam-induced changes in the (a) carbon, (b) oxygen, (c) tin, and (d) calcium Auger signal intensities on a float glass surface; the electron beam current density was 7.5 mA/cm^2 (From Ref. 189).

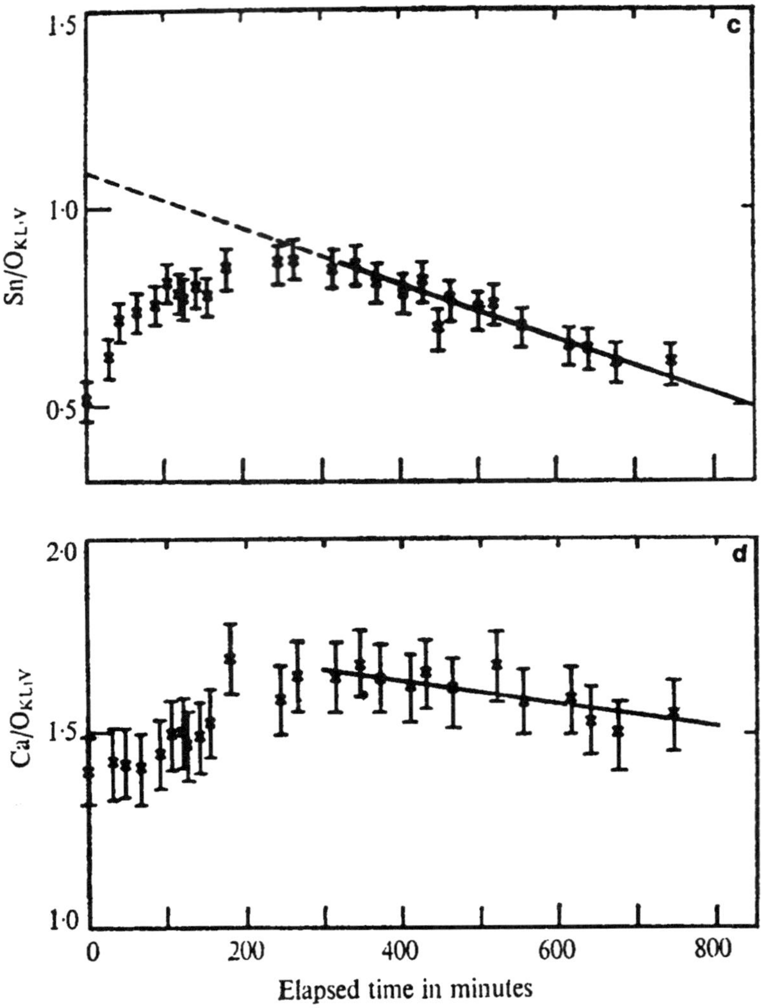

Figure 25. Continued.

but, if properly interpreted, the electron beam interactions can be beneficial in the characterization.

Figure 25 shows the changes observed in Auger peak intensities during the analysis of tin-oxide contamination on, and in, the surface of float glass. A key question in this analysis was the concentration and oxide stoichiometry of the tin introduced by the float bath during manufacturing. Initially, O, Si, Ca, Mg, K, Sn, Fe, C, and S were present on the surface, but irradiation caused a decrease in the

C, S, and K intensities. During this period, the intensity of the other elements increased. The data in Fig. 25a, when compared to Figs. 25b, 25c, and 25d, suggest that the ESD of a carbon overlayer reduces the attenuation of Auger electron emission in the substrate and thereby results in an increase in the O, Ca, Sn, Si, etc., Auger signals. Subsequently, the intensities of the Ca and Sn signals decrease. It was proposed that electron beam heating of the glass surface led to redistribution of the Ca and Sn via electromigration. Nevertheless, the complexity of the changes, especially in the case of oxygen, preclude any definitive quantitative analysis of the tin-oxide stoichiometry or surface composition.

Finally, Fig. 26 shows the effects of electron irradiation upon a deposited Pd–Ni thin film. Here, Stoddart *et al.*[190] observed the segregation of Ni in the thin-film catalysts during AES analysis. They noted that films deposited and immediately examined in the UHV system were stable under electron irradiation. However, exposure of the films to air at 133 Pa for 300 s resulted in the beam-induced changes shown in Fig. 26. These investigators proposed that C is removed by ESD and that the Ni diffuses to the surface through the Pd film where it is oxidized to form a layer of NiO.

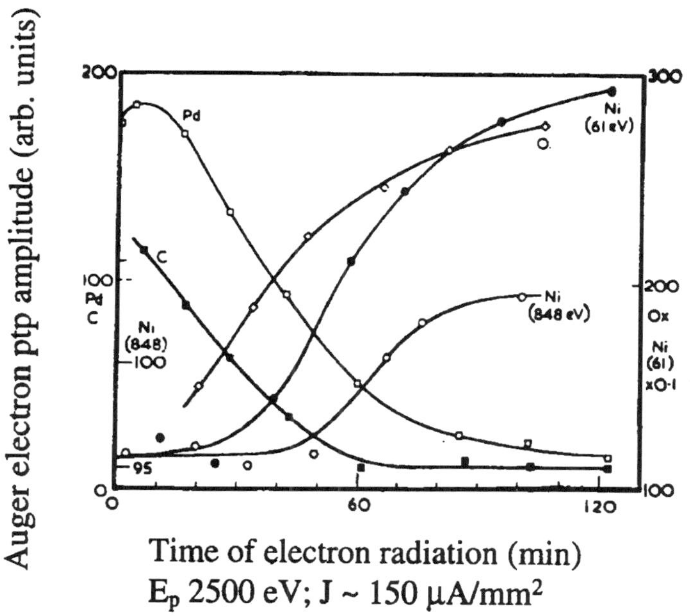

Figure 26. Electron-beam-induced changes in Auger signal intensities of a Pd–Ni alloy thin film; this film was deposited and exposed briefly to air. The films that were deposited and analyzed *in situ* did not undergo beam damage (From Ref. 190).

3.3. Polymers

AES analysis of polymers and other organic materials is not recommended, primarily because of their susceptibility to electron beam damage. The cross sections for ESD in polymers can be 10 to 1000 times higher than that for inorganic materials (Table 2), and this rapidly results in a macroscopic decomposition problem. The large band gap in these materials contributes to charge trapping, so relatively few neutralization mechanisms are available. The beam damage mechanisms in polymers are rather complex, and no attempt will be made here to review them. Interestingly, the fundamental interactions between energetic electrons and polymers have been studied extensively in the recent past because they are advantageous in electron beam lithography; a number of excellent treatments of this topic are available.(225) Here, we simply want to document the susceptibility of polymers to electron beam damage.

Figure 27 shows that even in a conducting polymer (ClO_4-doped polypyrrole), a significant electron beam effect is observed. The two AES spectra in Fig. 27 reveal that, even at beam current densities as low as 0.1 mA/cm^2, significant influence upon the surface composition can be observed. The visible deterioration of this polymer could be perceived at 40 μA/cm^2, and at 100 μA/cm^2 the damage was rapid and clearly visible.

Even polymers with inorganic thin-film coatings can be susceptible to electron beam damage. Figure 28 reveals the blistering and cracking of an In-doped SnO_2 film on Kapton resulting from excessive electron beam irradiation. The investigators also reported some melting of the Kapton. At more modest electron beam currents, the physical damage is less apparent, but the beam-induced segregation of Sn is very dramatic.

3.4. Glasses

The combined effects of charging, electromigration, electron-stimulated desorption, decomposition, and often heating, are especially evident during surface analyses of glass. The glasses of most interest are generally multicomponent, and the surface concentrations and in-depth profiles of alkali and alkaline-earth metal ionic species are of most interest. However, these ionic species are very mobile in glasses while being susceptible to ESD. The presence of mobile ions in many silicate glasses often provides sufficient conductivity and secondary-electron emission to eliminate difficult charging problems. However, the open structure of these frozen liquids and the absence of microstructural heterogeneity result in rather severe electromigration effects. The complications associated with the electromigration effects per se, in combination with desorption of the species that migrate to the surface, offer a real challenge for modeling the time-dependent change in Auger

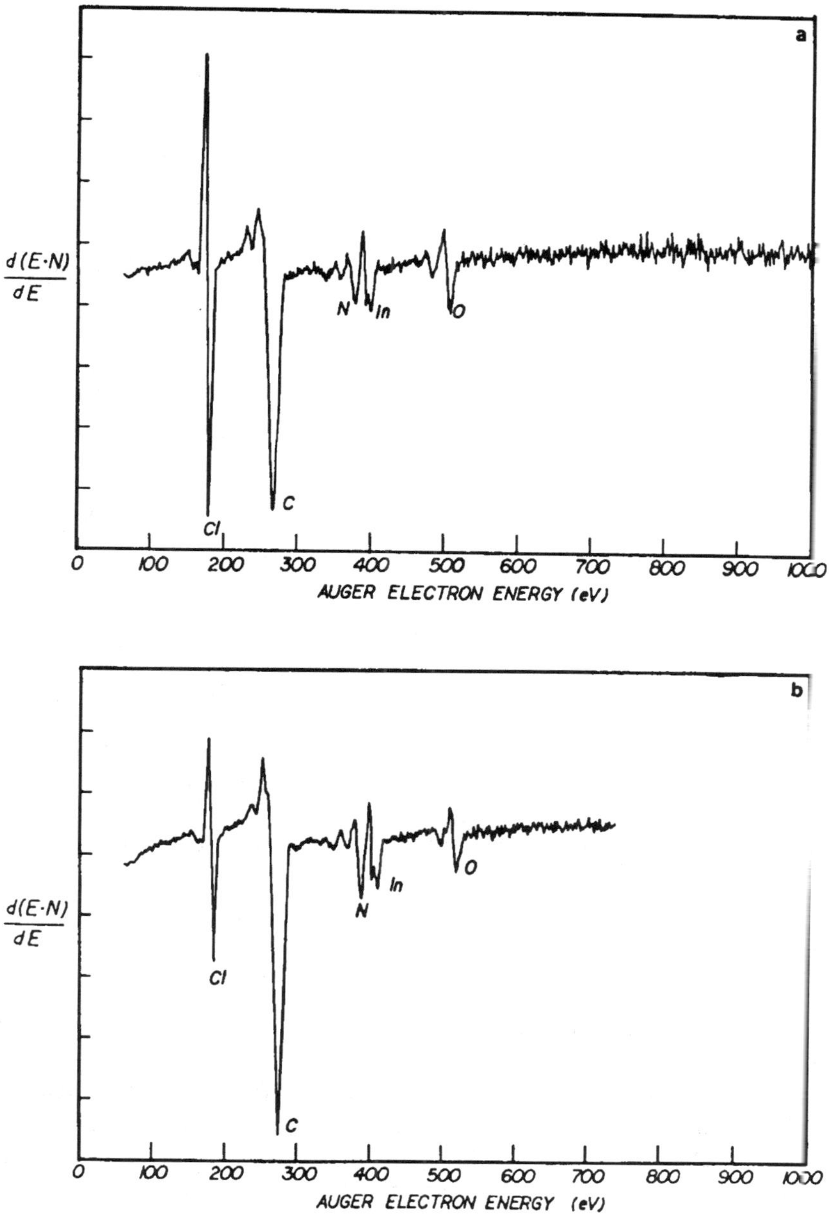

Figure 27. Auger spectra of a conductive polymer film (polypyrrole doped with ClO_4) (a) at ~20 μA/cm^2 and (b) at ~100 μA/cm^2; the film in (b) was visually damaged (From Ref. 52).

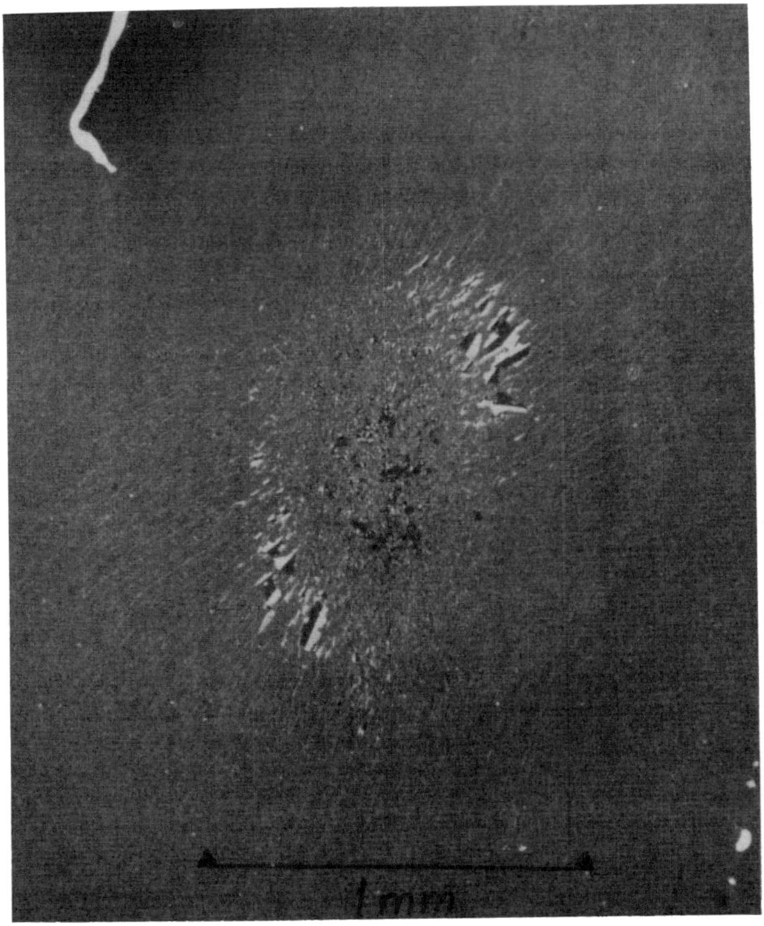

Figure 28. Electron micrograph of a polymer (Kapton) coated with In-doped tin oxide (ITO); the damage occurred after several minutes exposure to a 7-mA/cm^2, 3-keV electron beam (From Ref. 210).

signal intensity. In addition, the possible decomposition of the oxide glass matrix, usually silica, further complicates interpreting the AES results.

Figure 29 shows the time-dependent mass spectrometer signal obtained during the electron irradiation of a soda–silica glass. A clean surface was prepared by fracturing a glass rod in vacuum. These data verify that ESD of various species, including Si, is responsible for electron beam damage at the surface. The figure also shows the Si LVV Auger spectra that were measured simultaneously. The beam-induced decomposition of SiO_2 to Si is evident in the peak shape change. Presumably, the ESD of Si^+ results from the formation of elemental Si. At short irradiation times (low electron dose), where SiO_2 decomposition has not occurred, little or no Si^+

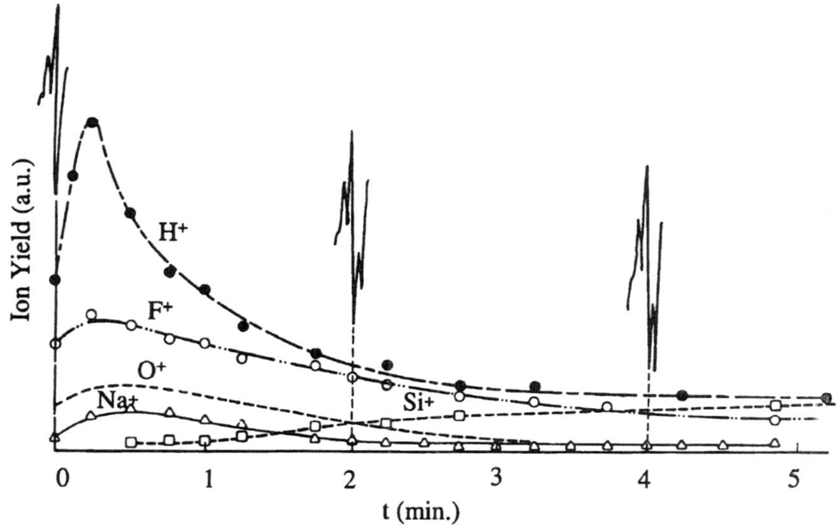

Figure 29. Mass spectrometer signals obtained as a function of electron irradiation time for a clean soda–silica glass fracture surface. The Si LVV Auger peaks were also monitored; note that the decomposition of SiO_2 (revealed by the appearance of the 92-eV elemental Si peak) correlated with the desorption of Si^+ (From Ref. 71).

desorption is observed. The increase and saturation of the Si^+ ESD signal correlate with the Si LVV peak shape change.

Figure 30 shows the measured change in the Na^+ ESD signal as a function of the electron irradiation dose. Again, the clean fracture surface of a soda–silica glass is considered. It can be seen that the beam damage is a function of the dose only. This result is consistent with the idea that electronic excitation by primary electrons is rate limiting to the beam damage process. Ohuchi and Holloway[173] modeled the data in Fig. 30 in terms of both electron-beam-enhanced diffusion and ESD. It was proposed that the primary electrons produce an intermediate excited state of Na—via bond breaking between the sodium and nonbridging oxygen in the glass structure—and then the Na can desorb or diffuse. After the data are fit to a differential equation that coupled the diffusion and desorption processes, a cross section for the ESD step of 10^{-18} to 10^{-19} cm^2 was obtained.

Miotello and Mazzoldi[175] suggested that the ESD contribution to the signal change during irradiation of glasses containing sodium is a secondary effect. Figure 31 shows the Na KLL Auger signal decay and the predictions of their model. Their expression for the signal change also considers the combined effects of diffusion and desorption, but they concluded that the data can be fit to the model with $Q_{ESD} = 4 \times 10^{-19}$ cm^2 or without ($Q_{ESD} = 0$) a contribution resulting from ESD.

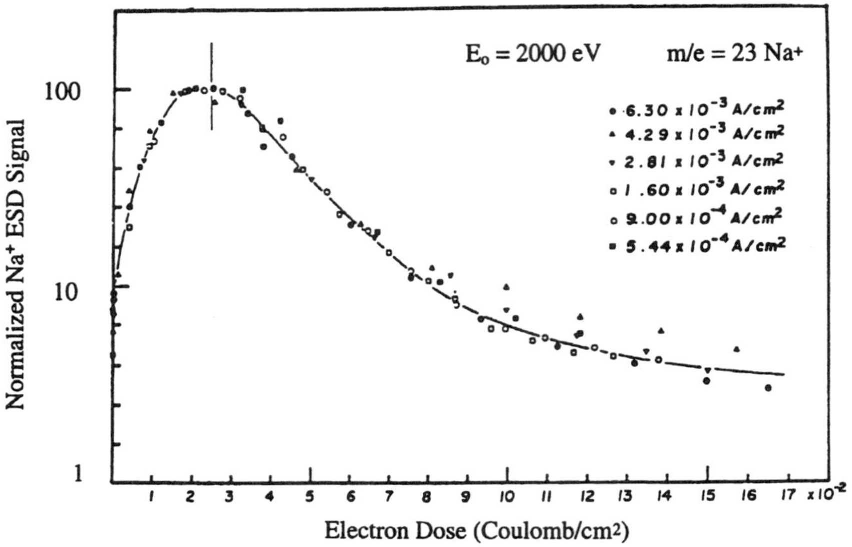

Figure 30. Normalized Na^+ ESD signal measured versus the electron dose for a clean soda–silica glass fracture surface; the measurements were made over a range of electron beam current densities (From Ref. 68).

It is likely that the relative contributions of migration and desorption in the beam damage process and the way they influence the signal intensity are functions of many factors. Most important are glass composition, surface condition, beam voltage, Auger electron escape depth of the mobile specie (relative to the inversion point in the positively charged surface layer), and beam heating effects. The extent of ESD depends upon the rate at which the mobile specie diffuses to the surface. The diffusion rate is a function of glass composition and temperature. Moreover, the escape depth associated with the Auger signal being monitored determines the relative sensitivity to ESD and out-diffusion versus diffusion into the bulk.

The condition of the surface and glass composition may be responsible for the apparent discrepancy in the models used to interpret the data in Figs. 30 and 31. Ohuchi and Holloway[173] used clean fracture surfaces of a binary Na_2O–SiO_2 glass, whereas the data modeled by Miotello and Mazzoldi[175] were obtained on surfaces of commercial Na_2O–CaO–Al_2O_3–SiO_2 glasses, which were ion-beam-sputter-cleaned at liquid-nitrogen temperature. Figure 32a shows a complete set of sodium Auger signal decay curves for the aforementioned commercial Na_2O–CaO–Al_2O_3–SiO_2 sputter-cleaned glasses. Note that the sodium signal always decreases and that a plateau appears as the beam current density or specimen temperature is reduced. Conversely, the Auger signal data reported for clean fracture surfaces of

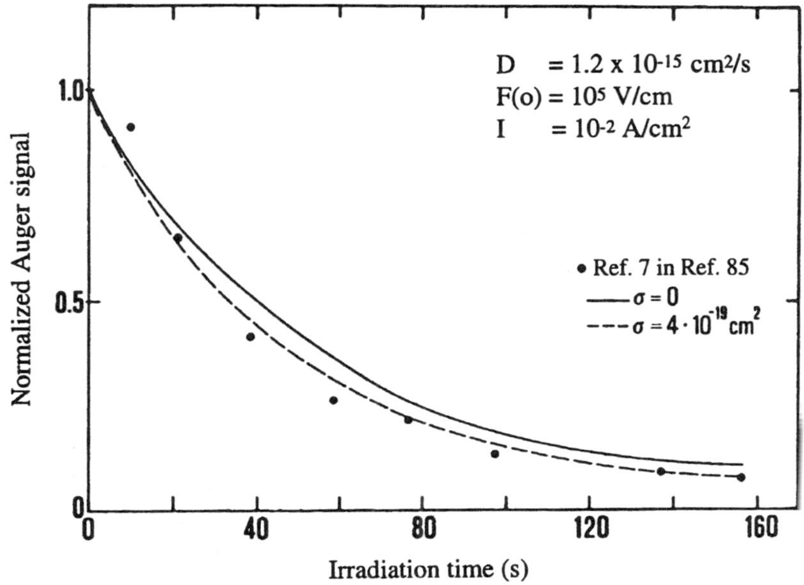

Figure 31. Calculated Na Auger signal decay resulting from electron beam irradiation at 0.1 mA/cm^2; the solid line assumes decay resulting from electromigration only, while the dotted line considers both electromigration and desorption. The data points were measured on an ion-beam-cleaned surface of soda–lime–silica glass (From Ref. 175).

Na_2O–SiO_2 in Fig. 32b shows an initial increase in signal followed by a decrease. No attempt will be made to explain the origin of this difference. It is simply pointed out that, because of the complexity of the phenomena, the dependence upon surface condition is not unexpected.

The data of Gossink *et al.*(170) are noteworthy in that they examined some rather complex glasses that contained mixed alkali species or alkali–alkaline earth species. Their time-dependent Auger signals are presented in Fig. 33 and show a decay of the alkali signals and an increase in the alkaline-earth signal. The different decay rates for the Na and K signals are not surprising because they have different diffusion coefficients and different Auger–electron escape depths. However, the increase and stabilization of the Ca signal, which have been observed in other studies, could not be explained.

It is noteworthy that the thickness of glass thin films influences the nature of electromigration. In most cases, this effect can be explained by the electric field, whose strength and uniformity depend on the proximity of a metallic or semiconducting substrate. Cauzaux,(146) in fact, modeled the thickness dependence of the electron-beam-induced electric field in thin-film insulators. Qualitatively, more of the primary electrons penetrate the metal substrate with decreasing glass thickness,

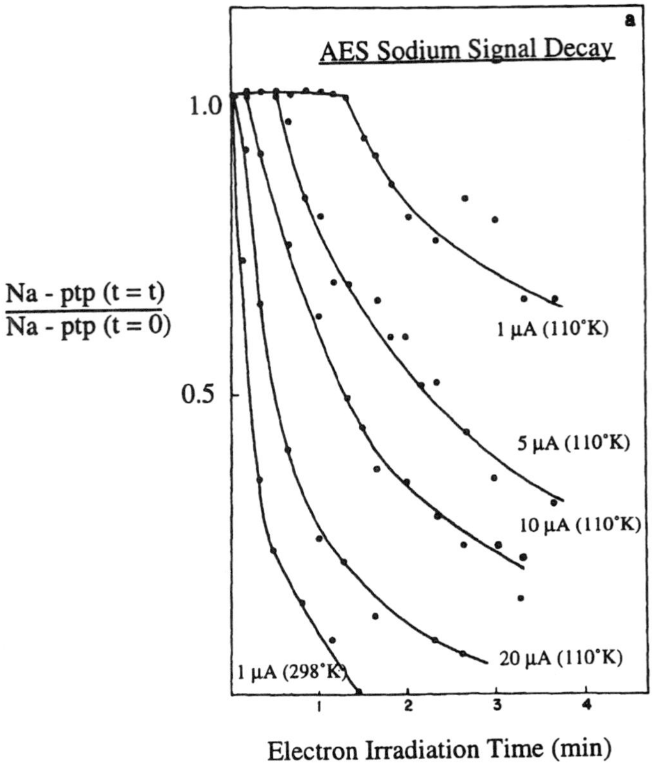

Figure 32a. Sodium signal decay measured during AES analyses of ion-beam-cleaned soda–lime–silica glass surfaces (From Ref. 167).

and the electric field, which drives the migration, is thus reduced. This effect is evident in the sodium signal decay curves of Fig. 34. The bulk glass shows the most rapid signal decay because of the high field gradient; the thin films show a less rapid decay that decreases further with smaller film thickness.

3.5. Sputter Depth Profiles

The surface and in-depth effects of electron beam damage not only influence the qualitative and quantitative interpretation of the Auger spectra but also the quality of a sputter profile (see Chaps. 3 and 5, this volume). Naturally, the acquisition of depth profiles requires repetitive exposure of the surface to electron irradiation. This exposure exaggerates any of the beam damage mechanisms that exhibit a time-dependent in-depth effect. The initial exposure of the specimen to

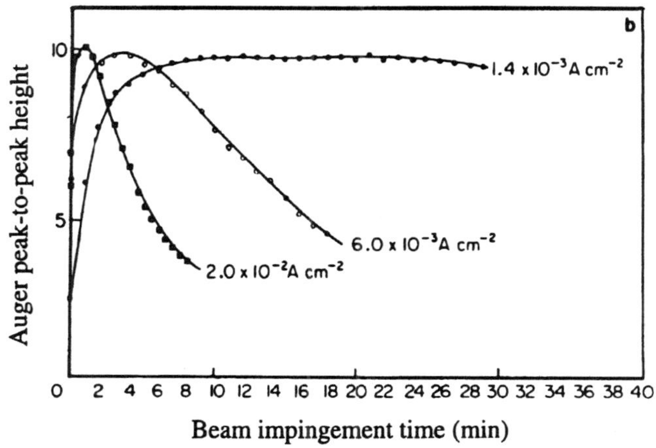

Figure 32b. Sodium signal changes measured during AES analyses of clean fracture surfaces of soda–silica glass (From Ref. 169).

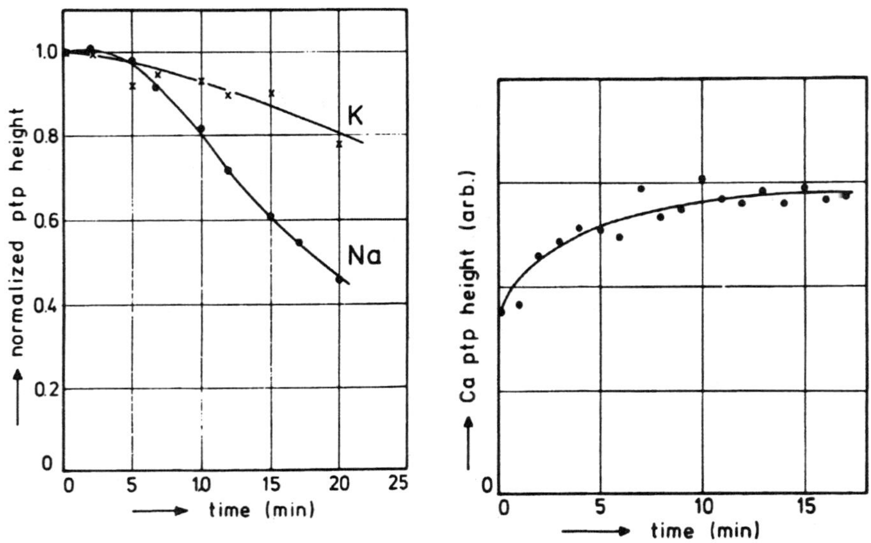

Figure 33. Electron-beam-induced changes in Auger signal intensities for K and Na in a Na_2O–K_2O–SiO_2 glass, irradiated at 3 keV and 10 μA (left) and Ca, in a Na_2O–CaO–SiO_2 glass irradiated at 1.5 keV and 2 μA (right); in the Na_2O–CaO–SiO_2 glass, the Na signal decay was essentially instantaneous (From Ref. 170).

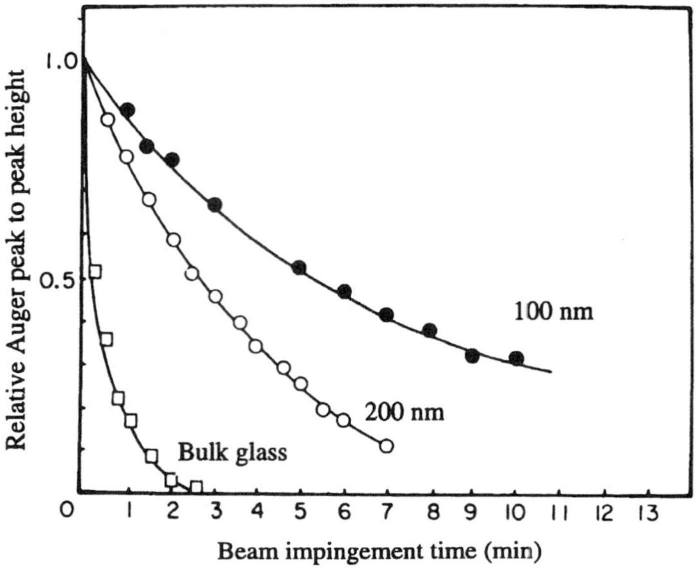

Figure 34. Effect of glass "thickness" upon the decay rate of the sodium Auger signal: the surface of bulk and thin-film soda–silica glasses were irradiated using a 3-keV electron beam at 5 mA/cm^2 (From Ref. 169).

the electron beam can even introduce an in-depth effect that distorts the true composition profile.

The Auger analyses for minor constituents and impurities when sputter-depth-profiling bulk insulators or insulating thin films represent a situation where exaggerated beam damage arises. It has been shown that electromigration in bulk glasses modifies the in-depth profile (Fig. 18). Inoue[198] showed that the distribution of phosphorous in phophosilicate thin films can be dramatically altered by electron irradiation of the sample prior to depth profiling. Naturally, these effects accumulate during depth profiling itself.

In chlorinated SiO_2 on Si, the electron beam effect is dramatic. Chou *et al.*[186] showed that initial irradiation of these films causes chlorine to migrate the film surface. In fact, they showed that virtually all of the chlorine in the film could be "distilled off" via electron-beam-induced migration and stimulated desorption. Some of their data are presented in Fig. 35. This plot of the Cl KLL Auger signal versus electron irradiation time shows an increase (because of migration and pile-up of the Cl at the film surface) followed by a decrease (resulting from ESD of Cl). The time (or electron dose) required to segregate and desorb the Cl scales with the film thickness. This example dramatically indicates that the in-depth composition

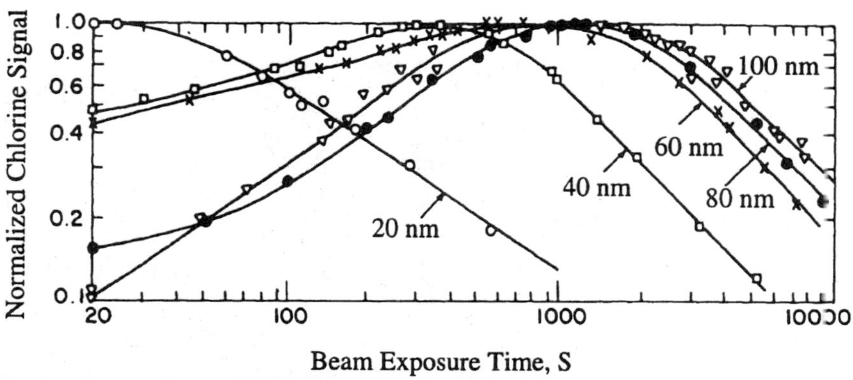

Figure 35. Effect of electron beam irradiation upon the Cl Auger signal in chlorinated SiO_2 films from 20 to 100 nm thick. The increase in signal results from migration of Cl to the film surface, and the decay is due to its desorption (From Ref. 186).

of a thin film may change during the acquisition of Auger data during sputter depth profiling.

Another effect of beam damage upon sputter depth profiles was reported by Ahn *et al.*[191] They showed that the erosion rate is substantially higher in the electron-irradiated region of the sputter-etch crater in SiO_2 films and soda–lime glasses. This irradiation created a pit—whose shape and size were an imprint of the electron beam—within the sputter-etched crater (see also Chap. 3). Obviously, electron-beam-induced pitting complicates calibration of the depth scale and degrades the depth resolution. The effect was attributed to ESD of oxygen and consequent decomposition of the SiO_2. This decomposition left a layer of silicon under the beam whose sputter yield is appreciably higher than that for silicon dioxide and, thereby, created an "electron-beam-induced pit" within the ion beam crater.

3.6. Microanalyses

Electron beam effects can be especially troublesome in Auger microanalyses because of the high current density of the beam. Figure 36 shows line scans for O, Ga, and As on a GaAs wafer after a lengthy exposure to a static beam at 0.2 A/cm^2. The vacuum system had a background pressure of water vapor equal to 6×10^{-8} Pa. It is clear ESA of oxygen, and probably in-depth oxidation, has occurred locally.

The high electron beam current density used in most microanalyses results in an exceedingly rapid accumulation of beam damage. Figure 37 shows the time-dependent changes in the Auger spectra of a thick oxide film on steel. The electron beam current density was 2 A/cm^2. Note that the S and Ca peaks undergo a doubling of their intensity in the time required to scan 0–500 eV. The data on the far right in

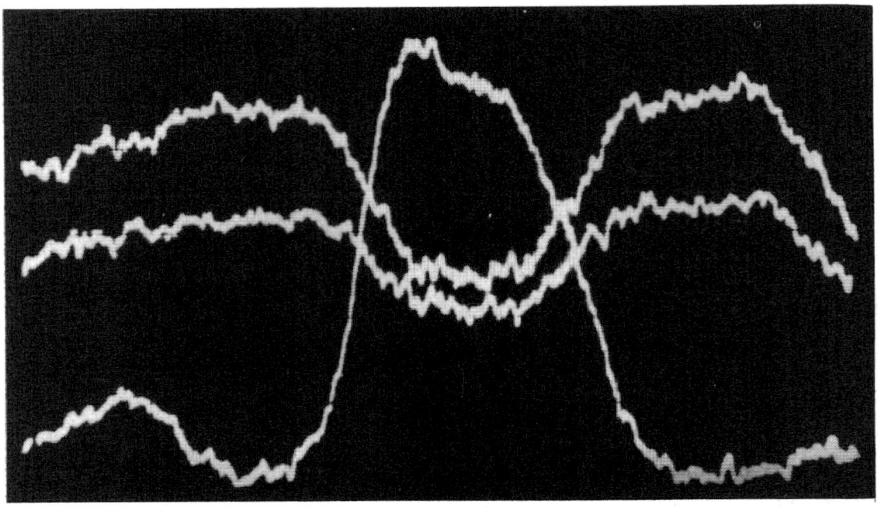

Figure 36. Line scans of the Ga (upper trace), As, and O (lower trace) Auger signals across a GaAs surface after electron beam irradiation with a static beam at 5 keV and 200 mA/cm^2 for several hours. Electron-stimulated adsorption of oxygen occurred under the static beam because of the presence of 6×10^{-8} Pa water vapor in the analysis chamber (From Ref. 9).

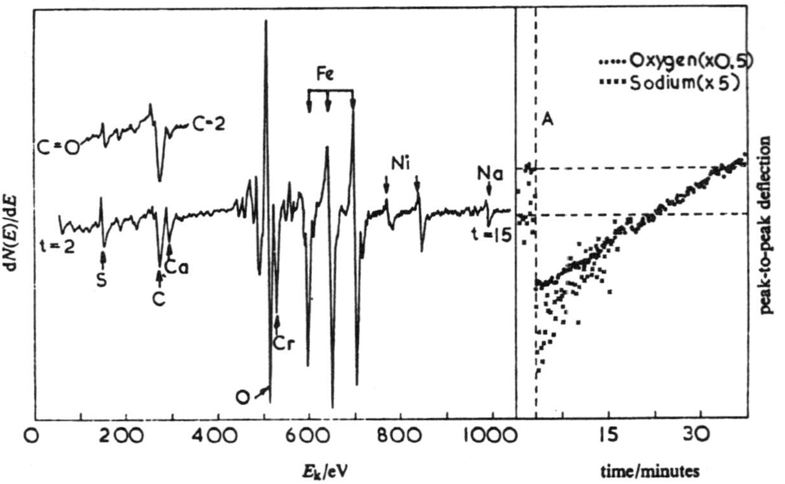

Figure 37. Changes in Auger spectra and signal intensity that occurred over the time scale required to obtain the AES microanalysis at 2 A/cm^2) of an oxidized steel surface. Note that the S, Ca, O, and Na signals are modified during the initial 2-min scan (From Ref. 8).

the figure reveal that the sodium peak recorded in the spectra (on the left) had increased in intensity by a factor of 2 to 3 during acquisition of the spectra. Some additional discussions of electron beam damage effects during high-current-density, small-spot AES analysis are available.[3,9,13]

4. Recommendations

Most of the mechanisms and consequences of electron beam damage during AES and other surface analyses have been described. These effects can seriously influence qualitative and quantitative analyses, whether are for routine, practical samples or for ideal model systems used in surface science. In most cases, beam damage cannot be eliminated, but it is often possible to reduce the rate of beam damage to a tolerable level. At the very least, one should determine whether, and to what extent, beam damage has occurred during the analysis.

The first consideration in the execution of any electron-beam-excited surface analysis should be the material and nature of the surface. The bulk materials most susceptible to electron beam damage are organic polymers and multicomponent glasses. Any surface with adsorbed species, especially water, halogens, organic molecules, and carbonaceous species, is susceptible to ESD. Clean surfaces, especially metals been prepared in vacuum, are exceedingly reactive and, thereby, are prone to ESA. Table 1 and the references provide a more complete compilation of materials susceptible to electron beam damage.

The second consideration is to determine whether electron beam effects are influencing the analysis and, if so, the rate at which beam damage occurs. In general, the acquisition of sequential spectra or peak scans during continuous electron irradiation provides a ready indication of the existence and degree of sample damage (unless, of course, the damage is so rapid that the changes occur before adequate data can be acquired). Usually, beam damage is noticed by a time-dependent increase or decrease in one or more signal intensities, a change in peak shapes, or a shift in peak position. In most cases, the observed change provides insight about the beam damage mechanism, particularly because Auger analysis provides information on the surface composition.

The rate of beam damage is evident in the time dependence of the observed spectral changes. Ideally, one strives for those conditions where beam damage is barely perceptible in the time period required to record a spectrum. If Auger sputter profiling is to be performed and the accumulation of in-depth beam damage is possible, a lower rate of sample damage may be required. Otherwise, common sense, relative to the objectives of the analysis, must be exercised to define the tolerable rate of sample damage.

It is recommended that the extent of sample damage be quantitatively treated and that the parameters be reported when publishing the results. Most beam damage mechanisms–ESD, ESA, decomposition, oxidation, and even combinations thereof—obey first order kinetics as in Eq. (1). Thus, it is a simple matter to plot the natural logarithm of the observed signal change versus irradiation time. The slope of the line can be used to calculate the cross section for damage if the beam current density is known. This procedure can be used, in turn, to define the beam conditions or electron doses necessary to reduce the damage rate to a tolerable level. The plots frequently reveal that some level of sample damage must be tolerated to perform the analysis with the desired sensitivity or spatial resolution. Of course, if a cross section for damage has already been reported, or if one assumes a cross section based on dissociative ionization or stimulated desorption, the first-order reaction in Eq. (1) can be used to define the beam parameters for the initial experiments.

Approaches exist for minimizing the rate of electron beam damage. Unfortunately, most of them compromise or complicate the analysis because they reduce the signal-to-noise or spatial resolution for the analysis. There is no question that the electron dose exerts the strongest influence on the rate of sample damage. This conclusion is virtually independent of the beam damage mechanism. Thus, a reduction in current density should always be the first approach for reducing electron beam damage. If there are no constraints placed on the spatial resolution of the analysis, this reduction is readily achieved by defocusing or rastering the beam. The beam current can also be reduced, especially if the maintenance of beam size is critical, but this will reduce the signal to noise. Unfortunately, there are scenarios where beam damage is still intolerable at beam currents that are inadequate to give spectra or signals with the desired sensitivity.

In those materials where sample damage has been verified, and *especially where an acceptable reduction in beam current density (or signal-to-noise) is insufficient* to bring the damage rate to the level required, the following approaches may also be required:

(i) Move the sample or beam to a new spot before the scan is initiated,
(ii) Turn off the electron beam except during the acquisition of spectra; this procedure is recommended at all times, but especially when combining Auger analysis with sputter depth profiling,
(iii) Use energy windows to scan and record only the peaks of interest,
(iv) Reduce the electron beam energy,
(v) Reduce the partial pressure of reactive background gases when electron-stimulated adsorption is suspected,
(vi) Change the sample mounting, a procedure that is of the most benefit for insulators and powders,
(vii) Mount the specimen on a cold stage.

Similar recommendations are also included in an ASTM standard guide for minimizing unwanted electron beam effects in AES.[226]

References

5.1. Review of Beam Damage in Electron Microscopy

1. R. M. Glaeser, Radiation Damage with Biological Specimens and Organic Materials, in: *Introduction to Analytical Electron Microscopy,* Plenum Press, New York (1979), pp. 423–436.
2. L. W. Hobbs, Radiation Effects in Analysis of Inorganic Specimens by TEM, in: *Introduction to Analytical Electron Microscopy,* Plenum Press, New York (1979), pp. 437–480.
3. M. S. Isaacson, Limits of detection sensitivity due to electron beam radiation damage, *Ultramicroscopy* **28**(1–4), 320 (1989).
4. M. I. Buckett, Electron irradiation damage in oxides, *Ultramicroscopy* **29**(1–4), 217 (1989).
5. J. Rickards, Electron beam damage in high Tc superconductor materials, *Ultramicroscopy* **30**(3), 425 (1989).
6. S. Kumar, Electron beam damage in high temperature polymers, *Polymer* **31**(1), 15 (1990).
7. C. J. Humphreys, 100 keV electron beam damage of metals, ceramics and semiconductors—implications for microanalysis and nanolithography, in: *Institute of Physics Conference Series* **119**, 319 (1991).

5.2. General Discussions of Electron Beam Damage in Surface Analysis

8. J. P. Coad, M. Gettings, and J. C. Riviere, Beam effects in AES revealed by XPS, *Discuss. Faraday Soc.* **60**, 269 (1975).
9. A. van Oostrom, Some aspects of Auger microanalysis, *Surf. Sci.* **89**, 615 (1979).
10. J. M. Fontaine, J. P. Duraud, and C. LeGressus, Electron beam effects in Auger electron spectroscopy and scanning electron spectroscopy, *Surf. Interface Anal.* **1**, 196 (1979).
11. C. G. Pantano and T. E. Madey, Electron beam damage in Auger electron spectroscopy, *Appl. Surf. Sci.* **7**, 115 (1981).
12. G. Pignatel and G. Queirolo, Electron and ion beam effects in Auger electron spectrometry, *Radiat. Eff. Defects Solids* **64**, 109 (1981).
13. L. L. Levenson, Electron beam-solid interactions: Implications for high spatial resolution Auger electron spectroscopy, in: *Scanning Electron Microscopy III*, 1982, p. 925.
14. G. Pignatel and G. Queirolo, Electron and ion beam effects in Auger electron spectroscopy on insulating materials, *Radiat. Eff. Defects Solids* **79**, 291(1983).
15. J. Cazaux, The influence of radiation damage (microscopic causes) on the sensitivity of Auger electron spectroscopy and x-ray photoelectron spectroscopy, *Appl. Surf. Sci.* **20**, 457 (1985).
16. T. E. Madey, Radiation damage in Auger electron spectroscopy and X-ray photoelectron spectroscopy, in: *Proc. of the Microbeam Analysis Society,* 1987.

5.3. Fundamentals of Electron-Stimulated Desorption

17. M. L. Knotek and P. J. Feibelman, Ion desorption by core-hole Auger decay, *Phys. Rev. Lett.* **40**, 964 (1978).
18. M. L. Knotek and P. J. Feibelman, Stability of ionically bonded surfaces in ionizing environments, *Surf. Sci.* **90**, 78 (1979).
19. N. H. Tolk, M. M. Traum, J. C. Tully, and T. E. Madey, eds. *Desorption Induced by Electronic Transitions: DIET I*, Springer Series in Chemical Physics, Vol. 24, Springer, Berlin (1983).
20. M. L. Knotek, Electron and photon stimulated desorption, *Rep. Prog. Phys.* **47**, 1499 (1984).

21. Gar B. Hoflund, Review of electron stimulated desorption (ESD) phenomena: energy analysis of desorbing ions, Proceedings of the Annual Scanning Electron Microscope Symposium, *Scanning Electron Microscopy IV*, Scanning Electron Microscopy, Elk Grove Village, IL (1985).
22. *Desorption Induced by Electronic Transitions, DIET II*, Springer Series in Surface Science, Vol. 4, Springer, Berlin (1985).
23. D. Menzel, Desorption induced by electronic transitions, *Nucl. Instrum. Meth. Phys. Res., Sect. B* **13**, 507 (1986).
24. T. E. Madey, Characterization of surfaces through electron and photon stimulated desorption, *Science* **234**, 316 (1986).
25. K. J. Snowdon, Molecule excitation in sputtering, scattering and electron or photon induced desorption, *Nucl. Instrum. Meth. Phys. Res.,Sect. B* **14**(4–6), 1986.
26. Theodore E. Madey, Mechanisms and applications of electron stimulated desorption in surface science, *Vacuum* **37**(1–2) 1987.
27. *Desorption Induced by Electronic Transitions, DIET III*, Springer Series in Surface Science, Vol. 13, Springer, Berlin (1988).
28. E. M. Williams, Electron stimulated desorption of ions from surfaces, techniques, methodology and some recent findings with water at metals and semiconductors, *Vacuum* **9**(7–8), 633 (1989).
29. *Desorption Induced by Electronic Transitions, DIET IV*, Springer Series in Surface Science, Vol. 18, Springer, Berlin (1990).
30. W. Husinky, Ion, electron, and photon-induced desorption of particles by electronic processes. Fundamentals and application, *Vacuum* **40**(1–2), 230 (1990).
31. J. Fine, Electron stimulated desorption mechanisms in insulators. *Interdisciplinary Conference on Dielectrics, Properties, Characterization, Applications*, Antibes, France (1992).
32. S. S. Elovikov, Application of electron spectroscopy and mass-spectrometry methods for electron-stimulated desorption investigation, *Poverkhnost* **6**, 99 (1992).
33. O. V. Rakhovskaya, Electron stimulated desorption process from ionic compound surface, *Surf. Sci.* **274**(1), 190–198 (1992).

5.4. Studies of Electron Beam Interactions at Solid Surfaces

34. D. Menzel and R. Gomer, Desorption from metal surfaces by low-energy electrons, *J. Chem. Phys.* **41**, 3311 (1964).
35. P. A. Redhead, Interaction of slow electrons with chemisorbed oxygen, *Can. J. Phys.* **42**, 886 (1964).
36. P. W. Palmberg and T. N. Rhodin, Surface dissociation of potassium chloride by low energy electron bombardment, *J. Phys. Chem. Solids* **29**, 1917 (1968).
37. R. G. Musket and J. Ferrante, Auger electron spectroscopy study of electron impact desorption, *Surf. Sci.* **21**, 440 (1970).
38. H. Tokutaka, M. Prutton, I. G. Higginbotham, and T. E. Gallon, (100) surface of alkali halides, II. Electron stimulated dissociation, *Surf. Sci.* **21**, 233 (1970).
39. J. P. Coad, H. E. Bishop, and J. C. Riviere, Electron-beam assisted adsorption on the Si (111) Surface, *Surf. Sci.* **21**, 253 (1970).
40. B. A. Joyce and J. H. Neave, Electron beam-adsorbate interactions on silicon surfaces, *Surf. Sci.*, **34**, 401 (1973).
41. R. E. Kirby and D. Lichtman, Electron beam induced effects on gas adsorption utilizing Auger electron spectroscopy: CO and O_2 on Si. I. Adsorption studies, *Surf. Sci.* **41**, 447 (1974).
42. R. E. Kirby and J. W. Dieball, Electron beam induced effects on gas adsorption utilizing Auger electron spectroscopy: CO and O_2 on Si. II. Structural effects, *Surf. Sci.* **41**, 467 (1974).
43. Y. Margoninski, D. Segal, and R. E. Kirby, The interference of an electron beam with the surface reaction between oxygen and germanium, *Surf. Sci.* **53**, 488 (1975).

44. J. Verhoeven and J. Los, The influence of an electron beam on oxidation of polycrystalline nickel surfaces, monitored by disappearance potential spectroscopy (DAPS), *Surf. Sci.* **58**, 566 (1976).
45. J. Dresner and B. Goldstein, Dissociation of MgO films under heat and electron bombardment and its effects on secondary emission, *J. Appl. Phys.* **47**, 1038 (1976).
46. G. F. Dionne and J. F. Fitzgerald, Effects of high-intensity electron bombardment on the secondary-emission characteristics of a MgO/Au-Cement film, *J. Appl. Phys.* **48**, 3028 (1977).
47. M. Housley and D. A. King, Electron beam interactions with CO on W, *Surf. Sci.* **62**, 81 (1977).
48. H. G. Tompkins, The electron stimulated interaction of H_2O with a nickel surface, *Surf. Sci.* **62**, 293 (1977).
49. B. Carriere and B. Lang, A study of the charging and dissociation of SiO_2 surfaces by AES, *Surf. Sci.* **64**, 209 (1977).
50. J. C. Fuggle, E. Umbach, P. Feulner, and D. Menzel, Electron induced dissociation of CO on Ru (001), *Surf. Sci.* **64**, 69 (1977).
51. A. G. Knapp and J. R. Hughes, The effect of electron bombardment on evaporated films of sodium aluminum fluoride, in: *Proceedings of the 7th International Vacuum Congress and 3rd International Conference on Solid Surfaces*, Vienna (1977).
52. M. L. Shek, S. P. Withrow, and W. H. Weinberg, Electron beam induced desorption and dissociation of CO chemisorbed on Ir (111), *Surf. Sci.* **72**, 678 (1978).
53. R. P. Holstrom, J. Lagowski and H. C. Gatos, Electron beam induced decomposition and desorption: ZnO, *Surf. Sci.* **75**, L781 (1978).
54. T. Sasaki, R. S. Williams, J. S. Wong, and D. A. Shirley, Radiation damage studies by x-ray photoelectron spectroscopy, I. Electron irradiated $LiNO_3$ and Li_2SO_4. II. Electron irradiated Li_2CrO_4 and Li_2WO_4, *J. Chem. Phys.* **68**, 2718 (1978); **69**, 4374 (1978).
55. J. Fine and R. Gorden, Jr., Visual ion-beam images produced by electron and ion-beam interaction on surfaces, *J. Appl. Phys.* **49**, 1236 (1978).
56. T. T. Lin and D. Lichtman, AES studies of chemical shift and beam effects on molybdenum oxides, *J. Vac. Sci. Technol.* **15**, 1689 (1978).
57. T. T. Lin and D. Lichtman, Electron irradiation effects on oxidized Nb foil and NbO, *J. Mater Sci.* **14**, 455 (1979).
58. T. T. Lin and D. Lichtman, The effects of electron bombardment on the surface composition of WO_3, oxidized W and Ta Foils, *J. Appl. Phys.* **50**, 1298 (1979).
59. P. H. Holloway, T. E. Madey, C. T. Campbell, R. R. Rye, and J. E. Houston, Electron spectroscopy of condensed multilayers: Line shape changes due to beam damage and excitation mode, *Surf. Sci.* **88**, 121 (1979).
60. R. G. Gossink and T. P. A. Lommen, Secondary ion mass spectrometry (SIMS) analysis of electron bombarded soda-lime-silica glass, *Appl. Phys. Lett.* **34**, 444 (1979).
61. V. M. Bermudez and V. H. Ritz, Investigation of the silica surface via electron-energy-loss spectroscopy, *Phys. Rev. B* **20**, 3446 (1979).
62. M. Lichtensteiger, C. Webb, and J. Lagowski, electron stimulated adsorption: surface activation and preferential binding of oxygen to sulfur on CdS, *Surf. Sci.* **97**, L375 (1980).
63. Y. Shapira, Electron beam induced dissociation and conductivity of SnO_2 films, *J. Appl. Phys.* **52**, 5696 (1981).
64. V. H. Ritz and V. M. Bermudez, Electron-spectroscopic studies of $LiNbO_3$ and $LiTaO_3$ surfaces, *Phys. Rev. B* **24**, 5559 (1981).
65. C. L. Strecker, W. E. Moddeman, and J. T. Grant, Electron-beam-induced decomposition of ion bombarded calcium fluoride surfaces, *J. Appl. Phys.* **52**, 692 (1981).
66. V. T. Sotnikov and O. V. Tuchin, Surface dissociation of CsI monocrystal upon electron irradiation, *Sov. Phys. Tech. Phys.* **26**, 231 (1981).
67. J. M. Fontaine, O. Lee-Deacon, J. P. Duraud, S. Ichimura, and C. LeGressus, Electron beam effects on Oxygen exposed Aluminum Surfaces, *Surf. Sci.* **122**, 40 (1982).

68. F. Ohuchi and P. H. Holloway, Contribution of electron stimulated desorption to electron beam damage in oxides, *Scanning Electron Microsc. IV*, 1453 (1982).
69. A. Livshits and M. Polak, Electron-induced segregation to surfaces of the superionic conductor Na-Alumina studied by AES and XPS, *Surf. Sci.* **119**, 314 (1982).
70. C. Webb, J. Lagowski, and M. Lichtensteiger, Electron stimulated adsorption on semiconductor surfaces, *Surf. Sci.* **138**, 399 (1983).
71. Y. X. Wang, F. Ohuchi, and P. H. Holloway, Mechanisms of electron stimulated desorption from Soda-Silica Glass surfaces, *J. Vac. Sci. Technol.* **A2,** 732 (1984).
72. S. Nagai and Y. Shimizu, Electron and Ar^+ ion impact effects on SiO_2, Al_2O_3 and MgO, *J. Nucl. Mater.* **128/129,** 605 (1984).
73. V. T. Sotnikov, V. F. Gaidukov, and S. S. Dobrotvorskii, Radiation stimulated defect formation in ruby surface layers, *Phys. Chem. Mech. Surf.* **2**, 1043 (1984).
74. P. Collot, Low-pressure oxidation of silicon stimulated by low-energy electron bombardment, *Philos. Mag. B* **52**(6), 1051 (1985).
75. M. Q. Ding and E. M. Williams, The interaction of electron beams with water admitted at aluminum single crystal (100) studied by ESD and AES, *Surf. Sci.* **160**, 189 (1985).
76. A. Glachant and D. Saidi, Kinetics of Si(100) nitridation first stages by ammonia: electron-beam-induced thin film growth at room temperature, *J. Vac. Sci. Technol. B* **3**, 985 (1985).
77. R. Beigang, Electron stimulated desorption of excited OH radicals from SrF_2 surfaces, *Nucl. Instrum. Meth. Phys. Res., Sect. B* **13**(1–3) 541 (1986).
78. R. F. Haglund, Mechanisms of electron and photon-stimulated desorption in alkali halides,, *Nucl. Instrum. Meth. Phys. Res., Sect. B* **13**(1–3), 525 (1986).
79. D. R. Baer and M. T. Thomas, H_2O adsorption on Fe: Disorder and beam effects, *Appl. Surf. Sci.* **26**, 150 (1986).
80. I. L. deCeballos, M. C. Munoz, J. M. Goni, and J. L. Sacedon, X-Ray photoemission spectroscopy study of the activated oxidation of GaAs (111), *J. Vac. Sci. Technol. A* **4**, 1621 (1986).
81. S. Aduru, S. Contarini, and J. W. Rabalais, Electron, X-ray and ion stimulated decomposition of nitrate salts, *J. Phys. Chem.* **90**, 1683 (1986).
82. S. Contarini, S. Aduru, and J. W. Rabalais, Decomposition of chromate salts stimulated by ion and electron irradiation, *J. Phys. Chem.* **90**, 3202 (1986).
83. V. M. Bermudez, T. M. Parrill, and R. Kaplan, Electron stimulated desorption of positive ions from hexagonal α-SiC, *Surf. Sci.* **173**, 234 (1986).
84. J. H. Craig, CO on Pt (111); TDS and ESD study, *Appl. Surf. Sci.* **25** (3), 333 (1985).
85. J. A. Kelber, Surface analysis of LiF films by electron stimulated desorption and x-ray photoelectron spectroscopy, *Appl. Surf. Sci.* **27**(2), 151 (1986).
86. J. L. Sacedon, Electron beam stimulated reactions on semiconductors, *Vacuum* **37**(5–6) (1987).
87. S. Rey, ESD threshold and local bonding of dissociated species from H_2O on niobium, *Vacuum* **37**(5–6) 1987.
88. J. H. Craig, ESD energy distributions for CO and O_2 on Rn(111), *Appl. Surf. Sci.* **28**(3), 323 (1985).
89. K. Saiki, Y. Sato, K. Ando, and A. Koma, In-Situ observation of defect formation in CaF_2 (111) surfaces induced by low energy electron bombardment, *Surf. Sci.* **192**, 1–10 (1987).
90. X. Chen, Electron stimulated F+/ desorption: A sensitive method for detection of fluorine, *Surf. Interface Anal.* **12**(1–12), (1988).
91. K. Suzuki, Cross sections of electron stimulated desorption under technical vacuum conditions at 1 to 3 keV electron energies, *Appl. Surf. Sci.* **33–34**, 325 (1985).
92. Tsuneo Yasue, Electron stimulated desorption ion angular distribution (ESDIAD) from LiF surface, *Appl. Surf. Sci.* **33–34**, 167 (1988).
93. L. Cota, Electron stimulated desorption study of irradiated poly(vinyl chloride), *Appl. Surf. Sci.* **35**, 213 (1985).

94. S. Rey, Thermal desorption and surface and layer characterization of H_2O adsorbed on Nb(110) at 240 K by TDS, ESD and ELS, *Vacuum* **39**(7–8), 665 (1989).
95. M. Andritschky, Damage of oxide layers on an Al-alloy by electron bombardment, *Vacuum* **9**(7–8), 649 (1989).
96. M. Q. Ding, Electron stimulated desorption of gases at technological surfaces of aluminium, *Vacuum* **39**(5), 463 (1989).
97. G. B. Hoflund, Use of electron stimulated desorption to study hydrogen at catalytic surfaces, Preprints—Division of Petroleum Chemistry, American Chemical Society **34**(3), 535 (1989).
98. J. A. Kelber, Electron stimulated desorption of CH_3OH and $(CH_3)_2NH$ from partially oxidized copper, *Appl. Surf. Sci.* **40**(1–2), 129 (1985).
99. S. Rey, ESD and TDS study of the adsorption of methanol on Nb(110), *Vacuum* **41**(1–3), 54 (1990).
100. S. L. Bennett, ESD study of the initial stages of oxide formation at the Si-H_2O interface, *Vacuum* **41**(1–3), 220 (1990).
101. K. Ueda, Initial oxidation study of titanium by dynamic measurement of TOF, electron stimulated desorption, *Vacuum* **41**(1–3), 11 (1990).
102. S. A. Joyce, Influence of adsorbed potassium on electron stimulated desorption of PF_3 on Ru(0001), *Vacuum* **41**(1–3), 738 (1990).
103. T. Gotoh, Observation of ion desorption from MgO(001) by electron bombardment, *Vacuum* **41**(1–3), 213 (1990).
104. T. Yasue, Electron stimulated desorption from hydrogen and alkali adsorbed Si(111) surface, *Vacuum* **41**(1–3), 561 (1990).
105. H. Daimon, ESDIAD with a new display-type spherical mirror analyser, *Vacuum* **41**(1–3), 215 (1990).
106. S. Baunack, Electron beam-induced decomposition of MBE grown CaF_2 films, *Vacuum* **41**(4–6), 1003 (1990).
107. A. Takano, Study of the oxygen adsorption process on a titanium single crystal surface by electron stimulated desorption, *Surf. Sci.* **242**(1–3), 450 (1991).
108. T. D. Wu, Experiments relative to the desorption of negative ions during low-energy electron impact on surfaces, *Nucl. Instrum. Meth. Phys. Res., Sect. B* **58**(3–4), 496 (1991).
109. O. Kreitschitz, Desorption kinetics of Li atoms from LiF under electron bombardment, *Nucl. Instrum. Meth. Phys. Res., Sect. B* **58**(3–4), 490 (1991).
110. S. L. Bennett, Evidence of multiple H^+ ion desorption pathways with ESD of chemisorbed and multilayer water, *Surf. Sci.* **251–252**, 857 (1991).
111. Y. B. Zhao, Electron stimulated desorption of CO from Pd/W(110), *Surf. Sci.* **250**(1–3), 81 (1991).
112. M. Szymonski, Electron stimulated desorption of neutral species from (100) KCl surfaces, *Surf. Sci.* **260**(1–3), 295 (1992).
113. J. Dirks, Electron stimulated field desorption of field adsorbed neon and helium from tungsten, *Surf. Sci.* **266**(1–3), 62 (1992).
114. J. Kolodziej, Temperature dependent dynamic ESD processes in alkali halides, *Nucl. Instrum. Meth. Phys. Res., Sect. B* **65**, 507 (1992).
115. U. Kazuyuki, Studies of hydrogen adsorption on silicon (100) surfaces by means of time-of-flight type electron stimulated desorption spectroscopy (TOF-ESD), *Vacuum* **43**(8), 795 (1992).
116. T. Hsu, Hydrogen desorption on various H-terminated Si(100) surfaces due to electron beam irradiation. Experiments and modeling, *Appl. Phys. Lett.* **61**(5), 580 (1992).
117. F. J. Palomares, Formation of SiO_2/GaAs(100) interfaces by electron stimulated oxidation of ultrathin Si overlayers. Subcutaneous oxidation processes, *J. Vac. Sci. Technol. B* **10**(5), 2201 (1992).
118. N. Seifert, et al. Simultaneous measurements of transmission optical absorption and electron stimulated Li desorption on LiF crystals, *Nucl. Instrum. Meth. Phys. Res., Sect. B* **72**(3–4), 401–408 (1992).

119. F. H. Eisen, Ion irradiation damage in n-type GaAs in comparison with its electron irradiation damage, *J. Appl. Phys.* **72**(12), 5593 (1992).
120. A. Szabo, Width of particle beams desorbed in electron stimulated desorption.O^+ and metastable CO from CO/Pt(111), *J. Chem. Phys.* **98**(1), 689 (1993).
121. M. R. Davidson, ESD of oxygen atoms and ions from an oxidized Ag-Zr alloy surface, *Surf. Sci.* **281**(1–2), 111 (1993).
122. I. V. Gontar', The investigation of electron stimulated processes in ionic oxides, *Izv. Akad. Nauk. Ser. Fiz.* **57**(2), 143 (1993).
123. D. E. Weibel, Electron stimulated desorption from the surface of solid Ne: kinetic energy and angular distributions of desorbed metastables, *Surf. Sci.* **283**(1–3), 204 (1993).
124. U. Kazuyuki, Desorption study of a proton from H/Si(100) by electron stimulated desorption spectroscopy, *Surf. Sci.* **283**(1–3), 195 (1993).
125. A. Takano, Electron stimulated desorption of O^+ from CO adsorbed on a Ni(110) surface, *Surf. Sci.* **283**(1–3), 199 (1993).
126. Y. B. Zhao, Electron stimulated desorption of hydrogen from W(110), *Surf. Sci.* **286**(3), 290 (1993).
127. M. Petravic, Electron stimulated desorption of positive and negative ions from SiO_2/Si surfaces, *Nucl. Instrum. Meth. Phys. Res., Sect. B* **78**(1–4), 333 (1993).
128. C. L. Greenwood, ESD and ESDIAD investigation of TiO_2/(110)-SO_2, *Surf. Sci.* **287–288**, 386 (1993).
129. U. Kazuyuki, Hydrogen termination study of silicon dangling bonds by electron stimulated desorption spectroscopy (TOF-ESD), *Surf. Sci.* **287–288**, 506 (1993).
130. J. H. Craig, Electron desorption study of HF etched Si(100), *J. Vac. Sci. Technol. A* **11**(3), 554 (1993).
131. H. J. Jaensch, Electron stimulated surface migration of CO on Pt(335). First spectroscopic evidence for a new phenomenon, *J. Chem. Phys.* **99**(1), 721 (1993).
132. M. V. Ascherl, ESD kinetic energy distribution study of trimethylsilane adsorbed on Si(100), *Appl. Surf. Sci.* **74**(1), 121 (1994).
133. N. Seifert, Influence of defects and defect clusters on alkali atom desorption stimulated by low energy electron bombardment of alkali halides, *Nucl. Instrum. Meth. Phys. Res., Sect. B* **84**(1), 77 (1994).
134. J. Kotodziej, Thermally assisted desorption processes in electron bombarded alkali halides, *Vacuum* **45**(2–3), 353 (1994).
135. N. Tanaka, Energy dependence and depth distribution of electron beam-induced damage in GaAs/AlGaAs heterostructures, *J. Electron. Mater.* **23**(3), 341 (1994).
136. K. Sakamoto, Electron-stimulated desorption (ESD) of the O_2/Si(111) surface, *Surf. Sci.* **306**(1–2), 93 (1994).

5.5. Charging of Insulators Resulting from Electron Beam Irradiation

137. B. Lang and S. Goldsztaub, Charging of insulators in LEED, *Surf. Sci.* **32**, 473 (1972).
138. A. Nelson and A. K. Green, Method to control specimen charging for SIMS and AES of insulators, *J. Vac. Sci. Technol. A* **17**, 855 (1980).
139. P. J. K. Paterson, P. H. Holloway, and Y. E. Strausser, Energy Referencing of Auger Electron Spectra from Charging Samples by Implanted Inert Gas Spectra, *Appl. Surf. Sci.* **4**, 37 (1980).
140. A. H. Clark, T. L. Michalka, and J. Maley, A simple method to minimize charging and thermal damage during AES analysis of paper and paper coatings, *J. Vac. Sci. Technol. A* **20**, 254 (1982).
141. H. W. Werner and N. Warmoltz, Beam techniques for the analysis of poorly conducting materials, *J. Vac. Sci. Technol. A* **22**, 726 (1984).

142. M. G. Vasilyev, D. V. Klyachko, V. G. Krigel, and I. V. Razumovskaya, The effect of charging on the reproducibility of measurements of Auger electron spectra in semiconductors, *Phys. Chem. Mech. Surf.* **2**, 1525 (1984).
143. P. J. Moller and J. He, Electron beam induced charging of Cu/MgO surfaces, *Nucl. Instrum. Meth. Phys. Res., Sect. B* **17**, 137 (1986).
144. J. Cazaux, Electromigration and charging effects during Auger and XPS analysis of insulators, *Surf. Interface Anal.* **9**, 133 (1986).
145. J. Cazaux, Electrostatics of insulators charged by incident electron beams, *J. Microsc.* **11** 293 (1986).
146. J. Cazaux, Some considerations on the electric field induced in insulators by electron bombardment, *J. Appl. Phys.* **59**, 1418 (1986).
147. J. P. Moller, Electron beam induced charging of Cu/MgO surfaces, *Nucl. Instrum. Meth. Phys. Res., Sect. B* **17**(2), 137 (1986).
148. M. Brunner, Charging effects in low-voltage SEM metrology, *Scanning Electron Microscopy II*, 1986.
149. S. Ichimura, Reduction of charging in surface analysis of insulating materials by AES, *Surf. Interface Anal.* **14**(5), 250 (1989).
150. C. Le Gressus, Charging phenomena on insulating materials: mechanisms and applications, *Proceedings of the 1st European Workshop on Modern Developments and Application in Microbeam Analysis*, University of Amsterdam (1989).
151. J. Cazaux, Surface potential and electric field induced by electron bombardment of insulators, *IEEE Trans. Electrical Insulation* **24**(6), 995 (1989).
152. M. Tence, Electron irradiation effects. A time-energy representation, in: *Proceedings of the Institute of Physics Electron Microscopy and Analysis Group and Royal Microscopical Society Conference* (1990).
153. J. Cazaux, Phenomena relating to charge in insulators: macroscopic effects and microscopic causes, *Scanning Microscopy* **5**(1), 17 (1991).
154. B. G. Atabaev, Influence of electron irradiation on surface charging of alkali halide crystals. *Poverkhnost* **1**, 88 (1992).
155. J. Cazaux, Electrostatic, electrokinetic and secondary emission properties of insulators submitted to *Electron Irradiation, Interdisciplinary Conference on Dielectrics, Properties, Characterization, Applications* (1992).
156. S. Hofmann, Charging and charge compensation in AES analysis of insulators, *J. Electron Spectrosc. Relat. Phenom.* **59**(1), 15 (1992).
157. L. Reimer, Charging of bulk specimens, insulating layers and free-supporting films in scanning electron microscopy, *Optik* **92**(1), 14 (1992).
158. D. P. Russell, Auger studies of damage and charging in NaCl, in: *Proceedings of the Fifth International Workshop on Desorption Induced by Electronic Transitions: DIET V*, Springer, Berlin (1993).
159. J. Cazaux, Some physical descriptions of the charging effects in insulators under irradiation, *Proceedings of a NATO Advanced Research Workshop on Ionization of Solids by Heavy Particles* (1993).
160. D. S. H. Chan, K. S. Sim, J. C. H. Phang, Simulation model for electron irradiation induced specimen charging in a scanning electron microscope, *Scanning Electron Microscopy* **7**(3), 847 (1993).

5.6. Electron Beam Damage in Glasses

161. J. L. Lineweaver, Oxygen outgassing caused by electron bombardment of glass, *J. Appl. Phys.* **34**, 1786 (1963).

162. A. K. Varshneya, A. R. Cooper, and M. Cable, Changes in composition during electron microprobe analysis of K_2O-SrO-SiO_2 glass, *J. Appl. Phys.* **37**, 2199 (1966).
163. M. P. Borom and R. E. Hanneman, Local compositional changes in alkali silicate glasses during electron microprobe analysis, *J. Appl. Phys.* **38**, 2406 (1967).
164. L. F. Vassamillet and V. F. Caldwell, electron-probe microanalysis of alkali metals in glasses, *J. Appl. Phys.* **40**, 1637 (1969).
165. R. Hayami and H. Wakabayaski, Compositional changes during electron microprobe analysis of alkali silicate glasses, *Yogyo Kyokai Shi*, **78**, 138 (1970).
166. P. J. Goodhew and J. E. C. Gulley, The determination of alkali metals in glasses by electron probe microanalysis, *Glass Technol.* **15**, 123 (1974).
167. C. G. Pantano, D. B. Dove, and G. Y. Onoda, Jr., AES compositional profiles of mobile ions in the surface region of glass, *J. Vac. Sci. Technol.* **13**, 414 (1976).
168. F. Ohuchi, D. E. Clark, and L. L. Hench, Effect of crystallization on the Auger electron signal decay in a Li_2O-$2SiO_2$ glass and glass-ceramic, *J. Am. Ceram. Soc.* **62**, 500 (1979).
169. F. Ohuchi, M. Ogino, P. H. Holloway, and C. G. Pantano, Electron beam effects during analysis of glass thin films with Auger electron spectroscopy, *Surf. Interface Anal.* **2**, 85 (1980).
170. R. G. Gossink, H. Van Doveren, and J. A. T. Verhoeven, Decrease of the alkali signal during Auger analysis of glasses, *J. Non-Cryst. Solids* **37**, 111 (1980).
171. P. G. Withrop, Sodium depletion on glass surfaces during Auger analysis, *Surf. Sci.* **110**, 261 (1981).
172. D. M. Usher, Sodium ion migration in glass on electron beam irradiation, *J. Phys. C: Sol. State Phys.* **14**, 2039 (1981).
173. F. Ohuchi and P. H. Holloway, General model of sodium desorption and diffusion during electron bombardment of glass, *J. Vac. Sci. Technol.* **20**, 863 (1982).
174. A. Miotello and P. Mazzoldi, Numerical analysis of field-assisted sodium migration in electron-irradiated glasses, *J. Phys. C: Sol. State Phys.* **15**, 5615 (1982).
175. A. Miotello and P. Mazzoldi, Alkali signal decay during Auger analysis of dielectric solids: secondary effect of desorption process, *Radiat. Eff. Defects Solids* **83**, 271 (1984).
176. P. Mazzoldi and A. Miotello, Radiation effects in glasses, *Radiat. Eff. Defects Solids* **98**, 39 (1986).
177. A. Miotello and F. Toigo, Microscopic mechanisms governing alkali-metal transport in electron irradiated glasses, *Phys. Rev. Lett.* **56**, 1940 (1986).
178. M. Corchia, P. DeLogou, R. Giorgi, P. Mazzoldi, and A. Miotello, Auger electron spectroscopy in glasses: correlation effects, *Nuclear Instr. and Methods in Physics* **B32**, 318 (1988).
179. A. Miotello, G. Cinque, P. Mazzoldi, and C. G. Pantano, Alkali-metal segregation at glass surfaces during electron irradiation, *Phys. Rev. B: Condensed Matter* **43**(5), 3831 (1991).
180. U. Rothhaar, V. Rupertus, and H. Oechsner, XPS and AES-investigations of electron and ion beam induced effects on Sk16 glass surfaces, *Fresenius J. Anal. Chem.* **346**(1), 96 (1993).

5.7. Electron Beam Damage during Surface Analysis

181. C. C. Chang, Silicon on sapphire epitaxy by vacuum sublimation-LEED-Auger studies and electronic properties of films, *J. Vac. Sci. Technol.* **8**, 500 (1971).
182. H. Poppa and A. G. Elliot, The surface composition of mica substrates, *Surf. Sci.* **24**, 149 (1971).
183. I. Janossy and M. Menyhard, LEED Study of Quartz Crystals, *Surf. Sci.* **25**, 647 (1971).
184. H. H. Madden and G. Ertl, Decomposition of Carbon Monoxide on a (110) Nickel Surface, *Surf. Sci.* **35**, 211 (1973).
185. T. N. Taylor and P. J. Estrup, Carbon Monoxide adsorption on Ni (110), *J. Vac. Sci. Technol.* **10**, 26 (1973).
186. N. J. Chou, C. M. Osburn, Y. J. van der Meulen, and R. Hammer, Auger analysis of Chlorine in HCl or Cl_2-grown SiO_2 films, *Appl. Phys. Lett.* **22**, 380 (1973).

187. S. Thomas, Electron-Irradiation Effect in the Auger Analysis of SiO_2, *J. Appl. Phys.* **45**, 161 (1974).
188. W. C. Johnson and L. A. Heldt, Composition of Pt anode surfaces by Auger spectroscopy, *J. Electrochem. Soc.* **121**, 34 (1974).
189. R. A. Chappell and C. T. H. Stoddart, An Auger electron spectroscopy study of float glass surfaces, *Phys. Chem. Glasses* **15**, 130 (1974).
190. C. T. H. Stoddart, R. L. Moss, and D. Pope, Determination of the surface composition of Palladium-Nickel Alloy film catalysts using Auger electron spectroscopy, *Surf. Sci.* **53**, 241 (1975).
191. J. Ahn, C. R. Perlebery, D. L. Wilcox, J. W. Coburn, and H. F. Winters, Electron-beam effects in depth profiling measurements with Auger electron spectroscopy, *J. Appl. Phys.* **46**, 4581 (1975).
192. M. P. Hooker and J. T. Grant, Auger electron spectroscopy studies of CO on Ni: Spectral line shapes and quantitative aspects, *Surf. Sci.* **55**, 741 (1976).
193. Y. E. Strausser and J. S. Johannessen, An Auger electron spectroscopy study of Silicon spectra from Silicon Monoxide, Silicon Dioxide and Silicon Nitride, in: *Surface Analysis for Silicon Devices*, NBS Special Publication, Vol. 400, 125 (1976).
194. J. S. Johannessen, W. E. Spicer, and Y. E. Strausser, An Auger analysis of the SiO_2 interface, *J. Appl. Phys.* **47**, 3028 (1976).
195. Z. Szklarska-Smialowska, H. Viefhaus, and M. Janik-Czachor, Electron spectroscopy analysis of in-depth profiles of Iron in Cl-containing solutions, *Corros. Sci.* **16**, 649 (1976).
196. J. C. Buchholz and G. A. Somarjai, The surface structures of Phthalocyanine monolayers and vapor-grown films: A low energy electron diffraction study, *J. Chem. Phys.* **66**, 573 (1977).
197. H. J. Mathieu, J. B. Mathieu, D. E. McClure, and D. Landolt, Beam effects in Auger electron spectroscopy analysis of titanium oxide films, *J. Vac. Sci. Technol.* **14**, 1023 (1977).
198. T. Inoue, Electron irradiation effect in the quantitative Auger analysis of PSG, *Jpn. J. Appl. Phys.* **16**, 851 (1977).
199. W. N. Unertl, L. C. DeJonghe, and Y. Y. Tu, Auger spectroscopy of grain boundaries in calcium-doped sodium beta-alumina, *J. Mater. Sci.* **12**, 739 (1977).
200. W. P. Ellis, Electron-beam decomposition of UF_4 optical interference layers, *J. Vac. Sci. Technol.* **14**, 1316 (1977).
201. T. E. Madey, C. D. Wagner, and A. Joshi, Surface characterization of catalysts using electron spectroscopies: results of a round-robin sponsored by ASTM committee D-32 on Catalysts, *J. Electron Spectrosc. Relat. Phenom.* **10**, 359 (1977).
202. R. G. Musket, Studies of clean and oxidized tellurium surfaces, *Surf. Sci.* **74**, 423 (1978)
203. T. E. Madey and J. T. Yates, Jr., The adsorption of cycloparaffins on Ru (001) as studied by temperature programmed desorption and electron stimulated desorption, *Surf. Sci.* **74**, 397 (1978).
204. P. T. Dawson, O. S. Heavens, and A. M. Pollard, Glass surface analysis by Auger electron spectroscopy, *J. Phys: Condensed Matter* **11**, 2183 (1978).
205. J. J. Wagner and C. W. Wilmsen, Auger analysis of ultrathin SiO_2 layers on silicon, *J. Appl. Phys.* **50**, 874 (1979).
206. G. T. Burstein, Effects of electron bombardment during Auger electron analysis of corroded metal surfaces, *Mater. Sci. Eng.* **42**, 207 (1980).
207. C. G. Pantano and T. N. Wittberg, Surface studies of the medicus nickel matrix cathode, *Appl. Surf. Sci.* **4**, 385 (1980).
208. G. Remmond, P. H. Holloway, and C. LeGressus, Electron spectroscopy and microscopy for studying surface changes of mechanically prepared pyrite and quartz, *Scanning Electron Microscopy* (1981).
209. G. B. Hoflund, G. L. Woodson, D. F. Cox, and H. A. Laitinen, Surface characteristics of antimony-doped tin-oxide films, *Thin Solid Films* **78**, 357 (1981).

210. I. Krainsky, W. L. Gordon, and R. W. Hoffman, Electron beam damage in ITO coated Kapton, *Appl. Surf. Sci.* **9**, 39 (1981).
211. S. Bouquet, J. Bergner, J. LeHericy, and J. P. Langeron, Chlorine migration under electron bombardment observed by AES, *J. Electron Spectrosc. Relat. Phenom.* **26**, 247 (1982).
212. V. G. Lifshits and A. A. Saranin, Electron irradiation effect on the surface composition of Ar^+ ion bombarded Si-nitride and Si-oxynitride, *Solid State Commun.* **50**, 925 (1984).
213. W. H. Hocking, P. J. Hayward, D. G. Watson, and G. C. Allen, Scanning Auger microscopy study of lanthanum partitioning in sphene-based glass-ceramics, *Phil. Mag. A* **49**, 637 (1984).
214. F. Fransen, R. Vander Berghe, R. Vlaeminck, M. Hinoul, J. Remmerie, and H. E. Maes, Electron and ion beam degradation effects in AES analysis of silicon nitride thin films, *Surf. Interface Anal.* **7**, 79 (1985).
215. W. D. Jennings, HREELS and Auger studies of conducting polymers, *Appl. Surf. Sci.* **21**, 80 (1985).
216. F. A. Chambers, G. W. Zajac, and T. H. Fleisch, An Auger electron spectroscopy, x-ray photoelectron spectroscopy, secondary ion mass spectrometry and bulk analysis of pyrolytic boron nitride crucibles after vacuum baking, *J. Vac. Sci. Technol. B* **4**, 1310 (1986).
217. L. L. Kazmerski, N. A. Burnham, A. B. Swartzlander, A. J. Nelson, and S. E. Asher, Electron beam effects in the analysis of compound semiconductors and devices, *J. Vac. Sci. Technol. A* **5**, 2814 (1987).
218. B. Lang, Build-up and annealing of damage induced by ion and electron beams at SiO_2 surfaces: An AES study, *Appl. Surf. Sci.* **37**(1), 63 (1989).
219. H. Bender, Damage and recovery by electron and ion beam irradiation during AES analysis of silicon oxynitrides, *Surf. Interface Anal.* **15**(1), 38 (1990).
220. W. Chen, Electron and ion beam irradiation effects in AES analysis of silicon nitride and oxynitride thin films, *Chinese J. Semicond.* **11**(6), 441 (1990).
221. A. Fuchs, Electron and ion radiation damage of a-C:H films detected by energy loss spectroscopy of reflected and transmitted electrons, *Thin Solid Films* **217**(1–2), 48 (1992).
222. L. Viscido, Auger electron spectroscopy electron-beam induced damage of a H_2O-covered Si(100) surface, *J. Vac. Sci. Technol. A* **11**(1), 175 (1993).
223. A. Fuchs, Determination of sp_2/sp_3 carbon bonding ratio in a-C:H including irradiation damage by factor analysis of Auger electron spectra, *Thin Solid Films* **232**(1), 51 (1993).
224. L. G. Pittaway, The temperature distributions in thin foil and semi-infinite targets bombarded by an electron beam, *Br. J. Appl. Phys.* **15**, 967 (1964).
225. J. Pacansky, A. Gutierrez, and R. Kroeker, Electron beam induced chemistry of lithographic materials, in: *Electron Beam Interactions with Solids*, SEM, Inc., AMF O'Hare, Chicago (1984).
226. ASTM standard E 983-89, *1994 Annual Book of ASTM Standards,* Vol. 3.06 ASTM, Philadelphia (1994).

3

Ion Beam Bombardment Effects on Solid Surfaces at Energies Used for Sputter Depth Profiling

L. S. Dake, D. E. King, J. R. Pitts, and A. W. Czanderna

1. Introduction

In this section, we provide an overview of the entire chapter, definitions and nomenclature used, and an overview of ion–surface and ion–solid interactions. Commonly used acronyms in surface science that relate to this chapter are defined in Appendix 1. The reader may also be helped by reviewing our summary and concluding remarks (Sec. 9) before and after studying Sec. 1.

1.1. Overview

Ion beam bombardment of solids is of interest to scientists for its effects on chemical reactions, adhesion, implantation or doping, cleaning, and sputtering for sputter depth profiling. The scope of this chapter focuses almost exclusively on ion beam effects resulting from bombardment with ion beams of inert gases in sputter depth profiling. First, we summarize briefly the importance of compositional depth

---

L. S. Dake, D. E. King, J. R. Pitts, and A. W. Czanderna • National Renewable Energy Laboratory, Golden, CO 80401. Present address (J. R. Pitts): National Center for Photovoltaics and Center for Basic Sciences.

Beam Effects, Surface Topography, and Depth Profiling in Surface Analysis, edited by Czanderna *et al.* Plenum Press, New York, 1998

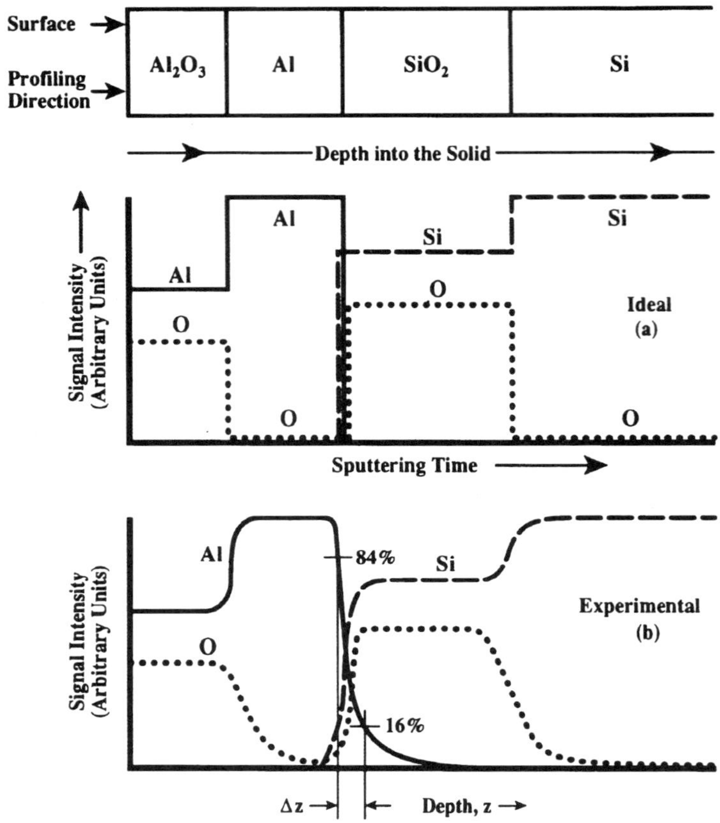

Figure 1.1. Depth profiles of thin-film multilayers (a) ideal, (b) schematic experimental of ideal, and (c) SIMS detection of $^{18}O^-$ from Ta_2O_5 multilayers in which the central 100 nm was enriched with ^{18}O. The 16% and 84% ^{18}O signal levels represent ±2 standard deviations from the half-maximum value and serve to evaluate the depth resolution, Δz. Parts (a) and (b) adapted and reprinted with permission from [1] A. W. Czanderna, in: *Ion Spectroscopies for Surface Analysis*, edited by A. W. Czanderna and D. M. Hercules, (Plenum Press, New York, 1991), pp. 1–44. Copyright 1991 Plenum Press, New York; part (c) adapted and reprinted with permission from[2] J. A. McHugh, in: *Methods of Surface Analysis*, edited by A. W. Czanderna (Elsevier Scientific Publishing Company, Amsterdam, 1975), pp. 223–278, and with kind permission of Elsevier Science - NL, Sara Burgerhartstraat 25, 1055 KV Amsterdam, The Netherlands.

profiling, mention the advantages and limitations of ion beam bombardment for sputter depth profiling to obtain compositional depth profiles, and illustrate a typical compositional depth profile. In the rest of the introduction, we list the ion beam–solid topics not covered in our review, define a number of terms we use, and provide a brief overview of ion beam–solid interactions.

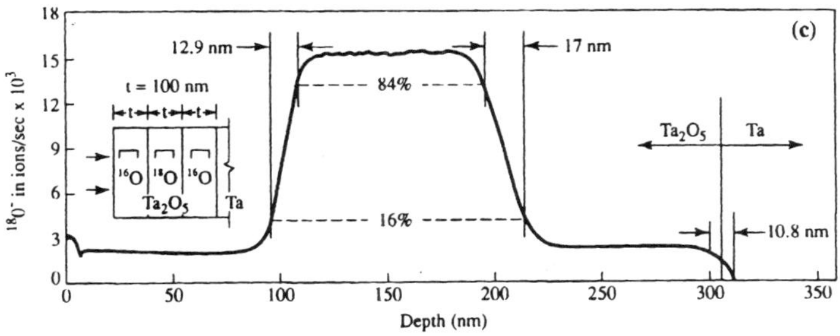

Figure 1.1 Continued.

The goal of compositional depth profiling is to determine the concentration of elements present as a function of the perpendicular distance (z direction) from the surface (the x–y directions). Sputter depth profiling, which is the most widely used method for obtaining a compositional depth profile, is compared and contrasted with other destructive and nondestructive methods in Chap. 5 of this volume. Many studies in science and technology require determining the depth distribution of elements in a wide range of specimen materials, particularly layered samples. Surface-sensitive probe beams can be used for this purpose in combination with the most commonly used method of etching by ion beam bombardment (i.e., sputter depth profiling), which is also known as ion erosion, ion milling, sputter etching, compositional depth profiling, depth profiling, and (atomic) layer-by-layer microsectioning. Ideal and schematic experimental depth profiles are shown in Figs. 1.1a and b for a multilayer stack of partially oxidized aluminum on partially oxidized silicon.[1] The experimental profile of oxygen-18 in Fig. 1.1c clearly does not provide one atomic layer of depth resolution at the interface,[2] and one of the purposes of this chapter is to explain why atomic-layer resolution cannot be obtained in practice. The *depth resolution*, Δz, is the depth (or thickness) range over which a signal increases or decreases between 15.87% and 84.13% when profiling an ideally atomically sharp interface between the two media.[3] The quoted limits correspond to $\pm 2\sigma$ limits for an assumed Gaussian broadening function described by a broadening parameter σ; the limits are frequently rounded to 16% and 84%.

Sputtering, the process of removing material by bombarding the surface with an energetic ion beam (typically 0.5–10 keV argon ions), in principle provides a means for atomically microsectioning the solid, and the sputtering time can be related to the depth.[4] The goal of microsectioning is to obtain an analysis at each atomic layer into the solid. The available "drill" is an ion beam that ideally produces a Gaussian-shaped crater in the static mode; the ion beam is rastered in the x–y plane to remove an area of material that may range from 1 to 100 mm^2 and to produce a crater with a flat bottom (ideally).[1] The composition is obtained from analyzing the surface in the central portion of the crater by any of the well-known methods of surface analysis,[5] e.g., XPS, AES, SIMS, or ISS.

The advantages of sputtering for in-depth analysis are considerable.[4] Physical sputtering is an atomic process, so, in principle, atomic depth resolution is possible. Sputter depth profiling can be performed *in situ* in vacuum, which means surface compositional analysis and sputtering can generally be performed simultaneously. When pure inert gases are used, clean surfaces may be produced, and reoxidation and recontamination can be avoided. Reasonable erosion rates are achieved, e.g., 0.1–100 nm/min, so microsectioning to depths of 1000–2000 nm is accomplished within reasonable periods of time.[4]

The sputtering process itself limits these considerable advantages, however, as we can see by comparing the ideal and experimental profiles in Fig. 1.1. A zone of mixing, which broadens the depth resolution (Sec. 6), is created from knock-on and other effects (Sec. 2). Differential sputtering (Secs. 2 and 5) may occur, and surface roughness (Sec. 4) may develop. Structure may be destroyed (Sec. 3) and new chemical states (Sec. 5) may be formed. Bulk and surface diffusion of the target may be enhanced during ion beam bombardment (Secs. 3 and 5). Matrix effects resulting from compositional and structural differences may change the rate of erosion as the process proceeds. The bombarding gas ions are implanted into the solid (Sec. 2); this is particularly serious when reactive gases such as oxygen or nitrogen are used. Redeposition (Section 2) of sputtered material occurs and complicates the analysis when it occurs on the surface being analyzed. Enhanced adsorption of residual gases from the vacuum may also occur, placing more stringent requirements on the vacuum level maintained during in-depth analysis (Sec. 6). Finally, the depth resolution degrades as greater microsectioning depths are reached (Sec. 6). The variation of Δz with sputtered depth is illustrated in Fig. 1.2.[6] The experimental results in Fig. 1.2 show some of the best results obtained for Δz, which is 0.7 nm for the Ta_2O_5/Ta interface; it is more typically about 2 nm at depths below about 300 nm. The actions of the ion beam, sample properties, and other factors we discuss in Sec. 6 result in depth resolutions larger than the best ones in Fig. 1.2. In actual practice, then, sputter depth profiling is not a technique in which successive single atomic layers are removed for convenient analysis of a newly exposed bulk layer. However, the general models that assume this is the case are useful for then deducing the various mechanisms that result in a departure from the ideal.

Most current approaches for simulating physical sputtering are based on a model in which a sequence of binary elastic collisions occurs (Fig. 1.3); i.e., the ion interacts with one target atom at a time.[7] During the multiple-collision process, the energy of the incident ion is transferred to lattice atoms until the incident ion is implanted or manages to escape from the solid. Winters[7] has compared theory and experiment regarding sputtering. As higher-energy ions are used, more energy is transferred to the lattice and the mean escape depth of the target atoms increases.[7,8]

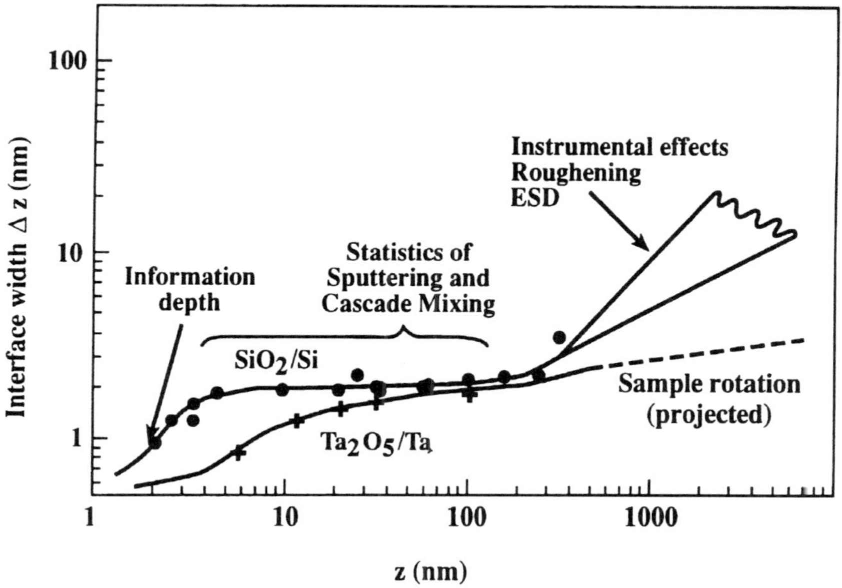

Figure 1.2. Depth resolution as a function of interface depth using 1 keV Ar ions for a National Physical Laboratory (UK) Ta_2O_5 on Ta reference metal (+) and the SiO_2 on Si thermal oxide (•) of Cook et al. (adapted from Ref. 6) The information depth is based on XPS or AES analysis; the other three processes cited will be discussed in Sec. 2, 4, and 6, respectively. Adapted and reprinted with permission from[6] M. P. Seah and C. P. Hunt, *Surf. Interface Anal.* **5**, 33–37 (1983).

For ideal depth profiling, the mean escape depth should be only from the outer monolayer, which cannot even be accomplished at low erosion rates. To achieve high erosion rates for reaching depths of 10,000–20,000 nm, the increased amounts of energy that must be transferred to the lattice per unit time result in many of the limitations mentioned in the previous paragraph. Most of the incident-ion energy does not result in the ejection of lattice atoms, but is, instead, transformed into increases in the local temperature (Sec. 7) of the solid and into the ejection of electrons and photons (Sec. 7).

The erosion (or sputtering) rate during ion beam bombardment of solids is given by $dz/dt = 0.00624SJ^{+}A/\rho$, where dz/dt is in nm/min, S is the sputtering yield in atoms per incident ion, J^{+} is the ion beam current density in $\mu A/cm^2$, A is the molecular weight of the target in g/mol, and ρ is the density of the target in g/cm^3.[(4)] *Erosion rate*, which is the time-dependent removal of material during ion beam bombardment, is a term preferred to *sputtering rate* because it is easy to confuse sputtering *rate* and sputtering *yield*.[(8)] According to Wehner,[(4)] when 2-keV Ar-ion bombardment at 0.15 mA/cm^2 is used, the erosion rates for stainless steels, Ta, SiO_2, and Ta_2O_5 range from 7 to 13 nm/min; for Pt, the rate is 23 nm/min; and

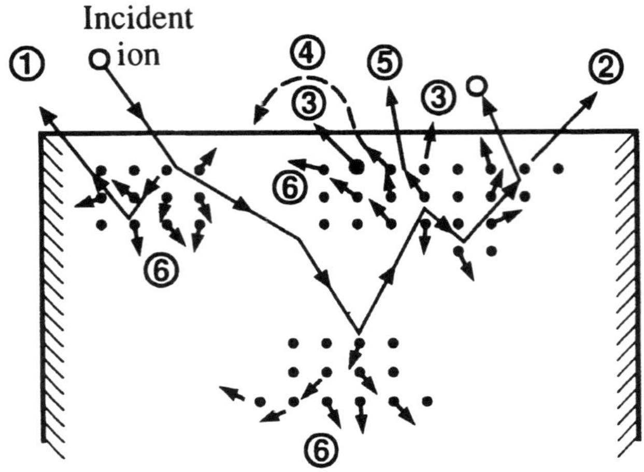

Figure 1.3. Several of the possible collision processes that occur during ion bombardment. Surface atoms are ejected after being hit from (1) above or (2) below; some do not receive enough energy to leave the surface (3), or they leave and fall back (4). Occasionally, bulk atoms receive enough energy to escape (5); others simply receive energy (6). Adapted and reprinted with permission from[7] H. F. Winters, in: *Radiation Effects on Solid Surfaces, edited by M. Kaminsky (American Chemical Society, Washington, DC, 1976), pp. 1–29. Copyright 1994 American Chemical Society.*

for Cu, Au, and Ag, the rates range from 30 to 40 nm/min. Werner[(8)] developed a nomogram relating dz/dt to J^+ for various values of S at an A/ρ value of 10 and published other useful nomograms for selecting various beam parameters and erosion rates for use in sputter depth profiling as well as for SIMS. (See also Chap. 5.)

1.2. Ion Beams and Solids: Topics Not Covered

Ion beams are widely used in science and technology, and many important topical areas of study are necessarily omitted from our consideration of the available literature. The topics omitted include Rutherford backscattering and nuclear reaction analysis,[(9)] secondary-ion mass spectrometry (SIMS),[(10)] ion scattering spectroscopy (ISS),[(11)] medium-energy ion scattering, other ion beam methods in surface analysis, including field ion microscopy,[(1)] ion implantation for materials modification, ion-beam-assisted deposition, reactive ion etching, magnetron sputtering, radio-frequency (rf) sputtering, other sputtering to prepare thin films, ion milling in TEM sample preparation, detailed consideration of sputtering yields, velocity of energy distributions of sputtered particles, sputtering by MeV ions, radioactive ion beams, electron–cyclotron resonance, ions in gas jets, plasmas, chemical vapor deposition, ion nitriding and variants, laser deposition of films, and many other applications areas we have not cited. We also excluded from consideration bombardment with molecular and cluster ions.

In searching the literature from 1980 to 1996, more than 20,000 citations were identified before exclusionary barriers were introduced to narrow the search to the roughly 1000 publications of specific interest about ion beam bombardment effects related to sputter depth profiling. Our literature search methodology is available on request. We are convinced that the articles referenced provide a good sample of the available information, but we readily admit that our survey has not identified every paper about the subject.

The topics in this chapter are of special interest to those using SIMS, ISS (also called LEIS), and ion beams for cleaning solids as well as those who only combine ion beam bombardment for microsectioning a solid and analyzing the exposed surface with XPS, AES, SIMS, or ISS.[1] In general, we have restricted our coverage to ion beam energies of 500 eV up to 10 keV, although some examples are given for energies as low as 20 eV and as high as 20 keV to illustrate important concepts or results. For the topics covered, we found the articles particularly helpful[4,7,10–27] for which the emphasis is ion–surface interactions,[4,7,10–16] implantation or trapping,[17,18] structure,[12,19,24] development of topography,[12,20–25] differential sputtering,[24–26] and limitations of depth resolution.[6,10,27]

1.3. Definitions and Nomenclature

We refer to the bombarding ion as an "ion" throughout its entire path through the solid, even though it will be neutralized and become an energetic atom after it penetrates the solid. Thus, "ion" is used for distinguishing it from atoms in the substrate, and not because it is scientifically exact. All ions are monovalent unless otherwise specified, e.g., in 3-keV Ar-ion beam, Ar^+ is meant. All ion beam angles of incidence are referenced to 0° for the surface normal to 90° for near grazing incidence to the plane of the solid. Because implantation is typically thought to involve processes with ion beam energies of more than 50 keV, we frequently, but not exclusively, use "trapping" to describe the implantation of bombarding ions at energies below 10 keV. We refer exclusively to the roughness of a surface as topography and do not use morphology, which we reserve for describing the nature of domains inside thin solid films or in solids. We refer to the rate of removal of material to expose the solid interior as an erosion rate, e.g., in nm/min, rather than an etch rate or sputter rate. The latter terms frequently are either confused with chemical etching or with sputtering *yields*,[8] so our purpose here is to avoid using terms that are known to be confusing. For the same reasons we use "ion-beam-bombardment-enhanced" instead of "radiation-enhanced" diffusion (RED), segregation, or reactions, because the widely used term "radiation enhanced" is ambiguous, whereas the term "ion-beam-bombardment-enhanced" is not. We use ion fluence and (total) ion dose, both with the units of number of ions/cm^2 interchangeably as is done in the literature, but reserve ion flux for the number of ions/cm^2-s and current density as a current/cm^2, e.g., μA/cm^2, for describing the ion beam. We list the acronyms and abbreviations used in surface science in Appendix 1; we also define less commonly used terms when they are first used in the text.

For the terms ion fluence and (total) ion dose, we have used the same terms as used by the authors of the paper(s) we referenced. We note definitions currently before the ISO/TC 201 Subcommittee on Terminology* have been *proposed.* "*Ion fluence* (F) for a parallel beam of particles is dN/dA, i.e., F = dN/dA, in which dN is the number of particles of a specified type *incident on an area dA* at right angles to the direction of the beam." Clearly, dA is the areal vector relative to the axis of the beam as normally defined in physics; a statement of the coordinate system used to describe the beam may also be required. "Dose, areic (D) or dose density (deprecated) is dN/dA, i.e., D = dN/dA, in which dN is the number of energetic particles of a specified type *introduced into a solid* through a surface area dA." In this case, the vector of the area dA is in the direction normal to the surface. Thus, "For the undeflected beam, areic dose equals fluence times cos θ (D = F cos θ) in which θ is the angle of incidence of the beam to the surface normal." Further confusion in the literature may result because "dose" is not always used for areic dose; some authors have used dose to mean "implanted areic dose" "nominal areic dose," "non-implanted areic dose," "received areic dose," "retained areic dose," or "sputtered areic dose." When the exact term used is important from our citation of fluence of dose, we urge the reader to secure the original published work we reference and deduce the author's intent.

For defining the ion beam itself, we expected all authors to state the ion used for ion beam bombardment, and the ion beam energy, current density, angle of incidence at the bombarded solid, and size, i.e., full width at half-maximum (FWHM) for a Gaussian distribution, e.g., an ion beam of 10 μA/cm^2 of 1-keV Ar ions with a 1-mm FWHM at 0° (normal incidence) was used. Most of the papers we reference do not provide a complete description of the ion beam, and the results from many papers of the 1000 we reviewed are not in this chapter because the description of the ion beam was not even marginally complete. In some of our text, we refer to the relative ion beam energy or the fluence as low, medium, and high. Whenever possible, we have indicated numerical ranges for the terms or provided specific examples that provide quantitative measures. In some cases, this was not possible because no mention of specific values was given in the original text. For the convenience of the reader, we broadly interpret these energies to be <1 keV (low), 1 to 5 keV (medium), and 5 to 10 keV (high), and broadly interpret the

*ISO is the International Standards organization; M. Seah (see Foreword) is currently the chairman of the Subcommittee on Terminology and C. Powell is currently a member of this subcommittee.

fluences to be $<10^{15}$ ions/cm^2 (low), 10^{15} to 10^{17} ions/cm^2 (medium), and $>10^{17}$ ions/cm^2 (high). In many cases specific values are highly dependent on the substrate and ion beam and other experimental conditions, so the terms "high" and "low" are relative and valid ranges must be determined for the specific sample and experimental conditions. If the exact energies or fluences are desired by the reader, the references cited in the relevant text should be sought and studied.

1.4. Overview of Ion–Surface and Ion–Solid Interactions

In this section, we present a broad overview of ion–solid interactions; more detail is given in Sec. 2. In Fig. 1.4, many of the ion–surface and ion–solid interactions or effects are shown schematically for *normally* incident ions.[12] The impact of the impinging ions results in sputtering, backscattering of the incident particle, and emission of photons, X-rays, and electrons. While some may be trapped (implanted), the impinging particles generally result in topographical surface roughening, structural damage and changes, and chemical changes, including differential sputtering. Changes in properties, e.g., electrical, optical, mechanical, chemical, and physical, take place in the damaged zone. Detailed summaries of ion–surface interactions are available.[6,10–16,20,23]

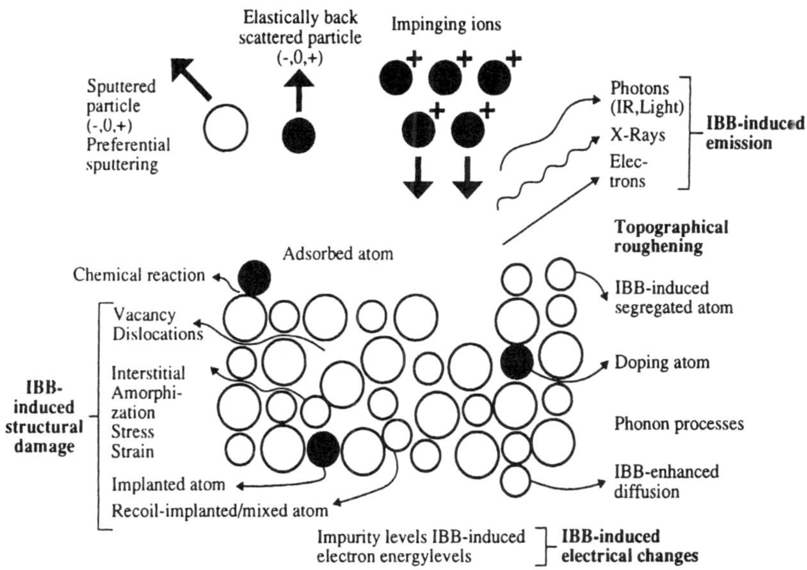

Figure 1.4. Some of the possible interactions that can take place during ion beam bombardment (IBB) of a solid. Adapted and reprinted with permission from JB. Malherbe, *Critical Reviews in Solid State and Materials Sciences* **19**(2), 55–127 (1994). Copyright 1994 American Chemical Society.

Many factors determine which effects will dominate. These factors include the solid itself, such as the masses and types of substrate atoms, the microcrystalline orientation and texture, and the ions in the ion beam, especially their mass, energy, current density, total fluence, and angle of incidence. For example, a damage map is shown in Fig. 1.5 for Ar-ion bombardment of GaAs (001) at current densities between 0.1 and 0.75 mA/cm^2 with energies from 20 to 95 eV.[(28)] An increased number of defects results from increasing *either* the current density or the ion beam energy, and this is typical for all types of materials. The lowest ion beam energies typically used in sputter depth profiling are 300 to 500 eV with comparable current densities, so the formation of defects in the near-surface regions is virtually ensured for all reported sputter depth profiles.

As an impinging ion approaches and strikes the surface, neutralization or an electron exchange occurs between the ion and the solid in one of two regimes. In

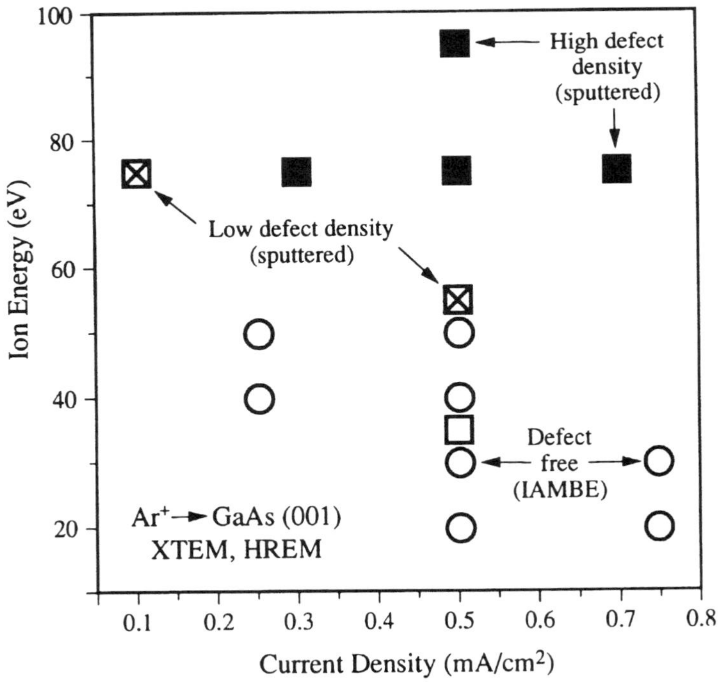

Figure 1.5. Degree of damage on GaAs (001) at 615°C as a function of Ar-ion energy an current density. The defect-free layers are represented by open symbols, low-defect-density layers by crosshatched symbols, and high-defect density layer by solid symbols. Sputter etched layers are shown by squares and ion-assisted MBE (IAMBE) by circles Adapted and reprinted with permission from[28] J. M. Millunchick, et al., *J. Vac. Sci. Technol., A* **13**(3), 1155–1159 (1995). Copyright 1995 American Vacuum Society.

the first regime, electron exchange occurs within a fraction of a nanometer from the surface by rapid ($t < 10^{-15}$ s) processes, e.g., resonance neutralization or ionization, Auger neutralization or deexcitation, nonadiabatic tunneling, or radiative neutralization.[1,29,30] The second regime is in the solid in which the electron orbitals of the collision partners overlap and evolve into molecular orbitals of a quasimolecule. A collisional reionization of the neutral species is the ultimate result. These electron processes are classified as potential energy processes and are relatively independent of impact energy.[31] Because the dominant processes are those that neutralize the ion beam, most of the collisions below the surface are energetic neutral–solid collisions. Some recent work on charge transfer processes in ion beam–solid collisions includes a review focused on resonant tunneling mechanisms,[32] a theoretical study of resonant transfer of conduction-band electrons into Rydberg states,[33] the calculation of Auger capture rates in which a free-electron gas is assumed for the metal,[34] an ARXPS study of the autoionization of He- and Ne-ion projectiles impinging on Al and oxidized Al surfaces,[35] and a discussion of excitation and loss processes of ions at surfaces based on experimental results for grazing incident ions with energies from 100 eV to 10 keV.[36]

As the neutralized ion collides with surface or subsurface atoms, several processes are possible that all depend on transferring momentum and energy. The bombarding ion may be backscattered, cause dissociation of an atom or molecule in the solid, react with the solid, become adsorbed, or penetrate the target lattice. The probability of the bombarding ion becoming implanted (trapped) in the solid increases with increasing ion energy, decreasing ion mass, and at normal incidence.[17,18] Prior to implantation, the bombarding ion loses energy via inelastic electronic interactions between the incoming ion and the solid and by nuclear interactions in which energy and momentum are transferred to the atoms in the solid. The energy loss process via nuclear interactions is primarily responsible for setting the substrate atoms into motion that results in sputtering, structural changes, and other motion-dependent damage. The energy transfer results in a collision cascade that is energy dependent and may be a ballistic series of binary collisions (lower energy, linear cascade) or an "instantaneous" event that sets all atoms into motion in the local volume (higher energy, spike regime). The low- and high-energy regimes are discussed in Sec. 2.4. The spike regime consists of atomic displacement, thermal, shock-wave, and ionization spikes.[37] For short periods of time the spike regime may result in pressures of ~1 GPa, temperatures higher than 3500 K, and atomic displacements even for Ar ions in the keV energy range. The ion-beam-induced movement of atoms develops in different time intervals for the ballistic, spike, and ion-beam-bombardment-enhanced diffusion regimes (the times and terms are discussed in Sec. 2).

In the ballistic regime the incident ion transfers most of its energy to the solid in a sequential series of binary collisions. When the bombarding ion comes to rest, it may be implanted, or it may migrate, adsorb, or react with the host atoms. The solid atoms are excited by a collision with the incident ion or from collision cascades with other target atoms until the energy is thermalized within about 10^{-13} s. The primary and secondary collisions, which are knock-on and recoil processes, result in forming defects, mixing of substrate atoms and layers, and in the recombination of defects. Thus, the atomic distributions are broadened and structural defects are introduced into the solid. The collision cascades are the main sources of the energy required for physical sputtering and erosion of the solid.

As the collision cascade subsides, the energy from the impacting ion beam is predominantly dissipated in the solid by phonon processes until total thermal equilibrium is established. During this process, ion-beam-enhanced diffusion may take place over a volume considerably larger than the spike or cascade volumes. New phases and uniformly mixed layers may result from the chemical driving forces involved in the process.

As part of the energy loss processes, ion beam impact also results in the ejection of secondary electrons and photons, as well as sputtering of the solid as atoms, ions, and molecular particles. Czanderna summarized these processes and their potential use as surface analysis techniques.(1) Ion-induced photon emission,(30,38–44) electron emission,(30,43,45–53) Auger electron emission,(29,54–56) and desorption(57) have been discussed in recent papers. Hofer's review is particularly comprehensive for ion-beam-induced electron emission processes.(45) We also note that electron tunneling processes are important for forming secondary ions, which is important in SIMS.(44,58) The fraction of ions that retain their charge during a collision process is important in ISS.(11,59)

2. Ion Beam–Solid Interactions

2.1. Introduction

This section provides a brief review of the fundamental ion beam–solid interactions. The purpose of this discussion is to provide a practical overview of the interactions, including important variables and parameters, as well as some proposed mechanisms, to assist in understanding and interpreting the experimental results in the rest of this chapter. The damaging effects of ion beam bombardment on a sample are considered to a limited extent in this section, but in depth throughout the rest of the chapter. A comprehensive theoretical treatment is beyond the scope of this work, but some indication of proposed mechanisms is provided, along with references to more thorough discussions of theoretical treatments.

For the range of ion beam parameters we are considering, most of the ion beam–solid interactions can be regarded as a series of binary collision events between an incoming ion and an atom of the substrate;(15,20) bombardment by clusters or molecular ions is not considered. Our treatment is limited to the physical sputtering process, as opposed to chemical processes, so reactive ion beam etching is not discussed. Chemical changes in the sample that are induced by sputtering are presented in Sec. 5, although some of the chemical changes resulting from purely ballistic interactions, such as differential sputtering and mixing, are discussed in this section. Topography development from ion beam bombardment is covered separately in Sec. 4.

When an ion beam impacts a solid surface, it may be backscattered, reflected, or adsorbed onto the sample surface, or it may penetrate to a depth dependent on the ion beam energy and mass, as well as various substrate properties. The transfer of energy and momentum from the ion beam to atoms in the sample results in the movement of substrate atoms, which induces the rearrangement, mixing, and sputtering of some of the substrate atoms. Figure 2.1 is a simplified schematic showing the implantation and rearrangement of both the incoming ions (solid circles) and substrate atoms (open circles) as well as the ejection of atoms and molecules, which is the desired result of sputter depth profiling.[60] Further details of these processes are given later. First we consider what happens to the ion beam,

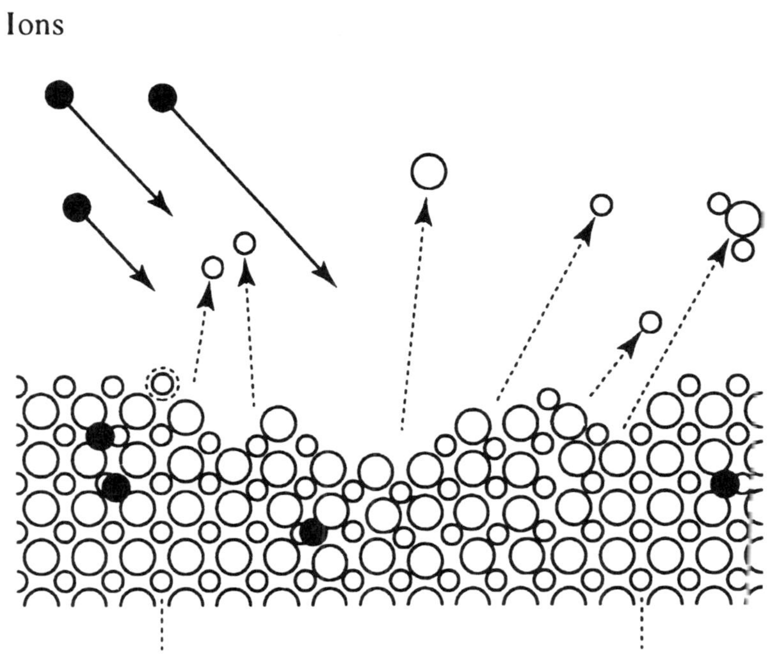

Figure 2.1. Ion-beam-bombardment-induced changes in the surface zone as illustrated by Benninghoven in 1977. Adapted and reprinted with permission from[60] A. Benninghoven, in: *Chemistry and Physics of Solid Surfaces*, edited by R. Vanselow and S. Y. Tong (CRC Press, Cleveland, 1977), pp. 207–233. Copyright 1977 CRC Press, Inc.

and then we discuss the nature and effects of the ion–substrate interactions. Although there is some understanding of most of the individual basic processes involved, there is only limited (and generally only qualitative) understanding of how these processes interact and of their combined effect on the sample.[61]

2.2. Ion Beam

Ions directed toward a solid surface backscatter (reflect), forward scatter, adsorb onto the surface, or penetrate the solid, depending on the specifics of the ion–substrate interaction. For the inert-gas ions and ion beam energies and substrate temperatures considered here, adsorption of the incoming ions is essentially negligible. The amount of scattering is largely a function of the ion beam angle of incidence because of the angular dependence of the scattering cross section, the relative masses of the bombarding ion and the substrate atoms, and, to a lesser extent, the ion beam impact site. If the ion penetrates, it will most likely be trapped inside the substrate. The concentration of trapped particles increases, and trapped particles may diffuse out of the sample or agglomerate into clusters. As the surface is eroded during sputtering, some of the trapped particles are exposed and released.[18,20]

As an ion approaches a (conducting) surface, electron exchange processes occur that depend on the nature of the surface and the energy of the ion.[12,20] The ion usually becomes neutralized (especially for energies below 1 keV) as it approaches the solid surface,[15] because the work function of most solids is lower than the ionization potential of the ion.[12] Thus, the backscattered and implanted primary ions are generally neutralized, although exceptions occur, especially for insulating or dielectric substrates. Guillemot *et al.*[62] conducted experiments in which a Si substrate was bombarded with Ne atoms and ions and reported similar charge fractions in the scattered species for both the neutral beam and ion beam bombardment. These experiments confirm that the observed ionized species are not simply reflected incoming ions, but are reionized neutrals produced in the final stage of the interaction, as the particle is emitted or reflected from the surface.[62]

If the ion penetrates the surface, its kinetic energy is dissipated by elastic (nuclear) and inelastic (electronic) collisions with substrate atoms.[16,20] The ion slows down and, if not reemitted at some point in the collision process, eventually comes to rest in the substrate.

2.2.1. *Reflection/Backscattering*

At low ion beam energies and near-grazing incidence, the interaction of the ion with the surface potential is largely repulsive, and reflection of the ion beam dominates. The reflection may result from a single binary collision or a sequence of a few collisions, but energy transfer to the substrate is small.[15] The particle reflection coefficient decreases with increasing bombardment energy, from about unity at a few electron volts to less than ca. 10^{-4} at 1 MeV.[15] Adsorption is also possible at low energies, but for the energies typically used for ion beam cleaning and sputtering the ion is most likely to penetrate the surface and enter the bulk of the substrate. There it undergoes a series of collisions with substrate atoms.[20] During this multiple-collision process, some of the incoming ions are scattered back out of the sample and contribute to the total backscattered flux. The total magnitude of the backscattered flux increases with increasing incidence angle of the ion beam with respect to the substrate normal. The backscattered flux also increases with decreasing ion beam energy and with decreasing ratio of the ion to substrate atomic masses.[20]

2.2.2. *Penetration and Trapping*

Once the ion penetrates the surface, it undergoes a series of collisions and deflections with substrate atoms until it is either reemitted or comes to rest. Classical trajectory simulations for inert ion bombardment of graphite indicate that the penetration probability rapidly increases from 10 to 100 eV, and that the probability of being trapped is ~1 for Ne and Ar ions above 100 eV in energy.[18] Below 10 eV, most of the incident ions backscatter, primarily as a result of head-on collisions with substrate atoms. Above 100 eV, penetration generally occurs even for near head-on collisions. The penetration probability is somewhat dependent on whether the impact point on the substrate is an occupied site or vacant. For He ions, the threshold energy for penetration is lower because of the smaller scattering cross section. Furthermore, calculations indicate that the penetration probability for He is never quite unity, because some backscattering will result from head-on collisions.[18]

The ion loses energy to the substrate via electronic (inelastic) and nuclear (elastic) interactions. In the 0.1- to 5-keV energy regime, interactions between the ion beam and substrate are generally treated as a series of independent binary collisions. The nuclear and electronic loss processes are treated separately and then summed to obtain the total energy loss rate (dE/dz) in the substrate.[15,20,63] The electronic interactions include Coulomb interactions of the ion with electrons in the substrate and can result in the emission of photons and electrons.[63–65] Electronic processes are only briefly mentioned here; electronic interactions induced by ion-induced ionization of core or valence levels of substrates (known as the Menzel–Gomer–Redhead and Knotek–Feibelman processes) may contribute to sputtering, especially for insulators,[66] alkali halides,[67,68] and some semiconductors.[69] The energy losses via nuclear collisions are of greater interest here, because these are primarily responsible for increasing the energy of the substrate atoms and for causing sputtering.[12]

Interactions of the ion nucleus with the nuclei of the substrate atoms serve to transfer momentum and energy from the ion to the substrate. The energy loss of the incoming ion increases roughly linearly with incident energy for energies below several kiloelectron volts;(15,16) it reaches a plateau somewhere from 10 to 100 keV and decreases at higher energies.(9,16,70) The energy loss also increases with increasing atomic number of the ion and substrate atoms.(16) Feldman and Mayer(9) express the energy loss from nuclear collisions $[dE/dz]_n = NS_n(E)$, where N is the atomic density of the sample, and $S_n(E)$ is the nuclear stopping cross section. For 1-keV Ar ions on Cu, they calculated an energy loss of approximately 1240 eV/nm.[9] The ion range R in the solid is

$$R = \int_{E_0}^{0} \frac{1}{dE/dz}\, dE \tag{1}$$

where E_0 is the incident-ion energy. Assuming that energy losses from nuclear collisions dominate and are independent of the energy, Eq. (1) becomes $R = E_0/(dE/dz)_n$. On the basis of these approximations, and using the value of energy loss estimated for nuclear collisions between Ar and Cu, Feldman and Mayer give 1 nm/keV of ion beam energy as a "rule of thumb" for the ion range in an amorphous sample. Further details of the calculation and example problems are available.(9)

Because each ion trajectory is different, the ion range is spatially distributed both laterally and in depth. The mean ion ranges generally increase with increasing ion energy and decrease with increasing ion mass.(20) The mean projectile ranges calculated from Schiott's formulas have been tabulated for energies from 0.2 to 500 keV for 14 typical projectiles and Al, Si, Ti, Fe, Ge, Nb, Ag, In, Sn, W, Au, and Bi targets.(10) The ion range also depends on the substrate mass, electronic properties, and, very importantly, the substrate crystallographic structure. If the ion beam is directed parallel, or nearly parallel, to crystallographic atomic rows or planes in a single-crystal substrate, the ions may penetrate up to tens of nanometers into the substrate. This process is known as channeling.(9,20)

The probability of trapping the incident ion in the substrate is higher for light ions and increases with ion beam energy.(12) Choi *et al.*(18) measured the trapping behavior of inert ions on graphite as a function of ion dose and ion energy, and they determined the desorption behavior by using AES and thermal desorption spectroscopy. The amount of trapping qualitatively follows the trends calculated for the penetration probability, although the trapping probability is lower, because not all ions that penetrate are trapped.(18) In particular, for Ar only about 20% of the atoms that penetrate the graphite substrate become trapped. Figures 2.2a and 2.2b illustrate the energy dependence of the number of trapped atoms and trapping probability for inert ions incident on graphite at a dose of 1.8×10^{15} ions/cm^2 and a higher dose of 1.1×10^{16} ions/cm^2 for ion beam energies between 10 and 500 eV. For all ion beam species the trapping probability increases rapidly from 10 to 100 eV and levels off thereafter. The number of trapped ions increases as a function of ion species in the order Ar < Ne < He, while the threshold energy for trapping increases slightly in the reverse order. The authors[18] did not provide an explanation for the He trapping probability being greater than unity in Fig. 2.2a; we presume the reason is the uncertainty in their measurement. For higher ion doses (Fig 2.2b), the number of trapped ions decreases significantly. Helium shows very different trapping behavior at higher doses, while Ne and Ar do not. At higher doses, the trapping probability continues to increase with increasing energy for He ions, but approaches a plateau for Ne and Ar. This situation likely results from the increased penetration depth of He ions with a corresponding decrease in desorption or reemission probability. Ne- and Ar-ion beams heavily damage the graphite surface, primarily by removal of C atoms and rarely penetrate deeper than three atomic layers, so the ions may more easily diffuse to the vacant sites and desorb.[18] This process is most efficient for shallow ion penetration depths, so it may not be effective for He, which penetrates significantly deeper than Ne or Ar. Thermal desorption data show that the peak temperature for desorption increases in the order of He (90–100°C) < Ne (120–160°C) < Ar (200–300°C), as do the desorption peak widths.[18] However, the intensity of the He peak was reduced to ~60% of its original value after 9 h in UHV at room temperature, but the Ar peak intensity was largely unchanged. The authors[18] suggest that the Ar atoms are bound between the basal C planes, while He atoms are relatively mobile at room temperature. Another study showed that the trapping of deuterium in TiC_x films depends on the composition of the film, as illustrated in Fig. 2.3.[71] The trapped deuterium concentration decreases with increasing carbon content in the TiC_x film. Seminal work by Kornelsen revealed that the number of binding states for trapped inert ions in single-crystal W, and their binding energy, depends on the ion mass and the exposed crystal face of the sample; it does not depend on the ion beam energy in the range of 40 eV to 5 keV.[17]

The areal density of trapped atoms can depend on the total ion dose, as shown in Fig. 2.4.[72] Many studies show a saturation of trapped ions at doses of 10^{16} to 10^{17} ions/cm^2. However, this work indicates that the number of trapped ions continues to increase at doses $\geq 10^{17}$ ions/cm^2 for both 100-eV Ar on Mo(110) and for 1-keV Ar on Si(111). Thermal desorption experiments on these samples suggest that the Ar is trapped more strongly at these higher doses and requires heating to higher temperatures for its release.[72]

A trapped atom constitutes a defect (Sec. 3), and it may dissolve interstitially or substitutionally into the substrate lattice or thermally diffuse along diffusion paths in the bulk.[20] At low concentrations, the trapped atoms are likely to form isolated defects, while at higher concentrations, neighboring defects may interact and segregate and agglomeration of the trapped atoms may occur. A significant number of the insoluble atoms may accumulate below the sample surface, especially for the case of He-ion bombardment. Bubbles can form, and blistering and

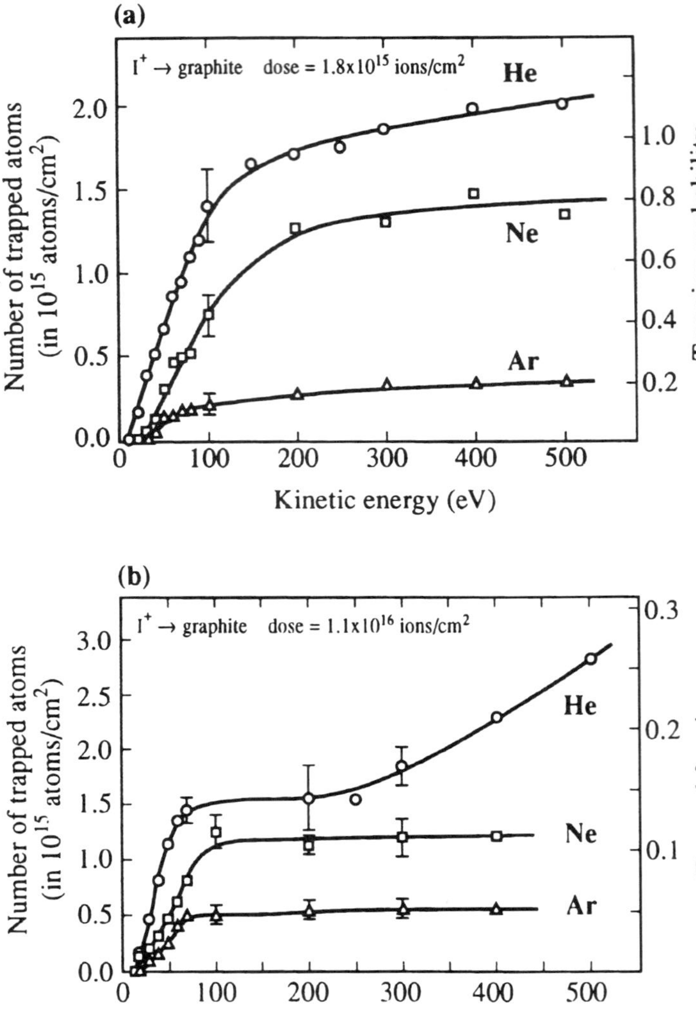

Figure 2.2. Trapping behavior of He, Ne, and Ar ions in graphite as a function of ion-beam energy at (a) low ion dose (1.8×10^{15} ions/cm^2) and (b) higher ion dose (1.1×10^{16} ions/cm^2). Adapted and reprinted with permission from[18] W. Choi, et al., *Surf. Sci.* **281**, 323–335 (1993), and with kind permission of Elsevier Science - NL, Sara Burgerhartstraat 25, 1055 KV Amsterdam, The Netherlands.

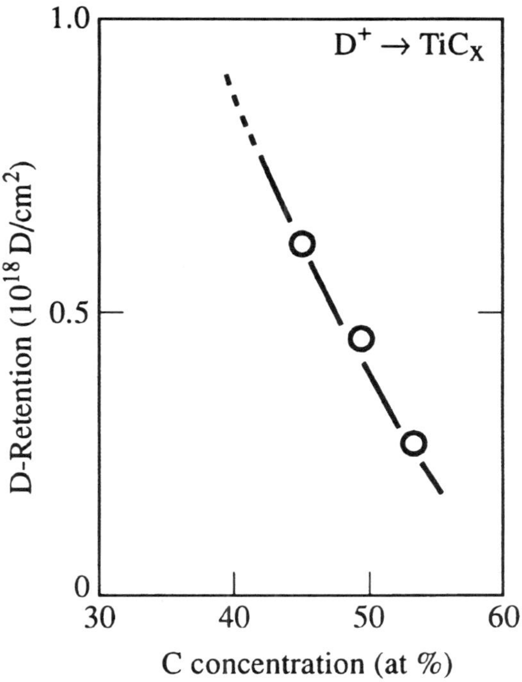

Figure 2.3. Trapping of deuterium in TiC films as a function of carbon concentration. Adapted and reprinted with permission from[71] M. Kitajima, et al., *Thin Solid Films* **156**, 231–238 (1998), and with kind permission of Elsevier Science - NL, Sara Burgerhartstraat 25, 1055 KV Amsterdam, The Netherlands.

flaking may occur as pressure inside the bubble increases and as the sputter front crosses the blister boundary.[(20,73)] A detailed description of blistering and flaking is presented in Ref. 74, but these processes usually occur for most solids when bombarding at energies greater than 5 to 10 keV.

Eventually, the rate of sputtering competes with the rate of trapping, so at high ion beam doses, the trapped fraction generally saturates at some steady-state level. Sputtering may not remove the substrate atoms and implanted ions at the same rate,[(20)] and ions trapped in the outermost layers of the sample are more efficiently sputtered. For trapping in graphite, Choi *et al.*[(18)] found that sputtering considerably reduced the trapped fraction of Ne and Ar ions but had much less effect on the He ions because of their deeper penetration.

For the surface analyst, trapped atoms from ion beam bombardment in the near-surface regime of the solid may occur at concentrations high enough to be detected by methods that are not exclusively surface monolayer sensitive, e.g., XPS and AES; trapped atoms are not detectable with ISS. We encountered a number of

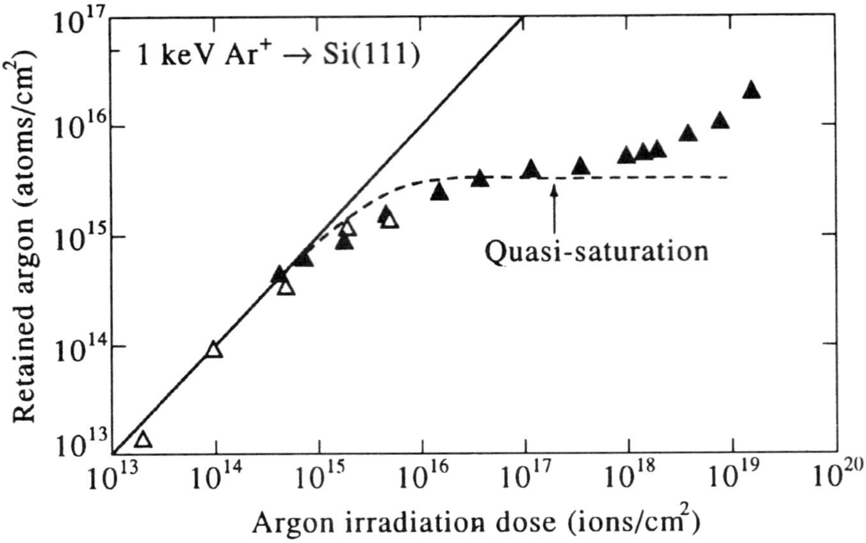

Figure 2.4. Trapping behavior of 1 keV Ar ions in Si (111) as a function of total ion dose, showing continued increase in retained Ar at ion doses in excess of 10^{18} ions/cm^2. Adapted and reprinted with permission from[72] H. A. Filius, et al., *Radiat. Eff. Defects Solids* **108**, 1–8 (1989). Copyright 1989 Gordon and Breach, Science Publisher, Inc.

papers with useful information and further details about trapped or implanted atoms from ion beam bombardment with rare-gas ions(17,18,69,70,75–81) and some that mention the well-known trapping found in SIMS studies.(82,83) The principal results are summarized in Table 2.1.

2.3. Ion–Substrate Interactions

Now we consider additional interactions of the ion beam with the substrate. The interactions result in direct recoils, cascade mixing, removal of material from the sample, and the implantation of material into the sample, as illustrated in Fig. 2.5.(84) The shaded circles represent the incoming ions or implanted atoms, and the solid circles are substrate atoms. The figure shows implantation (trapping) of the primary ions, discussed in Sec. 2.2; recoil implantation of both trapped atoms and substrate atoms; cascade mixing, which is isotropic and can result in both mixing and sputtering; and the sputter removal of substrate atoms and previously implanted atoms. Various diffusion and segregation processes also take place that are not shown in Fig. 2.5. An altered layer in the sample results from the implantation and mixing of atoms from the substrate and ion beam. Now we consider the fundamental

Table 2.1. Principal Results from Selected References about Implantation (trapping) of Bombarding Ions

Ref.	Ion$^+$	Energy (eV)	Target	Analysis methods	Principal results
17	Ar	50 to 4000	W	TDS	Seminal work by Kornelson showed the implanted Ar concentrations increased to ~10^{15} Ar/cm^2 for energies up to 4 keV and for fluences from 10^{11} to 10^{17} ions/cm^2.
71	D	7000	TiC	TDS	In addition to Fig. 2.3, blisters form at 4.5 to 5.5 × 10^{17} D/cm^2 for TiC or TiC with excess C, but disappear at 3.5 to 11 × 10^{19} D/cm^2.
18	He, Ne, Ar	10 to 600	Graphite	AES TDS	As shown in Fig. 2.2, the concentration of trapped atoms depends on the bombarding energy, ion used, and dose.
75	Ar	9400	Si	RBS, SIMS	Maxima in the atomic concentrations of Ar range from 10% for 5 × 10^{15} ions/cm^2 to 3% for 1.4 x 10^{18} ions/cm^2 with penetration depths of 10 nm and 35 nm, respectively. The results for Ar penetration into Be and Al are similar. The decreases at higher fluences result from redistribution or desorption of Ar at higher fluences.
76	He, H	300 to 500	Graphite	RGA	H that was implanted to saturation at 200 eV is released by He-ion bombardment. The depletion depth varied from 0.8 nm at 300 eV to 1.2 nm at 500 eV.
77	H, Ne, Ar	1 to 500	C(0001) Graphite	AES, TDS, STM	Hillocks formed at the position of the carbon atoms with 50-eV Ar or Ne are thought to represent point defects underneath one or two displaced surface atoms.
78	Ar	100	Mo(110)	—	As shown in Fig. 2.4, quasi-saturation of trapped gas occurs because of sputtering.

(continued)

Table 2.1. Continued

Ref.	Ion^+	Energy (eV)	Target	Analysis methods	Principal results
79	H, He, Ar	9100 to 9500	Several	RBS	The rms surface roughness increases from 100 to 250 nm for Ar doses from 10^{15} to 10^{19} Ar/cm^2. Ar concentrations of 10% to 15% saturate in Be or Al at a dose of $5–7 \times 10^{17}$ $ions/cm^2$.
82	O	50	Si	XPS	Stoichiometric SiO_2 is formed in 3–5 min when doses of 10^{17} O^-/cm^2 impinge on Si at 40 to 120 eV. At higher energies, the oxide thickness formed is self-limited by sputtering.
83	O_2, Cs	1–20 keV	Si	SIMS	Simulation of ion implantation during sputtering is a sum of superimposed Gaussian distributions for successive implants. Si^- yields vary with the type of silicide (Pt_2Si, Pt Si, Pd_2Si, Ni_2Si, NiSi, and Si and implanted Cs atomic fractions from 0.035 to 1.

processes and mechanisms that result in implantation, mixing, and removal of material.

2.4. Mixing and Implantation of Material

2.4.1. Ballistic Mixing

If enough energy and momentum are transferred from the ion beam to the substrate, the substrate lattice is disrupted; structural changes are discussed in detail in Sec. 3.2. Collisions at low energies may induce mobility of adatoms, defects, and surface atoms, as discussed in Sec. 3, and cause movement and displacement of substrate atoms at higher energies.[16] The movement of atoms can be divided into two phases: (1) ballistic or collision cascade and (2) cooling.[12] Extensive modeling of these phases has been done.[12,16,25,84,85]

In the ballistic phase (or cascade generation phase), the ion transfers most of its energy and momentum to the substrate in a series of collisions with the substrate atoms. If enough energy is transferred to a substrate atom to overcome its binding energy, it is displaced. The displacement energy for most metals is between 15 and

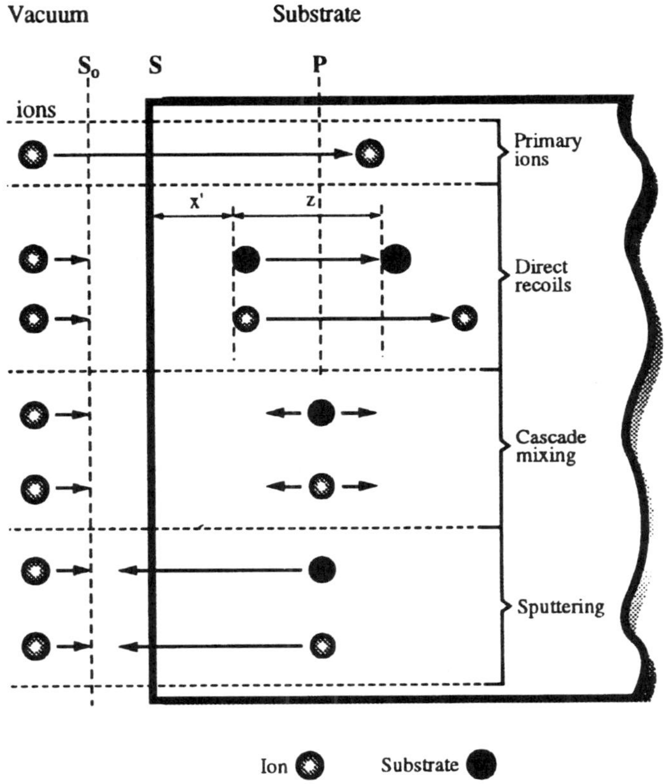

Figure 2.5. Ion beam-substrate interactions that result in implantation of primary ions, mixing by direct recoils or cascades, and removal of material. Adapted and reprinted with permission from[84] R. P. Webb, in: *Practical Surface Analysis, Volume 2-Ion and Neutral Spectroscopy,* edited by D. Briggs and M. P. Seah (John Wiley & Sons Ltd., Chichester, UK, 1992), pp. 657–704. Copyright 1992 John Wiley and Sons, Ltd.

40 eV.[16] The recoiling substrate atoms may, in turn, displace other substrate atoms. At high-enough impact energies (in the keV range), a large number of atoms are displaced in a volume about the original impact point and the sequence of collision events is called a collision cascade. For low-density cascades, the collisions can be treated as a sequence of separate binary events between a moving atom and a stationary atom, and these are known as linear cascades. High-density cascades, in which essentially all the atoms in a local volume are in motion at once, are called spikes.[12] The binary collision assumption breaks down for spikes because the motion of atoms in a spike can no longer be approximated by a series of independent binary collisions, and spikes are thus nonlinear events. The collision cascade continues until the recoiling atoms no longer have enough energy to become

displaced. They still have some excess energy, which is transferred to neighboring atoms by phonon processes, and this stage of the interactions constitutes the cooling phase. During thermal relaxation (cooling), locally enhanced vibrations and defect and diffusion processes can occur, which result in substantial atomic intermixing that may extend considerably beyond the collision cascade volume.[12,20]

The atomic displacements produce interstitial–vacancy pairs. Successive collisions give rise to a displacement chain, with a vacancy at the origin and an interstitial at the end.[16,61] Molecular dynamic simulations predict that many of the chains close upon themselves and annihilate the defect pair; however, a considerable number of atom displacements may occur prior to the defect annihilation.[16,61] This process is one mechanism for atomic mixing within the cascade volume.[16,20] Figure 2.6 shows a simulation of the defects produced at two different stages of a 0.5-keV collision cascade event with a Cu substrate.[16] The open circles represent vacancies, while the solid circles represent interstitials or displaced atoms. Figure 2.6a is a snapshot at the end of the collisional phase ($\sim 2 \times 10^{-13}$ s), at which time the number of Frenkel pairs is at a maximum. During the subsequent cooling phase, recombination of the defects occurs, and by the end of the cooling phase ($\sim 7 \times 10^{-12}$ s, depicted in Fig 2.6b) only a few defect pairs remain. However, the number of displacements per final defect pair is approximately 35. Figure 2.7 shows the simulation for a 5-keV collision cascade event. Figure 2.7a shows the defect pairs remaining at the end of the cooling phase. The defects form a central

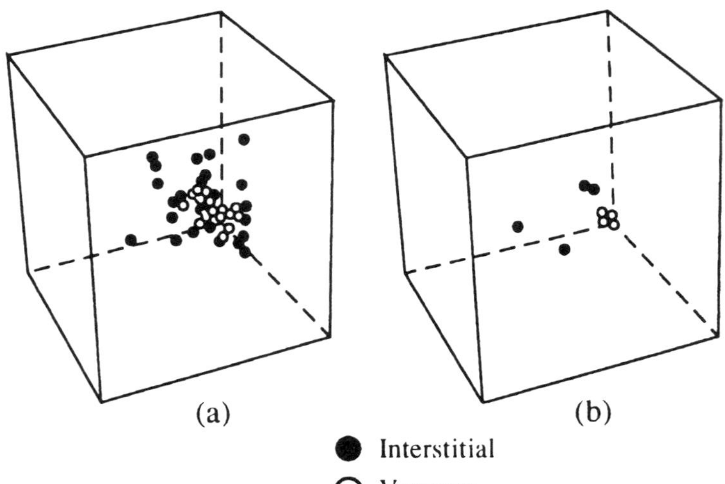

Figure 2.6. Simulation of defects produced in a 0.5 keV collision cascade event (a) at the end of the collisional phase, and (b) after the cooling phase. Adapted and reprinted with permission from[16] N. Q. Lam, *Scanning Microscopy Supplement* 4, **1990**, 311–352, and with permission from Scanning Microscopy International.

vacancy-rich core that is surrounded by interstitials. Figure 2.7b illustrates all of the intermediate displacements that occurred (crosses); these displacements provide an indication of the large amount of disordering or mixing in the central region of the cascade. In some cases, it appears that local melting takes place in the central region, which annihilates the majority of the defects but also induces considerable disorder.(16) A different simulation of the effect of normally incident 4-keV Ar ions on Cu indicates that lateral displacements in excess of 5 nm from the sputter beam axis may occur.(13) In insulators, it is possible for defects to have a range of charge states, especially for transition-metal constituents. Electric fields can be set up within the insulating material, and it may take days to attain equilibrium.(66) These different charged states may be detectable with XPS. Care must be exercised in interpreting XPS spectra of solids because the ion beam bombardment may result in forming charged states that are not present before bombardment.

Both recoil implantation and cascade mixing occur during the collision cascade event, and together they are known as displacement mixing. Recoil implantation, or knock-on, is atomic motion that occurs in the direction of the incoming ion beam and results from direct ion–atom collisions and preferential momentum transfer in the direction of the ion beam. At low ion energies, heavier substrate species tend to undergo recoil implantation, while at high energies the lighter species is more likely to preferentially recoil.(16) Cascade mixing, on the other hand, is generally regarded as a random-walk process because of the randomization of the recoil directions,

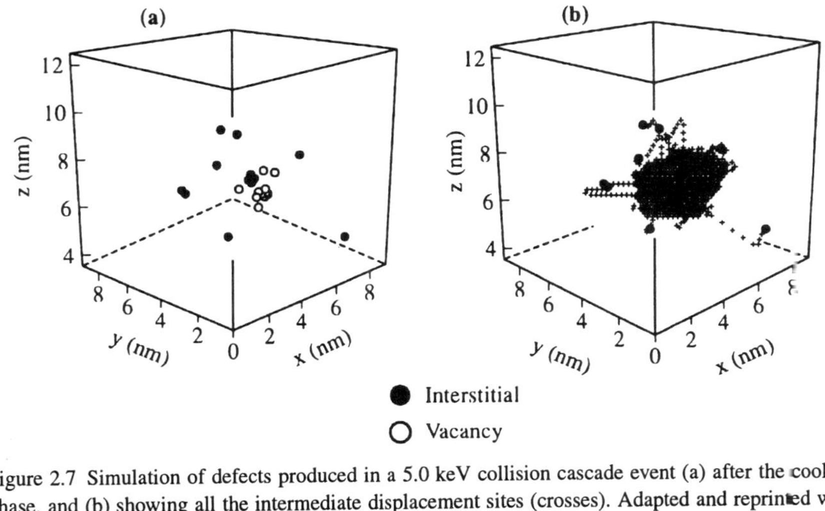

Figure 2.7 Simulation of defects produced in a 5.0 keV collision cascade event (a) after the cooling phase, and (b) showing all the intermediate displacement sites (crosses). Adapted and reprinted with permission from[16] N. Q. Lam, *Scanning Microscopy Supplement* 4, **1990**, 311–352, and with permission from Scanning Microscopy International.

and, thus, results in isotropic mixing. The effect of isotropic ballistic mixing generally far exceeds that of the recoil implantation.[61] However, recoil implantation can be an effective mechanism for formation of compounds between surface contaminants and substrate atoms. For example, 0.9 monolayer of hydrocarbon on the surface of Si, which is the typical coverage from minimal air exposure or even from the vacuum itself, is sufficient for a persistent carbide phase to be formed in the near-surface layer when bombarded with 1-keV Ar ions.[86]

2.4.2. Diffusional Mixing Processes

Displacement mixing is a ballistic process and is essentially independent of temperature. Diffusional mixing processes also occur during and just after ion bombardment, and these processes do exhibit a temperature dependence. In fact, the two types of mixing are often distinguished by their different temperature behavior. All of these mixing processes can be characterized by diffusion coefficients, and Fig. 2.8 shows the effect of the substrate temperature on the diffusion coefficients for the various mixing processes.[61] Below $0.2T_m$, where T_m is the substrate melting temperature, ballistic displacement mixing dominates and has a constant diffusion coefficient because the defects generated are relatively immobile. Between 0.2 and $0.6T_m$, defects are primarily formed by the ion beam and have very high mobilities; the diffusion coefficient generally depends on the defect concentration,[16] and ion-beam-enhanced diffusion dominates. Above $0.6T_m$, recombination of bombardment-induced defects is rapid (essentially an annealing process) and thermal vacancy concentrations are greater than those produced by ion bombardment, so pure thermally induced diffusion dominates.[13,16,61] The dominant mixing mechanism also depends on the substrate material. For metal systems, thermodynamics plays a dominant role even at room temperature, because defects are mobile at room temperature in most metals.[85] In oxide materials, the dominant mechanism may be difficult to predict because vacancies tend to be quite immobile at room temperature and the mobility of interstitials varies considerably for different materials.[85]

In the intermediate temperature range, ion-beam-enhanced segregation effects may become significant in an alloy.[13,61] Defect mobility is very high, and defects may migrate over considerable distances before annihilation. This process can result in defect fluxes. The preferential association of a particular defect type with one of the alloy components (e.g., an interstitial with a grain boundary) can result in considerable spatial redistribution of elements. At both lower and higher temperatures, defects do not migrate over such long distances, either because of immobility or enhanced recombination.[61] The substrate atoms tend to mix isotropically from ion-beam-enhanced diffusion; compositional gradients can be generated to depths much greater than the ion beam range from ion-beam-enhanced segregation.

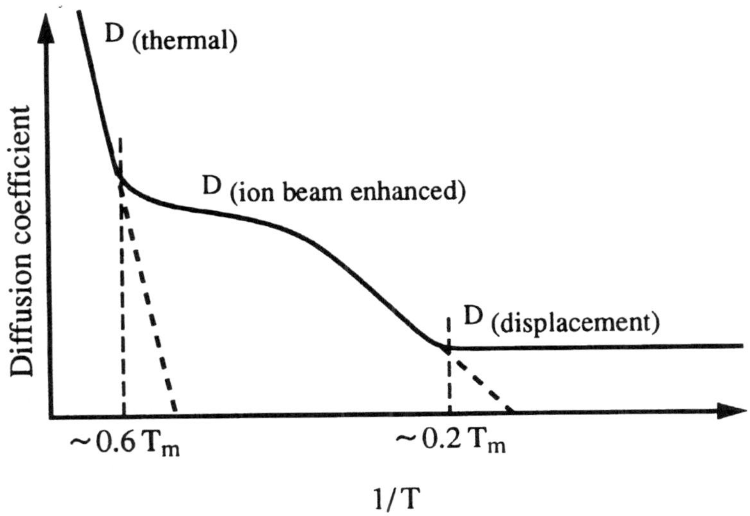

Figure 2.8. Diffusion coefficients and primary diffusion processes that contribute to mixing as a function of substrate temperature. Adapted and reprinted with permission from[61] N. Q. Lam, *Surf. Interface Anal.* **12**(1-12), 65–77 (1988). Copyright 1988 John Wiley and Sons, Ltd.

Gibbsian segregation at alloy surfaces occurs when the surface free energy can be minimized by segregation of one of the system components. This process results in compositional changes near the surface.[(16)] Gibbsian segregation is thermodynamically driven at elevated temperatures, but can also be driven at low temperatures by the collisional movement of the substrate atoms resulting from ion beam bombardment.

2.5. Removal of Material

Removal of material from the substrate is the most important process of interest for sputter depth profiling (as well as for cleaning), and is generally the primary intent of ion beam bombardment. However, as will be seen, the removal of atoms rarely occurs in an ordered, layer-by-layer fashion. This, along with the mixing already discussed, can result in significant physical and chemical changes in the sample. Our primary concern in this treatment is the physical sputtering process, i.e., that resulting from the collision process and the associated diffusion processes. Chemical sputtering (or reactive-ion sputtering) occurs by the formation of volatile compounds that desorb, and will not be discussed here. Electronic sputtering, which occurs as a result of electronic rather than nuclear energy transfer, is generally a minor process; however, it can be significant during the sputtering of insulators and dielectrics in which the implanted ionic species can generate strong repulsive forces that cause the ejection of material.(14,66) In this section, we discuss the physical sputtering process, sputter yields, and differential sputtering that results from ballistic considerations. Finally, we discuss the altered layer or zone of mixing that forms as a result of the mixing and sputtering processes.

2.5.1. Physical Sputtering

Physical sputtering occurs when energy in excess of the binding energy and with momentum directed out from the sample is imparted to the substrate atoms at or near the surface. There are several regimes of sputtering, corresponding to the amount of energy transferred. In order of increasing energy transfer, primary recoil sputtering is followed by cascade sputtering, which may be followed by spike sputtering. Additionally, diffusion processes set into motion by the sputtering process can result in the migration of atoms to the surface, where they may desorb or be sputtered.

In the 0.1- to 1-keV energy regime, as well as for very light ions, especially on heavy targets,(25,87) primary recoil events account for much of the sputtering. In this process, the primary-ion beam either ejects surface atoms directly or after a few recoil events. Examples of the process are illustrated in Fig. 2.9.(87) In Fig. 2.9a, the initial collision of the ion beam, which is generally at an angle to the substrate, directly removes a substrate atom either on the first or a subsequent collision. In Fig. 2.9b,(87) the initial collision produces a small number of recoils that result in the ejection of a substrate atom. In Fig. 2.9c, the incoming ion undergoes several collisions inside the substrate, and a substrate atom is ejected as part of the secondary-ion collision process. The threshold energy for sputtering is the minimum ion beam energy required to cause ejection of atoms from the substrate.(12) The threshold energy for most materials is 5 to 40 eV, or about 2 to 100 times the substrate binding energy.(12,16,87) The threshold energy for sputtering also depends on the angle of incidence of the ion beam, the ion beam species, and the mass and atomic number of the substrate species.(12) Although sputtering with very low energy ions significantly reduces the damage to the sample and the extent of mixing because of lesser amounts of energy transfer and small penetration depths, it is not very efficient, and ion beams with energies of at least a few hundred electron volts are required for efficient sputtering.(63) Experimental measurements of the threshold energy are difficult to perform because of the very low sputter yields at the threshold; the sputter yield is the number of atoms removed per incoming ion. Theoretical predictions for threshold energies and yields, which are beyond the scope of this chapter, are not definitive. A review of the theoretical models and results is given by Malherbe.(12)

For the typical energies used in sputter cleaning and depth profiling, linear cascade sputtering is the dominant mechanism for physical sputtering.(12,25) From

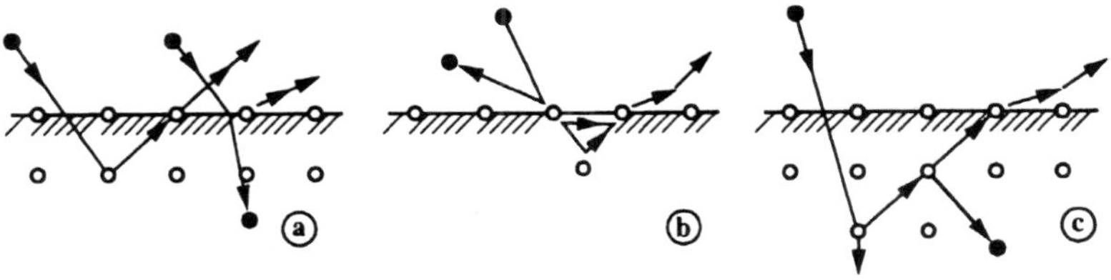

Figure 2.9. Some possible ejection processes for ion-beam bombardment at energies < 1 keV. See also Ref. 87.

the previous discussion of mixing, we recall that the collision cascade imparts motion to a large number of atoms in a local region about the ion impact site. This motion is isotropic, and atoms moving toward the surface can be ejected if their energy exceeds the surface binding energy when they reach the surface. Atomic motion generated in the spike regime may also account for removal of some substrate material, but this is not generally thought to be a major effect for most sputtering.[25] Spike-induced sputtering is often used as an explanation for much larger than expected sputtering yields. The probability of spike sputtering is increased by using heavy ions and higher energies.[12]

Most atoms removed by sputtering are ejected from the top few atomic layers of the sample.[13,16] The depth of origin of sputtered particles is a particularly important consideration for multicomponent targets, and depends on ion beam and target variables. The maximum in the energy spectrum for sputtered particles occurs at a few electron volts (see Fig. 16 in Ref. 16), which indicates that the atoms originate from the surface monolayer or one of the first several near-surface layers. The low energies of the sputtered particles also indicate that sputtering is not a very efficient process: the energy of the sputtered particles only accounts for 1% to 10% of the incoming beam energy.[13] Experimental results from Lam[16] show that for 3-keV Ne-ion sputtering of Ni–Cu, 65% of the sputtered particles are ejected from the surface monolayer; for 14-keV Ar-ion bombardment of a liquid GaIn eutectic alloy, 85% of the sputtered atoms originate from the outermost layer. Lam has found that, depending on the alloy, 50% to 95% of the sputtered atoms originate from the first surface monolayer.[16] We note that atoms in the surface monolayer may have been transported from greater depths by one or more of the various mixing processes.

2.5.2. *Sputter Yields*

The total sputter yield is defined as the average number of substrate atoms ejected for each incoming ion. In a multicomponent substrate, different constituents may have different yields, and the total yield then equals the sum of the sputter yields of the individual components.[12] Experimental sputter yields range between 10^{-5} and 10^{3} atoms per bombarding ion,[12] although more typical values range between 1 and 10 atoms/ion.[16] The sputter yield is an important quantity both in the theoretical description of sputtering and for interpreting experimental data, especially for composition and erosion depth determinations. Elemental yields are generally known, but yields for multicomponent materials are not as well known, and they can change considerably during, and as a result of, the sputtering process.(86) The yield depends on the ion beam parameters (energy, mass, and angle of incidence), sample parameters (e.g., mass, crystal planes exposed, stoichiometry, purity, phase, topography, temperature, and density), and on the surrounding environment (presence of contamination and neighboring material).(12,13,16,20,87) Many papers have been written about sputter yields; even a modest review of the results is beyond the scope of this chapter, so only some useful generalizations and observations will be presented. Tables of sputter yield data are compiled in Refs. 10, 88, and 89.

According to Sigmund theory, the total yield is directly proportional to the energy deposited by the incoming ions in the near-surface region of the sample.(12,20) Figure 2.10 shows calculated curves for the depth distribution of energy

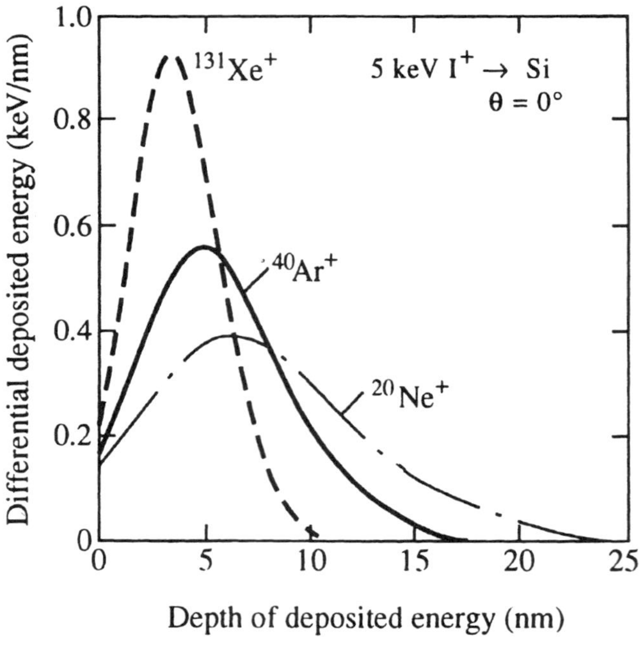

Figure 2.10. Calculated curves for the depth distribution of energy deposited into nuclear motion with 5-keV Xe, Ar, and Ne ions normally incident on Si. Adapted and reprinted with permission from[13] K. Wittmaack, in: *Practical Surface Analysis, Volume 2-Ion and Neutral Spectroscopy*, edited by D. Briggs and M.P. Seah (John Wiley and Sons, Ltd, Chichester, UK, 1992), pp. 105–175. Copyright 1992 John Wiley and Sons, Ltd.

deposited into nuclear motion with 5-keV Xe, Ar, and Ne ions normally incident on Si.(13) This figure shows that heavier ions deposit more of their energy closer to the surface and with less spread (in depth) than do lighter ions; thus, the yield is expected to be higher for heavy-ion bombardment.

The Sigmund theory is currently the basis for most theoretical descriptions of sputtering. It is a theory for elemental targets in the linear cascade regime and treats the sputter process as a series of binary collisions between a moving atom and a stationary one.(12,84) The Sigmund equation for the sputter yield Y can be written as(12)

$$Y = 0.042[\alpha(M_2/M_1, \theta)][S_n(E_0, Z_1, Z_2)]/U_0$$

where α is the fraction of energy available for sputtering and depends on the mass ratio of the substrate and ion beam, and on the ion beam incidence angle; S_n is the nuclear stopping power and is a function of the ion beam energy and the atomic numbers of the ion and substrate; and U_0 is the (local) substrate binding energy. The Sigmund equation predicts low yields for low ion energies and increasing yields

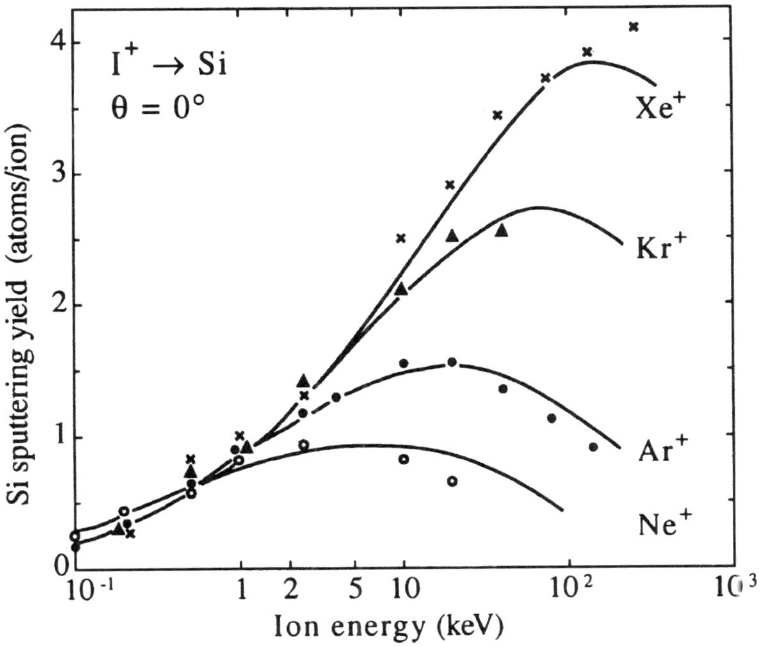

Figure 2.11. Silicon sputtering yield as a function of ion-beam energy for normally incident inert gas ions. Adapted and reprinted with permission from[87] P. C. Zalm, *Surf. Interface Anal.* **11**, 1–24 (1988). Copyright 1988 John Wiley and Sons, Ltd.

with ion energy up to a broad maximum in the range of 10 to 100 keV.[12] Figure 2.11 shows the Si sputtering yield as a function of ion beam species and energy.[85] The sputter yield also increases with increasing ion beam incidence angle, up to a maximum between 60° and 80°, because at these angles there is a higher probability of generating a collision cascade near the surface. At a more glancing incidence, less energy is deposited as the ion beam reflects off the surface, and the yields decrease significantly.[16]

Different choices for the values of the parameters in the Sigmund equation give large variations in the predicted yields.[12] One of the most difficult problems with yield predictions is how to choose the value for the surface binding energy. The sublimation energy is often used for elemental and metallic compounds, but the appropriate quantity to use for compounds is not generally known.[12] In fact, there are indications that the energy required to remove an atom from the surface exceeds the surface binding energy.[12]

Experimental data indicate that the total sputtering yield for an alloy depends on the alloy composition, as shown in Figs. 2.12 and 2.13 for Ag–Cu[90] and AlGaAs[13] samples. Generally, the total sputtering yield in alloys is found to be less than the yield of either pure component. The sputter yield for several metals

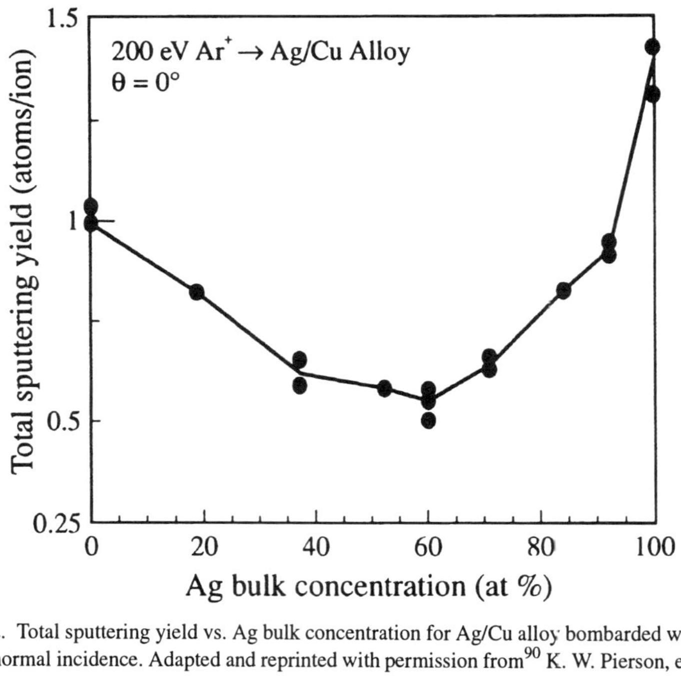

Figure 2.12. Total sputtering yield vs. Ag bulk concentration for Ag/Cu alloy bombarded with 200 eV Ar ions at normal incidence. Adapted and reprinted with permission from[90] K. W. Pierson, et al., *Nucl. Instrum. Methods in Phys. Res., Sect. B* **108**, 290–299 (1996), and with kind permission of Elsevier Science - NL, Sara Burgerhartstraat 25, 1055 KV Amsterdam, The Netherlands.

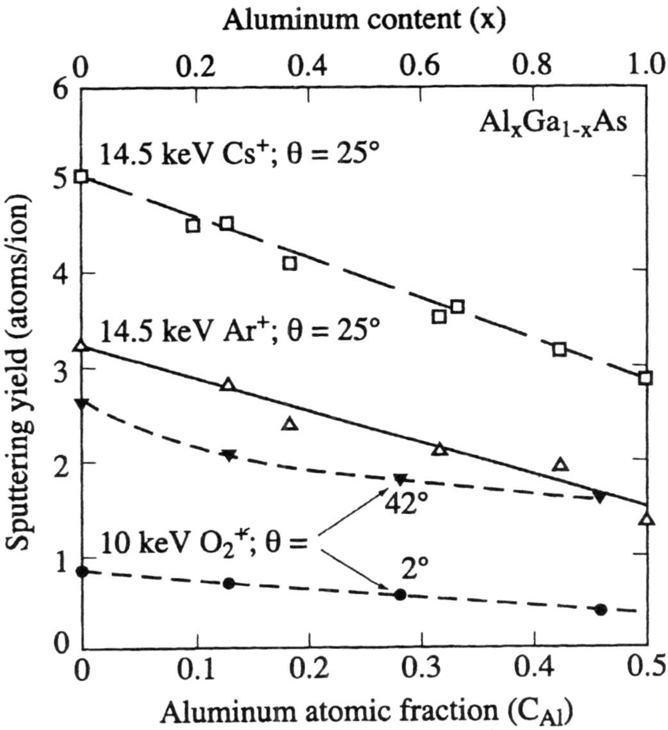

Figure 2.13. Total sputtering yield as a function of Al content in $Al_xGa_{1-x}As$ bombarded with Cs, Ar, or O_2 ions at different energies and incidence angles. Adapted and reprinted with permission from[13] K. Wittmaack, in: *Practical Surface Analysis, Volume 2-Ion and Neutral Spectroscopy,* edited by D. Briggs and M.P. Seah (John Wiley and Sons Ltd, Chichester, UK, 1992), pp. 105–175. Copyright 1992 John Wiley and Sons, Ltd.

and their oxides is shown as a function of Ar-ion energy in Fig. 2.14.[(91)] Metal oxide and insulator sputtering can be complicated by positive charge buildup at a dielectric surface, which can affect the homogeneity of the ion beam distribution on the surface and result in spatially and temporally nonuniform erosion conditions.[(91,92)] The yield may not be uniform across the surface of conducting samples because the binding energy may vary at different substrate sites (such as defects, steps, different crystal faces). The yield may even be a dynamic quantity that changes under the influence of sputtering as a result of composition and phase changes. Contamination can have a large effect on the yield, with chemisorbed species tending to reduce the yield. The yield may also depend on the substrate temperature and ion flux.[(12)] Channeling in single-crystal materials drastically decreases the sputter yield, since channeled ions deposit their energy deep inside the substrate and do not contribute to sputtering. The channeling effect can be

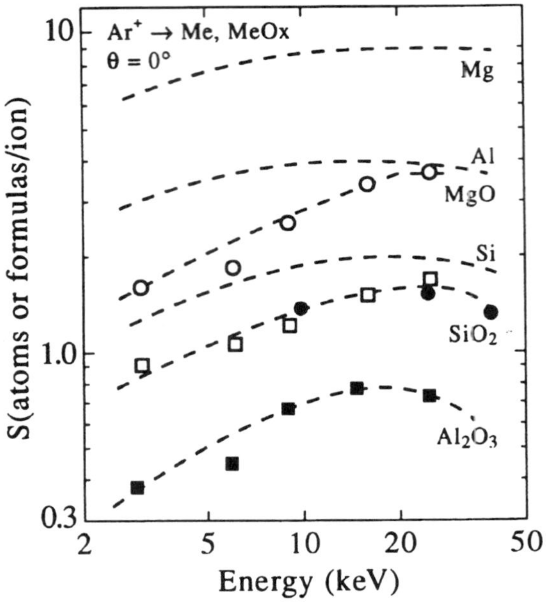

Figure 2.14. Total sputtering yield as a function of ion energy for several metals and metal oxides bombarded at different energies with Ar ions at normal incidence. Adapted and reprinted with permission from[91] T. Nenadović, et al., *Nucl. Instrum. Methods Phys. Res., Sect. B* **48**, 538–543 (1990), and with kind permission of Elsevier Science - NL, Sara Burgerhartstraat 25, 1055 KV Amsterdam, The Netherlands.

minimized by the proper choice of ion beam incidence angle. Also for crystalline materials, the yield may be highly anisotropic, as the ejected atoms show a strong directional preference. For ions normally incident on fcc targets, the yields generally increase as a function of the crystalline face in the order Y(110) < Y(100) < Y(111), as shown for Cu in Fig. 2.15.[13,16] In general, a greater yield is obtained as the interatomic spacing along the [*uvw*] direction increases,[87] and the same generality seems to hold for bcc, hcp, and diamond-type lattices.[16] Yields tend to be fairly isotropic for polycrystalline and amorphous samples.[12] The angular distribution varies with the energy of the ejected particle; atoms emitted along the preferred ejection directions also have higher average kinetic energies.[93]

An excellent description of the Sigmund model and other theoretical descriptions and treatments of the sputtering process, along with their limitations, is given in Ref. 84. Malherbe[12] provides a thorough comparison of different theories, including the effect of variation of parameters, such as binding energy, potential, etc., on the yield predictions, and compares the results with experimental data for semiconductors.

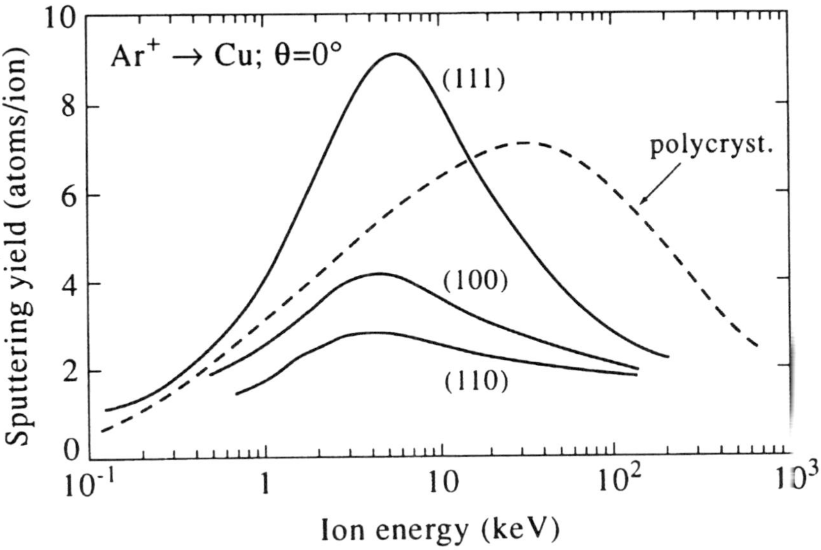

Figure 2.15. Variation in sputter yield with crystal face for Cu sputtered with normally incident Ar ions at different ion energies. Adapted and reprinted with permission from H. E. Roosendaal, in *Sputtering by Particle Bombardment*, edited by R. Behrisch, Vol. 1, Topics in Applied Physics Vol. 47 (Springer, Berlin, 1981) p. 219. Copyright 1981 Springer Verlag.

2.5.3. Differential Sputtering

Differential sputtering of the individual components is expected in most multicomponent materials; the yields generally vary for the individual components because of the different surface binding energies and the difference in energy transfer to substrate atoms of different masses.(13,16) In most cases, the binding energy effect is predicted to be larger than the mass-difference effect,(12,16,94) although the mass effect may dominate for sputtering of targets with components with large mass differences.(95) The collision-induced differential sputtering can drastically affect the composition, which may give rise to or be additionally influenced by diffusional processes. Generally more than one process occurs simultaneously (such as differential sputtering and ion-beam-enhanced diffusion), and the combined effects are not well understood or amenable to prediction but are usually inferred from experimental results. However, interpretation of data can be considerably complicated. The ballistic effects of differential sputtering are considered here. The chemical effects of differential sputtering are considered at length in Sec. 5.

The purely ballistic differential sputtering effects can be most reliably determined using multi-isotope studies, which avoid complications of chemical interactions. It is usually assumed that, in the absence of any diffusional processes, the low-mass component will be preferentially sputtered. In fact, if this is not observed, it is generally assumed that diffusional processes are dominant. However, the opposite effect of recoil implantation of the low-mass component may also occur. In an interesting study, Okano *et al.*(96) examined the variation in the isotope ratio as a function of bombarding time for oxygen bombardment of multi-isotope substrates of Cu, Ni, Ag, and Pd. They found that the values for the low- to high-mass ratio in the sputtered fluxes were low in the early stages of ion beam bombardment and increased with sputter time (tens of minutes) to constant values. Their possible explanations included preferential sputtering of the high-mass component or preferential recoil implantation of the low-mass component. For the conditions of their study, the authors concluded preferential recoil implantation of light species is the only explanation that could account for the magnitude of the effect they observed.(96)

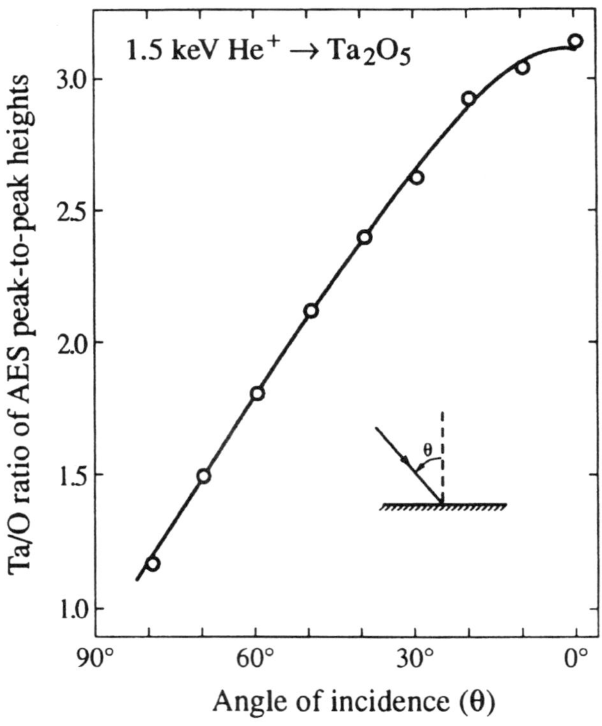

Figure 2.16. Steady state Ta/O ratios measured using AES for Ta_2O_5 samples bombarded with 1.5-keV He ions at different ion-beam incident angles. Adapted and reprinted with permission from[95] B. Baretzky and E. Taglauer, *Surf. Sci.* **162**, 996–1002 (1985), and with kind permission of Elsevier Science - NL, Sara Burgerhartstraat 25, 1055 KV Amsterdam, The Netherlands.

For a homogeneous multicomponent material, differential sputtering changes the surface and near-surface composition until a steady state is reached; achieving a steady state may require extensive sputtering corresponding to the removal of several tens of nanometers of material.[97] Because sputter yields depend on the angle of incidence, it is expected that the differential sputter yields are also angle dependent. Baretzky and Taglauer[95] bombarded Ta_2O_5 with 2.5-keV He ions and observed considerable variations in the steady-state surface composition as a function of the ion beam incident angle, as illustrated in Fig. 2.16. The Ta/O ratio, measured using AES, decreased with more glancing ion beam incidence angles, and the steady-state surface concentration was attained more rapidly at glancing incidence. Angular effects may result from variations in the composition gradients that quickly develop as a result of differential sputtering.[97]

Ion-beam-bombardment-induced Gibbsian segregation (e.g., of alloys) results in a surface-enriched component and may mimic differential sputtering. Gibbsian segregation is characterized by spikes in depth profiles and preferential perpendicular emission of the unsegregated species, which are concentrated in the subsurface layers. Compositional depth profiles measured using AES or XPS and ISS will be different for systems exhibiting Gibbsian segregation because of the different sampling depths of the techniques.[98]

2.6. Altered Layer (Zone of Mixing)

The result of the atomic sputtering and mixing processes is the formation of an altered layer (or layers) in the sample. The altered layer may differ from the pristine substrate in properties such as composition, density, and phase. In this section, we focus on the extent of the mixing zone and its effect on the measured depth resolution at an interface. We note that, in some cases, the damage zone may extend far deeper than the mixing zone (Sec. 3). Competition between thermodynamic and collisional processes may produce material that is far from its thermodynamic equilibrium state.[20] The displacement mixing and ion-beam-enhanced diffusion tend to mix atoms in the substrate and flatten out concentration gradients, while ion-beam-enhanced segregation has a stratifying effect.[12] Isotropic mixing broadens the depth resolution, and the extent of this broadening is proportional to the square root of the ion dose.[61] Differential sputtering and Gibbsian segregation alter the alloy composition in the first few atomic layers.[12] Recoil implantation tends to shift the centroid of a marker layer in the direction of the impinging-ion beam, and the magnitude of the shift is proportional to the ion dose.[61] Since material can be transported in both directions (towards and away from the original interface), concentration gradients are likely to be produced at otherwise sharp interfaces, and these gradients may be asymmetric.[99]

One way to measure the extent of mixing is to examine the broadening induced at an otherwise atomically sharp interface. The Andersen model gives an estimate of the broadening induced as a result of each main cascade collision event:[97]

$$\Delta = 2R\,[n(E)/Y]^{1/2}, \qquad n(E) = 0.42E/E_d \tag{2}$$

where Δ is the FWHM broadening of the interface in nanometers, $n(E)$ is the number of primary displacements within the collision cascade, Y is the sputter yield, R is the range of the recoils in the cascade (~1 nm), E is the primary-ion energy, and E_d is the displacement threshold energy (if unknown, 15 eV is often used for semiconductors and 25 eV for metals).[12] The Sigmund formula for the yield in atoms/ion can be expressed as[9]

$$Y = 0.042\alpha\rho S_n/\rho U_0 \tag{3}$$

where ρ is the atom density, S_n is the nuclear stopping cross section, U_0 is the surface binding energy, and α is a function of the mass ratio of the incident projectiles and the substrate. The units are nm/eV for $0.042/\rho U_0$ and eV/nm for ρS_n.[9] Substituting Eq. (3) into Eq. (2) gives

$$\Delta = 2R\left[\frac{0.42E}{E_d}\,\frac{\rho U_0}{0.042\alpha\rho S_n}\right]^{1/2} \tag{4}$$

Typical values for Δ are between 2 and 20 nm. Low-energy, high-mass projectiles (e.g., 1-keV Xe ions) give lower Δ values, while higher Δ values correspond to high-energy, low-mass projectiles (e.g., 20-keV Ne ions).[12] At low temperatures, ballistic displacement mixing dominates—a process that is not thermally activated—and the altered layer thickness should correspond to the mixing in the collisional phase, or roughly to the range of the primary ion.[9,83] At higher temperatures, the diffusion processes that occur primarily during the cooling phase can greatly enhance the defect diffusion distance and produce altered layers that extend very deeply into the substrate.[16]

The total thickness of the altered layer varies widely, depending on the ion beam parameters and the substrate properties. Direct measurements of the mixing regime are not always easy to perform or interpret, and care must be taken to ensure that a steady state has been attained. Changes in the composition of the altered layer that result from differential sputtering will eventually reach a steady state, in which the composition of the sputtered flux is equal to the bulk stoichiometry. However, this may occur only after a considerable amount of material has been sputtered, e.g., up to 100 nm.[87] Depth profiles of Si–Ta layers using AES analysis during bombardment with 3-keV Ar ions gave a mixing regime thickness of approximately 4 nm.[12] Another group used medium-energy-ion scattering to determine the range of displaced Si atoms after 0.5- and 1-keV normally incident Ar-ion bombardment to fluences of $10^{16}/cm^2$.[86] Silicon atoms were displaced an average of 1.7 and 3.1 nm by 0.5- and 1-keV ions, respectively. Bombardment of single-crystal GaAs substrates with 0.1-keV Ar ions produced an amorphous layer 2.3 nm thick, which resulted from direct damage from the incident ions, and a damaged diffused layer of simple defects approximately 200 nm thick. The diffused layer was the same thickness after bombardment with 2-keV ions, but the amorphous layer was 7.2 nm deep for the higher-energy ions.[12] $Hg_{1-x}Cd_xTe$ samples were bombarded with a neutralized 100–150-eV Ar-ion beam with current densities between 0.5 and 0.3 mA/cm^2; the bombardment produced an altered layer from tens to hundreds of nanometers thick and an electrically induced *n*-type layer 5 to 50 μm deep within the first 1–3 min of ion beam bombardment.[100] The dilution effect of mixing may also be an important experimental consideration, especially for buried interfaces, where the mixing occurs much faster than the interface can be measured with sputter-depth-profiling techniques. For a mixing zone of ~50 atom layers, an interfacial layer one monolayer thick will be diluted to an effective concentration of only 2% (over a much longer range).[83]

Occasionally, coalescence of one layer rather than intermixing is observed. The tendency of a layer to coalesce seems to be correlated with the free energy of the product. Mixed phases form when the free-energy change of mixing is negative; positive free-energy changes result in coalescence of the films. Metallic films that intermix with suitable thermal treatment also intermix with ion irradiation, while those that coalesce thermally also tend to do so with ion beam bombardment.[20,85]

The processes described in this section have been subjected to extensive scrutiny by simulations. As indicated by most authors, the major challenge is the proper choice of the interaction potential for the ion–atom collisions. We summarize the principal results and processes modeled from selected references (not previously cited in this section) in Table 2.2.[94,101–117] We do not summarize papers in which comparisons are made between various simulation methods (e.g., Refs.

Table 2.2. Selected References and Principal Results from Modeling of Various Processes Resulting from Ion Beam Bombardment of Solids

Ref.	Ion	Energy (keV)	Solid	Process simulated	Principal results
101	Ar	0.2 to 100	Si	Depth distribution Sputtering yield	Monte Carlo simulations based on Molière, universal, Kr–C, and Lenz–Jensen screened interaction potentials do not give good fits with experimental data.

(*continued*)

Table 2.2. Continued

Ref.	Ion	Energy (keV)	Solid	Process simulated	Principal results
102	Ar	5	Ge	Depth of origin Sputtering yield	Monte Carlo study of collisional mixing shows relocation depths of Ge from recoil implantation and cascade mixing. The depth of origin of sputtered particles ranges up to 2 nm. The lighter isotope is preferentially depleted close to the surface.
103	Ar	1	Cu (100)	Depth of origin Sputtering yield	Molecular dynamics simulation shows more vacancies are formed than ejected atoms/ion or target adatoms formed/ion.
104	He, Ar, Kr	0.4	W(100)	Ejection mechanisms Sputtering yield	Molecular dynamics simulation shows ejection by primary recoils is the dominate mechanism for He and to a much lesser extent for Ar; cascade mixing is dominant for Kr. The vast majority of sputtered atoms are from the surface layer.
105	Ar	0.5	*p*(2×2) C_xH_y on Pt(111)	Ejection mechanisms Hydrocarbons (HCs) Sputtering yield	Molecular dynamics simulation shows that the fragments of different hydrocarbons ejected are H or the intact adsorbate for small HCs; more fragmentation results when larger (e.g., C_5H_9) HCs are employed.
106	Ar	1	diamond-like (001) with four reconstructions	Ejection mechanisms for angular distribution	Molecular dynamics simulations show trajectories for causing ejection and the angular dependence of ejection from (1×1), (2×1), (2×4), and (1×4) reconstructions using the modified Tersoff interaction between Ar and Si. The yield dependence on structure is tabulated for the first four layers and azimuthal differences are illustrated.
107	Ar	1	Ag (100)	Ion beam mixing Sputtering yield	Molecular dynamics shows the most probable displacement is less than one-eighth of the equilibrium bond length for atoms in the first four layers. The direction of the sputtered atoms depends on the azimuthal angle at different polar angles and energies.

(*continued*)

Table 2.2. Continued

Ref.	Ion	Energy (keV)	Solid	Process simulated	Principal results
108	Ar	3	Ta/Si	Ion beam mixing	A model, based on linear transport theory in the diffusion approximation, is used to predict concentration profiles for multilayers of Ta and Si. Asymmetric shapes observed experimentally are predicted, but abrupt changes in profile shapes are not.
109	Ar	5	LuFe	Sputtering yield Preferential sputtering	The preferential enrichment of Lu and sputtering yields for Ar, Lu, and Fe are calculated for increasing fluences up to 4×10^{16} ions/cm^2 using the Monte Carlo TAMIX program.
94	Ar	0.5–5	M(100)	Preferential sputtering and yields	Classical dynamics calculations were used to study the effects of the target atom mass and surface binding energy for the binary systems O–Cu, Au–Cu, CuO–O, CuO–Cu, CuAu–Cu, and CuAu–Au. The results focus on sputtering yields.
110	Ar	5 1	CdHgTe SiO_2 Al/Si	Preferential sputtering O implants Sputter yield Erosion rate	Monte Carlo simulations predict different steady-state concentrations of CdHgTe than compared with experiment. Good fits with RBS data are obtained for O implants in SiO_2 and with SIMS data for the sputtering yield for Al deposited on Si. The simulation shows Ar + oxygen erodes Al_2O_3 and SiO_2 surfaces faster than oxygen alone, and that SiO_2 erodes faster than Al_2O_3; the fluence ranges up to 8×10^{17} ions/cm^2.
111	Ar	0.3 to 4	$NbSi_2$	Sputtering yield Preferential sputtering	A master equation is derived and used to obtain good agreement with experimental preferential sputtering yields from 2 to 4 keV for a model two-component system.

(continued)

Table 2.2 Continued

Ref.	Ion	Energy (keV)	Solid	Process simulated	Principal results
112	Ar	1 5	Cu(010) Cu(3 11 1)	Sputtering yields	The sputtering yields are markedly different from calculations for high- and low-index planes. For the (010) plane, 99% of the sputtered atoms are ejected from the surface and the next plane below it; for the Cu(3 11 1) plane, only 37.5% of the atoms are ejected from the first two planes and 10% still originate from the sixth plane.
113	Ar	0.1 to 10	pc-Mo in which pc is polycrystalline	Sputtering yields	Binary collision simulation shows isotopic enrichment of the $^{92}Mo/^{100}Mo$ ratio is independent of Mo concentration. A strong dependence on the emission angle of the yields is obtained with the ratio depending most strongly at the lowest bombarding energy (100 eV).
114	Ar	0.1	SiO_2	Sputtering yields	Dynamic Monte Carlo simulation show good agreement with experimental sputtering yields at the sputtering threshold; the simulation also considered surface binding energy and O^{2-} redistribution in the target.
115 116	Ar normal inci-dence	1	Ag(111)	Sputtering yields	The embedded-atom method interaction potential is used to show that 7.3% of the total yield of 4.66 atoms/ion consists of Ag_2. The experimental total yield is 4.5 atoms/ion. The energy distribution of sputtered silver has a maximum at about 3 eV; the maximum dimer ejection energy ranges from 4 to 2.5 eV for the planes in the order 110 > 100 > 111, respectively. The work is extended to larger Ag clusters in Ref. 115.
117	Ar	3.6	2 ML Cu on Ru(0001)	Depth of origin Sputtering yields	Experimental work shows 75% and 20% of the sputtered particles are from the surface and first bulk layer, respectively. TRIM calculations agree with measured Ru sputter yields for thickness up to 1 ML.

118–120) or papers in which self-ion bombardment is simulated (e.g., C^+ on C[121] and Cu, Ge, Au and, Pt ions on Cu, Ge, Au, and Pt targets, respectively).[122]

From this section, we conclude that sputtering, in general, is *not* a technique in which successive layers of the sample are removed for convenient analysis.[83] The sample is almost always significantly altered in a complex and difficult-to-predict manner by the ion beam bombardment and related processes.

3. Structural Changes Resulting from Ion Beam Bombardment

As a section overview, the structural changes from ion beam bombardment of solids are summarized in Table 3.1, in which the progression from 1 to 11 is generally in the direction of increasing fluence of the ion beam with sufficient energy to cause the damage. The extent and rate of damage to a particular material depend on the ion beam mass, energy, current density, total fluence, and angle of incidence. As will be seen from the selected references, validation of some of the items in Table 3.1 required carrying out studies at energies below 100 eV and with

Table 3.1. Structural Changes Resulting from Ion Beam Bombardment of Surfaces (for ion energies up to 10 keV)

1.	Implant atoms from bombarding beam
2.	Stretch bonds (surface relaxation) at low fluences; results in different bond lengths
3.	Break bonds; reconstruction may occur; reactivity enhanced
4.	Produce surface adatoms, adatom clusters, and subatomic bumps with an adatom yield in adatoms per ion
5.	Sputter most atoms from surface plane; form monolayer-deep vacancy islands
6.	Form surface and bulk vacancies, dislocations, interstitials
	• surface and bulk diffusion enhanced
	• more vacancies formed than ejected atoms
	• anion vacancies in compound semiconductors and halides
7.	Remove atoms layer by layer (fluence and temperature dependent)
8.	Form point, complex, and continuous defects plus interface defects (F-centers, Frenkel defects, etc.) (defects/ion quantified in some cases; increase with J and E)
9.	Amorphize the bombarded zone
10.	Develop damaged layers for thicknesses greater than the mean projectile range
11.	Induce changes in electronic states/properties; form electron traps and bound excitons; interband transitions

Structural damage changes optical, electrical, magnetic, density (of porous films), and high-temperature superconducting thin-film properties.

fluences below 10^{14} ions/cm^2. One ML on a solid is about 10^{15} atoms/cm^2, and the exact monolayer density depends on the exposed crystal plane and on the material.(123) We found the following articles particularly helpful for developing our summary of structural changes from ion beam bombardment.(12,19,24,31,78,103,124) As expected, the bombarded solids were typically single-crystal metals, semiconductors, or inorganic compounds, e.g., Cu(100), Pt(111), Ni(110), Ni(111), Au(001), Au(111), diamond, graphite, GaAs(100), InP(100), Si(001), Si(111), Ge(001), Ge(111), SiO_2, and LiF. The changes were generally determined by using the common structurally sensitive techniques such as RHEED, TEM, STM, ISS, and LEED, as well as by AES, EELS, and techniques sensitive to optical and electrical changes or chemical bonding. Stoneham has summarized defects and defect processes resulting from ion beam bombardment.(66)

In the modification of material properties by ion beams, defects and defect processes are important components. They are involved at several stages. First there are defects—some transient—that result from the incident beam. Once the bombarding particle energies have decreased to thermal energies, several defect processes may result in defect annihilation, migration, aggregation, or dislocation climb. More complex phenomena, e.g., amorphization, may occur by nucleation and growth in some cases; moreover, the effects of the ion beam bombardment may sensitize the material to subsequent processes, as in fission track formation, where etching becomes easier. At any stage, the defects present or impurities implanted can affect observed behavior. The electrical resistivity, the optical behavior (e.g., refractive index), and even mechanical properties may be altered. In describing representative defects and defect phenomena, Stoneham(66) notes that ion–surface interactions involve different time scales, energy densities, and especially differences in the distribution of energy between electronic and nuclear systems. When electronic excitation occurs, it can have very different consequences, depending on whether the core electrons or the valence electrons are excited. The mechanisms that result in the removal of surface species also depend on the chemical nature of the bombarding species (e.g., oxygen or cesium); moreover, some processes occur just outside the surface (e.g., charge exchange between the surface and a sputtered atom or ion after its ejection from the solid). Thus, we must consider a very complex, time-dependent system and try to draw from this complexity some general ideas of the key processes and how they operate.

3.1. Bond Stretching, Bond Breaking, and Surface Reconstruction from Ion Beam Bombardment

The principal results from selected references about bond stretching, bond breaking, and surface reconstruction caused by ion beam bombardment are summarized in Table 3.2.(125–128) The modifications of the In–P and P–P bond distances in an InP(100) surface shown in Table 3.3 resulted from normal–incident, 500-eV

Table 3.2. Principal Results from Selected References about Bond Stretching, Bond Breaking, and Surface Reconstruction

Ref.	Ion^+	Energy (keV)	Target	Analysis method	Principal results
125	Ar	0.5	InP(100)	PEXAFS	From 2 μA/cm^2 and up to 15 min, the P–In bond initially is shortened, and then returns to normal; the P–P bond length is increased.
126	H	1–3	a-Si a=amorphous	UPS	SiH_2 and SiH formed in H_2 on amorphized Si and preferentially loses SiH_2 under H^+ bombardment.
127	Ar	—	Si(111)	TEM	Annealing amorphous Si at 600°C yields Si(111)–7×7, Si(111) amorphized, SiC, stacking faults, and screw dislocations during electron annealing. Superlattice fringes from reconstructed Si(111)–7×7, SiC, and stacking faults are shown.
128	Ar	0.015–0.05	Si(001)–1×1	RHEED TEM	Ar-ion bombardment of 1.5×10^{16} ions/cm^2 at 190°C causes loss of surface dihydride and Si(001)–1×1 becomes amorphous; at 100°C and 65° incidence angle, Si(001)–2×1 is formed. The same reconstruction can be formed with 15-eV He ions from 190°C to 50°C at a fluence of 7.5×10^{16} He^+/cm^2.

Ar-ion bombardment at 2 μA/cm^2 for 5, 10, and 15 min.(125) The bond distances were obtained by processing the P 2*p* photoelectron EXAFS (PEXAFS) spectra, as illustrated in Fig. 3.1; the changes in the In 4*d* core-level spectra (Fig. 3.1a) show the growth of a metallic In shoulder at about 58 eV. The metallic In is formed into metallic islands with a concomitant depletion of P from the surface. For the P 2*p* PEXAFS data, the experimental EXAFS modulation functions are shown in Fig. 3.1b, and two dominant peaks are observed in the Fourier transform in Fig. 3.1c. The position of peak A changes in position with sputtering time, but the position of peak B does not change appreciably. The authors conclude that ion beam bombardment induces surface relaxation with the formation of In clusters and that the ion beam bombardment stretches and subsequently weakens In–P bonding.(125) Muller *et al.*(127) also showed that amorphized Si(111) and Si(100) wafers, which were hydrogenated in 1000 L of hydrogen at 600 K (monohydride formation)

Table 3.3. Bond Distances at the InP(100) Surface after 500-eV Ar-Ion Bombardment at 2 μA/cm²

P 2*p* PEXAFS	P–In (peak A)[a]	P–P (peak B)[a]
5-min ion sputtering	0.240 ± 0.004 nm	0.421 ± 0.005 nm
10-min ion sputtering	0.249 ± 0.004 nm	0.419 ± 0.005 nm
15-min ion sputtering	0.257 ± 0.004 nm	0.423 ± 0.005 nm
Bulk InP	0.254 nm	0.415 nm

[a]Peak A and peak B are shown in Fig. 3.1c. From Ref. 125.

followed by 500 L hydrogen at room temperature (dihydride formation), yielded UPS characteristic peaks at 5.3 eV (Si–H) and 6.2 eV ($Si–H_2$). Upon ion beam bombardment with 1.0-keV H, the dihydride peak was preferentially lost, leaving the monohydride peak; however, the monohydride was also removed after bombardment by 2-keV H.[(127)]

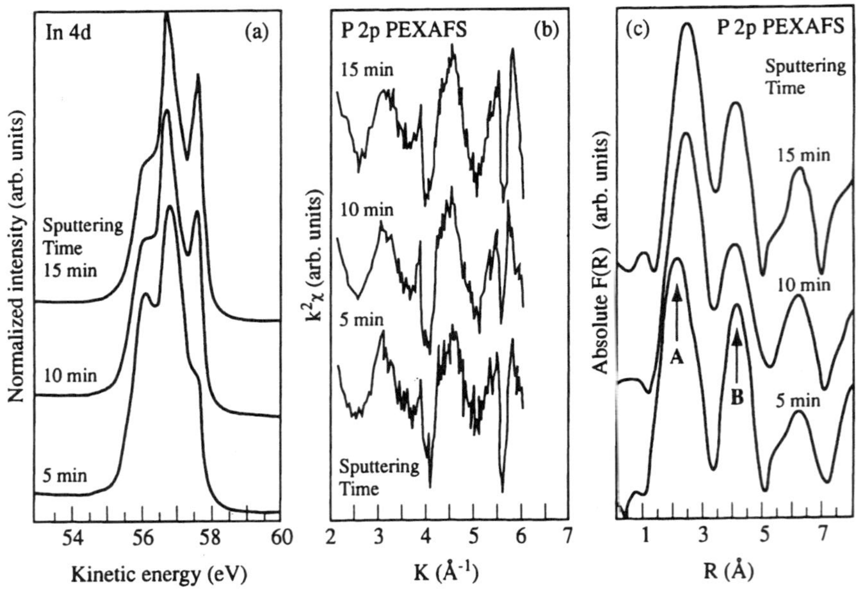

Figure 3.1. (a) In 4d core-level photoemission spectra at a photon energy of 80 eV for ion bombardments of 5, 10, and 15 min; (b) $k_\chi^2(k)$ function for InP (100) surface for the same ion bombardment times; and (c) the absolute part of the Fourier transform of the corresponding EXAFS function in (b). Adapted and reprinted with permission from[125] P. S. Mangat, et al., *Appl. Phys. Lett.* **63**(14), 1957–1959 (1993). Copyright 1993 American Institute of Physics.

For studying the temperature and fluence dependence of reconstructed surfaces, dihydride-terminated Si(001)–1×1 surfaces were prepared and the structure verified by RHEED.[128] The surface was bombarded with 50-eV Ar ions at 190°C and 100°C up to a fluence of 1.5×10^{16} and 1.2×10^{16} ions/cm^2, respectively A Si(001)–2×1 surface is formed from 190°C to 100°C by Ar-ion bombardment at an angle of incidence 65°, but the surface is amorphized at 50°C. However, 15-eV He-ion bombardment generated Si(001)–2×1 surfaces at 190°C and 50°C from the Si(001)–1×1 surfaces. The 2×1 reconstructions result from removing surface hydrogen by sputtering and recoil implantation of H into the substrate.[128] In a TEM study of microstructure evolution, Murty and Atwater[128] monitored the effects of Ar-ion bombardment and annealing of Si(111) in which the Si(111) is first amorphized. With a clean enough sample, a flat surface was obtained with some residual defects, including screw dislocations. Annealing this surface at 600°C resulted in the formation by epitaxial growth of the 7×7 reconstructed Si surface; the growth was not disturbed by a few isolated SiC islands. By annealing at 800°C, a $\sqrt{3} \times \sqrt{3}$ R 30° reconstruction is stabilized by boron adatoms that segregate on the surface.

3.2. Structural Changes as Nanotopography from Ion Beam Bombardment

Principal results from selected references about surface nanotopographical changes that resulted from ion beam bombardment and measured using STM are listed in Table 3.4.[19,129–137] For investigation of sputtered surfaces, the STM is limited to surfaces with a small amount of disorder; otherwise the unique capability for imaging atomic scale features is lost. Michely and Comsa[19] overcame this potential problem by (1) using limited doses of ion beam bombardment to disturb an otherwise perfect surface or (2) bombarding under conditions of competing conditions for damage creation and annealing, i.e., elevated temperatures and low ion fluxes. They used target surfaces with terrace widths of ca. 200 nm separated by monatomic steps. Their investigation ranged from studying the formation of adatom islands, vacancy islands, and subsurface bubbles (from the submonolayer removal of metal surface atoms) and replacement of them from the "bulk," to layer-by-layer removal of atoms, and finally to the removal of up to 50 ML that resulted in deep hexagonal-shaped, tiered pits.[19] Substrate temperatures as low as 190 K and as high as 910 K were used with Pt(111)[19,129,135] and Au(111).[130] Michely and Comsa[19] showed that the yield of target adatoms is greater than the sputtering yield for the same ion beam bombardment conditions. We summarize their results in this section as they relate primarily to nanotopographical changes in the monolayer regime. Although not all of their data are for ion beam bombardment of substrates at room temperature; Michely and Comsa's results provide insight about processes that occur at room and other temperatures, e.g., they elucidate structural damage by the removal of material to form vacancy islands one

Table 3.4. Principal Results from Selected References about Surface Nanotopographical Changes from Ion Beam Bombardment

Ref.	Ion$^+$	Energy (keV)	Target	Analysis method	Principal results
129	He	1	Pt(111)	STM	Hexagonal-shaped adatom islands formed at 753 K, bumps from subsurface He bubbles, and hexagonal-shaped 1-ML-deep vacancy islands.
132	Xe	0.2	Si(100)–2×1	STM	Generation and annealing of about 1/3 to 1/2 ML of defects from 350°C to 650°C
133	Ar	0.6	Cu(100)	STM	Adatom and 0.3 ML of vacancy islands formed after 4×10^{14} ions/cm^2 and formation of vacancy islands up to 3-ML-deep after 8×10^{14} ions/cm^2 at 300 K
134	Ar	0.6	Pt(111)	STM	One monolayer of adatom and vacancy islands formed after removal of 0.45 ML at 350 K, 450 K, 550 K, and 650 K
135	Xe	0.225	Si(111)–7×7	STM	Removal of up to 1 ML at 250°C that disrupts and reforms 7×7 reconstruction
136	Ne, Ar, Xe	0.04 to 5	Pt(111)	STM	Adatom yields at 150 K vs. ion energy range from 0.01 up to 66 adatoms/primary ion
19	Ar	0.6	Pt(111)	STM	Adatom and vacancy islands formed at 190 K and 240 K and topographs after removal of up to 50 ML at 625 K
137	Ar	0.3–1.2	Si(111)–7×7	STM	Adatom and vacancy islands at room temperature and partial recovery of reconstructed surface at 500°C or 730°C. Activation entropies were calculated for the recovery process.

(continued)

Table 3.4. Continued

Ref.	Ion+	Energy (keV)	Target	Analysis method	Principal results
130	Ne	2.5	Au(111)	STM	Triangular and hexagonal vacancy islands are formed at 400 K and bounded by monatomic steps oriented along <110> directions.
131	Ar	0.6	Au(100)	STM	Rectangular, 1-ML-deep, vacancy islands are formed at room temperature and fluences of 4×10^{14} ions/cm^2 to 1.1×10^{15} ions/cm^2.

ML deep and adatom islands one ML higher than the original surface monolayer. Adatom islands and vacancies are illustrated in the topographs in Figs. 3.2, 3.3, and 3.4. For an Ar-ion-bombarded (600 eV, 3.1×10^{14} ions/cm^2) Pt(111) surface, Fig. 3.2 shows an adatom island as a bright spot, vacancy islands as dark areas, and the original surface as a gray area after 0.45 ± 0.05 ML removal at 540 K. The line scan

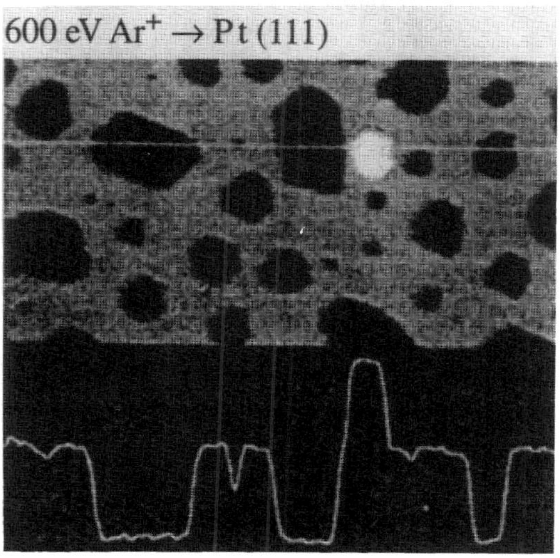

Figure 3.2. Topograph of Pt (111) after removal of 0.45 ML at a sputtering temperature of 540 K and a scan width of 40 nm × 20 nm. The line scan demonstrates that the *adatom island is one monolayer high.* Adapted and reprinted with permission from[134] T. Michely and G. Comsa, *Phys. Rev. B* 44(15), 8411–8414 (1991). Copyright 1991 by the American Physical Society.

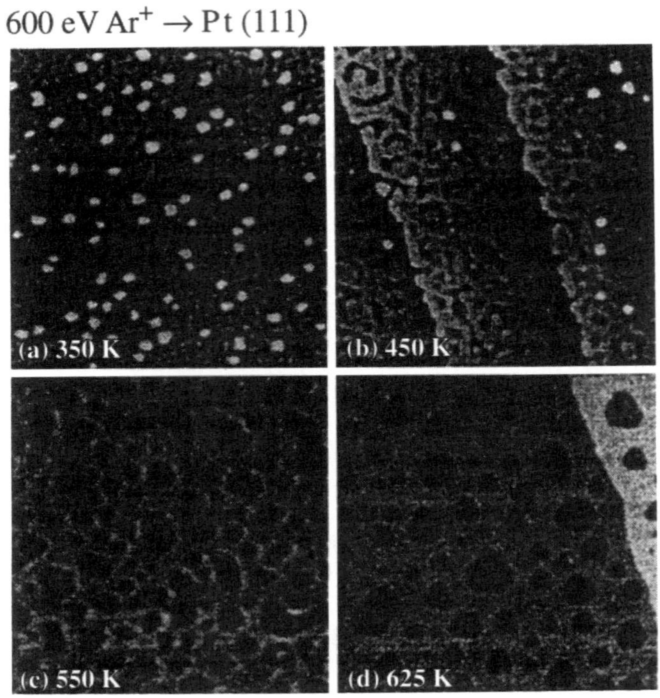

Figure 3.3. Topographs of Pt (111) after removal of 0.45 ML at (a) 350 K, (b) 450 K, (c) 550 K, and (d) 625 K, respectively. The scan width in each topograph is 70 nm × 70 nm and the same conditions of preparation and sputtering were used. Adapted and reprinted with permission from[134] T. Michely and G. Comsa, *Phys. Rev. B* **44**(15), 8411–8414 (1991). Copyright 1991 by the American Physical Society.

shows that the islands are one monolayer high or deep. With the same beam parameters to remove 0.45 ML of Pt, Fig. 3.3 shows decreasing amounts of adatom islands and increasing feature sizes for the vacancy islands at progressively higher temperatures of 350 K, 450 K, 550 K, and 625 K.[(134)] Adatom islands are not observed at 550 K because their rate of disappearance from surface mobility exceeds their rate of formation. Note the tendency for the islands to form hexagonal shapes in registry with the <110> substrate directions. Triangular and hexagonal vacancy islands are also formed at 400 K during 2.5-keV Ne-ion beam bombardment of Au(111); the islands are bounded by monatomic step heights oriented along the <110> directions.[(130)] In earlier work with Pt(111), Michely and Comsa studied the nanotopography of 1-keV He-ion beam bombardment at 735 K.[(129)] In addition to vacancy and adatom islands, they observed subatomic bumps with smooth rather than stepped-height profiles. They ascribed the bumps to strain fields resulting from subsurface helium bubbles, i.e., coalesced vacancies below the surface.[(129)]

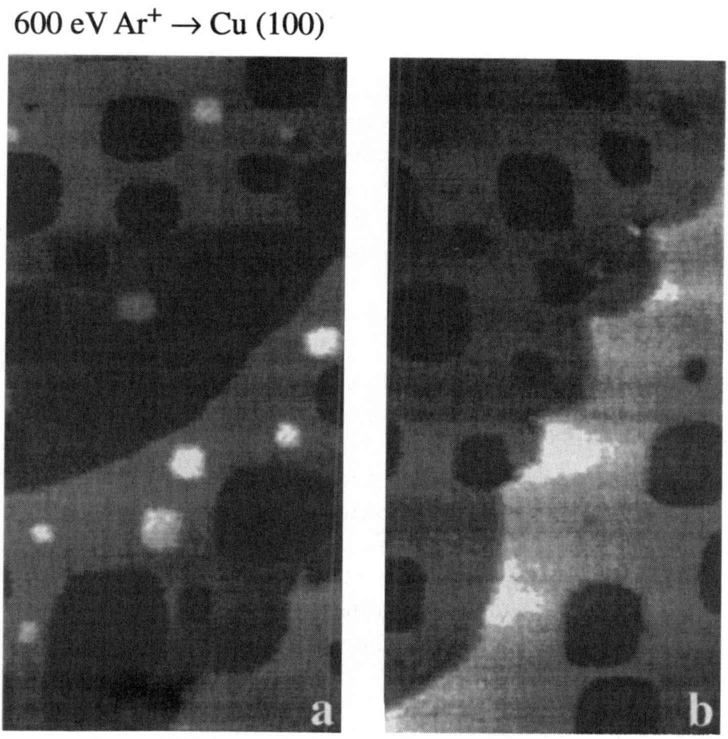

Figure 3.4. Images of the Cu (100) surface (scan area 63 nm × 117 nm) after bombardment at 300 K with 600 eV Ar. The bright areas are adatom islands and the dark areas are vacancy islands one atomic layer above or below the gray surface plane. The bombardment parameters were 1×10^{12} ions $cm^{-2}\ s^{-1}$ for 400 s (a) and 1×10^{14} ions $cm^{-2}\ s^{-1}$ for 4 s (b). The total area of the vacancy islands corresponds to 0.3 ML. Adapted and reprinted with permission from[133] J. C. Girard, et al., *Surf. Sci.* **302**, 73–80 (1994), and with kind permission of Elsevier Science - NL, Sara Burgerhartstraat 25, 1055 KV Amsterdam, The Netherlands.

Similar results were obtained for forming adatom islands and vacancies at 300 K for 600-eV Ar-ion bombardment of Cu(100), as shown by Fig. 3.4.(133) In both cases, 0.3 ML are removed with 4×10^{14} ions/cm^2, but different island sizes and distributions result from currents of 10^{12} ions/cm^2-s (Fig. 3.4a) or 10^{14} ions/cm^2-s (Fig. 3.4b). Both the adatom and vacancy islands are square with rounded corners and are identically oriented. By varying the bombardment parameters and Cu(100) temperatures, Girard *et al.*(133) showed the square feature is the *equilibrium* shape at room temperature; i.e., the surface mobility at room temperature is high enough for the equilibrium shape to be attained.(133) By analyzing the distribution of vacancy and adatom islands near the steps shown in Fig. 3.4 these authors also concluded that each step prevents the simultaneous vacancy nucleation process on the upper terrace and the adatom nucleation process on its lower terrace. The calculated vacancy coverage increased to 0.3 ML for terrace widths up to 30 nm, and then remained constant at 0.3 ML up to a width of 100 nm.[(133)] For the clean Au(100) surface that is known to reconstruct to form a Au(111) surface layer, rectangular, one-ML-deep vacancy islands are formed at room temperature and fluences of 4×10^{14} ions/cm^2 to 1.1×10^{15} ions/cm^2 of Ar ions.[(131)] The corners of the rectangles are defined more sharply than for Cu(100). The authors also noted that the vacancy islands disturb the reconstructed Au(111) surface and change to square shapes at the highest fluences.[(131)]

The adatom yields were determined for Pt(111) by using an STM and Ne-, Ar-, and Xe-ion bombardment from 40 eV to 5 keV at 150 K.[(136)] These data, plotted in Fig. 3.5, show that the adatom yield increases with ion energy and ion mass, although the yield from Ne-ion bombardment saturates as 5 keV is approached. The ion doses ranged from 2.4×10^{12} ions/cm^2 for 5-keV Xe ions to 2.8×10^{15} ions/cm^2 for 40-eV Ne ions; the ion fluxes ranged from 1.7 to 7.6 ions/cm^2-s. The adatom yields were the highest when the lower fluxes were used. The typical adatom island edge size formed is 1 nm, and these begin to coalesce at 80 K. Onset temperatures for other defect mobilities are 30 K for interstitials, 180 K for surface vacancies, and 500 K for bulk vacancies. By using low doses, the damage from single 5-keV Xe ions to Pt(111) was restricted to 24 impacts in a 77 nm × 77 nm STM viewing

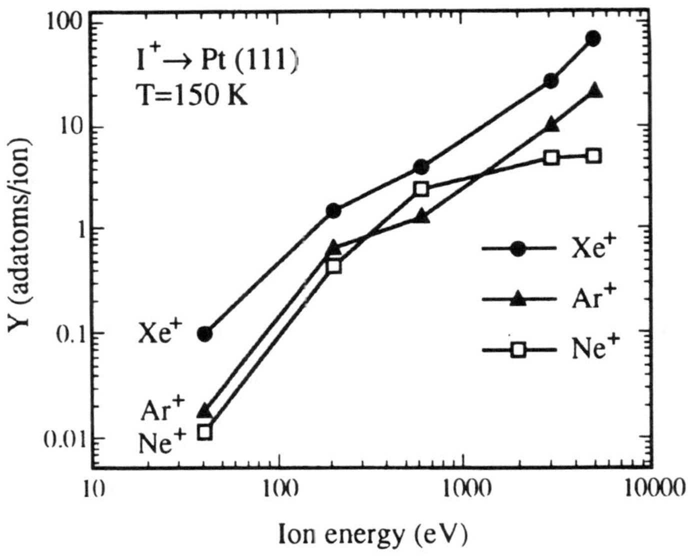

Figure 3.5. Adatom yields Y determined by STM for Pt (111) bombarded at 150 K and at normal incidence. Adapted and reprinted with permission from[136] T. Michely and C. Teichert, *Phys. Rev. B* **50**(15), 11156–11166 (1994). Copyright 1994 by the American Physical Society.

area. Each impact caused damage with a typical linear dimension of 5 nm but less than 10 nm. The features identified in each defect area were typically adatom islands, vacancy islands, and ringlike triangular corrugation line features (called loops for convenience). The distribution of these varied across the surface and is discussed in detail by Michely and Teichert.(136)

The development of vacancy islands for submonolayer-ion-beam bombardment removal of Si has been reported for Xe-ion bombardment of Si(100) from 350°C to 650°C.(132) Removing one ML of Si at 250°C with 225-eV Xe ions, disrupts the Si(100)–7×7 surface, but the 7×7 reconstruction is reformed.(135) In another study of the Si(111)–7×7 surface, adatoms and a distribution of vacancy islands are formed by 1-keV Ar-ion beam bombardment at room temperature, but only partial recovery of the reconstructed surface results from heating for up to 900 s at 500°C, 650°C, or 730°C.(137) The average *size* of the vacancy regions decreases with the annealing time, and large vacancy regions disappear because the smaller vacancy regions formed are preferentially stabilized at corner sites.(137)

3.3. Surface Defect Formation from Ion Beam Bombardment

The principal results from selected references about surface defects formed from ion beam bombardment are listed in Table 3.5.(28,103,138–143) Point, complex, and continuous surface and bulk defects are formed as surface and bulk vacancies, dislocations, and interstitials. The *rate* of defect formation, however, can be exceeded by their rate of annihilation, e.g., by diffusion as the substrate temperature is increased during ion beam bombardment. Computer simulation work by Webb and Harrison provides a basis for illustrating the severity of damage for ion energies from 1 to 20 keV.(103) They showed that pits are produced when the energy transferred by ion beam bombardment greatly exceeds that determined from the linear cascade theory; the maximum observed pit size increases with ion beam energy. Relatively large pits are found for simulation of 5-keV Ar-ion beam bombardment of the Cu(001) system. Figure 3.6 shows top views of the atom positions after 5-keV ion beam bombardment of a Cu(001) crystallite; the positions were calculated using a Bohr–Mayer "R" potential function that is similar to a Ar^+–Cu Moliere potential. The end of the atom ejection cascade (~5×10^{-13} s) is shown in Fig. 3.6a, and after cooling and coalescence (~10^{-11} s) in Fig. 3.6b. In these figures, white circles represent atoms near the original surface layer, and the degree of darkening indicates the relative displacement above or below the surface. Shaded circles on top of white circles are more than 0.1 nm above the surface. In the recrystallized state (Fig. 3.6b) shaded circles form part of a new layer resting on a reformed surface. Fragments of a second adatom layer occur occasionally The two pits formed (black areas, Fig. 3.6b), which are a connected set of vacancies, typically have less than 50 vacancies. The calculated sputtering yield is 4 atoms/Ar ion, and most of those are ejected from the surface plane. The adatoms do not

Table 3.5. Principal Results from Selected References about Surface Defects from Ion Beam Bombardment

Ref.	Ion$^+$	Energy (keV)	Target	Analysis method	Principal results
138	H^+	0.23	Si(100)	TEM	Extended defect depths ranged from 15 to 300 nm at ion bombardment for 2 min and 60 min and 150°C, 300°C, and 450°C.
139	Ar, Xe	0.07–0.5	Ge(001) 2×1 reconstr.	RHEED	Defect (surface only) yield per ion at 200-eV Ar- and Xe-ion bombardment ranged from 0.3 to 25 at 173 K but decreased to 1 above 350 K.
103	Ar	5	Model(001)	Computer	Top view of a simulation of a (001) surface after impact with a 5-keV Ar ion.
140	Ar	0.5–3	Ni(111)	Xe,XPS	Xe $3d_{5/2}$ XPS spectra of Xe adsorbed at 88 K at steps after bombardment from 100 to 800 K validates vacancy coalescence at higher temperatures.
28	Ar	0.02–0.095	GaAs(001)	XTEM HREM	Damage of surface layer mapped from 20 to 95 eV and for current densities from 0.1 to 0.75 mA/cm^2. Subsurface dislocation loops generated for $E \geq 55$ eV. No extended defects below 50 eV.
141	Ne	0.04–0.15	Graphite	AES EELS	Damage threshold is 35.8 eV and the number of defects/ion increases from 0 to 4.7 at 150 eV.
142	He, Ne Ar	3–30	Ni(110)	ISS	Bombardment-induced surface damage dependence on ion mass, energy, and temperatures from room temperature to 230°C are reported.
124	Ar	1–5	GaAs	EELS LEED AES	Amorphization fluence (LEED) decreases from 1 to 3 keV. The number of new bonds formed (EELS) and the Ga/As ratio (AES) increase for fluences up to 10^{16} ions/cm^2.

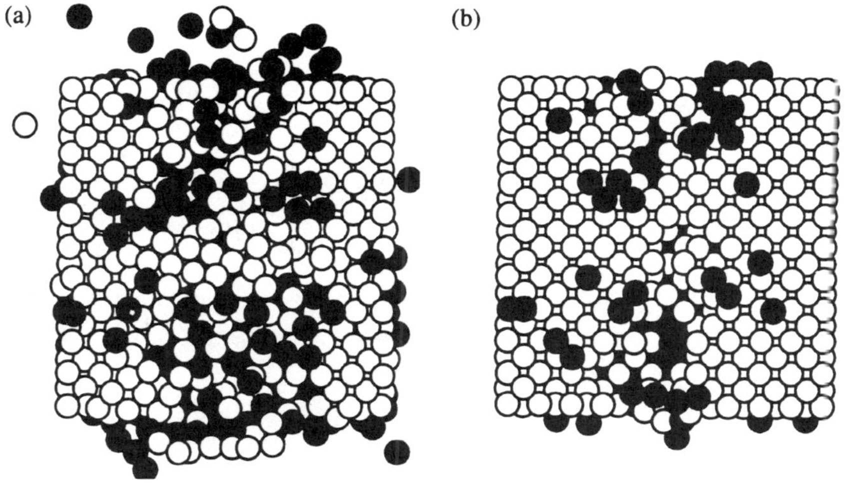

Figure 3.6. Top views of a model (001) crystallite calculated using masses and the lattice constant for an Ar ion impact on Cu. The open circles represent atoms in the surface plane, the shaded circles represent atoms on the surface plane, and the dark circles represent vacancies below the surface plane. The figure compares the results from 5.0-keV trajectories calculated with the same modified Bohr-Mayer "R" potential functions. The condition of the target at the end of the sputtering is shown in event (a) and of the target after cooling and coalescence by (b). Atoms appearing on the sides of the targets in (b) indicate that the targets are somewhat small for complete containment in the simulation. Adapted and reprinted with permission from[103] R. P. Webb and D. E. Harrison, Jr., *Phys. Rev. Lett.* **50**(19), 1478–1481 (1983). Copyright 1983 by the American Physical Society.

originate from the pits but typically have moved only one atomic layer, and the adatom yield is slightly greater than the sputtering yield. The number of vacancies is much larger than the combined adatom and sputtering yields. The frequency distribution of pit sizes has its maximum number for 10 vacancies or less, and the largest pit formed has 130 vacancies. The authors did not report the number of interstitials formed, but the number of occurrences are for a single Ar-ion trajectory used in their damage calculations.[103]

Verheij *et al.*[142] used ISS to measure the damage to a Ni(110) surface after He-, Ne-, and Ar-ion bombardment at different ion beam energies and target temperatures. Their data are plotted in Fig. 3.7, in which N_0 is the normalized height of the damage peak at saturation for some characteristic ion dose D_C. The damage peak is obtained from the number of ions N scattered from the atoms at steps into the detector; ions scattered from surface atoms in terraces are blocked from entering the detector by the choice of the angle of incidence and scattering angle. The authors chose scattering conditions so that elastically scattered ions were at an angle of only 6° relative to the surface plane. Under these conditions, ions scattering from an atom at a step arrive at the detector, but those ions from

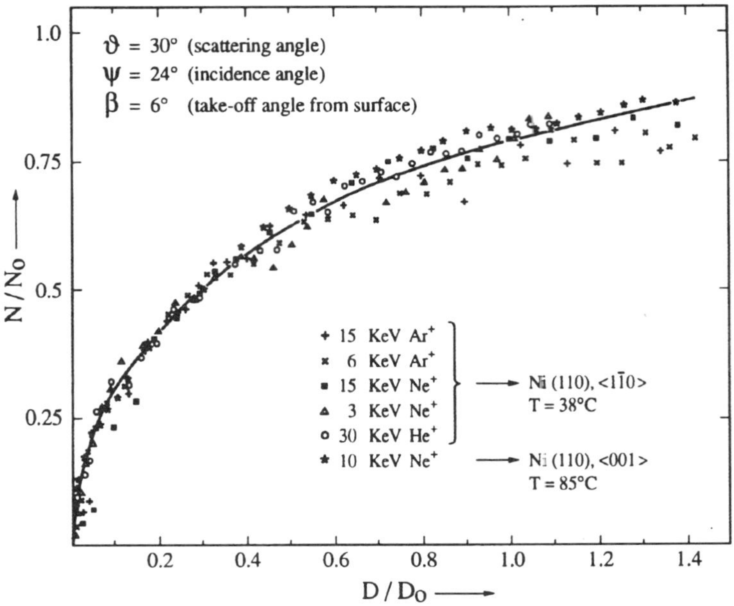

Figure 3.7. The dependence of the damage signal on the bombardment dose for different scattering conditions used for ISS analysis. The normalization factors N_0 and D_0 depend on the scattering condition. The solid line represents the function $[1 - \exp(-D/D_0)]^{1/2}$. Ion beam bombardment was along the <110> direction at T = 38°C and the <001> direction at T = 85°C. Adapted and reprinted with permission from[142] L. K. Verheij et al., *Nucl. Instrum. Methods* **168**, 595–599 (1980), and with kind permission of Elsevier Science - NL, Sara Burgerhartstraat 25, 1055 KV Amsterdam, The Netherlands.

terrace atoms undergo multiple collisions and are scattered away from the detector. The data in Fig. 3.7 have an ion dose of 10^{12} ions/cm^2 subtracted as the background level. The solid line is described by N (the ISS signal height from step atoms) = $N_0[1 - \exp(-D/D_0)]^{1/2}$, in which D is the dose at 10 keV up to a value of D_0 of 9.5×10^{16} ions/cm^2 for He, 6.7×10^{14} ions/cm^2 for Ne, and 5.7×10^{14} ions/cm^2 for Ar. The value of N_0 is constant for different masses and energies of the ion beam used and substrate temperature. However, the dose D_0 had to be increased by about a factor of 15 for a temperature increase from 38°C to 85°C for the Ni(110) crystal, a result that shows the damage is annealed at these relatively low temperatures; the annealing rate becomes very large above 100°C.[(142)] The authors' results can be explained in terms of clustering of vacancies formed by ion beam bombardment.

Defect generation at 615°C resulting from E = 20- to 95-eV Ar-ion beam bombardment of ion-assisted MBE grown GaAs(001) was evaluated using RHEED and cross-sectional transmission electron microscopy (XTEM).[(28)] The 2 × 4 reconstruction RHEED patterns are indicative of flat-ordered surfaces, but the XTEM showed subsurface dislocation loops are generated at ion beam energies >55 eV (Fig. 1.5). At ion beam energies below 50 eV and current densities of less than $J = 0.5$ mA/cm^2, stable extended defects are not produced. The loop densities, which ranged from <10^5/cm^2 (XTEM detection limit) to 10^9/cm^2, are primarily oriented along the $[1\bar{1}1]$ and $[\bar{1}11]$ planes. The dislocation loops are formed from supersaturated point defects that result from the ion beam bombardment and defects formed per ion. The point defect density is approximately proportional to EJ, the product of the energy E times the current density J. The critical value of EJ for defect production in Fig. 1.5 is about 10 eV mA/cm^2. A sputtering yield of 0.04 atoms/ion was measured at 50-eV Ar-ion beam bombardment and 0.5 mA/cm^2.[28)]

Steffen *et al.*[(141)] analyzed line-shape changes in AES to deduce the damage threshold in Ne-ion bombardment of graphite and estimated the number of defects formed per ion from π-plasmon shifts in the EELS spectra. The threshold for forming defects was measured to be 36 eV ± 1 eV at ion beam bombardment energies from 40 to 150 eV at room temperature and doses of $< 2 \times 10^{15}$ ions/cm^2. The number of defects per ion increases from 0 at 36 eV to 4.7 at 150 eV, with a tendency toward a saturation value at the higher bombardment energies.

Floro *et al.*[(139)] detected and quantified defect concentrations with *in situ* RHEED while using 70- to 500-eV Ar- or Xe-ion beam bombardment of Ge(001) and a maximum current density of 50 nA/cm^2 at temperatures from 175 to 400 K. The total defect (vacancies and interstitials) yields at 175 K increased almost linearly from 2 to 20 defects/ion for Ar-ion energies from 70 to 500 eV, but increased from about 5 to 28 defects/ion for Xe ions over the same energy range. The defect yield dropped dramatically between 200 and 350 K; e.g., yields from 200-eV Ar-ion beam bombardment decreased almost linearly from 10 defects/ion at 175 K to about 1 defect/ion at 350 K. The yield reduction is attributed to adatom and vacancy recombination produced in the same collision cascade.[(139)]

Malafsky[(140)] detected defect concentrations on Ni(111) surfaces by using the $3d_{5/2}$ XPS peak shift of 0.7 eV from Xe preferentially adsorbed at 88 K at the defect sites rather than at terrace sites. The Ni(111) crystal was sputtered with up to 0.06 ML of 500-eV to 3-keV Ar ions to yield Xe coverages of 0.13 to 0.18 ML of Xe after ion beam bombardment. Most of the defects were formed at a fluence of about 1×10^{13} Ar ions/cm^2 (0.005 ML). The Ar-ion beam bombardment of Ni(111) at increasing temperatures shows that the defect concentration decreases rapidly (by a factor of about 4) from 88 to 375 K, with little further decrease at temperatures up to 775 K. The loss in Xe coverage is attributed to changes in the size and shape of vacancy islands (from small and irregular shapes to larger hexagonal shapes) and not to the loss in defect sites from adatom vacancy recombination.[(140)]

Gnaser[(124)] summarized damage to GaAs(110), as monitored at room temperature by LEED, AES, and EELS, from 1- to 3-keV Ar-ion beam bombardment at fluences up to 10^{16} ions/cm^2 and flux densities of 1 to 16×10^{12} ions/cm^2-s. He also referenced work done on the damage to Si(111), Ge(001), and GaAs(110) from Ar-ion beam bombardment from 100 eV to 5 keV. An amorphization fluence, which was defined as the ion fluence required for complete extinction of the LEED pattern, was used to show that the fluence decreased from over 4×10^{15} Ar ions/cm^2 at 1 keV to 1.5 to 2.5×10^{15} Ar ions/cm^2 for Ar-ion energies of more than about 1.75 keV for all ion fluxes studied. The amorphization fluence required decreases with increasing current density. With EELS, Gnaser[(124)] reported that 1.5-keV Ar-ion beam bombardment produced dramatic decreases in the excitation of an As dangling-bond into a Ga dangling-bond state with fluences of 10^{13} to 10^{15} ions/cm^2 and a flux of 10^{12} Ar ions/cm^2-s. Other peaks that also decreased concomitantly are from the excitation of the 3*d* core electrons and the surface plasmon. The qualitative shapes of the peak intensity versus fluence curves are essentially independent of the ion beam bombardment energy to as low as 200 eV. The fluence, current density, and temperature-dependent changes in the surface concentration ratio (from AES) of Ga–As are also reported.[(124)] Additional detail is available in a previous publication.[(143)]

Using TEM, Liu *et al.*[(138)] detected extended defects in Si(100) from bombardment with a 230-eV H-ion beam at about 1.5×10^{14} H ions/cm^2-s for exposure times of 2 min or 60 min to produce defects from 150°C to 450°C. The depth of the defects ranged from 15 nm after 60 min and 450°C or 2 min and 150°C exposures to a maximum of 300 nm at a 60-min exposure at 150°C. For 2-min exposures between 150°C and 450°C, the defects are $\leq$20 nm below the surface at a concentration of $\geq 10^{12}$ defects/cm^2. The average surface roughness, surface and bulk minority lifetimes, and volume density of the defects are reported for six time–temperature exposure combinations.[(138)]

From this section, we conclude that (1) surface and bulk defects result from ion beam bombardment, (2) vacancies coalesce into extended defects at high concentrations, (3) the number of step defects increases with ion beam bombardment energy, (4) defects are annealed during and after ion beam bombardment of targets at increasing temperatures, (5) the ion beam bombardment threshold energy for forming defects is 36 eV for graphite, (6) the number of defects increases with increasing ion beam bombardment energy, total ion fluence, current density, and time of ion beam bombardment at a particular energy, and (7) during ion beam bombardment at a particular energy the defect density is limited by a competitive rate process that decreases the number of defects from various annihilation processes.

3.4. Damage Depth and Defect Density from Ion Beam Bombardment

The defects resulting from ion beam bombardment are dislocations, interstitials, vacancies, complexes, structural defects, and partial or complete amorphization. The principal results from selected references about damage depth and defect

Table 3.6. Principal Results from Selected References about Damage, Defect Density, and Damage Depth from Ion Beam Bombardment

Ref.	Ion+	Energy (keV)	Target	Analysis method	Principal results
144	Ne, Ar Kr	0.2 to 2	Si(111), (100) at T <40°C doped with boron	PL, RBS, TEM	Noble-gas (NG) defects are detected by photoluminescence (PL); point defects are produced to depths of 100 nm and deeper and increase in concentration with ion beam energy; Si vacancies and interstitials are formed; dislocations are formed; excitons are bound at B; and the energy threshold for forming NG defects is between 180 and 420 eV.
145	Ar	1, 3, & 4	GaAs(100) (c-GaAs) c is for crystalline	DLTS	Ion-bombardment-induced damage confined to a few tens of nanometers below the surface; electron traps are formed 0.32 to 0.76 eV below the conduction band with densities from 4×10^{13} to 2×10^{16} per cm.[(3)] The layers formed on bulk c-GaAs are a-GaAs, a-GaAs and microcrystalline GaAs, microcrystalline GaAs, and c-GaAs.
146	Ar	0.5, 1, 2, and 3.9	GaAs	Raman UVR	From ultraviolet reflectivity (UVR) and Raman scattering, structural damage was identified to a depth of 60 nm for 3.9-keV Ar ions. The damage layer consists of a very high density of point defects, but distinct nanocrystals of GaAs are not present.

(continued)

Table 3.6. Continued

Ref.	Ion^+	Energy (keV)	Target	Analysis method	Principal results
147	Ar	5.5	Si(111) Si(100)	TEM SEM	Cross-sectional TEM shows amorphous layers 5 nm (111) and 15 nm (100) thick near the surface and a severely damaged structure at >ca. 50 nm below the surface; the (100) surface has topographical protrusions (ca. 2 nm) surrounded by (111) planes smaller than 10 nm.
148	Kr	8	PbS	STM	Surface and in-depth damage maps show the effects of a single Kr^+-ion impact on PbS.
149	Au	10 or 20	Au(001)	Simulation	Defect production at the surface is different from bulk; transient localized melting is especially important at the surface.
150	Ar	0.5 to 5	GaAs(100)	DR	From differential reflectance (DR), ion-bombardment-induced structural damage changes the optical properties and electronic interband transitions; vacancies and interstitials are annealed and recombine in stages at 200°C, then more completely from 400°C to 450°C, and completely at 700°C.
151	Ar	0.01 to 5	LiF	Calculation	Ion penetration depths, damage profiles, and sputtering yields calculated for amorphous LiF; anion vacancies plus electrons yield F-centers that migrate to the surface; Li atoms evaporate. The Ar-ion interaction is described by resonant and Auger neutralization, vacancy formation, other inelastic energy loss channels, and sputtering.

(continued)

Table 3.6. Continued

Ref.	Ion$^+$	Energy (keV)	Target	Analysis method	Principal results
152	Ar H_2	0.3, 1.2 1.0	SiO_2	SE	From spectroscopic ellipsometry (SE), a 5-nm-thick (300 eV), 11-nm-thick (1.2 keV), or 36-nm-thick (1.0 keV) a-Si layer is formed from ion bombardment at normal incidence.
78	Ar, Ne Xe	0.05–0.5, 10 0.1, 0.2, 0.3	Si(001), Ni(110) Ge(001)	MEIS, ISS RHEED	Review article on ion bombardment damage gives examples of displaced Si atoms, defect concentration in Ni(110), and surface damage to Ge(001).
153	Ar	0.1 to 1.5	SiO_2/c-Si	Se	SE used to determine damage depth dependence on IB energy and dose to form layers of voids and a-Si and a-Si on c-Si up to a thickness of 13.8 nm for the damaged layer.
154	Ar	5	Ge(111)	IBSE	Inelastic backscattered electrons (IBSE) used to support model calculations of the thickness of a-Ge formed with increasing ion doses and four current densities at 190°C and 5-keV ion bombardment.
155	—	—	Amorphizable solids	—	Review of amorphizable insulators gives examples of solid–*electron* interactions but provides useful tables of nine solids subject to amorphizing from ion bombardment.
156	Ar	1	Diamond	AES C(KLL)	Diamond (c-sp^3) is converted into graphite-like C(a-sp^2) and then glassy C (disordered sp^2) with IB doses of 7×10^{12} ions/cm^2 to 6.2×10^{17} ions/cm^2 as monitored from changes in KLL Auger data. No C-sp^2 (graphite) is actually identified.

(continued)

density from ion beam bombardment are listed in Table 3.6.[78,144–156] By using Ar-ion doses from 10^{15} to 10^{16} ions/cm^2, Armour and Al-Bayati[78] showed that the number of displaced Si atoms per incident Ar ion increases from 3 to 30 for Ar-ion beam bombardment energies between 50 eV and 1 keV. Threshold energies for forming point defects (vacancies) from implanted noble-gas atoms are given in Table 3.7 for Si(111) and Si(100) surfaces after 1-keV Ar-ion beam bombardment at < 40°C; annealing at 350°C was used to coalesce the vacancies into the three or four needed for detection by photoluminescence (PL).[144] A highly damaged layer about 5 nm deep with a concentration of more than 10^{19} atoms/cm^3 was identified with RBS, and point defects were produced to depths of 100 nm as monitored with PL. The defect concentration in Si increases with ion beam energy from 0.2 to 1 keV of Ne, Ar, and Kr ions. The PL intensity decreases after annealing to more than 350°C and is eliminated at 550°C. The effects of annealing the damaged Si to 1050°C are also discussed in which 5-nm-diameter Ar bubbles (TEM), new defects from implanted noble gases (PL), and copper-pair defects (PL) are formed between 550°C and 1050°C.

The progression in damage depth up to tens of nanometers and the defect density increase with ion beam bombardment energy, current density, and total fluence as long as the substrate temperature is low enough to preclude annealing the defects. Table 3.6 provides some detail about the damage depths. Figure 3.8 shows explicitly that damage depths range up to 60 nm for GaAs(100) bombarded by 0.5- to 3.89-keV Ar ions.[146] The structure of the ion-beam-bombardment-damaged layer is distinctly different from Ar-ion-implanted GaAs and exhibits red shifts in the Raman spectra in addition to the broadening of the peaks.[146] The structural damage in GaAs(100) from 0.5- to 5-keV Ar-ion beam bombardment changes the optical properties and interband electronic transitions.[150]

Table 3.7. Threshold Energies for the Formation of Defects from Implanted Ne, Ar, and Kr after 1-keV Ion Beam Bombardment at 0.2 to 0.4 mA/cm^2 for 15 min at $T < 40°C$[a]

Gas	Si wafer orientation	Threshold (eV)
Ne	(111)	420±110
Ne	(100)	210±80
Ar	(111)	180±30
Ar	(100)	190±30
Kr	(111)	290±40
Kr	(100)	340±50

[a]Samples were annealed at 350°C to detect the defects with photoluminescence.[144]

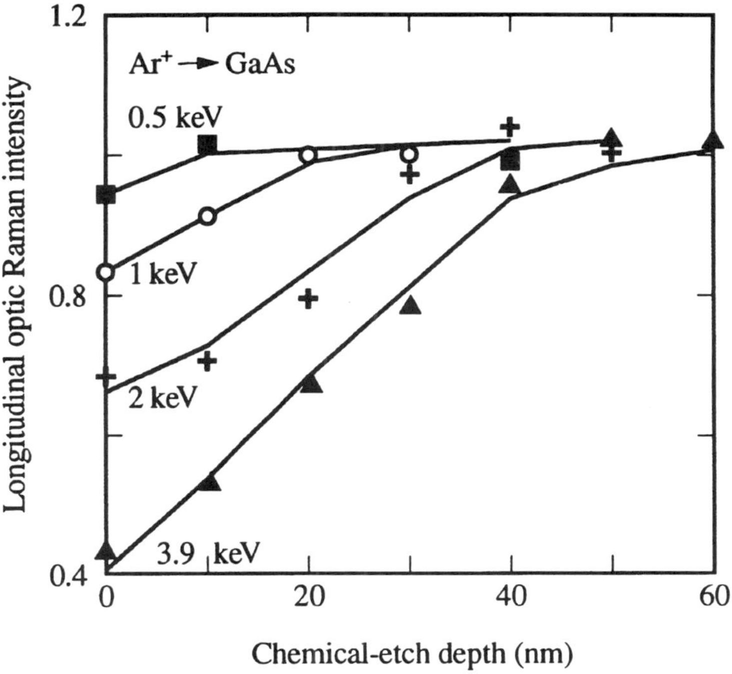

Figure 3.8. The longitudinal optical (LO) Raman intensity (normalized to a c-GaAs reference standard) of the Ar-bombarded GaAs samples following chemical etching. The symbols are the experimental data. The lines indicate the fits discussed in Ref. 146. For each sequence of chemically etched samples, the "fitted curve" consists of linear segments connecting points at which the fit was done at 10-nm intervals. Adapted and reprinted with permission from[146] G. F. Feng, et al., *Phys. Rev. B* **43**(12), 9678–9686 (1991). Copyright 1991 by the American Physical Society.

For Si(111) and Si(100), TEM shows an amorphized layer 10 nm thick and a damage zone in the crystalline structure at 50 nm below the surface (Fig. 3.9) from 5.5-keV Ar-ion beam bombardment at a current of 0.8 mA for 1 h.[(147)] For developing Fig. 3.10, Irene *et al.*[(152,153)] analyzed spectroscopic ellipsometry data for 0.1- to 1.5-keV Ar-ion beam bombardment of single-crystal Si wafers with damage layers consisting of a mixture of voids and a-Si and a layer of a-Si on the c-Si substrate. The damage progression shown resulted from increasing ion beam bombardment energies at current densities of 0.02 to 0.2 mA/cm^2 and total Ar-ion doses of 10^{14} to 10^{16} ions/cm^2.[(153)] Samples of SiO_2 on Si were also etched with Ar and H_2 under different conditions, as listed in Table 3.8. A model similar to Fig. 3.10 has also been shown for GaAs(110).[(145)]

Damage from ion beam bombardment was also reported for Ge(111),[(154)] LIF,[(151)] and diamond(110).[(156)] An initial growth of a-Ge on Ge(111) to thicknesses

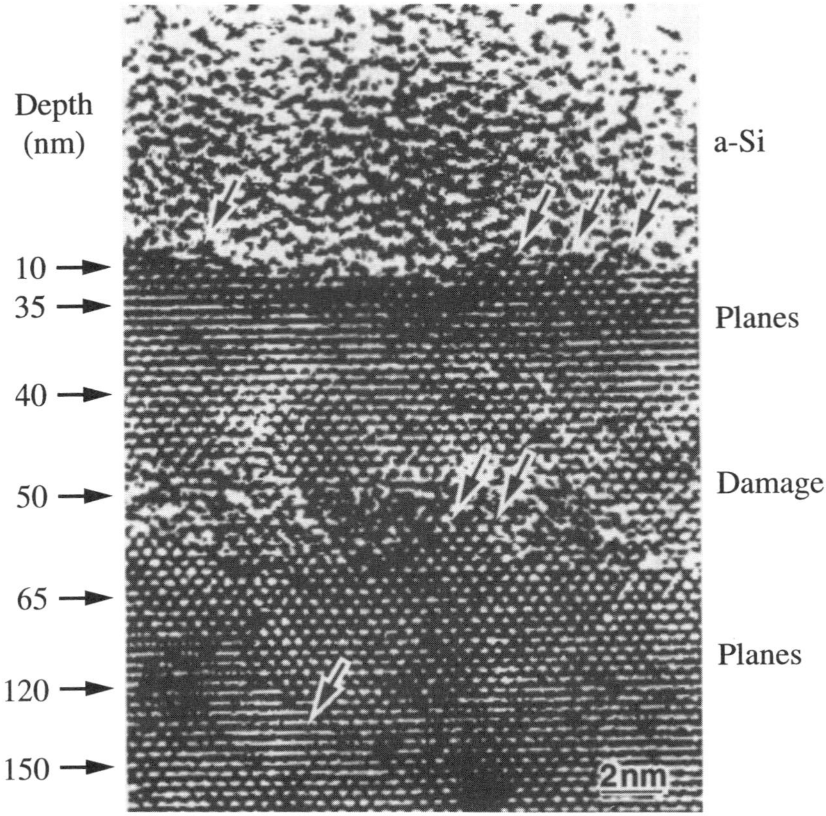

Figure 3.9. Cross-sectional high-resolution TEM image of a Si specimen with the (100) surface projected to the (110). The thickness increases from the top to the bottom; the arrows indicate the triangular image contrast resulting from protuberances. Adapted and reprinted with permission from[147] M. Song, et al., *Scanning* **12**, 29–32 (1990). Copyrighted and reprinted with the permission of SCANNING, and/or the Foundation for Advances of Medicine and Science (FAMS), Box 832, Mahwah, NJ 07430, USA.

of about 190 nm was determined after 5-keV Ar-ion beam bombardment with ion doses for up to about 1.8×10^{16} ions/cm^2 and current densities of up to 4 μA/cm^2 at 190°C.[(154)] Ion penetration depths and damage yields were calculated for Ar-ion beam bombardment of a-LiF in which ion vacancies and electrons yield F-centers that migrate to the surface and Li atoms evaporate.[(151)] The penetration depths for Ar ions was 0.36 to 5 nm between 0.1- and 5-keV ion beam bombardment energies. Considerable detail is presented about defect production, and calculated Ar-ion sputtering yields for LiF are given from 10 eV to 5 keV.[(151)] The transformation of type IIa diamond(110) induced by 1-keV Ar-ion beam bombardment for doses of

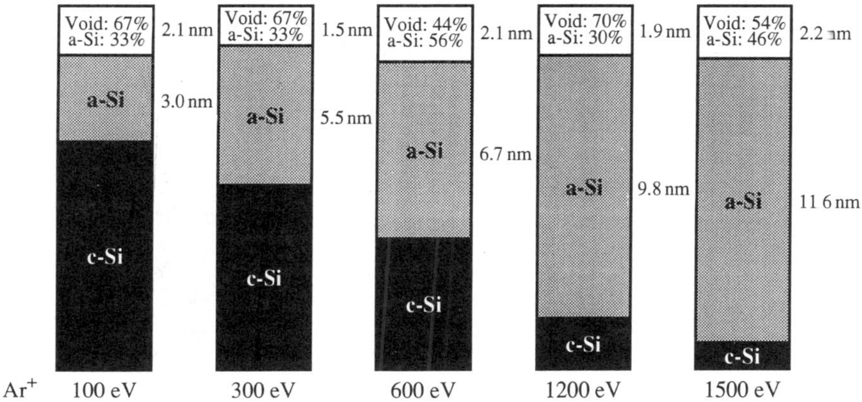

Figure 3.10. The damage products and thicknesses after Ar saturation bombardment for c-Si after using different bombarding energies. Adapted and reprinted with permission from[153] Y. Z. Hu, et al., *Nucl. Instrum. Methods Phys. Res., Sect. B* **59/60**, 76–79 (1991), and with kind permission of Elsevier Science - NL, Sara Burgerhartstraat 25, 1055 KV Amsterdam, The Netherlands.

7×10^{12} to 6×10^{17} ions/cm^2 was studied by AES, secondary-electron emission, and total secondary-electron yield measurements.(156) Below a critical dose of about 2×10^{15} ions/cm^2, diamond remains crystalline but with increasing levels of disorder; at a dose of 2 to 3×10^{15} ions/cm^2, the damaged diamond suddenly transforms into an amorphous carbon with sp^3 bonding and no long-range order. Further irradiation results in the a-sp^3 bonding gradually changing into an a-sp^2 carbon with a concomitant increase in the electrical conductivity. A micropolycrystalline graphite-like layer is produced at the highest doses.(156)

Molecular dynamics simulations of 10-keV Au-ion beam bombardment of the targets considered show that nearly all the damage results from local melting, viscous flow of a hot liquid on the surface, and defect production.(149) For one 10-keV event, 554 atoms flowed to the surface and resulted in the formation of a dislocation network that extended to about 6 nm below the surface. Pressures to 6 GPa and temperatures to 7000 K were calculated for 1 to 2 ps following ion impact. The cavity formed contracts by 9 ps.(149) In an STM study substantial damage was observed from the effects of a single 8-keV Kr-ion impact with a cleaved PbS solid.(148) The surface topograph showed displaced atoms, interstitials, vacancies, and an impact crater of approximately 6×6 atomic rows. Within the damage crater, the interstitials either recombine or segregate to the surface, and clusters of vacancies collapse into dislocation loops about three layers below the surface.(149)

Hobbs(155) discussed the topology and geometry of ion-beam-bombardment-induced amorphization of insulators. Readily amorphizable nonmetallic solids

Table 3.8. Comparison of Etch Rates, Damage Depths, and Composition after Ar and H_2 Ion Beam Etching of SiO_2 on Si under Different Conditions

Ion beam	Ar		H_2			
Ion energy (eV)	1200	300	1000	300	1000	300
Temperature (°C)	25	25	25	25	500	500
Ion incidence angle (0° is the surface normal)	0°	0°	0°	0°	60°	60°
Removal of oxide	Yes	Yes	No	No	Yes	Yes
Etch rate (nm/min)	15	10	None	Small	1.5	1.4
Damage depth (nm)	11	5	36	6.3	9.5	1.5
a-Si in the damaged layer	100%	100%	71%	41%	22%	95%
Voids in the damaged layer	—	—	20%	25%	11%	5%
c-Si in the damaged layer	—	—	9%	34%	67%	—

From Ref. 152.

Table 3.9a. Amorphizable Solids and Amorphizing Radiation

Compact solids	
Cu	none
CuTi	electrons
NaCl	none (radiolysis)
MgO	heavy ions (difficult)
Al_2O_3	heavy ions (high fluence)
$CaTiO_3$	heavy ions (high fluence)
Molecular solids	
$(-CH_2-)_n$	electrons (radiolysis)
$C_{14}H_{10}$	electrons (radiolysis)
Network solids	
SiO_2	electrons (radiolysis)
$NaAlSi_3O_8$	electrons (radiolysis)
Si	light ions (low fluence)
SiC	light ions (low fluence)
Si_3N_4	ions (moderate fluence)
$ZrSiO_2$	ions (moderate fluence)

From Ref. 155.

Table 3.9b. Amorphization Criteria

Atom (ion) size ratios (atom packing)
Ionicity (directed covalent bonds)
Melting point (bond strength)
Crystallization temperature (diffusion)
Structural complexity (number of cation sites)
Chemical complexity (number of atom types)
Combinatorial geometry (connectivity)

(Table 3.9a) are typically complex ionic compounds or covalently bonded solids to which a number of criteria (Table 3.9b) have been applied to predict the ease of amorphization. For Table 3.9a Hobbs did not define radiolysis, which appears to be connected with electron bombardment of solids. The amorphizing mechanism depends either on accumulating a critical density of point defects or reaching a critical energy density within a collision cascade (Table 3.9c).(155) Because Hobbs(155) did not define low and medium fluences, we think he means 10^{13}, 10^{15}, and 10^{17} ions/cm^2 for low, medium, and high fluences, respectively.

3.5. Enhanced Diffusion and Changes in Electrical Properties from Ion Beam Bombardment

The principal results from selected references about enhanced diffusion and changes in electrical properties resulting from ion beam bombardment are given in Table 3.10.(135,144,145,150,157–169) The high-defect densities formed during ion beam

Table 3.9c. Energy Density for Amorphization by Ion Implantation

Compound	keV/atom
MgO	>6
Al_2O_3	3.5
$CaTiO_3$	0.1
Si_3N_4	0.05
$ZrSiO_4$	0.04
SiC_4	0.02
Si	0.013
$NaAlSi_3O_8$	0.008
SiO_2	0.007

Table 3.10. Principal Results from Selected References about Enhanced Diffusion and Changes in Electrical Properties from Ion Beam Bombardment

Ref.	Ion$^+$	Energy (keV)	Target	Analysis method	Principal results
157	Ar	0.05–0.2	Al or Au on GaAs	*I–V* Schottky capacitance spectros.	Schottky diode (Al or As/GaAs) barrier heights decrease for *n*-type and increase for *p*-type GaAs with increasing bombardment energies but can be increased (*n*-type) or decreased (*p*-type) with a 400°C anneal.
158	Ar	1 or 4	Ag/Ni multilayers	SDP AES	The diffusion rate of Ag (3–4 nm) in Ni (50 nm) multilayers is enhanced by bombardment at room temperature but is enhanced less at 603°C (defects are annealed).
144	Ne, Ar Kr	0.3 to 1	Si(100) Si(111)	PLS	Noble-gas ions form defects as detected by photoluminescence spectroscopy (PLS). New defects are formed on annealing from 250°C to 550°C and their concentration generally increases with bombardment energy.
145	Ar	1, 3, & 4	GaAs	DLTS	Persistent photoconductivity, photo and thermal quenching of metastable defect complexes, new electron traps below the conduction band, and trap-assisted tunneling in Schottky diodes result from ion bombardment damage.
159	Ar	1	GaAs(100)	DLTS	Five different electron traps (defect centers) resolved much deeper than the mean ion range; the trap concentration is reduced by annealing at 500°C.
160	Ar	5	CdTe(110) ZnCdTe, or CdSeTe	PLS PR	Surface electronic defects are introduced with ion bombardment and are located in the band gap; alloys have differences; and the ion beam damage is much deeper than the mean ion range. Photoreflectance (PR) and PLS used.
150	Ar	0.5 to 5	GaAs(100)	DR	Differential reflectance (DR) shows structural damage is gradually annealed from 200°C and is completed at 550°C.

(continued)

Table 3.10. Continued

Ref.	Ion$^+$	Energy (keV)	Target	Analysis method	Principal results
135	Xe	0.225	Si(111)–7×7	STM	With increasing target temperatures, a transition occurs from disruption of the ordered surface from ion bombardment to preservation of the 7×7 reconstruction.
161	Ar	0.5 to 5	*n*-GaAs	*I–V*,*C–V* DLTS	Schottky diode barrier heights increase monotonically with ion bombardment energy from both discrete- and continuous-level defects introduced in the band gap.
162	Ar	0.4, 0.5	GaAs InP	PL	Changes in photoluminescence (PL) efficiencies show ion-bombardment-enhanced diffusion is enhanced further by He–Ne laser illumination.

bombardment results in enhanced diffusion, often termed radiation-enhanced diffusion (RED), and has been observed in many systems. We do not use RED because radiation is ambiguous and ion beam bombardment enhanced diffusion is not. For example, Marton *et al.* prepared Ag–Ni multilayers of five 3- to 4-nm-thick Ag and six 50-nm-thick Ni layers.[158] These samples were subjected to 1- or 4-keV Ar-ion beam bombardment from 299 to 800 K; AES spectra were simultaneously recorded during sputtering. The diffusion of Ag in Ni is induced by sputtering, and the diffusion rate D must be divided by the sputtering rate S because D is linearly dependent on S. Arrhenius plots of the data, shown in Fig. 3.11 (in which the ordinate contains the ratio D/S), yield activation energies of –0.04 eV and –0.038 eV at 4-keV and 1-keV ion beam bombardment energies, respectively. The results were interpreted in terms of RED with long-lived complex defects that can migrate from large distances, and that are themselves subject to annealing. Annealing of defects results in the negative activation energy.[158]

Bedrossian and Klitsner[135] used STM to study the vacancy-mediated layer-by-layer removal of Si(111) during 225-eV Xe-ion beam bombardment at elevated temperatures. At 250°C, a high density of small monolayer-deep vacancy islands is formed as the 7×7 reconstructed surface is randomly removed; the 7×7 surface disappears from view when one monolayer is sputtered. In contrast, ion beam bombardment at 580° yields vacancy islands of larger size because of the enhanced mobility of vacancies; in contrast to ion beam bombardment at 250°C, the 7×7 reconstruction is visible on all terraces of a stepped structure. At 400°C, some

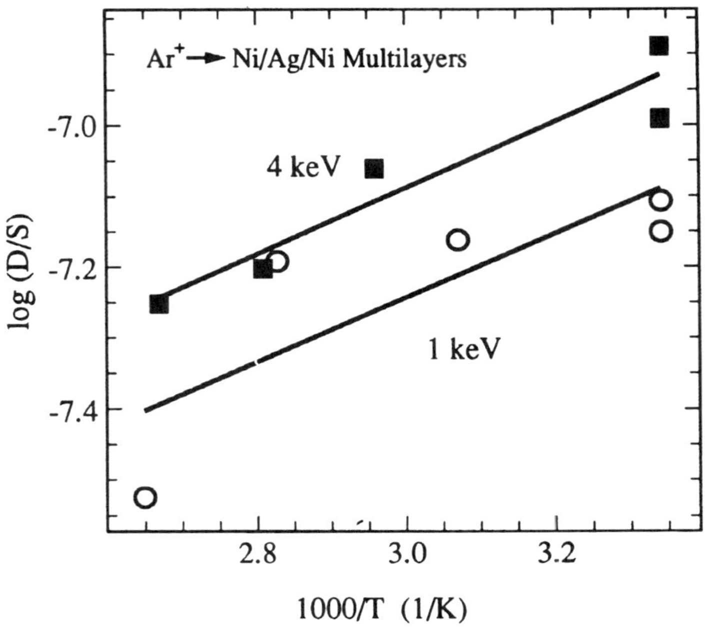

Figure 3.11. Arrhenius plots of the ratio of the ion-beam-bombardment-induced diffusion rate of Ag to the sputtering rate for 1 keV and 4 keV Ar ion bombardment of Ag/Ni multilayers. Each point corresponds to the average rate obtained from all of the 5-Ag layers measured in Ni/Ag/Ni etc. thin film multilayer samples. Adapted and reprinted with permission from[158] D. Marton, et al., *Phys. Rev. Lett.* **61**(23), 2697–2700 (1988). Copyright 1988 by the American Physical Society.

unreconstructed areas are visible near the monolayer-high steps in the surface.[(135)] Chen *et al.*[(162)] used the known ion-beam-bombardment-enhanced diffusion of defects (~10^{-15} cm^2/s) during 400-eV or 500-eV Ar-ion beam bombardment at 50 $\mu A/cm^2$ of GaAs or InP surfaces to determine if simultaneous laser illumination influenced the diffusion rate. They observed that the photoluminescence efficiency degrades dramatically with increasing intensities of a 4-mW/cm^2 HeNe laser in quantum well structures of both GaAs and InP. Corresponding losses are not observed with the laser alone; several possible mechanisms for the laser-induced enhancement are discussed.[(162)] In another study of GaAs, differential reflectance was used to monitor interband electronic transitions that are changed by the structural damage from 0.5- to 5-keV Ar-ion beam bombardment.[(150)] Vacancies and interstitials are formed during ion beam bombardment, but recombine at ~450°C.

Changes in the electrical properties of semiconductors as a result of ion beam bombardment have been extensively summarized.[(12,24)] From Table 3.10 we note

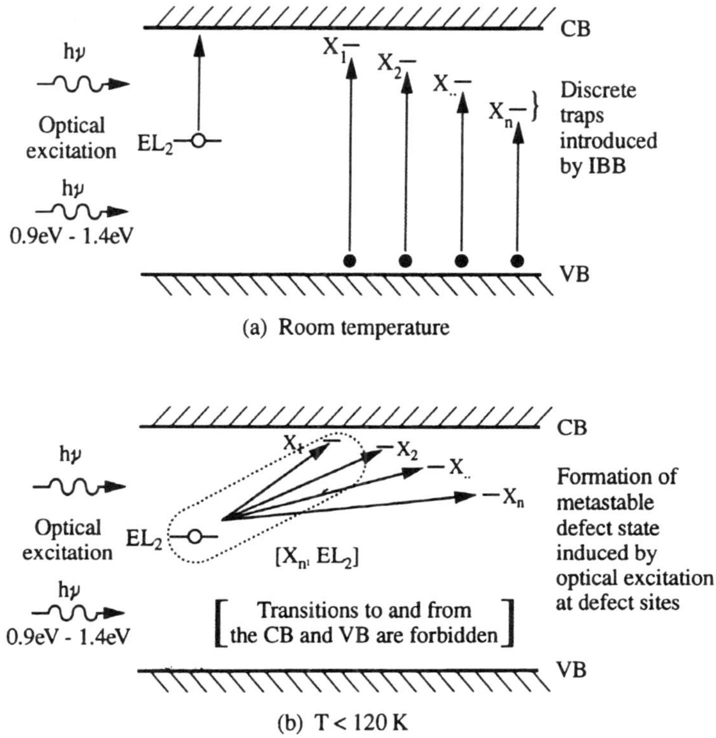

Figure 3.12. Phenomenological model for forming optically-generated metastable defect states by ion-beam etching: (a) at room temperature, transitions to and from the conduction band (CB) and valence band (VB) are permissible; (b) at $T \leq 120$ K, optically generated ion-beam-induced metastable states find no transition pathways. Adapted and reprinted with permission from[145] A. Vaseashta, et al. *Mater. Sci. Eng.* **B9**, 498–100 (1991), and with permission from Elsevier Science.

that Ar-ion beam bombardment results in changes in Schottky diode (Al or Au–GaAs) barrier heights,[(157)] new noble-gas defects during annealing,[(144)] new electron traps in GaAs(110) below the conduction band,[(145)] traps that are much deeper than the mean ion range,[(159)] surface electronic defects in the band gap of CdTe,[(160)] and monotonic increases in the Schottky diode (Au–*n*-GaAs) barrier height.[(161)] Figure 3.12 shows a phenomenological, flat-energy-band representation of the optical excitation of metastable defect states in GaAs(110) formed by 1- or 3-keV, 40-μA/cm^2 Ar-ion beam bombardment.[(145)] Table 3.11 is a summary of typical metal, semiconductor, oxide, and halide solids used for studying structural changes in the surface and the near-surface regions. The principal results listed are roughly in the same order as the structural changes given from 1 to 11 in Table 3.1.

Table 3.11. Examples of Targets Bombarded for Studying Structural Changes

Ref.	Target	Ion$^+$	Energy (eV)	Principal result(s)[a]
125	InP(100)	Ar	500	Bonds stretch before breaking
128	Si(001)–1×1	Ar	50	Reconstructs to Si(001)–2×1
129	Pt(111)	Ar, He	1 K	Point defects, dislocation lines, adatom islands
133	Cu(100)	Ar	600	V, D, and adatom islands
144	Si(111) B-doped	Ne, Ar, Kr	1 K	V, I, point defects from Ar trapped at Cu contamination, and boron-bound excitons
145	GaAs(100)	Ar	1–4 K	Electron traps formed
146	GaAs	Ar	0.5–4 K	D, V, electronic interbands
139	Ge(001)	Ar, Xe	70–500	D, D-recombination
151	LiF	Ar	10 to 5 K	Anion vacancies; Li evaporates
136	Pt(111)	Ne, Ar, Xe	40 to 5 K	D, V, I, bulk vacancies, adatoms
152	SiO_2	Ar	300 & 1.2 K	Amorphous Si formed

[a]V = vacancies; I = interstitials; D = defects.

4. Physical Effects: Ion-Beam-Induced Topography

4.1. Introduction

In previous sections we discussed the fundamentals of ion beam interactions with solid samples, along with the structural changes that occur in the initial stages of sputtering. In this section we consider the physical effects of more prolonged ion beam bombardment as typically used for cleaning and compositional depth profiling of solid samples, with the major focus on ion-beam-induced topography. Chemical changes from ion beam bombardment are discussed in Sec. 5. Topography development has a profound effect on the depth resolution in sputter depth profiling, because of the nonuniform erosion of the surface. Topographical changes of solids are also of interest because of their effect on the physical and chemical properties of the surface (crystallinity, surface area, composition, reactivity, strength, friction, etc.). We found several review papers particularly helpful for preparing this section.[12,20–25]

A wide range of sputter-induced topographical features have been observed that appear to depend in a complex manner on the sample composition, pretreatment, condition, and temperature as well as on a variety of ion beam parameters. Significantly different topographical features are often reported in the literature for the same material. Unfortunately, general conclusions cannot be derived from papers published on this topic.[23] Because it is impractical in this work to present an exhaustive listing of *all* the features observed on *all* materials, we have chosen to describe the range of features and behavior seen, along with some plausible mechanisms for their development. For any material, we recommend that a literature search be performed to determine what is known about ion-beam-bombardment-induced topography for the ion beam(s) and material of interest.

Topographical features observed include etch pits and craters, pyramids, cones, whiskers, and a wide range of corrugation and periodic ridge-type features. Both random and periodic structures are seen as well as microscopic and macroscopic roughness. In thin film and multicomponent systems, clustering and islanding have also been observed. Any of the topographical features can develop even on initially smooth surfaces. For polycrystalline samples, differential sputtering rates for different crystal faces perturb the planarity of the surface to roughly 10% of the sputtered depth, so these effects are significant for depth profiling and many other applications.[163] While ion bombardment generally induces roughness (both microscopic and macroscopic) on a sample, under certain conditions smoothing or polishing is also observed.

Ion beam parameters that may influence topographical development include the ion species, its purity, energy, total fluence, current density, uniformity, and angle of incidence of the beam on the sample surface. The angle of incidence is particularly critical, as is evident in Fig. 4.1, which shows the micron scale topography development on Ar-ion-sputtered Cr–Ni multilayers for ion beam incidence angles from 0° to 80°.[164] Sample parameters that may have an influence include the chemical matrix, preparation method, crystalline orientation, initial and ion-beam-induced defect structure, initial topography and cleanliness, and homogeneity. In addition, time and temperature effects have been reported, with some systems showing healing of some ion-beam-induced damage with time[155] or annealing treatments or both.[21]

4.2. Mechanisms for Topography Development

We cannot provide a detailed theoretical treatment of topography development in this work because the subject deserves to be treated critically and broadly. In general, predictive theoretical models are complex and incomplete. Furthermore, the predictions often depend critically on sample parameters that are not generally well known, and may be dynamic, such as the surface binding energy and local sputter yield. As yet, the models cannot be used to predict the topography development for an arbitrarily chosen sample, experimental apparatus, and ion beam. However, some success in modeling specific features has been achieved and may provide valuable insight into the operative mechanisms. We provide a brief discus-

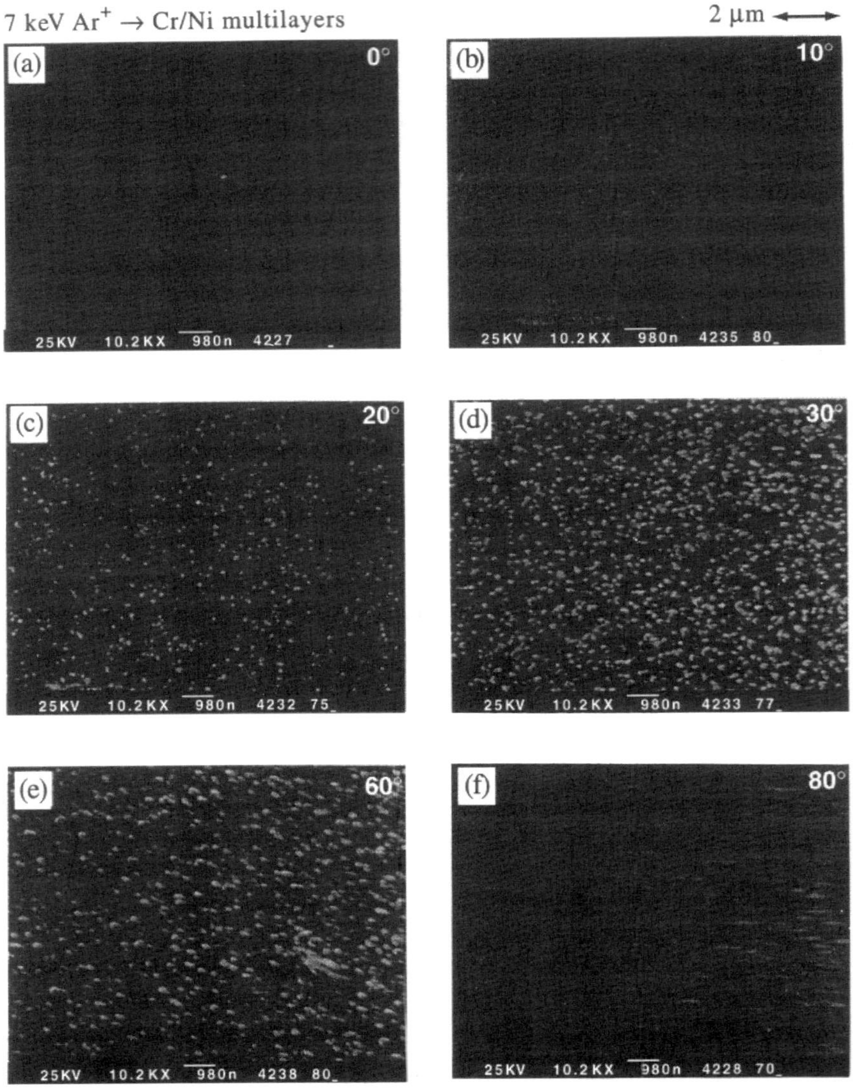

Figure 4.1. Ion-beam-induced topography development on Cr/Ni multilayers by 7-keV Ar ions at different incidence angles. Adapted and reprinted with permission from[164] D. W. Moon and K. J. Kim, *J. Vac. Sci. Technol., A* **14**(5), 2744–2756 (1996). Copyright 1996 American Vacuum Society.

sion of some proposed mechanisms, reference the theories that result in more specific predictions, and briefly discuss the theories in the following sections.

Both erosion and growth theories are invoked, sometimes in combination, to explain the development of ion-beam-induced topography. Differences in local sputter yields, and hence the variation in erosion rate across a surface, is the primary reason for sputtering-based topography development.[24,166] Different crystalline faces, phases, defects, and other inhomogeneities, including nonuniformities in the ion beam, can all affect the local sputter yield. Both lateral and in-depth spatial variations in the erosion rate mean that the surface is not uniformly eroded, thus leading to topography development, as illustrated schematically in Fig. 4.2a.[90] The darker regions represent areas with higher sputter yields; they erode faster and result in the projections of lower-sputter-yield material as shown. Furthermore, as discussed in Sec. 2, sputter yields are highly dependent on the angle of incidence of the ion beam as well as the local radius of curvature of the sample.[167] Thus, any initial sample roughness or ion-beam-induced microroughness that results in microplanes with different inclinations to the ion beam will propagate and tend to become more pronounced as a result of local differences in the *effective* ion beam incidence angle. Impurities on the sample surface, whether from surface contamination, deposition from the surroundings during sputtering, or diffusion from the bulk, can act as nuclei for topography development. This process is known as impurity seeding and is illustrated in Fig. 4.2b.[90] Impurity seeding will be discussed further in the section on sputter-cone formation. Bulk defects, which become exposed upon erosion of the overlaying surface, can also result in topog-

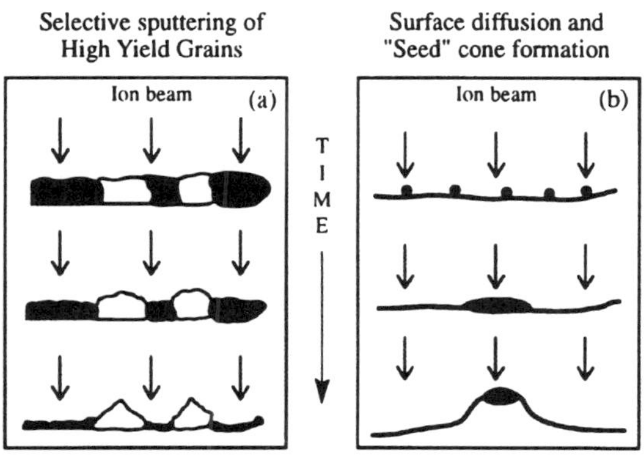

Figure 4.2. Ion-beam-induced topography caused by (a) selective sputtering of high yield areas (shown in grey) and (b) surface diffusion and subsequent seed cone formation. Adapted and reprinted with permission from[90] K. W. Pierson, et al., *Nucl. Instrum. Methods in Phys. Res., Sect. B* **108**, 290–299 (1996), and with kind permission of Elsevier Science - NL, Sara Burgerhartstraat 25, 1055 KV Amsterdam, The Netherlands.

raphy development because areas with high defect densities often have different sputter yields.

It may be surprising that growth theories are relevant to the erosion process, but many surfaces formed during ion beam etching or thin film deposition have some similar topographical features, and some aspects can be treated by the same theories.(20) Additionally, some true growth processes do occur along with ion beam erosion.(23) From Sec. 3.2, the initial stages of sputtering result in forming vacancy islands and adatoms on top of the surface. The diffusion of impurities (including defects) may be enhanced by interaction with the ion beam, and these impurities may nucleate to form islands (Fig. 4.2b).(90) Similarly, preferential sputtering of the more volatile component of a compound (Sec. 5) results in the surface enrichment of the lower-sputter-yield species, which may then migrate and coalesce and contribute to protuberance growth. Thus, migration and coalescence play a role in sputter-induced growth processes. These processes may be thermally or chemically driven, or both. A microregion model postulates the creation of subsurface microcrystallites within the amorphized zone (Sec. 3.4), which grow outward to the sample surface by some unknown growth mechanism to produce topographical changes.(24,168) Whisker growth and the growth of other features are also observed, although again the exact mechanisms are not known.(24,169) Redeposition of sputtered material from the substrate or the surroundings may also contribute to the growth of features, although this is not generally thought to be a dominating process in most cases.(24,170) Redeposition of ejected species from the substrate usually occurs on roughened surfaces with tall, steep features (Fig. 4.3),(90) but can also result from the electric field of the ion beam, which repels ejected positive ions back toward the surface.(171,172)

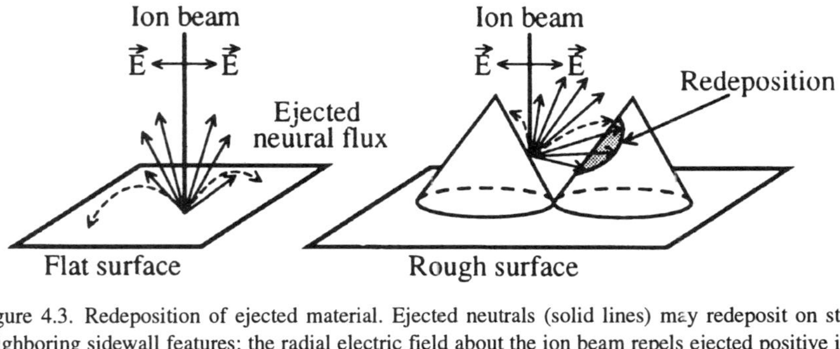

Figure 4.3. Redeposition of ejected material. Ejected neutrals (solid lines) may redeposit on steep neighboring sidewall features; the radial electric field about the ion beam repels ejected positive ions (dashed lines) towards the sample surface. Adapted and reprinted with permission from[90] K. W. Pierson, et al., *Nucl. Instrum. Methods in Phys. Res.*, *B* **108**, 290–299 (1996), and with kind permission of Elsevier Science - NL, Sara Burgerhartstraat 25, 1055 KV Amsterdam, The Netherlands.

Minimization of free energy is a driving force for topography formation. While thermodynamic considerations are important, kinetic processes may dominate, and both compete with the ballistic sputtering processes.[20] The topographical development during ion beam bombardment must be regarded as a balance between kinetically driven erosion processes, kinetically controlled atom transport processes, and thermodynamically driven growth processes.[21,23,167] Thus, when ion beam bombardment is terminated or the sample temperature is lowered, or both, the surface topography may be in a nonequilibrium metastable state, but one that cannot rearrange itself because of kinetic limitations.

In-depth reviews of various models and simulations of surface roughening and topographical development we found particularly helpful are Refs. 12, 20–24, and 173.

4.3. Microscopic and Macroscopic Roughness

Topography develops from microscopic (atomic) levels to macroscopic (micron and larger) scales during ion beam bombardment. Prior to the development of STM and AFM instruments, most experimental topographic determinations were done with SEM images, which could not generally resolve atomic scale features. Some TEM work has also been done, although usually in the replica mode. In one study comparing TEM and AFM images, more details are evident in the AFM images, even at similar magnifications.[174] Comparisons of SEM and AFM images at the same magnification give similar results,[175] but generally AFM can realize better atomic scale resolution. Recent STM and AFM work indicates that using these techniques provides valuable insight into the topography development at the atomistic level, as illustrated in Sec. 3.2, for the initial stages of ion beam bombardment and should aid in the evaluation and further development of theoretical models.

Microscopic topography evolution is inevitable (in the absence of diffusion) because of the fundamentally statistical nature of the sputtering process.[20,23 173] In Sec. 2, we discussed how ion beam bombardment can be considered as a sequence of binary collisions between individual incoming particles and atoms in the surface and near-surface regions of the sample. Within the sample irradiation area, individual incident ions randomly strike specific points on the surface; because there is a statistical variation in the atomic scale erosion from primary-ion sputtering, atoms are not removed in a layer-by-layer sequence. This randomness results in features approximately 0.1 to 1.0 nm. Toh *et al.*[176] calculated that the layer distribution density (i.e., the number of exposed atoms in each layer as a function of the number of atomic layers removed) conforms roughly to a Poisson distribution about the mean number of layers removed for a small total number of atoms (or clusters) sputtered and transforms to a Gaussian distribution with a large number of atoms sputtered (Fig. 4.4). For both distributions, the standard deviation about the mean increases as the square root of the number of atoms sputtered.[176]

The collision cascade, which results in emission via ballistic recoil events, is also generally statistical in nature because of the random-walk nature of the cascade. This process also produces random variations in the erosion rate on an even larger scale (~10 nm) than do primary-collision events.[20] The collision cascade may also smooth out some of the atomic scale topography by allowing for migration of underlying substrate atoms to the surface, especially for low-sputter-yield materials.[21] Macroscopic features (>100 nm) also develop from ion beam bombardment, but it is not known if there is an ordered evolution of features from microscopic to macroscopic size,[23] although some recent work does support such a progression.[148,177] In contrast, SEM images of the crater bottoms formed with 7-keV oxygen- and Ar-ion bombardment of Cr–Ni multilayers at normal incidence do not show development of visible features.[164] However, marked differences are evident on the AFM images with the smaller (height) scale (Fig. 4.5). For these Cr–Ni multilayers, oxygen-ion beam bombardment results in less roughness than Ar-ion beam bombardment if all other variables are the same.

The topography of sputtered multilayer structures has additional complications, particularly when the components of the layers have widely different erosion rates.[178] When the erosion rate of the substrate is much larger than that of the overlayer, it is expected that the height of the surface microtopography will be large, because rapid erosion of the substrate will occur while there are still significant

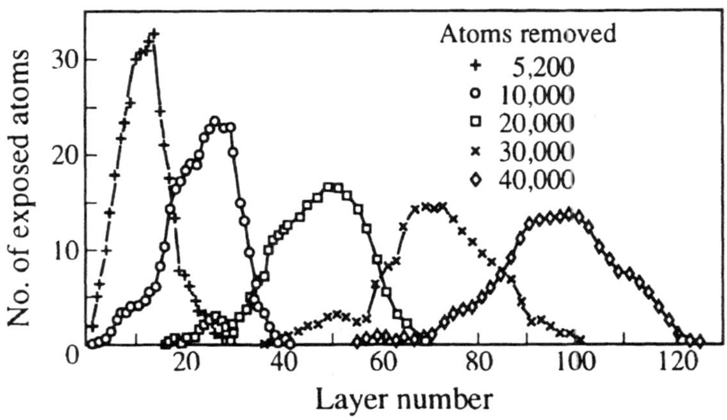

Figure 4.4. Number of atoms remaining in each exposed layer as a function of the layer number beneath the original surface, and for different total numbers of atoms removed. Adapted and reprinted with permission from[176] Y. S. Toh, et al., *Nucl. Instrum. Methods Phys. Res., Sect. B* **67**, 586–589 (1992), and with kind permission of Elsevier Science - NL, Sara Burgerhartstraat 25, 1055 KV Amsterdam, The Netherlands.

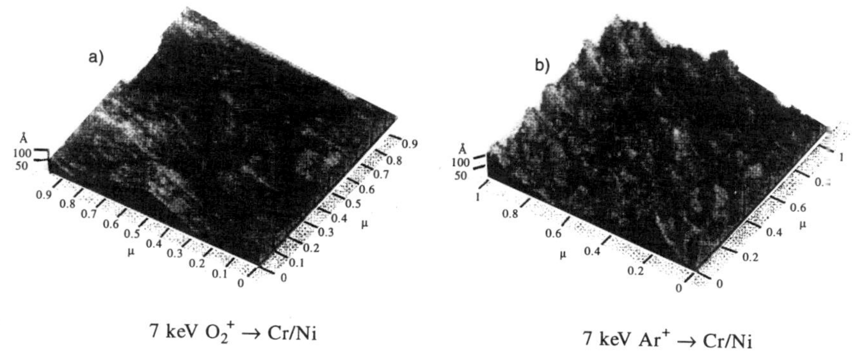

Figure 4.5. AFM images of crater bottoms of Ni/Cr multilayers bombarded with normally incident (a) oxygen and (b) argon ions, showing increased roughness for the argon-bombarded sample that is not evident in the SEM images recorded after oxygen-ion bombardment. Adapted and reprinted with permission from[164] D. W. Moon and K. J. Kim, *J. Vac. Sci. Technol., A* **14**(5), 2744–2756 (1996). Copyright 1996 American Vacuum Society.

amounts of overlayer atoms present. In the reverse situation, the feature height is expected to be smaller, because the overlayer will be removed relatively rapidly, thus allowing less time over which differential erosion rates may dominate.[(178)] These expectations agree with the experimental data in which ion bombardment of targets covered by low-erosion-rate films, such as some types of carbon contamination, results in the development of considerable topography, but substrates covered with films with much larger sputtering yields tend to develop less topography.[(178)] However, these expectations may be altered by the properties at the interface between the layers, i.e., the presence or absence of strong interactions, or bonding, between the components.

In the expression (Sec. 1) for the macroscopic erosion rate,[(4)] $dz/dt =$, $0.00624SJ^{+}A/\rho$ the erosion rate for a given material depends primarily on J (or $J \cos \theta$ for beams at other than normal incidence), the ion flux density, and on S, the local sputtering yield. Changing J or a change in S within the bombarded area results in variations in the erosion rate and, hence, topography. As discussed in Sec. 2, the sputtering yield varies slowly with incidence angle for θ near 0°, so the topography formed may be fairly stable (all other factors being uniform) for sputtering at near-normal ion beam incidence.[(20)] The sputtering yield may also depend on the crystal orientation, grain boundaries, and the presence of defects, including those produced as a result of the ion-beam-bombardment process, so complete *absence* of differential sputter yields across a surface area will be rare, even on a microscopic level. Locally differential erosion rates will also result from any localized amorphization of the solid (Sec. 3).[23] Lapujoulade has given a detailed treatment of the theory of roughening.[22]

Theoretical treatments indicate that the surface will eventually tend to erode to a configuration of minimum erosion rate (after all the high-yield areas are sputtered away), although high-erosion-rate crystal surfaces may develop initially.[20] Theoretical considerations also predict that circular cross-section conical structures are expected to form on amorphous surfaces, while crystallographically oriented pyramids, prisms, and etch pits should develop on surfaces that remain crystalline during ion beam bombardment. Under certain conditions, a randomly rough homogeneous surface is predicted to smooth out eventually with sputtering.[20] Often more than one level of roughness is formed by ion beam bombardment.[90,175,179,180] On multicomponent polycrystalline samples, a large-scale roughness is formed that roughly corresponds to the grain size in the material, but a smaller-scale roughness develops within each grain.[90,179] According to Carter,[175] the nanometer-scale roughness on Si does not increase in amplitude with increasing ion fluence; this result indicates that the purely statistical process may be moderated by short-range relaxation or diffusion processes. A steady increase in the roughness with depth will result from topography development governed only by the statistics of ion beam bombardment.[175]

4.4. Etch Pits

Etch pits are depressions that form during ion beam bombardment. Etch pits act as precursors for some of the other ion-beam-induced topography. A bulk impurity with an erosion rate greater than the surrounding matrix can produce local etch pits upon sputtering.[20] Dislocations are generally associated with lower atomic binding energies and are thus one source of etch pit formation.[20,23] Of course, dislocations may be present initially in the sample or may be sputter induced. Etch pits have been observed in the sputtering of polycrystalline Cu and on the (11 3 1) face of Cu.[20] In some cases, etch pits are produced in such abundance that they overlap and produce wavelike structures. An example of this type of evolution is illustrated in Fig. 4.6, which shows the effects of 7-keV Ar-ion beam bombardment on an initially smooth crystalline Si surface.[20] However, this type of topography did not form in other experiments on the same type of Si surface. In some cases, sputtering at elevated temperatures appears to minimize topography development, because the ion-beam-induced dislocations are annihilated by annealing before they can propagate into pits. However, for single-crystal Cu(11 3 1) faces, pit formation is independent of sample temperature from 77 K to 800 K.[20]

Another source of etch pits is secondary-sputtering effects resulting from the reflection of ions from the steep slopes of a surface onto other portions of the surface.[20,23,170] The ion reflection results in locally enhanced ion fluxes and erosion rates. The magnitude of this type of flux enhancement increases with

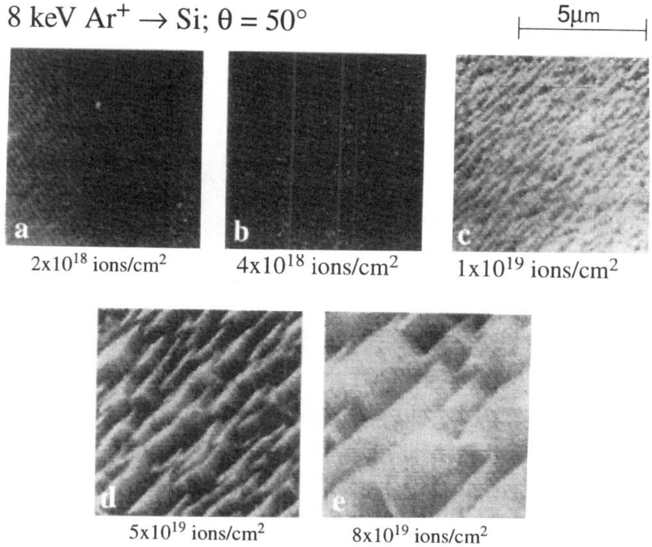

Figure 4.6. Etch pits on Si increase in density with increasing ion fluence. The etch pits eventually overlap and form a wave-like topography. Adapted and reprinted with permission from[20] G. Carter, et al., *Scanning Microscopy* **2**(3), 1293–1327 (1988), and with permission from Scanning Microscopy International.

increasing angle of incidence, decreasing ion energy, and decreasing ratio of the projectile-to-substrate masses[(20)]; in this case, rougher surfaces are formed using an ion beam at grazing incidence. Because the microplanes at different angles on the surface act as sources and sinks for redeposition, the process is highly dependent on the local topography.[(170)] Figure 4.7 shows how either enhanced erosion or deposition is obtained from sidewall sputtering, depending on the ion beam angle of incidence.[(23)]

Etch pits may also be formed from the coalescence of implanted atoms from the ion beam; the coalesced atoms form bubbles and then are breached by erosion of the surface.[(181,182)] Barna and Menyhard[(183)] reported that pits initially present in thin films generally spread laterally with sputtering and, with enough sputter time, eventually merge to give a smooth surface. However, the time required may be greater than that required to sputter away the layer.

On crystalline Cu substrates, the pit geometry and density are influenced by ion energy and incidence angle. They are not so much influenced by the ion species or sample temperature from 77 K to 800 K, although bombardment with Xe ions does cause an increase in the pit density.[(20)] Pits form in abundance on the fcc (11 3 1) face of Cu, Ni, Pd, Ag, Ir, and Au samples.[(20)] With continued sputtering,

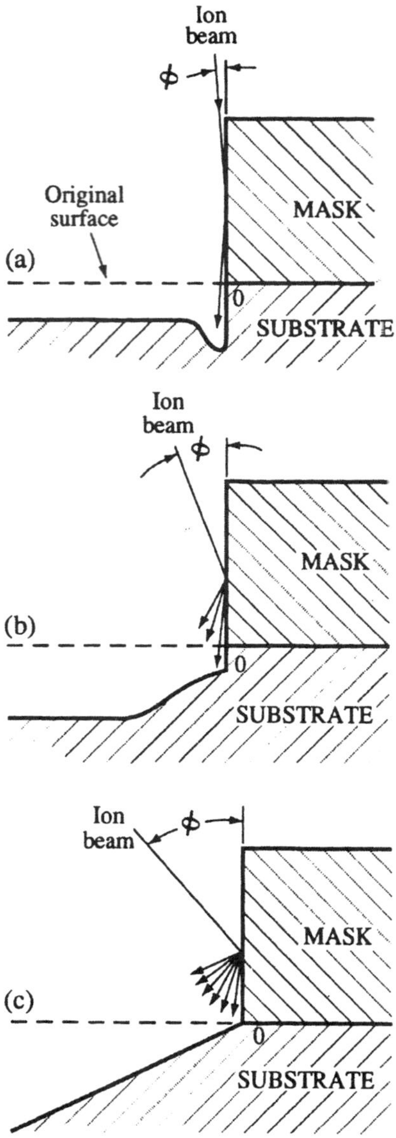

Figure 4.7. Schematic showing enhanced erosion (part a) and build-up of material (parts b and c) at the base of an ion-bombarded wall. Enhanced erosion (trenching) only occurs for grazing ion beam incidence along the sidewall. Adapted and reprinted with permission from[23] I. H. Wilson, in: *Materials Modification by High-Fluence Ion Beams,* edited by R. Kelly and M. Fernanda Da Silva (Kluwer Academic Publishers, Dordrecht, The Netherlands, 1989), pp. 421–466. Copyright 1989 Kluwer Academic Publishers.

the pit size and density increase until they overlap and produce regularly faceted structures or terraces that are often wavelike in appearance.(20,23) Pits have also been observed on amorphous elemental and compound semiconductor samples and on glass samples. Motohiro and Taga(184) examined the topography development on fused silica, borosilicate, and phosphosilicate glasses; they observed circular etch pits at normal incidence and crescent-shaped pits at a 22.5° incidence angle. At larger angles of incidence, microripples and bandlike structures developed, as indicated in the schematic of the pits that developed on fused silica (Fig. 4.8).

Theoretical calculations indicate that the standard deviation of the surface roughness increases with increasing cluster or etch pit size, and hence the depth resolution degrades when etch pits develop.(176) Because etch pits erode faster than the mean surface, an apparent outward shift of a depth-profile interface as well as broadening of the composition profiles will result when etch pits develop.(185) Experimentally, etch pits on Cu increase in density and size with increasing ion fluence.(163) Figure 4.9 shows the development of pits with increasing ion fluence,(163) and Fig. 4.10 shows the near linear dependence of pit width (in micrometers!) on the total ion dose.[163,185] The doses in these figures are 2×10^{19} ions/cm^2 and greater. For doses of 10^{17} ions/cm^2, it is estimated that the fraction of pitted surface area will only be between 10^{-4}% and 10^{-1}%.(163) This result suggests that

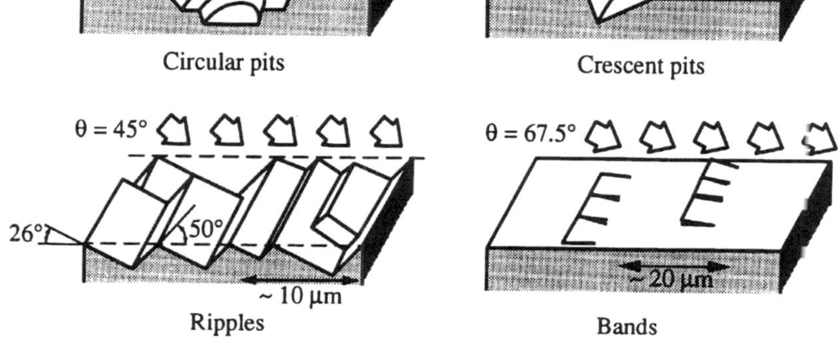

Figure 4.8. Schematic showing the change in pit shapes formed on fused silica as a function of ion-beam incidence angle. Adapted and reprinted with permission from[184] T. Motohiro and Y. Taga, *Thin Solid Films* **147**, 153–165 (1987), and with kind permission of Elsevier Science - NL, Sara Burgerhartstraat 25, 1055 KV Amsterdam, The Netherlands.

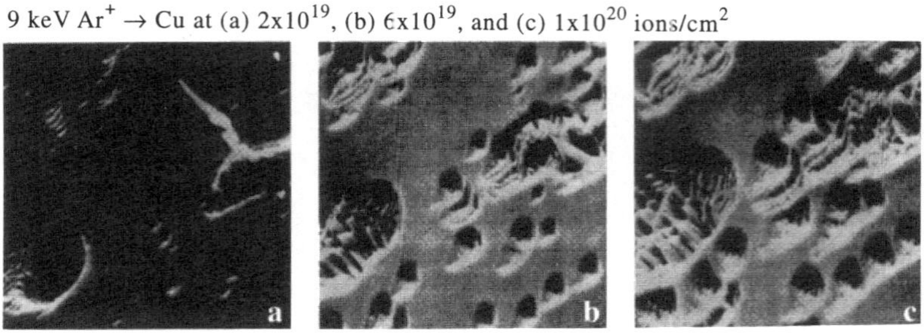

Figure 4.9. Pit development on Cu (11 3 1) as a function of increasing ion fluence. Adapted and reprinted with permission from[163] G. Carter, et al., *Mater. Sci. Eng.* **90**, 21–32 (1987), and with permission from Elsevier Science.

ion beam bombardment with low-energy ions might minimize large etch pit formation and reduce the degradation in depth resolution. Because ion beam bombardment at large angles of incidence i.e., grazing angles of incidence, increases the size of the pits along with the projected size of the ion beam on the surface, normally incident ion beam bombardment may also reduce the overall pit size, particularly in amorphous or amorphizable materials. However, grazing-incidence ion beam bombardment may result in changes in the erosion rates of the pits and the mean surface; the pits are eliminated if the surface erodes faster than the pits.[163,185] The azimuthal angle used for ion beam bombardment also has considerable influence for crystalline samples, and it is thought that ion incidence angles nearly parallel to pit-initiating dislocations will amplify the pits.[163] Pits formed on the (11 3 1) face of copper can be eliminated by changing the azimuthal

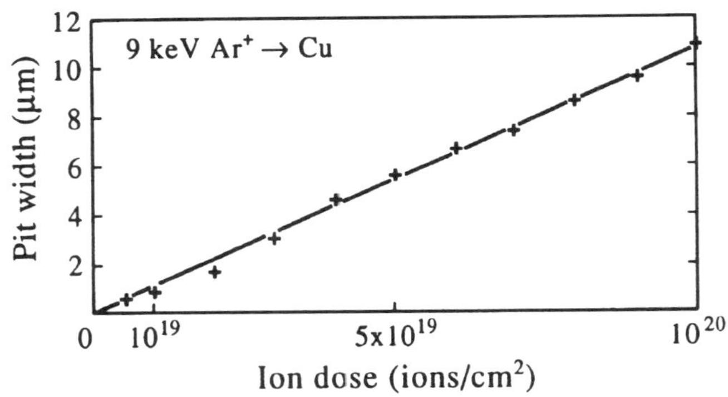

Figure 4.10. Increase in pit width on Cu (11 3 1) as a function of total ion dose. Adapted and reprinted with permission from[163] G. Carter, et al., *Mater. Sci. Eng.* **90**, 21–32 (1987), and with permission from Elsevier Science.

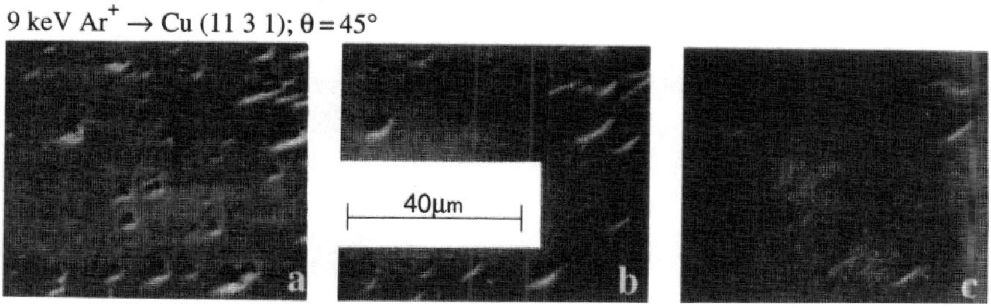

Figure 4.11. Smoothing of ion-bombardment-induced pits in Cu (11 3 1) by varying the ion beam azimuthal angle. The initial azimuthal angle in part (a) was arbitrarily chosen, and was rotated 90° for parts (b) and (c). Adapted and reprinted with permission from[20] G. Carter, et al., *Scanning Microscopy* **2**(3), 1293–1327 (1988), and with permission from Scanning Microscopy International.

incidence angle of the ion beam, as shown in Fig. 4.11.[(20)] The authors' explanation is that the ion beam was initially aligned close to parallel to the dislocation core that initiated the pits, thus enhancing pit formation. The change in azimuthal angle removed the alignment and smoothed the pits.

4.5. Pyramids

Pyramids differ from cones because pyramids are faceted, have specific crystallographic identity, and the apices of the pyramids do not extend above the plane of the surface. Pyramids are found embedded in pit structures, at the intersection of two or more pits, or at elevated grain boundary–pit intersections, and sometimes on the sidewalls of more elevated grains.[(21,23,63)] Pyramids occur in isolation and in crystallographic formations on a substrate.[(20,23)] They have not been observed on amorphous or amorphizable substrates, and it appears that some sort of defect population, either intrinsic or ion beam induced, is a necessary precursor for their development.[(20)]

Pyramid formation is not associated with contamination or phase separation (e.g., seed cones, Fig. 4.2b), and they are not thought to be associated with a growth-assisted process; instead, they appear to be associated with high-erosion-rate orientations.[(20)] The pyramidal habit is related to the matrix crystallography, though it is not known precisely how. Pyramids form selectively in pits with specific geometries and sometimes only for specific crystalline orientations of the surface. Pyramid apices often are found to erode at the same rate as the mean surface, although their height increases to 1 μm as the pit deepens. Continued evolution of pyramids into overlapping structures often results in forming a wavelike corrugation of the surface.[(20)] After extensive ion beam bombardment, the pyramids may start to erode and become smaller. Although not a subject of our review, chemically reactive ion sputtering has been shown to suppress the formation of some pyramidal structures, even though etch pits may still form,[163] and to obliterate existing structures formed by inert-gas-ion beam bombardment.[20,21] Pyramid structures, which appear to be an example of the emergence of preferred planes corresponding to sputtering-yield extrema, are the result of a variation of the sputtering yield with incidence-ion beam direction and crystallographic orientation.[21]

Pyramids have been observed on fcc (11 3 1) single crystals of Cu, Pd, Ir, Ag, Ni, and Au at doses of $>10^{19}$ ions/cm^2 and energies of 20 to 40 keV using Ar, Kr, and Xe ions.[186] On copper samples, pyramids were also formed using Cu-ion sputtering, thus indicating that ballistic processes are important.[186] Sputtering Cu with chemically active Fe, Ni, or Br ions produced no pyramids.[186] Argon-ion bombardment of a (11 3 1) oriented Ni–8%Cu alloy only produced etch pits at current densities of 60 μA/cm^2 and total doses of 1×10^{19} ions/cm^2, but some pyramids developed when the current density was increased to 150 μA/cm^2.[186] Pyramids were extensively formed at a current density of 60 μA/cm^2 when the total dose reached 2×10^{19} ions/cm^2.[186] These dose rates are lower than those necessary for pyramid formation on pure Cu or Ni. No sputter-induced segregation was evident in the alloy.

Both etch pit and pyramid formation depend on ion beam species and energy, substrate material and orientation, and ion fluence. As expected from the discussion about forming etch pits that serve as a precursor to pyramid development, topography development (at least on fcc metals) is expected to be minimal for fluences of 10^{17} ions/cm^2 or less.[163]

4.6. Cones

Cones rising above the mean surface plane are a commonly observed feature of ion-beam-bombarded surfaces. Cones may be isolated, exist in clusters, or be densely spread over the entire sputtered area. They may develop on certain crystal planes, but be absent from other parts of the surface. Cones can reach heights in excess of 1 μm. Cone formation is often viewed as a primary product of using ion beams to erode surfaces and to result from localized irregularities, impurities, defects, etc. The formation of cones is often explained in terms of variations in the sputter yield with ion incidence angle that permits stable inclines to form. However, it now appears that surface diffusion, ion reflection, and perhaps redeposition may also be contributing factors.[23,24,187,188]

One known mechanism for cone formation on metal surfaces is the initiation of growth by impurity seeding. An example of cones formed from impurity seeding is shown for InP in Fig. 4.12.[24] Impurity atoms having either a lower sputter yield, or a higher melting point than the surface constituents, act as nucleation sites for cone formation.[169] According to Wehner, sufficient amounts of low-melting-point materials can agglomerate and nucleate the growth of cones.[169] The rate of surface

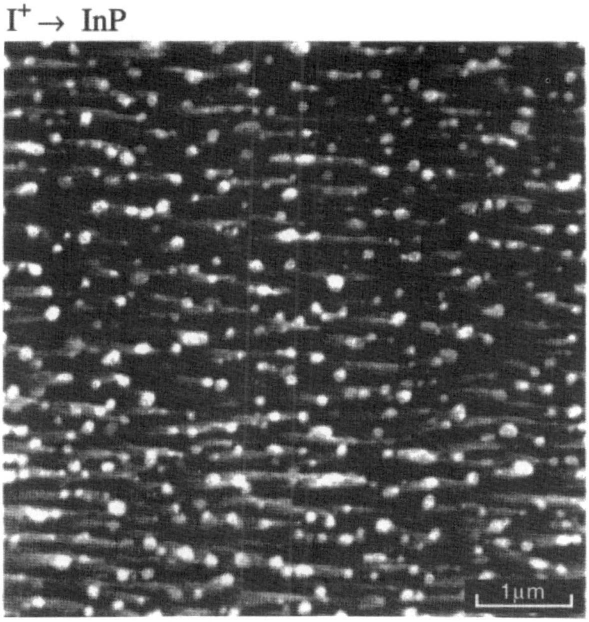

Figure 4.12. Seed cones on InP originating from impurities redeposited from the sample holder. Adapted and reprinted with permission from[168] W. H. Gries, *Surf. Interface Anal.* **14**, 611–618 (1989). Copyright 1989 by John Wiley, Ltd.

diffusion of the impurities into clusters may also be important, along with the rate of surface movement of the substrate atoms.[(24)] The impurity material shields the surface beneath it, and the surrounding material sputters away faster, thus leaving a surface protrusion beneath the seed material that acts as a starting point for the cone. The angular dependence of the sputter yield results in the formation of conical shapes. Once the seed material is eroded, the size of the cone will decay with prolonged sputtering, although this may take considerable time.[(20,24,185)] Rotating or rocking the sample can minimize the amplitude of protuberant features by undercutting.[(185)] Chini *et al.*[(189)] report that, despite deliberate seeding of Si surfaces, only very few cones were observed after sputtering with 30-keV Ar ions. These cones did not form because of impurities—they were also observed on the unseeded surface.

Cones may also form during ion beam bombardment in the absence of impurities, and often form preferentially on specific high-index faces of crystalline materials, such as the (11 3 1) face of Cu and other fcc metals. Cones from impurity seeding are formed on InP, but cones are also formed on pure InP, and it is hypothesized that they result from ion-beam-bombardment-induced recrystallization of microregions at the amorphous–crystalline interface; the microregions grow in a columnar fashion to the surface.[24,190] The surface of the sample consists of both amorphous and recrystallized regions, and the cones consist of crystalline InP that grow epitaxially on top of other crystalline InP regions.[190] Cones appear on InP at doses of approximately 3×10^{15} ions/cm^2 and grow with increasing dose (10^{15}–10^{16} ions/cm^2) to 10 to 100 nm. For beam energies below 10 keV, there is little influence of beam energy on roughness, although the size and density of the cones depend on the angle of incidence of the ions with the maximum topography development reported for incidence angles around 41°.[191] The cone formation is suppressed upon cooling the samples to 90 K, which indicates that erosion alone is not responsible for the cone formation, but that some other growth process is also involved. At angles greater than 45° for 0.5-keV Ar ions and 21° for 0.5-keV Kr ions, Demenet *et al.*[191] found that the cones merge and form ripples. The authors also found a strong correlation between the rms roughness and the experimental erosion rate as a function of incidence angle.[191] They suggest that, for incidence angles up to that for the maximum erosion rate, the topography development (primarily cones) results from the collision cascade process, which shifts closer to the surface layer up to the critical angle of incidence. Beyond this critical angle, direct knock-on collisions become increasingly important, and the topography changes to the ripple structure, the sputter yield drops, and the rms surface roughness decreases somewhat. At extremely glancing angles, most of the ions are backscattered, and very small cones develop—the surface becomes smoother at this point.[191]

Another group[192] found no cone formation on InP when sputtering at room temperature with 5-keV Ar ions at 30°; instead, the surface was covered with ~30 nm droplet-type features that persisted to eroded depths of 1.5 μm. They did observe dramatic cone formation when sputtering at 100°C. Although this was not experimentally verified, the group thinks this might be the result of surface enrichment of In, diffusion of In to form islands on the surface, and the islands acting as a seed for cone formation. Below 100°C, the In diffusion rate is presumed to be too slow to form islands.[192] Chini *et al.*[193] sputtered single-crystal InP at normal incidence to a total dose of 8×10^{18} ions/cm^2 and found distinct differences in the topography from the center of the sputtered region to the periphery, as illustrated in Fig. 4.13. They found isolated cones embedded in pits in the center region, more cones and pits (smaller pits) as they moved outward from the center, and dense spherically shaped granules of In 30–100 nm in size in the outer periphery. HRTEM results for the cones shown in Fig. 4.14 indicate that the cones consist of single-crystalline InP in the interior, polycrystalline InP on the outer surface, and In(101) at the tip. Some small In granules were also detected on the cone sidewalls and at the surface material surrounding the cones. The lack of cones in the periphery region, which was covered with In granules, indicates that the In granules do not act as seeds for cone formation.[193] These authors also postulate

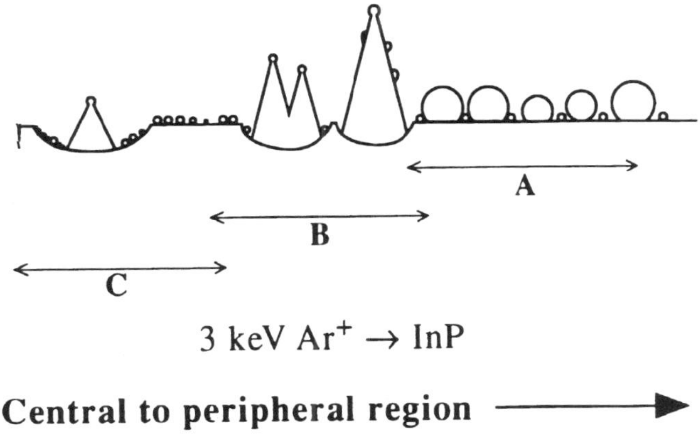

Figure 4.13. Schematic of cone, pit, and droplet features formed on InP from the periphery (A) to the center (C) of the ion-beam bombarded region. Adapted and reprinted with permission from[193] T. K. Chini, et al., *Nucl. Instrum. Methods Phys. Res., Sect. B* **95**, 313–318 (1995), and with kind permission of Elsevier Science - NL, Sara Burgerhartstraat 25, 1055 KV Amsterdam, The Netherlands.

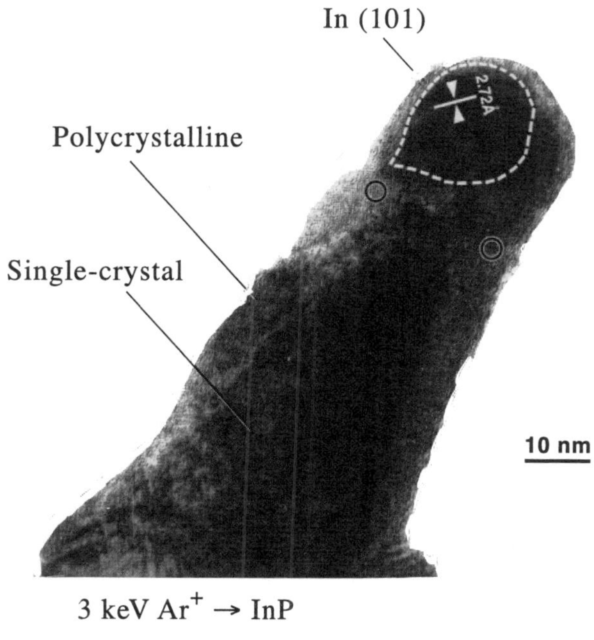

Figure 4.14. High-resolution TEM image of a cone formed on InP with ion-beam bombardment showing variations in crystalline phase and composition along the cone. Adapted and reprinted with permission from[193] T. K. Chini, et al., *Nucl. Instrum. Methods Phys. Res., Sect. B* **95**, 313–318 (1995), and with kind permission of Elsevier Science - NL, Sara Burgerhartstraat 25, 1055 KV Amsterdam, The Netherlands.

that enhanced mobility and diffusion of In, resulting from the local rise in temperature during sputtering, plays a vital role in the formations observed.[193]

Microcones have also been observed on polycrystalline thin films of Ag on Si after ion beam bombardment with 3-keV Ar ions incident at 60°.[187] Figure 4.15 shows high-resolution SEM images of the crater bottom after eroding to various depths. Before sputtering, the surface was nearly flat with some texture apparent. The grains appear to grow and roughen in the initial stages of sputtering and then develop into rod-shaped projections (Figs. 4.15c and d).[187] After sputtering to approximately 300 nm in depth, the rods were 50–150 nm in diameter and up to 300 nm high. They did not grow further in height, but transformed into cone shapes with rounded tips, as shown schematically in Fig. 4.15 below the images. The rod formation appeared to be suppressed by cooling the sample while sputtering, indicating that mobility of surface atoms is a factor in this growth.[187]

There is conflicting evidence as to whether sputtering with oxygen or halogen ions results in suppression of cone growth, although evidence suggests that the growth depends on the chemistry of the sample surface as well as on the defect structure.[24] At interfaces between dissimilar materials, cone growth may be observed because of redistribution and segregation of material across the interface, even for very pure materials.[185]

Micrometer-sized cones form on both Be-rich and Cu-rich surfaces of a 2% Be–Cu alloy sputtered in a carbon ambient. The carbon originated from a graphite backing sheet beneath the sample.[188] The dimensions of the cones depend strongly on Ar-ion energy (0.25–1.5 keV), total Ar-ion dose, and the magnitude of the ion flux used to reach the same total dose. The temperature during sputtering, previous alloy heat treatments, and the presence of Mo or C in the background ambient have less influence on the texture formed than the ion beam parameters. Small cones form initially, and these appear to mature into larger cones with increased ion dose, as evidenced by the images in Fig 4.16.[188] Cones formed at a current density of 0.5 mA/cm^2 or above can be erased if they are bombarded with ions at current densities of 0.1 mA/cm^2 or less; thus, there is a lower limit on the current density or ion beam flux required to maintain cone growth. Some ion erosion theories predict that the sidewall angle of stable cones is related to the angle of the maximum ion erosion yield, but this would not result in a lower ion beam flux bound for cone stability. Panitz[188] speculates that surface transport may also be involved, because larger ion current densities will increase the near-surface temperatures and, hence, the surface mobility.

Another characteristic of cone topography is a trench or groove at the base, which results from enhanced sputtering at the base of the cone from ions back-reflected from the cone walls.[20,24] Alternatively, material may be redeposited from the steep cone walls to the neighboring substrate, depending on the local ion incidence angle, as illustrated in Fig. 4.7. These phenomena were discussed in Sec. 4.4. Eventually, the cone size should decrease because the downward erosion of the

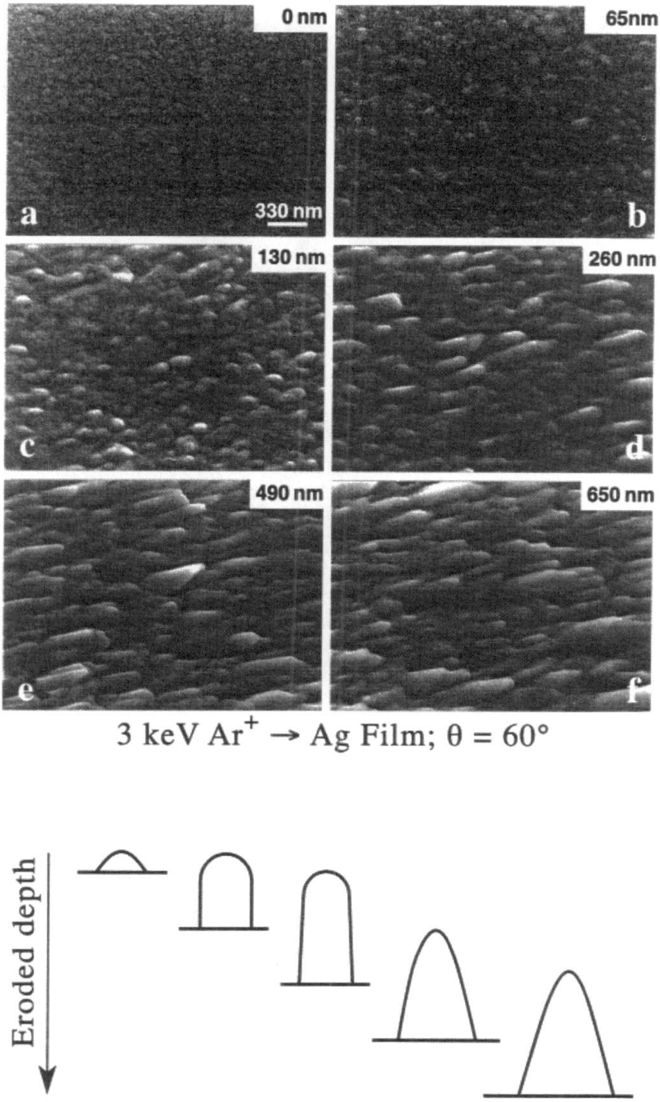

Figure 4.15. High-resolution SEM images of cones formed on the crater bottom of Ag films after ion-beam bombardment to depths from 0 nm to 650 nm. The schematic illustrates evolution of cone shape as a function of eroded depth. Adapted and reprinted with permission from[187] M. Tanemura and F. Okuyama, *Nucl. Instrum. Methods Phys. Res., Sect. B* **47**, 126–132 (1990), and with kind permission of Elsevier Science - NL, Sara Burgerhartstraat 25, 1055 KV Amsterdam, The Netherlands.

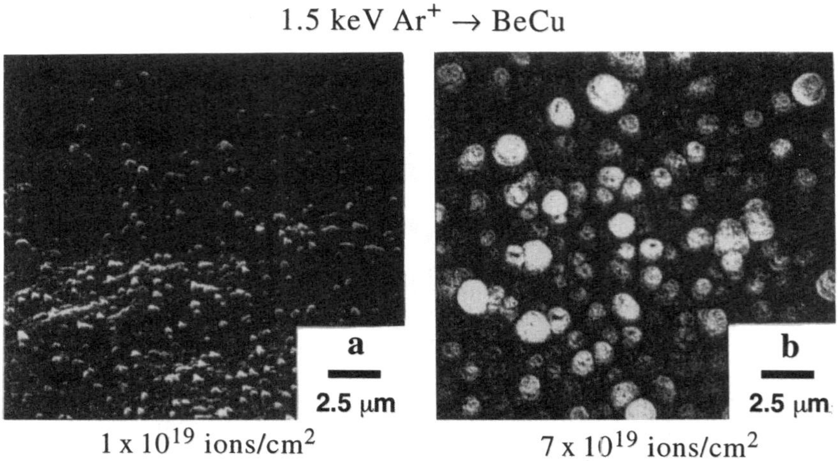

Figure 4.16. Increase in the size of cones formed on 2% BeCu bombarded with 1.5-keV Ar in a C ambient as a function of ion dose. Adapted and reprinted with permission from[188] J. K. G. Panitz, *J. Micromech. Microeng.* **1**, 52–59 (1991), IOP Publishing Limited.

steeply angled cone sides will be faster than the surroundings, and with enough sputtering the cone should disappear, leaving a shallow pit.[(20)] Evidence for this process is shown in Fig. 4.17, where the deepening of the trench pit at the base of a cone, along with erosion of the cone, is evident with prolonged sputtering. In some cases, a lip forms in the vicinity of the base of the cone or sometimes along the top of the trench. Lip formation is indicative of redeposition and is, as with trench formation, more likely to develop in the vicinity of steep-sided features.[(20)] Ion reflection may also occur from outside the sample; for example, ion beams may reflect from the sample holder or surroundings and affect sample topography.

Ion beam parameters, especially the ion beam incidence angle, can have a large effect on cone development. For cones or any protuberant structure, there is

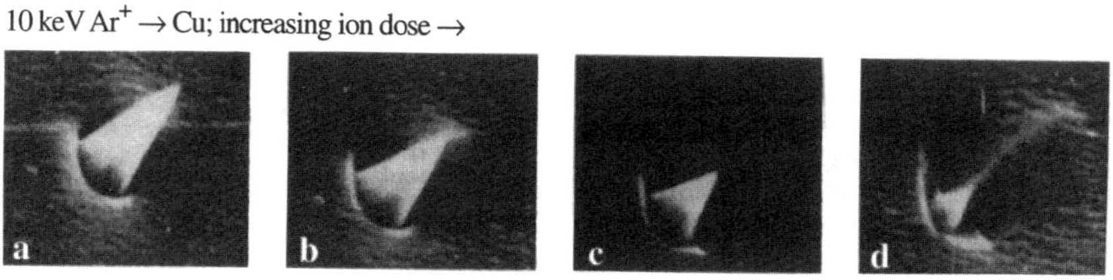

Figure 4.17. Erosion of cone and evolution of the surrounding etch pit on Cu with increasing Ar ion fluence. Adapted and reprinted with permission from[21] G. Carter, et al., *Defect and Diffusion Forum* **57–58**, 97–126 (1988).

evidence that near-normal ion incidence minimizes topography development, and ion beam bombardment at ~1 keV and oblique angles with a light-ion species (e.g., He) is the worst case for developing topography. Rocking or rotating the sample may help minimize topography development.(20) Further discussion of the effects of instrumental parameters, including sample rotation, is given in Sec. 6.

4.7. Whiskers

Whiskers, or filamentary growth, are sometimes observed on metal surfaces after ion beam bombardment. Whiskers that protrude above the surface are thought to be primarily a growth feature, and some are extremely resistant to sputter erosion. They appear to point in all directions; i.e., they have no preferred alignment with either the substrate or ion beam bombardment direction.(169) Whisker-like features have been observed on Be, Mg, and graphite, as well as InSb.(23) Some whiskers have been formed because of seeding, and whisker growth may be enhanced at ion beam energies close to the threshold for sputtering (Sec. 2, <100 eV).(23,169) Whiskers are converted into seeds for cone formation with higher-energy ion beam bombardment, but some controversy exists as to whether this is a general mechanism for cone development in metals or whether it occurs only in some circumstances.(23,169) Whiskers may also be associated with the formation of a second phase because of contaminants.(23) On InP, the whisker features are associated with In-rich regions that develop during ion beam bombardment.(23) Whisker growth has also been observed on InP seeded with Ni during sputtering. The whiskers were about 10 nm in diameter and consisted of polycrystalline InP in the shank area, but with an amorphous tip area containing large amounts of Ni.(194)

4.8. Ripples and Corrugation

Ripples are periodic, wavelike features that develop on the sample surface, sometimes only on surfaces with selected orientations.(21,24) Ripples typically have wavelengths on the order of 0.1 to 1 μm, although shorter wavelengths have also been reported. These wavelengths are up to two orders of magnitude larger than the surface component of the ion range.(195) The periodicity is at least an order of magnitude larger than the typical dimensions for the collision cascade.(20) The formation of ripples can result in significant changes in the erosion rate and secondary-ion yield.(196) One example of such ripples formed on GaAs is shown in Fig. 4.18.(24) At near-normal ion beam incidence, the ripples are aligned perpendicular to the incidence direction of the ion beam. However, near grazing incidence, the ripples are parallel to the ion beam direction.(21,24) Ripple formation was accompanied by an abrupt change in the sputter rate; the wavelength of ripples formed on GaAs were also found to depend on the substrate temperature, ion dose, energy, current density, and angle of incidence. The AFM image in Fig. 4.19 shows the ripples formed by 20-keV Xe-ion bombardment of Si to a total dose of 3×10^{18} ions/cm^2.[175] The ripples have an average peak-to-peak height of 16 nm and a wavelength of 440 nm. Some ripples form from the eventual overlapping of high-density etch pits formed on the sample surface during ion beam bombardment. Other ripple features may originate from a sputtering-induced ordered dislocation array. Ion-beam-bombardment-induced surface deformation has been proposed to contribute to the ripple formation.[(24)] The interaction between the eroded surface and redeposited species, along with sample and sputtering parameters, may also contribute to the development of wavelike features.[(20)] More rapid erosion of valleys than the upper part of hills on the surface is expected from local radius-of-curvature-dependent sputtering yields.[(177,182,195)] However, smoothing processes such as diffusion, atomic reordering, and viscous flow may counteract the roughening.[(182,196)] Other potential factors include reflection and shadowing effects, variations in the sputter yield along the surface, and collision cascade effects.[(197)] Bradley and Harper[(195)] developed a model for ripple formation in amorphous solids, which predicts that ripples should be perpendicular to the ion beam direction for incidence angles close to normal, but should be parallel for grazing incidence. Their theory is based on surface diffusion and the Sigmund

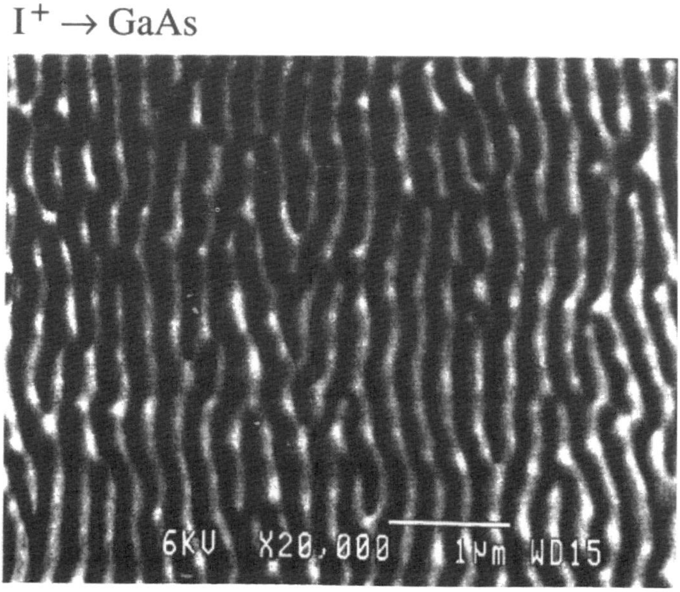

Figure 4.18. Ripples on GaAs formed by oxygen- or cesium-ion bombardment. Adapted and reprinted with permission from W. H. Gries, in: *Proceedings SIMS VIII*, edited by A. Benninghoven, K. T. F. Janssen, J. Tümpner, and W. H. Werner (John Wiley & Sons, Chichester 1991), p. 323. Copyright 1991 John Wiley and Sons.

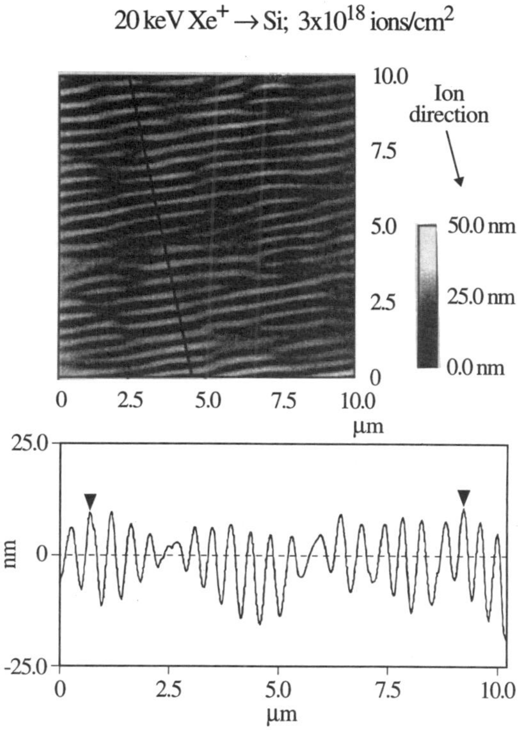

Figure 4.19. AFM image of ripples formed on Si by 20 keV Xe ion bombardment. The ripple height scale ranges from 0 to 50 nm from the dark areas to the white areas, respectively. Adapted and reprinted with permission from[175] V. Vishnyakov, et al., *Vacuum* **46**(7), 637–643 (1995), and with permission from Elsevier Science.

theory for variations in sputter yields induced by changes in the surface curvature.[(195)] They developed an equation for the wavelength variations at high temperatures as a function of temperature, ion beam bombardment flux, and the activation energy for surface self-diffusion.[(195)] In another model, ripple formation is attributed to ion-beam-induced defects that migrate to form a regular dislocation array that has an enhanced sputter yield along the defect array.[(20)] In yet another model, the evolution of a step into a ridge and furrow, which propagates into additional ridges and furrows, is considered.[(197)] In some cases, ripple formation has been observed with self-ion-beam bombardment (i.e., on Cu sputtered with Cu ions),[(21)] so inert-gas implantation is not required to initiate ripples. Other models and simulations are discussed by Vajo *et al.*[(196)] and Carter and Nobes.[(177)]

On InP bombarded with 0.5-keV Ar ions at 71° and with 5.0-keV Ar ions at 41°, cones were formed at doses in excess of 10^{15} ions/cm^2 and were transformed into ridges at higher doses. For an incidence angle of 71°, the ripples were discernible after a dose of 5×10^{15} ions/cm^2, but at 41° the dose required for ripple formation was 2×10^{18} ions/cm^2.(174) However, the surface roughness parameters were larger for the lower energy and lower incidence angle.

The InP samples, which were discussed under cone growth, develop ripples with Ar-ion sputtering after initial formation of the cones. Ripples were not formed from ion beam bombardment at –50°C; this result may indicate that surface diffusion is required for cone formation.(24) GaAs samples also develop ripples when sputtered with oxygen or cesium but without cones as precursors.(24) One group(192) found that ripples did not form on GaAs at 100°C or below –30°C, and proposed that at 100°C GaAs undergoes a phase change that affects the rate of surface diffusion and that the lower diffusion rate at –30°C suppresses ripple formation.

In some cases, the ripples are either found to be, or evolve into, faceted structures. The origin of these features is unknown, but they may result from some underlying dislocation network or from differential thermal or mechanical stresses induced in the sample surface.(20,167) Implantation of the ions results in a trapped gas and may also contribute to ripple formation.(182) It is not known whether any of the faceted structures (pits, pyramids, ripples) are primarily a result of orientation-dependent sputtering processes or atomic migration processes that result in surface energy minimization, or both,(20) or even whether a common mechanism is present in all cases.

Vishnyakov and Carter(181) found the ripple development on Si depended strongly on ion species, energy, fluence, incident angle, and substrate temperature. Their results are difficult to explain theoretically. No major topography developed at room temperature when using Ne ions at between 5 and 40 keV at a 45° incidence angle, or Ar ions between 5 and 10 keV, and for fluences less than 5×10^{18} ions/cm^2.(181) At higher fluences, a few isolated pit structures were observed. At 20 keV, Ar ions produced ripples and etch pits, thus indicating a threshold of somewhere between 10 and 20 keV for Ar-ion ripple formation. This threshold was also observed using Xe ions(175) and cannot be explained with current theories.(181) Furthermore, after prolonged ion beam bombardment with 20-keV Ar ions at high ion flux densities, the wave structure was replaced with a low density of faceted etch pits. Xenon-ion sputtering resulted in extensive ripple production, but of a different character than that observed using Ar ions. Cooling the substrate to 203 K and 123 K allowed for ripple production using Ne ions and resulted in more rapid topography development at lower fluences for all ion species. This result might be explained by a competition between statistical roughening and smoothing by atomic migration; migration is suppressed at lower temperatures.(181) In another study using 7-keV Ar ions incident at 45° and fluences in excess of 10^{19} ions/cm^2

to a total erosion depth greater than 1 μm, faceted ripples were observed, and the facet dimensions increased linearly with eroded depth. For incident beam angles between 0° and 55°, the ripples are transverse to the ion flux direction, but for larger incidence angles the ripples are oriented parallel to the incident beam. At an ion beam incidence angle of 30°, only isolated pits were formed after fluences of 8.8×10^{19} ions/cm^2.[185] For large eroded depths, ion beam bombardment at normal incidence is preferable for obtaining the best depth resolution.[198]

Ion beam bombardment of Si substrates with Si ions did not result in ripple development; instead, only hillocks a few monolayers high and an occasional pit were formed after sputtering to depths of 10^4 atomic layers.[175] The Si surface was completely amorphized to depths of tens of nanometers (see Sec. 3.4) at doses that were three orders of magnitude less than that where ripples were observed; this amorphization depth is similar to that of the ripple depth. Also, the accumulation of implanted ions is expected to saturate at fluences of 10^{17} ions/cm^2. In all these studies on Si, the doses at which ripples were noticed are in excess of 10^{17} ions/cm^2.

In other studies, topographically featureless surfaces are obtained for Si samples sputtered with 1.5-keV Ar ions at a 40° incidence angle to depths of 4.8 μm and a fluence of 1×10^{19} ions/cm^2 and for 7-keV Ar sputtering at a 45° incidence angle to depths of 46 μm and a fluence of 6×10^{19} ions/cm^2.[196] Although the mechanism for ripple formation is still unclear, Vishnyakov *et al.*[175] suggest using light (Ne) ions or lower-energy Ar ions (10 keV or less) for sputtering Si to help prevent ripple formation and subsequent faceting.[181,182] Ion beam bombardment using Ne or Ar eliminates ripples formed from Xe sputtering.[181] In another study,[196] Ar sputtering also destroyed the ripples resulting from oxygen ion beam bombardment of Si.

4.9. Sputter-Induced Recrystallization

Recrystallization was mentioned in the discussion of cone development on InP. Evidence for such recrystallization is also present for GaAs samples. The recrystallization is thought to be localized in microregions at the amorphous–crystalline interface. Recrystallization proceeds from this interface to the surface and is thereby a mechanism for growth.[24]

4.10. Coalescence to Form Islands

Multilayer systems show an interplay between mixing of layers and coalescence upon ion beam bombardment, driven by ballistic processes and ion-beam-enhanced diffusion.[20] Segregation at the surface of the slower-sputtering component can initiate contaminant-like processes for topography development. In some cases, changing the bombarding species or residual-gas environment results in the formation of a stabilizing layer that suppresses topography formation. In the case of thin metal films, those metals that intermix coherently with the substrate under appropriate thermal treatment also do so under ion beam bombardment. Those metals that coalesce with thermal treatment also do so under ion beam bombardment. The similar behavior is predictable from the free energy change in the reaction. If the change in the free energy of mixing is negative, thermodynamics favors mixed-phase formation. If the free energy change is positive, the formation of islands is favored. These effects have been observed for metal films on SiO_2, Au films on Si substrates, and in W–Cu bilayer systems.(20)

4.11. Swelling

As described in Sec. 2.2.2, ion beam bombardment results in the implantation of the primary ions. The implanted ions may cause density changes within the sample, and either swelling or compaction of the surface, with accompanying changes in the sputter yield. At low doses, light (e.g., He) inert-gas ions tend to dissolve interstitially in metals, but the heavier inert-gas ions (e.g., Ar, Kr, and Xe) often favor substitutional sites.(21) At higher doses, clusters and even pressurized gas bubbles may form.(21) Bubble formation is usually observed when an insoluble species is used in the ion beam, such as H, He, and the other inert gases.(23) The implanted ions can aggregate by defect-enhanced bulk diffusion and may form bubbles and blisters in the sample if they do not diffuse to the surface.(23) The extent of bubble formation depends on the implanted ion and target species, temperature, and ion dose and energy. At higher temperatures, voids may form in the solid(21) and ordering of the ion-beam-induced defects may occur. Soluble trapped species may form alloys with the substrate material. As sputtering erodes the surface, a bubble may be breached and result in a pitlike feature, which can generate further topographical roughness with continued sputtering. However, surface diffusion may occur rapidly enough, especially in metals to fill the void caused by bubble formation. TEM studies of 5-keV Ar-ion bombardment of Si show a dense shallow layer of bubbles of ~1 nm radius that trap about 50% of the Ar and that may be responsible for the topography observed upon sputtering.(23)

4.12. Smoothing

For an arbitrarily chosen sample, either roughening or smoothing may occur with ion beam bombardment, depending in a large part on the ion beam parameters. The stochastic nature of the sputtering process results in roughening at the atomic scale. However, this process is mediated by local differential erosion rates so that the roughness is reduced and an initially rough surface may undergo smoothing.(199) If a material and its surface are homogeneous and structureless, smoothing should preferentially occur, and even an initially rough surface can be polished to some extent by using the right ion-beam-bombardment parameters.[200,201] Carter predicts that under the right conditions, initially rough, homogeneous surfaces will tend to be smoothed because the high-surface-area asperities will erode faster, as will the areas with higher sputter yields.[20,199] Thus, the ultimate surface topography should conform to that of the lowest sputter-yield phase. Peaks will tend to erode and narrow and valleys will broaden until the roughness is minimized.[201] Carter provides a model for this smoothing process and shows how it can be applied to polishing both large-area and small-area surfaces.[201] Another mechanism for smoothing involves the ion-beam-induced generation of mobile defects. Migration of the defects results in redistribution of material and can moderate the surface topography; it is predicted this process will help smooth a rough surface.[20] The driving force may be minimization of surface area, which reduces the surface free energy. The ion flux and sample temperature are also expected to influence this process. On a small scale, some smoothing might be observed, but features ~1 μm in size may show only a tendency toward smoothing because a large amount of material transport is required for complete smoothing.[23] In addition, minimization of surface free energy can result in a diffusion-induced, modified topography.[23] Rounding or flattening of small protrusions and pits reduces the surface area and minimizes the free energy (Fig. 4.20). Faceting also occurs, which is a well-known result from minimizing the free energy. In addition, plastic deformation and twisting may occur in response to surface stresses[23]; this mechanism has been proposed for the formation of drooping and twisted cones, although some suggest that redeposition may also play a role.[169]

Sputtering atomically flat Au substrates with 3-keV Ar ions with a total dose of ~10^{15} ions/cm^2 initially produces small faceted islands and pits about 2 to 10 nm in lateral dimension and a few monolayers in height.[165] At room temperature, however, considerable reordering of the surface takes place over a period of hours after sputtering is stopped. STM images taken 30 min and 36 h after ion beam bombardment show that some sort of short-range diffusion takes place. The diffusion results in an increase in size of the islands, a decrease in the number of the islands, and a filling in of the pits. The rms surface roughness diminishes significantly. The diffusion processes seem to differ on the (100) versus (111) faces.[165] Vishnyakov *et al.*[165] indicate that small groups of gold atoms 2–3 nm in lateral dimension and 1–2 monolayers in height detach from the adatom islands and diffuse across the surface to a pit where they lodge.

Smoothing of an initially rough surface has been observed by sputtering the rough back side of a Si wafer. The proposed mechanism in this case is an annihilation of macrosteps, e.g., steps with an area larger than that of the collision cascade.[200] According to the theory, a critical parameter is the ratio of the perpendicular erosion rate to the lateral displacement speed of the macrosteps. However, the formation and displacement or occurrence of the macrosteps depend on the sample matrix and sputtering parameters; they depend particularly on the

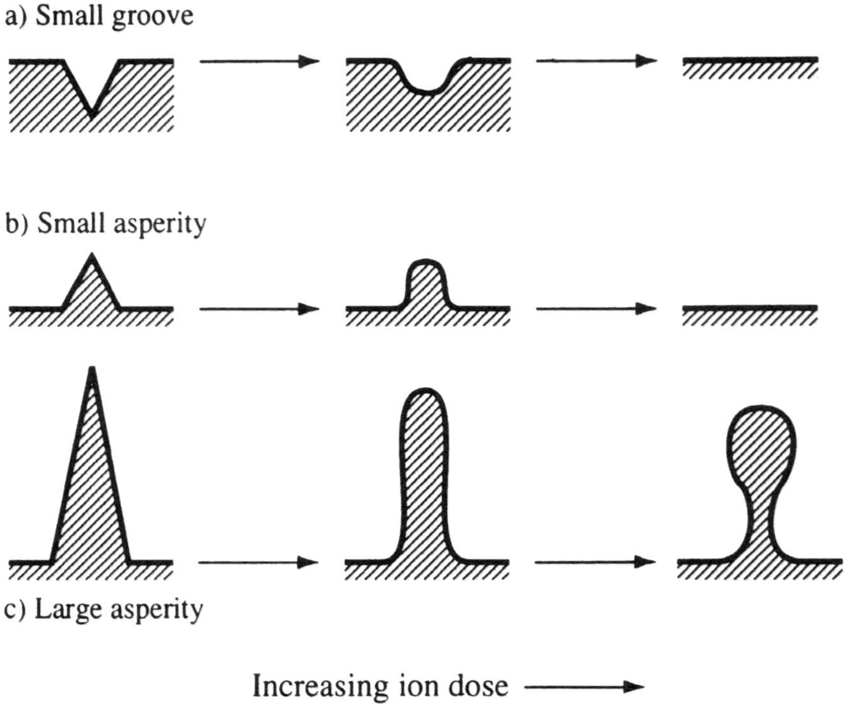

Figure 4.20. Expected rounding and smoothing of surface features with increasing ion beam bombardment, driven by the minimization of surface area and free energy. Adapted and reprinted with permission from[23] I. H. Wilson, in: *Materials Modification by High-Fluence Ion Beams,* edited by R. Kelly and M. Fernanda Da Silva (Kluwer Academic Publishers, Dordrecht, The Netherlands, 1989), pp. 421–466. Copyright 1989 Kluwer Academic Publishers.

ion beam angle of incidence (grazing incidence preferred) and are more favorable with sample rotation (Sec. 6). For homogeneous single-crystal samples, polishing can occur if the displacement speed of the macrosteps is sufficiently rapid and the sample is rotated. If the annihilation of macrosteps is slow, roughness will be preserved. If macrosteps are continuously created, such as in polycrystalline samples or otherwise inhomogeneous samples, smoothness may be improved but will not be perfect; predictions of the limiting value of surface smoothness have been obtained.(202) Smoothing was also observed on single-crystalline Si and on Al, results that indicate that amorphization of the surface by ion beam bombardment is not a necessary condition for polishing.(202) The theory was successfully used to predict the optimal ion beam bombardment conditions for minimizing roughness of sputtered Ni–Cr thin films.(200) Similar results and a similar explanation were reported by Venables and Makh[203] for polishing lithium niobate by Ar-ion etching at 600 eV and room temperature.

Chason and Mayer[204] found that SiO_2 surfaces are roughened on the nanometer scale by sputtering with 1-keV Xe ions, with a nearly linear dependence of the roughness on the eroded depth, a result consistent with a statistical roughening process. Sputtering the Xe-roughened surfaces with 0.5- to 1.0-keV H ions resulted in considerable smoothing, the extent of which depends on ion energy and fluence. The authors speculate that the H reacted with the ion-beam-induced surface defects and resulted in rearrangement and relaxation of the surface and minimization of the surface area, energy, and strain.[204] Vishnyakov and Carter[181] reported similar smoothing of transverse waves generated by Xe-ion bombardment on Si by using Ne or Ar ions at room temperature, but reduced smoothing effects were observed at lower temperatures. Csahok *et al.*[206] report smoothing for Ni films deposited on Si wafers over lateral dimensions of several tens of nanometers, but they observed increased lateral roughening of about 1 nm for 1-keV Ar-ion sputtering at an 86° incidence angle and room temperature. They attributed the fine-scale roughening to the statistical nature of the sputtering process.

4.13. Miscellaneous Results

4.13.1. Topographical Differences in the Same Sample

A variety of topographical features is often found on a single sample after sputtering. Different grains may produce different features. Sputtering of Ag–Cu two-phase alloys has been found to produce a surface covered with cones, ridges, and pebblelike features.[90] Targets with the smallest grain sizes had the largest sputtering yields. Targets consisting of mostly either Ag or Cu had the largest average grain sizes, while targets near the eutectic composition had smaller grain sizes. Topographical features were developed more rapidly in samples with smaller grain sizes.[90] Metal alloys[90] and multicomponent polycrystalline samples[179] often develop two levels of surface features: large rounded bumps that are micrometers in size, and smaller submicrometer hills or cones. The work on alloys suggests that the larger features are associated with the grain size of the material. The enhanced or preferential sputtering of Ag along with Cu diffusion and seed cone formation result in the growth of submicrometer hills or cones (Fig. 4.2).

*Kapton is a DuPont trademark name for polyimide. Many types of polyimides are available for application from ca. 4 K to 673 K. *Teflon is also a DuPont trademark for polytetrafluoroethylene (PTFE). Unfortunately, Teflon is used for fluorinated ethylene propylene (FEP). If the exact composition of either polymer needs to be known, the reader is urged to contact the author(s) we cite.

4.13.2. Topography of Kapton® and Teflon®* after Atom Beam Bombardment*

The effects of atom beam bombardment on the surface topography of Kapton-H films was studied[206] with 6-keV Xe atoms for doses between 1.6×10^{17} and 7.5×10^{18} atoms/cm^2. At the lowest dose the surface was covered with submicron-sized protrusions, which progressed to semicolumns with bases aligned toward the incoming atoms. Separation of the columns resulted in microcracks. Further atom beam doses resulted in forming short columnar structures about 0.25 μm in size, further breaking up of the semicolumns, and more extensive microcracking. After doses of 1.3×10^{18} atoms/cm^2, the surface consisted of 3 μm × 0.25 μm fibril-like columns that grew to 5 μm × 0.5 μm after further ion beam bombardment. Holes in the polymer surface were also seen. After a total dose of 7.5×10^{17} atoms/cm^2, the columns transformed into shorter and sharper cones, along with alternating regions of holes and newly formed semicolumns. The holes were approximately 1 μm deep. At this stage approximately 5 μm of the surface had been eroded.[206] From similar studies, Teflon has a much higher erosion rate than Kapton-H, a faster topography development, and a different progression of topography formation. The initial topography of the Teflon surface consists of a very dense microcrack network, which developed into long, thin, filament-like structures, upon Xe atom bombardment. After further atom beam bombardment, the individual filaments fused to form conical structures.

4.13.3. Sputtering with Non-noble-Gas Ions

Non-noble-gas ions are used in sputtering for a variety of reasons. Although we have not included reactive-ion sputtering or chemical sputtering in our review, we are compelled to mention some alternative ion beam species in common use. In SIMS, sputtering with oxygen and cesium ions is used to enhance the yields of some species. Sputtering with oxygen ions is sometimes used in an attempt to maintain sample stoichiometry or to oxidize a metal surface to enhance the detectability of trace elements. Small amounts of oxygen in the background gas facilitate removing some contaminant species, such as carbon. Sputtering with oxygen or nitrogen ions is also used on occasion to suppress topography formation or to reduce damage. However, the results vary widely from sample to sample. Malherbe reports[24] that sputtering with O_2^+ or Cs ions suppresses the topography development in InP and InSb, but gives more pronounced topography than Ar-ion sputtering for many other compound semiconductors. Karen *et al.*[207] report on the topography developed for several III–V semiconductors with oxygen-ion sputtering. They found ripples on GaAs, together with changes in the secondary-ion yield; no ripple generation on GaP; ripple formation on InP, but no change in the secondary-ion yield; and roughened surfaces on GaSb and InAs.[207] They also reported successfully suppressing both ripple development and the ion yield changes by rotating the sample during sputtering (Sec. 6). Cooling a GaAs sample to –30°C or increasing the temperature above 100°C suppresses ripple formation from Cs-ion bombardment.[192] Cooling also suppressed the ripple formation on InP (Cs ions), but heating to 80°C resulted in pronounced cone formation.[192]

AFM images of crater bottoms on an AlAs–GaAs superlattice sputtered with 3-keV O_2^+ andAr ions at a 60° incidence angle show clearly the enhanced topography developed with O_2^+ sputtering (Fig. 4.21).[164] The feature height on the oxygen-ion-bombarded sample was about 10 times that observed on the argon-ion-bombarded sample. However, Cr–Ni samples sputtered with 7-keV O_2^+ or Ar ions and at normal incidence develop a somewhat smoother surface with oxygen ions, as shown in Fig 4.5. However, the depth resolution in sputter depth profiles of oxygen-bombarded samples deteriorated when the ion beam incidence angle was increased to 20° and greater. Figure 4.22 shows the corresponding XPS data for Cr and Ni multilayers sputtered with oxygen ions at different ion beam incidence angles. At normal incidence, a metal oxide is clearly formed, as evidenced by the broad, multiple peaks. The oxide features diminish in magnitude with an increase in the ion beam incidence angle. Spectra for the clean metallic surfaces shown for comparison at the top of Fig. 4.22 were obtained after Ar-ion sputtering, so they do not show the presence of a native surface oxide.[164] Similar results for the decrease in depth resolution of Cr–Ni multilayer samples at increasing oxygen-ion incidence angle were reported by Zalar *et al.*[208] They explained that the thickness of the oxide layer formed with oxygen-ion bombardment decreases with increased ion beam incidence angle, so the smoothing effects in absence of rotation are not as

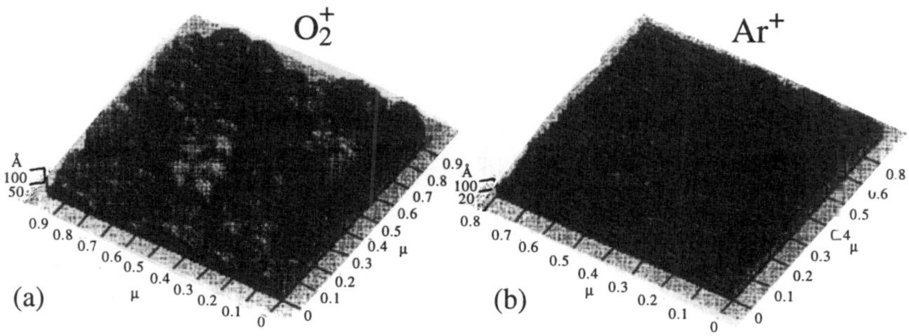

Figure 4.21. AFM images of crater bottoms formed on an AlAs/GaAs superlattice bombarded with 3-keV (a) oxygen and (b) argon ions at a 60° incidence angle showing decreased roughness with argon ion bombardment. Adapted and reprinted with permission from[164] D. W. Moon and K. J. Kim, *J. Vac. Sci. Technol., A* **14**(5), 2744–2756 (1996). Copyright 1996 American Vacuum Society.

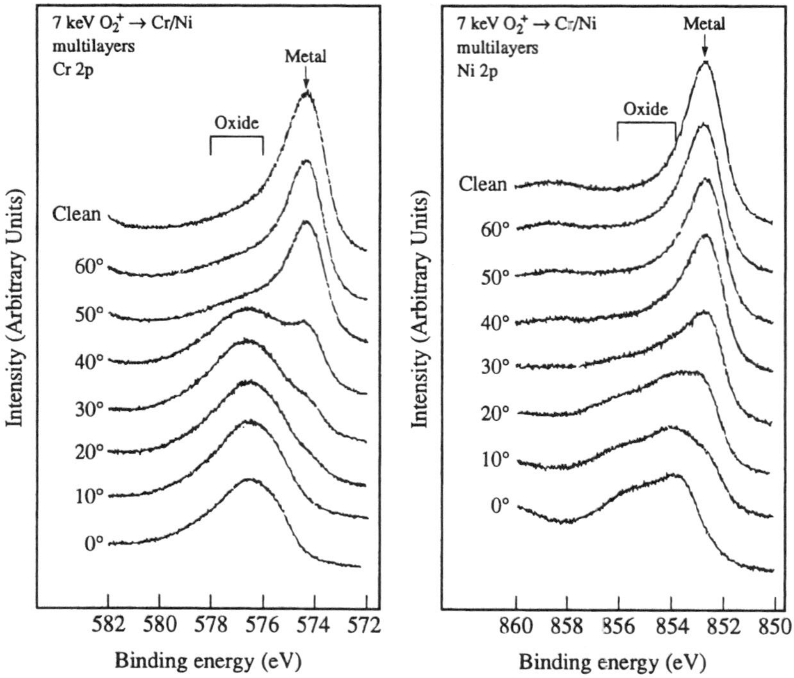

Figure 4.22. XPS Cr 2p and Ni 2p spectra as a function of oxygen ion beam incidence angle, showing that less oxide is formed as the incidence angle increases. Adapted and reprinted with permission from[164] D. W. Moon and K. J. Kim, *J. Vac. Sci. Technol., A* **14**(5), 2744–2756 (1996). Copyright 1996 American Vacuum Society.

pronounced. They found that for ion beam incidence angles of 64° and 77° the metallic surface was only slightly oxidized to depths of 1 nm or less, whereas oxide thicknesses of 4 nm and 2 nm were found for incidence angles of 33° and 52°, respectively.(208) Ripples have been observed on GaAs and Si surfaces sputtered with oxygen ions. On Si, the ripples appeared to be ion dose dependent but not ion flux dependent.(196)

Pamler and Wangemann(209) compared the depth resolution in sputter depth profiles for Al–1.0 wt% Si and Al–1.0 wt% Si–0.17 wt% Ti sputtered with oxygen and nitrogen, both of which are predicted to react with the Al to form amorphous layers; they obtained the best results with oxygen ions at incidence angles between 0° and 40°. The results using oxygen ions were also considerably better than those obtained with ions of inert gases. When oxygen ions are used, detection of oxygen contamination in the sample is problematic, but the authors found that results nearly as good could be obtained by presputtering up to 90% of the metal film using oxygen, followed by Ar-ion sputtering through the interface. Chassé *et al.*[210] found considerably improved depth resolutions and smoother crater bottoms on GaAs–GaAlAs heterostructures sputtered with nitrogen ions compared to those with Ar ions.

Some attribute the effects of non-noble-gas-ion sputter smoothing to a chemical reaction with the substrate that results in formation of an amorphous oxide or nitride layer, which sputters more evenly and reduces channeling effects.[209] Others claim that the layer formed chemically inhibits migration and segregation.[20] Thus, it is not surprising that different samples give varied results, especially because of the varying magnitudes of the different ion-induced effects.

4.13.4. Annealing Sputter Damage

Sometimes sputtering is performed at elevated sample temperatures, or the sample is annealed after sputtering in an attempt to remove the ion beam damage (Sec. 3.5). Again, the success of this strategy is highly material dependent. For InGaAs, for example,[24] a postsputtering heating to 400 to 500°C results in annealing the simple defects into more extended defects. The amorphous layers generally do not recrystallize completely, leaving a high degree of residual disorder. Annealing may result in decomposition of compounds, selective loss of one component such as Hg in $Hg_{1-x}Cd_xTe$, and rapid diffusion or segregation, or both.[24] Annealing heavily sputtered (5×10^{17} ions/cm^2) Si at 1300 K significantly, but not completely, flattened the surface.[211]

4.14. Concluding Remarks

In this section we have described the growth and evolution under ion beam bombardment of a large number and wide range of topographical features. Topographical roughening can have an adverse and large effect on depth resolution (Sec. 6). Some insight into the mechanisms for topography development are given, but as yet no model can accurately predict the type and extent of topography development for an arbitrarily chosen sample. Ion beam parameters can be optimized to minimize the extent of topography development, but the optimal parameters depend on the sample, so generally some experimentation is necessary to determine the best operating conditions. In Sec. 6 we further discuss the optimization of experimental parameters for ion beam bombardment especially for improving the depth resolution.

5. Compositional Changes and Chemical Effects

5.1. Introduction

Low-energy ion beams provide a convenient source of energy and are routinely used in a variety of vacuum and surface science applications. Ion beams are used to clean samples to remove adsorbed contaminants, as an integral part of surface analysis techniques such as ISS, in sputter depth profiling to ablate material to reach buried interfaces, and to remove material for direct analysis as in SIMS. The composition at any eroded depth is determined with analytical techniques such as XPS, AES, SIMS, and ISS. Some degree of care is usually exercised to use gases of sufficient purity to avoid sample and vacuum contamination, but the effects of the ion beams on the chemistry of the sample are often ignored and most often not understood. Sample composition, atomic mass, and surface binding energy play a significant role in how sputtering can affect the sample composition.

There has been a great interest in the determination of sputtering effects: simple diffusion, ion-beam-bombardment-enhanced diffusion (also termed RED), and increased or nonequilibrium segregation of components. While ion beams provide a method to obtain compositional depth profiles, great care must be exercised to interpret correctly the data generated from ion-beam-bombarded samples. Obvious changes in the sample during ion bombardment were discussed in Secs. 3 and 4. More subtle chemical changes also take place and, if undetected, can result in erroneous interpretation of the data.

Chemical changes are extremely sample dependent and may be affected by the ion beam energy, the ion species used to sputter the sample, the angle of incidence of the beam, and the sample temperature. Complex processes begin at the onset of irradiation and may even continue after reaching what is considered a conventional equilibrium state. In general, for a particular sample, the ion beam energy and mass govern the depth of penetration (Sec. 2.2.2) and the degree of sample modification (Secs. 3 and 4).

Ion-beam-bombardment-induced chemical effects and compositional changes are discussed in other parts of this chapter. These include using ion beam bombardment for cleaning surfaces, depth profiling, and surface analysis in Sec. 1; implanting ions at various depths (Sec. 2.2.2); causing cascade sputtering (Sec. 2.4.1); producing an altered layer in a zone of mixing (Sec. 2.6); producing defects and structural changes (Sec. 3); amorphizing crystalline materials (Sec. 3.4); and enhancing surface and bulk reactivity from the formation of broken bonds (Sec. 3.1). Redeposition of ejected particles was discussed in Sec. 4.3, annealing of defects and depleted surfaces in Sec. 3.5, and differential sputtering in Sec. 2.5.3. Instrumental, sample, and beam effects will be discussed in Sec. 6. In this section, we focus exclusively on chemical effects and compositional changes resulting from ion beam bombardment.

We present an overview with some specific examples for organic materials, alloys, semiconductors, oxides, and compounds. We include examples that illustrate differential sputtering and steady-state surface composition; ion-beam-bombardment-enhanced diffusion, segregation, and reactivity; and chemical-state changes involving reduction of the surface. We also indicate the importance of obtaining accurate XPS data with narrow lines to minimize chemical interference with the analysis.[212] Thus, we approach differential sputtering of semiconductors and binary targets[26] differently from that taken in two recent and useful review articles.[12,26] In both of these papers, detailed listings are given showing the extent of deviations from stoichiometry following ion bombardment.

The authors of other review papers[12,24,26] discuss theories and models of differential sputtering, unresolved differences in theoretical and analytical models, ion beam effects, and comparisons between surface compositions determined by AES and ISS. In contrast with AES and XPS, ISS is uniquely sensitive to the outer monolayer and provides a direct measurement of the outer monolayer composition. Figure 5.1 shows the measured surface concentration of Au as a function of temperature as determined with both ISS and XPS for two AuPd alloys.[13] ISS measures elemental concentrations in the outermost exposed surface layers, whereas XPS measures elemental concentrations averaged over the information depth of the near surface (~10 ML for the Au 4*f* signal). The large differences in the ISS and XPS spectra indicate that Au is surface enriched in the surface layer (ISS) as the temperature increases. The XPS data suggest that the surface enrichment of Au is balanced by the Au depletion in the underlayers, so the average concentration over the XPS analysis depth is largely unchanged.[13]

In Ref. 26, tables are given that show the sputtering conditions, surface enrichment, and whether ISS or AES was used for the binary systems CdTe, SiC, TiC, WC, TaC, HgTe, AuCu, Au–Ag, AuPd, CuNi, CuTi, CuPd, PtNi, PtPd, In_xSn_y, InP, LuFe, Ta_2O_5, ZnNi, and silicides of Co, Cr, Fe, Mo, Nb, Ni, Pd, Pt, Ta, Ti, Tb, and V. Malherbe[12] focused his review on noble-gas sputtering of semiconductor systems, especially GaAs and InP, and CdS, CdTe, CdSe, Cu_2Te, GaN, GaP, GaSb, GeSi, HgTe, InAs, InSb, PbTe, SiC, SiGe, and SnGe.

5.2. Organic Materials

Carbonaceous materials such as graphite and simple aliphatic polymers pose an interesting problem for the surface analyst. Adventitious carbon is a ubiquitous contaminant on many specimens, and its removal is most often the focus of extensive efforts of surface cleaning prior to analysis and processing of samples in vacuum. Organic materials are occasionally ion-bombarded, for the specific purpose of altering the surface chemistry for subsequent processing either by metalization[213] or sometimes for biocompatability enhancement.[214] Kowalski[214] reviewed the interaction of low-energy Ar ions on biopolymers such as bioelectric polyurethane, cross-linked polyurethane, ultrahigh-molecular-weight polyethylene with and without carbon fibers, polyoxymethylene, carbon-impregnated polyolefin, silicone rubber, and polytetrafluoroethylene (PTFE). All except carbon-impregnated polyolefin and Teflon showed significant incorporation (trapping) of Ar in the polymer. In general, the ion beam bombardment process increased the C/O ratios of the polymers slightly; chemical changes were minimal, but

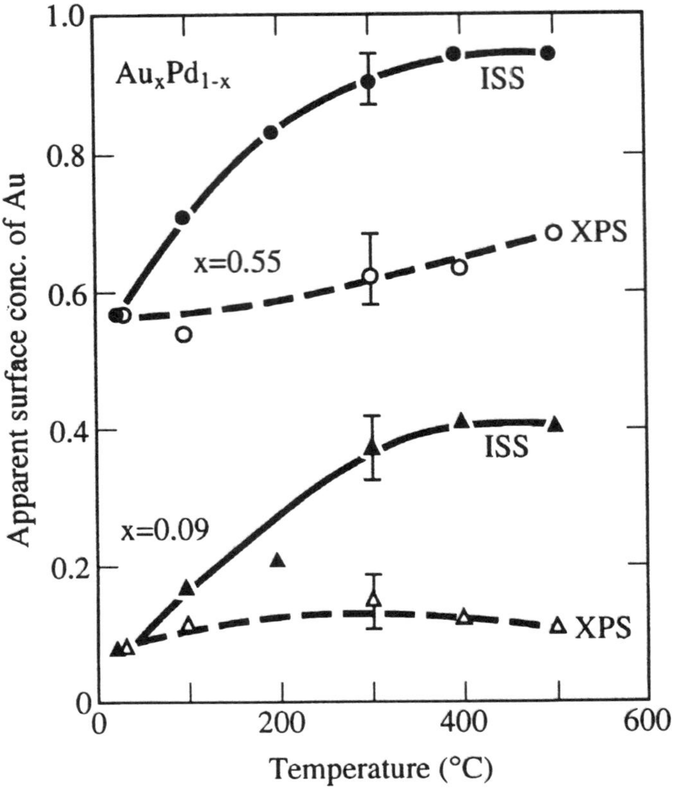

Figure 5.1. The surface composition as a function of temperature is shown for two AuPd alloys measured with either XPS or ISS. ISS clearly shows Au segregation to the surface, whereas XPS, because of its larger sampling depth, does not. Adapted and reprinted with permission from[13] K. Wittmaack, in: *Practical Surface Analysis, Volume 2 - Ion and Neutral Spectroscopy*, edited by D. Briggs and M. P. Seah (John Wiley and Sons, Ltd. Chichester, UK, 1992), pp. 105–175. Copyright 1992 John Wiley and Sons, Ltd.

polymer topography changes were significant. The responses of tissues to the chemical and physical changes of these biopolymers are not yet known.

Delcorte *et al.*[(215)] found that no true static regime exists for sputtering simple aliphatic polymers. TOF–SIMS shows that large molecular ions are ejected from low-density polyethylene, polypropylene, and polyisobutylene after bombardment with Xe ions of less than 4-keV energy. A continuous dynamic transformation occurs as the polymers are sputtered. Decomposition of the polymers is found to occur in several ways; dehydrogenation is a key mechanism, but cross-linking, segmentation, and aromatic-ion formation also play a role. Simple free-radical degradation of PVC, poly(vinylchloride) and PMMA, poly(methyl methacrylate) occurs with low-energy ion beams, but at energies greater than 4 keV the formation of aromatic ions is also found. These data indicate that for some polymers the energy of the ion beam plays a significant role in the way the polymer decomposes and in the composition of the resulting surface.(216)

Bombardment-induced differences can be significant between a charged beam and a neutral one in inert-gas or ion sputtering of polymers. A SIMS study comparing the interaction of 2-keV Ar ions and an Ar neutral beam with PET [poly(ethylene terephthalate)] and PTFE illustrates this point.(216) With PET, the sputtering process can be considered as an example of causing simple physical degradation by the random segmentation of the polymer. In addition to the random segmentation mechanism, another chemical mechanism is apparently operating for PTFE. As sputtering proceeds, there is a preferential loss of F as noted by an increase in SIMS peaks with a low F/C ratio. No preferential F loss occurs when a neutral

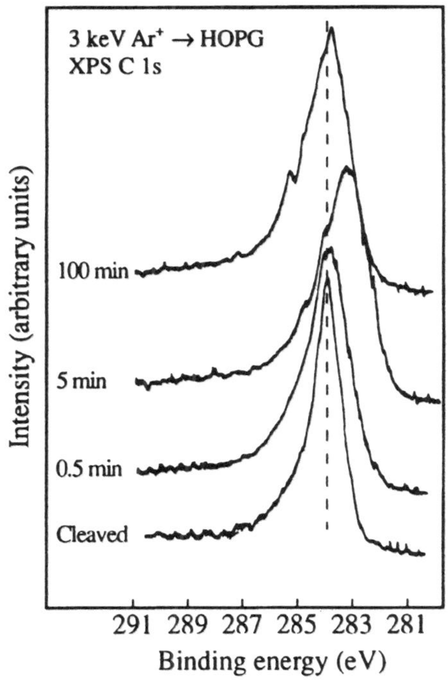

Figure 5.2. Carbon 1*s* XPS spectra of highly oriented pyrolytic graphite (HOPG) after various Ar ion sputter times at 300 K. Shifts in the position of the Ar 2p photoline do not follow the changes in the C 1s peak and do not match Ar imbedded in polycrystalline samples of Al and Au. The data indicate significant surface chemical changes from the ion bombardment. Adapted and reprinted with permission from[217] R. Schlögl, *Surf. Sci.* **189/190**, 861–872 (1987), and with kind permission of Elsevier Science - NL, Sara Burgerhartstraat 25, 1055 KV Amsterdam, The Netherlands.

beam is used, suggesting that the mechanism is an electronic interaction at the surface. Thus, the charge of the beam may play a significant role in the degree of ablation and changes in the surface composition of bombarded polymers.[216]

Sputtering highly oriented pyrolytic graphite with 3-keV Ar ions also results in changes to the material, as observed by Schlögl.[217] An insulating disordered phase is formed that has a significantly modified valence structure. In addition, surface chemical changes result in shifts in the C 1*s* XPS peak of up to 0.9 eV that do not necessarily result from the modified electric properties of the material, as shown in Fig. 5.2.[217]

5.3. Alloys

Metal alloys provide researchers with a convenient medium for studying the complexities of low-energy ion beam interactions with solids. With techniques such as ISS, AES, and SIMS, workers have examined ion-bombarded alloys and compounds and determined the nature of the beam-induced compositional changes in these materials. Kelly *et al.*[25,98,218,219] have extensively reviewed the literature in several publications and discussed the general effects of ion beams on alloys and oxides. Surface compositional changes resulting from ion bombardment have been attributed to several different physical and chemical effects: surface chemical bonding, electronic processes, mass-dependent effects, and diffusion, all of which result from temperature and ion-beam-induced effects. Early research suggested that elemental mass differences in the samples were the primary factor in which the lighter element was lost preferentially during sputtering, but more recent research has shown that this explanation is inadequate for results with many different systems. A somewhat better understanding has been obtained by using chemical bonding as a better factor for predicting compositional changes of ion-bombarded solids, with oxides having the best correlation.[220] Nevertheless, the steady-state composition of ion-bombarded alloys can only be understood in terms of all of the effects listed along with bombardment-induced Gibbsian segregation.[98,218,219] Kelly also suggested there is a weak chemical driving force of 0.06–0.52 eV/atom that plays a significant role in the compositional changes resulting from ion bombardment of solids. He explains this force in terms of a fraction of the ion's ballistic trajectory terminating within the outer two atomic layers of the target, followed by low-energy, chemically guided steps.[213,221] Lam[61,222] advanced a model that combines the synergistic effects of ion beam–target interactions and of temperature to explain general compositional changes in ion-bombarded alloys.

Galdikas and Pranevičius[223] applied a model based on differential sputtering and mutual diffusion for the redistribution of atoms in the altered layer (Sec. 2.6) to explain compositional changes in two- and three-component alloys. Good qualitative agreement with experimental data was obtained when differential sputtering of the surface coupled with mutual diffusion in the near-surface layers was included.

Baretzky[224] compared ISS and AES results on ion-bombarded Cu–Li alloys (also Ta_2O_5) and found good agreement between the two techniques. The resulting surface composition and depth profiles of the ion-bombarded tantalum oxide were found to be the result of collisional processes, whereas the CuLi alloy showed Li segregation to the surface with the final composition determined by the balance between the differential sputtering and the segregation processes.

In Cu–Ni alloys, copper has the lowest surface free energy, so it will segregate to the surface if Cu atom mobility is possible. From computer simulations compared with experimental data, Miteva *et al.*[225] were able to distinguish temperature and ballistic effects in 5-keV Ar-ion-bombarded Cu–Ni alloys of different stoichiometries. Variable-angle XPS depth profiles of the altered layers showed Cu atoms segregate to the surface at room temperature and that the final surface composition depends somewhat on the initial stoichiometry of the alloy.[225] Similar results were obtained by Lam *et al.*[226] with ISS of Ni–Cu alloys bombarded with 3-keV Ne ions. They also determined that the steady-state copper-rich surface composition was temperature dependent above 400°C (Fig. 5.3). In both cases, defects formed by ion beam bombardment provide the means for Cu transport, so defect-enhanced segregation takes place at the lower temperatures. Between 400 and 550°C, the segregation results from the usual diffusion process,[226] and the

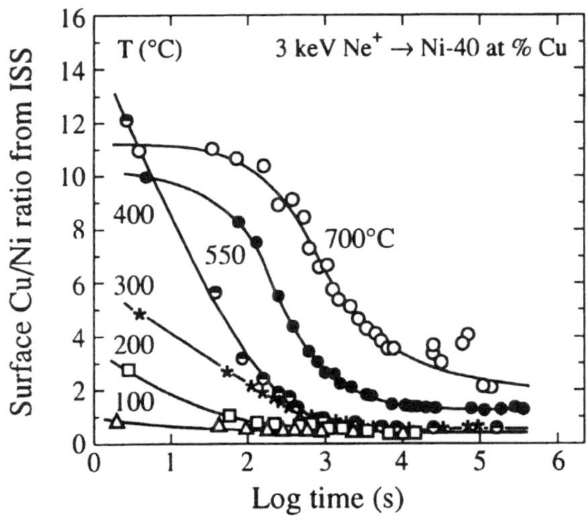

Figure 5.3. Time evolution of the Cu/Ni ISS intensity ratio measured during 3-keV Ne-ion bombardment flux of 3.75×10^{13} ions/cm^2-s of a Ni 40 at % Cu alloy at temperatures from 100 to 700°C. Adapted and reprinted with permission from[226] N. Q. Lam, et al., *Surf. Sci.* **149**, 517–536 (1985), and with kind permission of Elsevier Science - NL, Sara Burgerhartstraat 25, 1055 KV Amsterdam, The Netherlands.

defects formed by the ion beam become less important (Secs. 2.4.2 and 3.5). Using elevated temperatures and time-of-flight sputtered neutral mass spectrometry, Sekine *et al.*[(227)] found that the transport of Cu to the surface region during 3-keV Xe-ion sputtering of Cu–Ni alloys occurs along grain boundaries. The activation energy for the Cu grain-boundary diffusion process was determined to be 54 kJ/mol.[(227)] Jiang *et al.*[(228)] performed AES compositional depth profiles on Cu–Ni alloys as a function of ion fluence. Samples were profiled with 0.7-keV Ar ions after bombardment with 7-keV Ar ions. Kurokawa *et al.*[(229)] obtained good agreement between experimental data and their Monte Carlo simulations of surface compositional changes in ion-bombarded Cu–Ni alloys.

Li and Sun[(230,231)] found with AES that 2-keV Ar-ion bombardment of Au–Cu alloys at substrate temperatures below 300°C results in severe gold depletion at the surface. Above 300°C, however, thermal segregation and diffusion dominate and the gold-depleted layer is partly repopulated. AES sensitivity factors were also developed from scraped samples of different Au–Cu alloys and used to calibrate the measured intensities of low-, medium-, and high-energy Auger peaks. The calibration was then used to quantify the amount of preferential sputtering in these alloys.[(230–232)] Li *et al.*[(233)] also detected some post-ion-bombardment segregation of Au in Au–Cu alloys at room temperature. The segregation is attributed to the redistribution of atoms to lower the surface free energy in the disordered surface that results from 0.5- to 3-keV Ar-ion bombardment. Kang *et al.*[(234)] used mixed beams of 1.5- to 5-keV Ar and He ions to simultaneously ion-bombard and collect ISS compositional depth profiles for Au–Cu alloys. They found that the outer layer of the alloy is Au rich because of the preferential sputtering of Cu and that a depletion layer of Au atoms lies beneath the outermost atomic layer resulting from the ion-beam-enhanced surface segregation. The monolayer sensitivity of ISS allowed the detection of these surface and near-surface compositional changes; AES analysis frequently fails to detect these subtle changes in surface compositions.

Marton *et al.*[(158)] found that the temperature-dependent radiation-enhanced diffusion rates for Ag in Ni decrease with increasing temperature. They interpreted these data in terms of a general model of radiation-enhanced atom diffusion that involves long-lived complex defects, which can migrate for long distances but are subject to annealing. Marton *et al.*[(235)] determined the rate of ion-beam-bombardment enhanced diffusion in multilayered thin films of Ni–Ag under 1 to 44-keV Ar-ion bombardment by studying layer broadening in AES depth profiles.

Lam and Hoff[(236)] used ISS to measure changes in the surface and the revealed subsurface compositions of Ni–Si alloys bombarded with 3-keV Ne and Ar ions between 200 and 700°C. During sputtering at elevated temperatures, the surface concentration of Si decreased to a low steady-state limit. This concentration during sputtering was found to be nearly independent of temperature, suggesting that the contribution from the second atomic layer to the sputtered atom flux was nearly zero in these alloys.

Van Wyk *et al.*[237] determined surface composition changes with AES and ISS for 2-keV Ar-ion-bombarded amorphous Cu–Ti alloys. Copper was preferentially sputtered as it segregated to the surface during the ion bombardment. A model to explain the ion-beam-bombardment-enhanced surface segregation was used to fit the experimental data. The depth of the altered layer was found to be 4 nm, which agrees well with the depth of defects calculated to occur in these materials with the Ar-ion bombardment. Bondarenko *et al.*[238] bombarded Al and Al–Li alloys with 2.5-keV Ar ions and He ions at 300°C. Lithium segregated to the surface and decreased the sputtering coefficient by a factor of 1.4 as compared with Al. Applying a –50-V bias to the sample resulted in a further decrease in the sputtering coefficient by a factor of 4. The authors attribute this effect to radiation-stimulated segregation of Li.[238] Dilute alloys of lithium in copper have been studied[239] under 3-keV Ar-ion bombardment for use as self-sustaining low-atomic-number components in nuclear reactor environments. As Li segregates to the surface, the erosion rate of copper drops by up to 40%.[239]

Kang *et al.*[240,241] studied 5-keV Ar-ion-bombarded Ni–Zr and Cu–Zr alloys with AES, XPS, and ISS. They found with ISS that the outermost atomic layer is Zr rich, and that the subsurface is also Zr rich from the XPS and AES data. This finding is in contrast to other results from Ni–Zr alloys that show a subsurface layer depleted of Zr. Differential sputtering and ion-beam-bombardment-induced surface segregation effects are cited to explain this behavior.

Fan and Yang[242] developed an analytical model incorporating differential sputtering, ion-beam-bombardment-induced segregation, and ion-beam-bombardment-enhanced diffusion effects to predict surface composition and compositional profiles in alloys. This method was found to agree with published experimental results for Au–Cu, Cu–Ni, and Au–Pd alloys.

Lead–tin alloys (40 at% Pb) are enriched in the lighter component during Ar-ion bombardment, but quickly return to the presputtering value after the ion bombardment is terminated.[243] The steady-state composition is attributed to the higher surface binding energy of Sn, which results in the preferential sputtering of the heavier Pb. This effect was also detected in other Sn–Pb alloys.[244]

Hammer and Shemenski[245] found with XPS and AES that bombarding Zn–Cu (brass alloys) with 1- to 4-keV Ar and Xe ions results in surfaces depleted in zinc for their sputtering conditions. However, significant differences in the final surface composition was noted for the two phases studied. The composition of the fcc surface depended on ion energy, whereas the composition of the bcc phase depended on both the ion energy and the mass of the incident ion, showing a decreasing surface Zn concentration for increased ion energy and mass. Darque-Ceretti *et al.*[246] studied surface concentrations of fcc brass alloys bombarded with 5-keV Ar ions. Measurements of the number of sputtered atoms with thermal-ion mass spectrometry and AES analysis were used to develop improved methods for quantitative Auger analysis.

Du Plessis *et al.*[247] compared AES and ISS analyses of 1-keV Ar-ion-bombarded Pd–Pt alloys and found that the Pd segregates to the surface and is preferentially sputtered. Application of a model to fit the experimental data resulted in calculated values for the surface segregation energy of 2–8 kJ/mol and for the ion-beam-bombardment-enhanced diffusion coefficient of $1–8 \times 10^{-20}$ m^2/s.

Jeng and Holloway[248] examined compositional changes in Ni–Cr(15%)(110) single crystals with AES during 2-keV Ar-ion bombardment at temperatures from 200 to 650°C. They found the data are consistent with the prevailing theory of differential sputtering. Segregation of Ni to the outer atom layer led to an enrichment of Cr in the subsurface region for all temperatures except at 650°C. Incremental temperature studies suggest that the composition of the second atomic layer may also depend on the composition of the layers below it. Lill *et al.*[249] found that Sn segregates to the surface of Sn–Ga alloy clusters bombarded by a 4-keV Ar-ion beam and analyzed with time-of-flight mass spectrometry.

Sharma and Hofmann[250] studied amorphous and crystallized FeB alloys with 1- and 3-keV Ar-ion bombardment. XPS and Auger analysis of the sputtered samples showed a B-rich surface layer in all forms of the alloy studied. Surface binding-energy effects are thought to play a significant role in the sputter-induced compositional changes in these materials. Differences in the altered layer formed on the different crystalline forms of the alloy are also discussed.[250]

5.3.1. *Ternary Alloys*

Zhe[251] developed a general expression for the steady-state compositional changes in ion-bombarded Ag–Au–Pd alloys based on differential sputtering effects. Dudonis and co-workers[252,253] studied the surface composition of 2-keV Ar-ion-bombarded Ag–Au–Pd alloys with different bulk compositions. They found that the depth of the altered layer is equal to the ion penetration depth and that surface binding energies are more important than mass effects for describing the differential sputtering of these materials. For these ternary alloys, the ratio of component sputtering yields depends on the bulk composition of the alloy, in contrast to binary alloys. Tang *et al.*[254] and Lam *et al.*[255] studied the effects of 3-keV Ne-ion bombardment on Ag–Au–Cu alloys at different temperatures with ISS. During sputtering at elevated temperatures, the surface concentrations of both Cu and Au increase. Theoretical modeling of the ion-beam-induced modification of the surface was compared with the experimental results, and good agreement was found.

Yacout *et al.*[256] theoretically investigated the surface compositional changes in Fe–Cr–Ni alloys from bombardment with 3-keV Ne ions at elevated temperatures. They found that ion beam effects dominated the changes in the surface composition below 200°C. With increased temperature, thermal activation processes become increasingly important for the final surface composition. Betz *et al.*[252] studied the Ag–Cu–Pd alloy system and found that the addition of a third element does not effect the concentration ratio of the original two elements under steady-state conditions. Zhe[257] developed a graphical method for determining the steady-state composition of ternary alloys from the constituent binary systems under ion bombardment. This method gave good agreement with experimental results for the Ni–Cu–Pd ternary system.

5.4. Semiconductors

Many of the process steps in the manufacture of semiconductor devices involve subjecting their surfaces to ion bombardment. Often, ion bombardment energies at less than 1.5 keV are used to minimize the formation of defects. These low-energy ion beam treatments may still result in changes in surface composition, bulk morphology (Sec. 4), crystal structure (Sec. 3), and electronic properties (Sec. 3.5). These changes then ultimately influence the final device properties and thus are of general interest. Because the optical properties of most semiconductors are changed by ion beam bombardment, beam-damaged systems are often investigated by using optical techniques such as Raman spectroscopy, photoluminescence, UV reflectivity, ellipsometry, and infrared spectroscopy. Similarly, ion beam damage results in changes in the electrical properties of semiconductors. For this reason electrical methods are also used to assess ion-beam damage and changes in device performance (Sec. 3.5). Damage at low ion doses can often be detected by optical or electrical methods before any changes are evident from techniques such as TEM, XPS, AES, and RBS. As mentioned in Sec. 5.1, Malherbe[12,24] compiled an excellent and comprehensive review of ion beam damage in many semiconductors. His review includes tables of the semiconductors studied, the ion beam composition(s) used, and results from bombardment by inert-gas ions. References for ion beam interactions with oxide semiconductors such as In_2O_3, BaO, CdO, HgO, and ZnO are also tabulated.

Some discussion of the effects of ion beam bombardment of Si is warranted because of its use in the manufacture of semiconductor devices. Thus, the Si surface has been investigated in many surface analytical studies and is used extensively for device manufacture. Ion beams are often used to clean and prepare silicon wafers for subsequent manufacturing and processing steps. However, not much work has been done specifically on pure silicon. Silicon has been modified with Ar ions in the presence of Br to produce surfaces containing bromides.[258] Todorov *et al.*[82,114,259] used mixed 100-eV Ar–O-ion beams to oxidize Si selectively in vacuum to a self-limiting thickness of 4 nm. The self-limiting effect is attributed to an equilibrium between removal by sputtering and growth of the oxide induced by the ion beam. In another study,[260] 500-eV Ar ions decomposed monolayers of

XeF_2 on silicon, yielding SiF_x species. The ion beam shows some chemical specificity as the trifluoride is preferentially removed, leaving a SiF_2 surface. The efficiency of removing thin films of Cu on Si was examined with XPS and RBS by Menzel and Wittmaack,[(261)] who compared the differences in sputtering for 12-keV Ne-, N-, and O-ion beams at various angles of incidence.

The native oxide of silicon also plays a significant role in the manufacture of semiconductor devices. As with most oxides, ion beam bombardment results in the preferential loss of oxygen from the surface. Collart and Visser[(262)] have shown that a 1-keV Ar-ion beam can deplete oxygen from SiO_2 to a depth of 8.5 nm. Changes in the oxide thickness were monitored with a laser interferometer, and the data were compared with a computer simulation. Increasing the Ar-ion beam energy can result in more rapid loss of O from SiO_2 surfaces.[(263)] In addition to conventional XPS work, synchrotron sources have been used to characterize the damaged layer with a depth resolution of 0.4 to 0.9 nm, and have been useful for characterizing Ar-ion artifacts in SiO_2 complementary metal–oxide–semiconductor systems. In other insightful work, Mizutani[(264)] found that, as with some polymers, the interaction of an energetic neutral beam has significantly different chemical consequences on the surface of SiO_2 than does a comparable ion beam. He showed that 0.5-keV Ne-ion beams cause preferential loss of oxygen from the surface, whereas neutral Ne beams appear to densify the oxide layer and also render the surface more resistant to subsequent charged beam ablation. The results of his work[(264)] suggest that the dominant mechanism of sputtering SiO_2 is electronic in nature (Sec. 2.5).

Hydrogen-ion beams have been used at elevated temperatures to remove up to 10-nm-thick silicon oxide layers, whereas room temperature ion bombardment was not effective.[(265)] AES data showed no oxygen is detectable on the surface after bombardment at beam energies from 0.5 to 1 keV and substrate temperatures of 450–500°C. Ion beam sputtering with He was ineffective for removing the oxide under the same conditions.

InP has received a lot of attention because of its extensive use in producing devices for fiber optic communications. Ion-beam-induced compositional changes in InP result from pure collisional effects (i.e., differential sputtering; see Sec. 2.5.3).[(24)] Ion bombardment of InP crystals with 0.5 to 5 keV and Kr ions result in an In-rich surface that is virtually independent of sample crystal orientation and ion energy. However, the preferential sputtering of P is strongly dependent on the angle of incidence of the ion beam. At grazing angles of bombardment, preferential sputtering of P is minimized.[(266–269)] Similar preferential sputtering of Sb in InSb crystals has also been observed with 500-eV Ar-ion bombardment.[(269)]

Depth profiles of thin Al films on InP are often distorted by bombardment with noble-gas ion beams. Using a N_2^+ beam can alleviate some of the interfacial mixing and, in general, improve the interfacial resolution in compositional-depth profiles. Skinner *et al.*[(270)] provide a general discussion about effects of ion beam species, energy, and incidence angle for the analysis of thin Al films on InP. They propose that sputtering with nitrogen ions generates an amorphous layer of AlN that sputters more uniformly because some of the interfacial mixing induced by other probe beams is eliminated.

Several studies with GaAs have shown that sputter-induced compositional changes result in bombardment-induced segregation of As at the surface, with an As-depleted region formed just below the surface.[24] Other work has shown that As is depleted at the surface of GaAs samples sputtered with 0.65–3-keV Ar ions. The effects are independent of sample crystal orientation and ion beam orientation, but they do depend on ion beam energy, with higher beam energies increasing the amount of As depleted at the surface. The depletion effects have been observed to extend up to 0.1 m into the bulk of the sample with 5-keV Ar-ion beams.[271–273]

Chemical and electronic changes are not usually desired in semiconductor devices. Nevertheless, sputtering does not always result in decreased performance, but it has been used to improve the performance of some fabricated devices. For example, the barrier height of Au Shottkybarrier diodes constructed on *n*-type GaAs was found to decrease monotonically with an increased Ar-ion beam dose, thus improving the final device performance.[273] Khalki *et al.*[274] found that nitrogen ions with energies between 15 and 40 eV did the most damage to GaAs surfaces. Other researchers have explored using ion-induced desorption and metal overlayer interactions on GaAs.[275,276] Evdokimov *et al.*[277] found that Sb rapidly diffuses to the surface of InSb(100) during 4.8-keV Ar-ion bombardment. The surface composition was found to oscillate under continuous sputtering. This behavior is attributed to layer-by-layer sputtering through an Sb-depleted crystal lattice and radiation-assisted transport of Sb to the surface.[277]

Kazmerski *et al.*[278] determined that the surfaces of sputtered $CuInSe_2$ (CIS) crystals remain essentially stoichiometric for Ar-ion beam energies up to 1 keV. When the ion beam energy is increased further, however, the CIS surface becomes Cu rich as In and Se are preferentially ejected. The authors report[278] that differential sputtering is minimized by using large-ion-beam angles of incidence and beam energies of less than 3 keV. Niemi and Stolt[279] found that the surface of 4.5-keV Ar-ion-sputtered CIS thick films is partially restored after a vacuum anneal at the deposition temperature of the film.

5.5. Metal Oxides

5.5.1. Simple Metal Oxides

Metal oxides are technologically significant materials for many commercial applications. Metal oxides are used as ceramic precursors for preparing high-voltage and high-temperature insulators, ceramic dielectrics for multilayer capacitors, sensors, resonators, high-temperature superconductors, electronic packages and circuit boards, and many other specialized materials and components. For many of these applications, the ceramic material is sputter cleaned in vacuum prior to metalization or bonding. Sputtering of most metal oxides results in a preferential loss of surface oxygen and sometimes a concomitant reduction of the metal to a lower oxidation state. Malherbe *et al.*[(280)] examined the altered layer of more than 20 different metal oxides bombarded with noble-gas ions and developed a model to compare the measured results with theoretical calculations. The data suggest that the dominant factor in sputtering simple metal oxides is the mass difference between oxygen and the metal. This difference results in the preferential loss of oxygen and an enrichment of the metal at the surface. However, his model does not fit the experimental data as well as a model that considers both the mass difference and surface binding energy. In Fig. 5.4, the effect of surface binding energies obtained by using Pauling electronegativities to calculate covalent bond energies for each oxide, coupled with the mass difference of the metal and oxygen, is

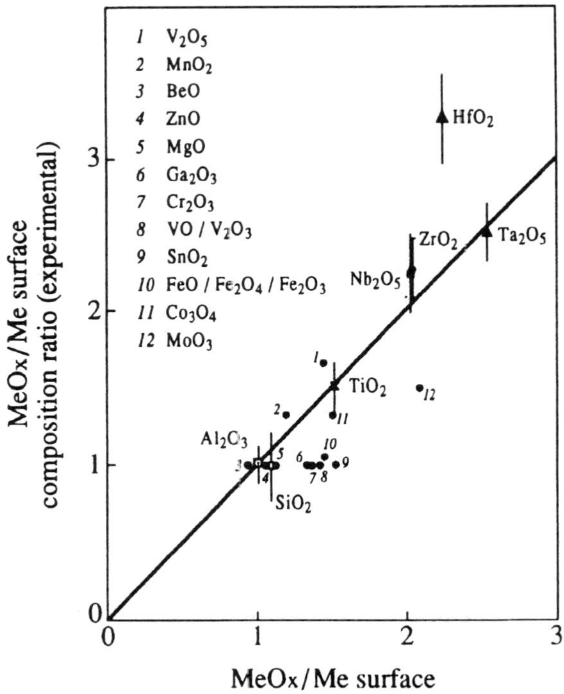

Figure 5.4. Surface composition ratios of oxide to metal plotted against the predicted ratio for a variety of metal oxides taking into account both mass differences and surface binding energy effects of the constituent atoms. Adapted and reprinted with permission from[280] J. B. Malherbe, et al., *Appl. Surf. Sci.* **27**, 355–365 (1986), and with kind permission of Elsevier Science - NL, Sara Burgerhartstraat 25, 1055 KV Amsterdam, The Netherlands.

compared with experimental sputtering data. By considering the mass difference of the atoms and the surface binding energy, better agreement with the experimental data is obtained than with either factor alone. In addition to the mass difference in oxygen and the metal, other factors found to be significant in the decomposition of the metal oxide surfaces studied include incident-ion mass in relation to metal cation and oxygen anions as it affects classical momentum transfer, the heat of formation of the oxide from its elements, the possible existence of suboxides and the differences in the stability of the suboxides, and the difference in surface binding energies of the cation and anion.

Mitchell *et al.*(281) exhaustively studied at 20° sputtered metal oxides—both powders and oxidized metals—and concluded that metal oxides can be divided into two broad categories: those that reduce under ion bombardment and those that do not. The authors report for 3-keV Ar-ion bombardment the following extent of sputter reduction: Al_2O_3, CaO, Cu_2O, Ga_2O_3, In_2O_3, La_2O_3, MgO, Sc_2O_3, SiO, Y_2O_3, and ZnO (0%); B_2O_3, BaO, BeO, CdO, Cr_2O_3, GeO_2, SiO_2, and V_2O_5 (≤1.2%); CeO_2, Co_2O_3, Co_3O_4, Fe_2O_3, Fe_3O_4, FeO, Mn_2O_3, Nb_2O_5, PbO, SnO_2, Ta_2O_5, ThO_2, TiO_2, and ZrO_2 (>1.2%≤5.2%); CuO, HfO_2, and NiO (>5.8%<10%); and MoO_3 and PdO (>15%≤20%). None of the five properties considered to be important, either alone or in combination, have given a satisfactory model that explains the experimental data. For the metal oxides that are reduced, the ion beam energy, mass, and the erosion rate were found to have little effect on the differential sputtering of the oxide surface.(280)

Bertóti *et al.*(282) studied the effects of ion beam bombardment on amorphous Cr–Si–O. XPS data indicate that the bombardment of this complex oxide with 1- to 4-keV Ar ions results in Si^0 formation. This effect appears to saturate at an ion dose of 10^{16} ions/cm^2. Bombardment of the oxide with nitrogen ions results in the formation of metal nitrides, and no free Si is detected. Varga and Taglauer(283) used AES and ISS to study the depth of the altered layer that resulted from Xe-, Ne-, and Ar-ion bombardment of Ta_2O_5. ISS compositional-depth profiles of the altered layer formed by the bombardments were then determined by using 0.7- to 3-keV He-ion beams. The depth profiles indicate that the depth of the altered layer corresponds to the mean range of the bombarding ion in the target oxide. Hamilton *et al.*(284) have shown that Raman spectroscopy can be used to study the altered layer formed by the 1-keV Ar-ion bombardment of Fe_2O_3 single crystals, oxidized iron foil, and oxidized stainless steel. The effects of the loss in surface oxygen on the Raman spectra are discussed.

Karulkar(285) found that preferential oxygen sputtering from NbO, NbO_2, Nb_2O_5 produces vacuum-stable reduced oxide species (Fig. 5.5). He found that 500-eV Ar-ion bombardment was more effective in reducing the surface than a 2-keV bombardment, and attributed this energy dependence to differences in the sputtering mechanisms at the two energies. Two target processes are thought to occur during ion bombardment: first, an Ar-ion–O atom collision and subsequent reflection of O atoms from Nb atoms; second, sputtering of O atoms by Ar ions scattered from Nb atoms. These two mechanisms can operate in the low-energy regime. At higher energies the Ar ions penetrate deeper into the surface, and the effectiveness of both of these mechanisms is reduced.

Fang *et al.*(286) found significant chemical and electronic changes in the ion-bombarded surfaces of SnO_2. He used 2- to 3-keV Ar, He, and H ions to bombard the tin oxide surfaces and UPS and XPS to characterize the altered layers. Loss of oxygen resulted in changes in the work function, and band bending occurred in the sputtered polycrystalline oxide. Morant *et al.*(287) studied the compositional

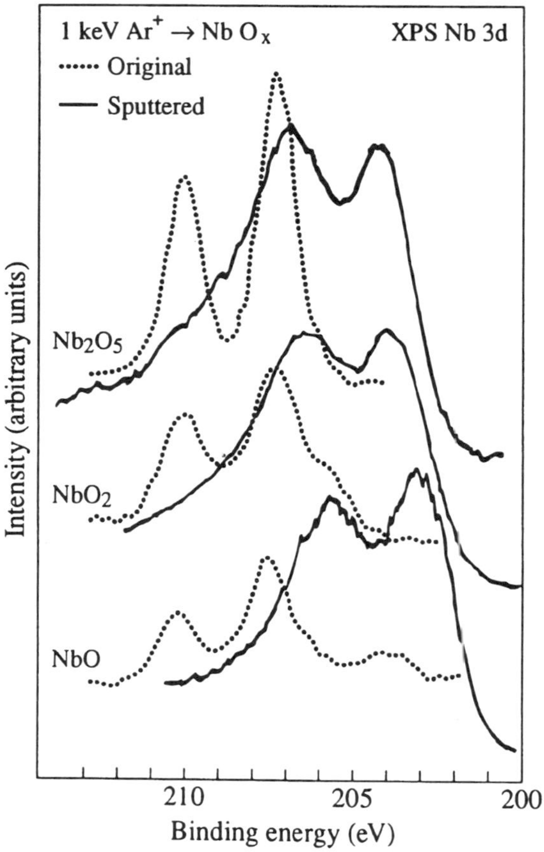

Figure 5.5. Nb 3*d* XPS spectra of different niobium oxides sputtered to a steady state with 1-keV Ar ions. The initial spectra are shown as dotted lines and the solid lines show the spectra after sputtering. The different Nb oxide species are reduced toward Nb^0 in different amounts by the ion beam. Adapted and reprinted with permission from[285] P. C. Karulkar, *J. Vac. Sci. Technol., A* **18**(2), 169–174 (1981). Copyright 1981 American Vacuum Society.

and electronic changes resulting from 3-keV Ar-ion bombardment of ZrO_2. An altered layer 1.2 nm thick was found to be depleted of oxygen and the Zr was reduced to an average stoichiometry of $ZrO_{0.44}$. These oxidation state changes were also reflected in drastic changes in the valence-band spectrum of the bombarded surface. González-Elipe *et al.*(288) found that ion bombardment of mixed oxide systems consisting of CeO_2 and TiO_2 or SiO_2 resulted in the almost complete reduction of Ce^{4+} to Ce^{3+} and significant mixing of the remaining constituents. In many cases, it is desirable to be able to sputter-depth-profile a surface without concern for the problems associated with differential sputtering effects. Moon and Kim(289) found that an undistorted depth profile of Ta_2O_5 can be obtained with 3-keV O_2^+ ions at an angle of 30°. Beams of Ar ions at comparable energies and angles result in the typical reduction of the metal oxide and preferential sputtering of oxygen as detected using XPS analysis. Kim *et al.*(290) also depth-profiled $SrTiO_3$ films with oxygen-ion beams and successfully extracted chemical-state information on these complex oxide films. The formation of metal oxides and changes in sputter yields using mixed-ion beams of O_2 and Ar have been discussed by Saidoh and Yamada.(291)

Christy(292) used XPS and X-ray-induced Auger transitions to study Ar-ion bombardment effects on various oxides, ceramic precursors, oxidized metal surfaces, and ceramic surfaces. Effects of the different materials and matrices of the oxides and the sputtering time on the surface composition were examined, and an example is shown in Fig. 5.6. In some cases, complex oxide surfaces were generated (Fig. 5.7). The effects of instrumental parameters, such as the use of an electron flood gun on the XPS data, were also addressed.

5.5.2. *Complex Oxides: Perovskites*

Ion beam interactions with perovskites result in surfaces of these complex oxides that are altered both chemically and physically. Nefedov *et al.*(293) found that 7-keV Ar ions caused an altered layer in $YBa_2Cu_3O_x$ (YBCO) and annealed $Bi_2Sr_2CaCu_2O$ high-T_c ceramic superconductors (HTCS). Sensitivity factors for both surface and in-depth analysis with the use of Ar-ion sputtering were developed, and it was determined that the ion beam irradiation results in preferential sputtering of the constituent atoms in the order Cu > Ba > Y. Matsui *et al.*(294) also electrically characterized YBCO sputtered with 5-keV Ar ions and found a degradation in electrical properties of the material resulted from the thick insulating altered layer formed by ion beam interactions. Other insulating perovoskites have also been studied. In numerous electronic devices, these oxides are sputtered prior to metalization by physical vapor deposition techniques so the influence of the ion beam dose on the surface properties of these materials is of interest. Mukhopadhyay and Chen(295) used 0.2-keV and 3-keV Ar ions to bombard surfaces of $SrTiO_3$ and $Pb(Zr,Ti)O_3$ (PZT). The lower-energy bombardment had no detectable influence

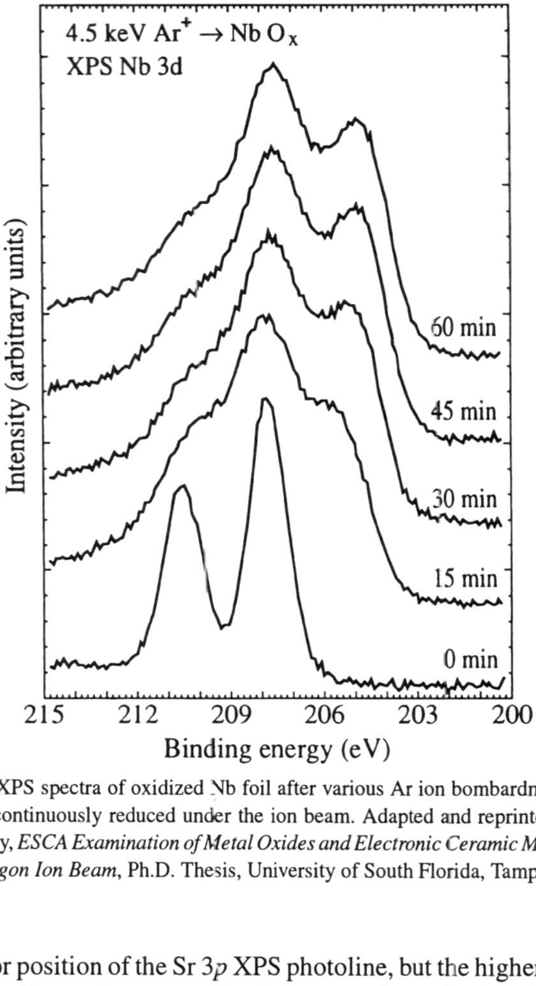

Figure 5.6. Nb 3d XPS spectra of oxidized Nb foil after various Ar ion bombardment times showing that the surface is continuously reduced under the ion beam. Adapted and reprinted with permission from[292] C. N. Christy, *ESCA Examination of Metal Oxides and Electronic Ceramic Materials: The Effect of a Low Energy Argon Ion Beam*, Ph.D. Thesis, University of South Florida, Tampa, 1989.

on the FWHM or position of the Sr 3*p* XPS photoline, but the higher-energy Ar-ion beam caused significant damage to the surface, as shown by broadening of the Sr XPS-peak in Fig. 5.8. Interestingly, the XPS-determined stoichiometric ratios remained constant with sputtering, although the surface chemistry changed. Argon-ion bombardment of the lead-containing PZT resulted in a depletion of lead at the surface and a reduction of the Pb^{2+} to Pb metal. Titanium was also reduced from the 4^+ state to lower valences, but the oxygen-to-cation ratio was preserved.

5.5.3. Complex Oxides: Glasses

As with most oxide samples, ion beam bombardment results in chemical and physical changes in glass surfaces. Brow[(296)] studied the modification of sodium silicate and phosphate glasses by 2.5-keV Ar ions. In the silicate glasses, sodium is

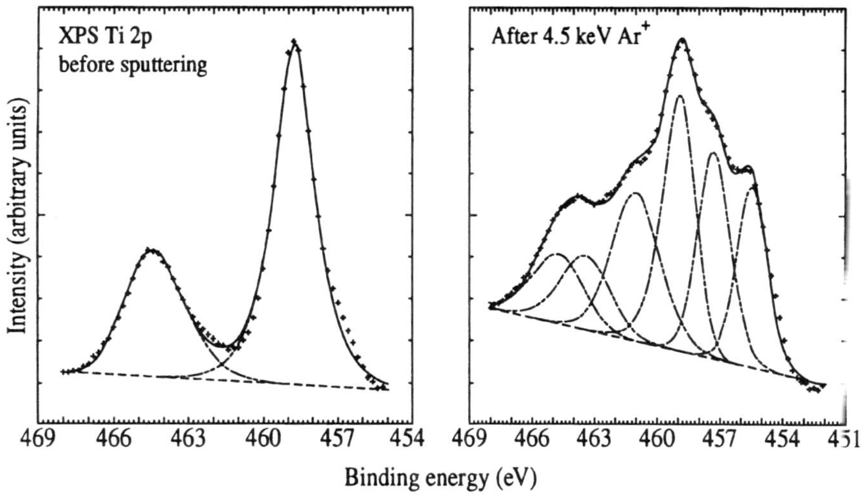

Figure 5.7. Curve-resolved XPS data showing features of the complex oxides generated by the Ar ion-beam-induced reduction of oxidized Ti metal. Adapted and reprinted with permission from[292] C. N. Christy, *ESCA Examination of Metal Oxides and Electronic Ceramic Materials: The Effect of a Low Energy Argon Ion Beam*, Ph.D. Thesis, University of South Florida, Tampa, 1989.

depleted to a 9-nm depth following ion beam bombardment; the loss of sodium is attributed to field-induced migration. Alterations in the surface composition were also detected by changes in the XPS O 1*s* photoline. Brow attributed these XPS-detectable changes to a repolymerization of the glass in the sodium-depleted, ion-modified layer. These chemical and structural changes are observed after as little as 10 min of sputtering at a current density of only 0.1 $\mu A/cm^2$. Higher current densities resulted in faster loss of sodium from the glass surface (Fig. 5.9). Coulombic migration of sodium is the accepted mechanism. As an ion approaches the glass surface, electron transfer from the surface creates a positive surface layer on the glass. This positive region then repels mobile sodium ions deeper into the glass matrix and results in a sodium-depleted zone. XPS photolines recorded during the sputtering process are shifted to apparent higher binding energies, confirming that the bias on the sample surface is positive. However, the positive bias generated by photoelectrons leaving the sample during XPS analysis does not result in significant sodium migration. Other than the depletion of sodium from the near-surface region, little change is evident in the O/Si ratios during bombardment, supporting the argument that for the most part only the bonding in the glass surface changes with ion bombardment.

In contrast to the silicate glasses, phosphate glasses show a surface enrichment in sodium while the phosphorus content drops with Ar-ion bombardment.[(296)] This

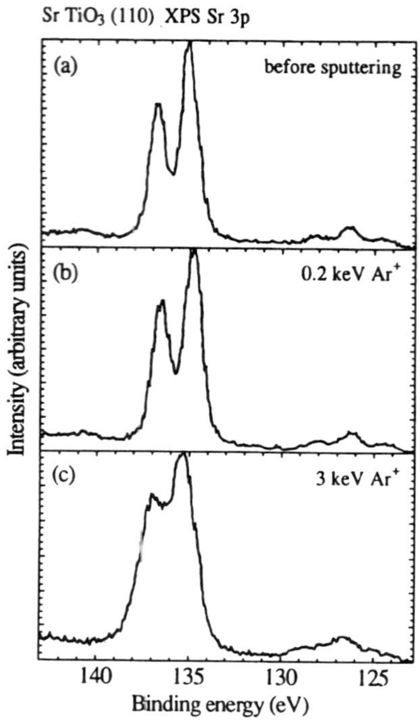

Figure 5.8. Sr 3*p* XPS spectra from a $SrTiO_3$ (110) sample (a) as received, and (b) after the surface carbon contamination is removed with 0.2-keV Ar-ion bombardment. Note that the peak width is unaffected by this low-energy ion cleaning; (c) the same sample after 3.0-keV Ar-ion bombardment. Surface chemical changes are evident from the broadening of the XPS peaks leading to loss in peak resolution. Adapted and reprinted with permission from[295] S. M. Mukhopadhyay and T. C. S. Chen, *J. Appl. Phys.* **74**(2), 872–876 (1993). Copyright 1993 American Institute of Physics.

finding may be attributed to differences in surface binding energy of the two species. From a thermodynamic viewpoint the P–O bond is weaker than the Na–O bond; thus, P is expected to be removed at a greater rate than Na and results in a sodium-rich surface. Other effects may contribute to the final state of the surface after ion bombardment, but this simple explanation provides some insight into the mechanism of chemical changes in these glasses.

Studies of glasses doped with other metals, such as Ti-doped silicate glasses, also have been reported.[(297)] Sputtering these glasses with 1- to 3-keV Ne or He ions results in an enrichment of titanium at the surface because of the preferential loss of Si. Differences in the surface oxygen bonding between the metals was detectable with XPS, but no discussion of the mechanism of the differential sputtering was given. Hara and Itoh[(298)] analyzed arsenic-containing silicate glasses with XPS after noble-gas-ion bombardment and found that arsenic is reduced to

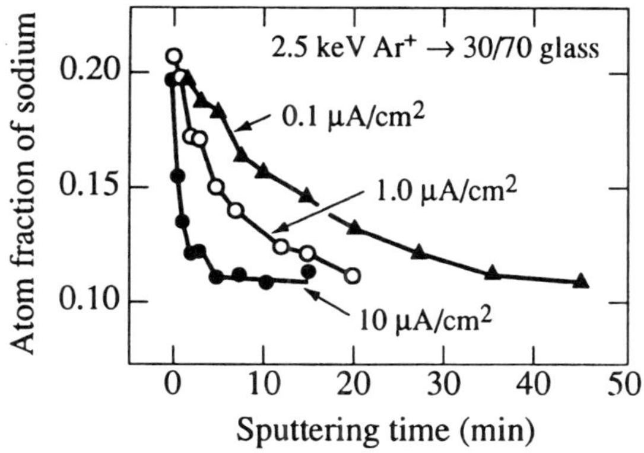

Figure 5.9. Changes in the surface Na concentration (determined by XPS) of a 30%Na_2O·70%SiO_2 silicate glass surface bombarded with 2.5-keV Ar-ion beams with different current densities. Adapted and reprinted with permission from[296] R. K. Brow, *J. Non-Cryst. Solids* **107**, 1–10 (1988), and with kind permission of Elsevier Science - NL, Sara Burgerhartstraat 25, 1055 KV Amsterdam, The Netherlands.

the metallic form. The thickness of the reduced layer depends on ion energy, current density, angle of incidence of the beam, and ion mass.

5.6. Compounds

Lince *et al.*[(299)] studied damage to MoS_2(0001)–1×1 surfaces by 1-keV Ne ions with photoelectron spectroscopy, using a synchrotron source. The ion bombardment produced metallic Mo interspersed in a MoS_x matrix (Fig. 5.10). Within this depleted sulfur layer, a highly amorphous polysulfide species $(S–S)^{2-}$ was also detected. The ordered surface of the crystal could not be regenerated by thermal treatments.

Modification of $NiSi_2$ and Cr_3Si at 400°C with 2- to 5-keV Ar-ion beams results in AES-detectable Si enrichment at the surface.[(300)] The same type of segregation is also found at even higher temperatures without ion bombardment because of thermally induced segregation. Valeri *et al.*[(301)] examined the time, temperature, and dose dependence of the effects of 0.7- to 4.5-keV Ar ions on $TaSi_2$ with AES and found significant modification of the surface extending to a depth of 10 nm into the sample. They suggest that the synergistic effects of differential sputtering and ion-beam-bombardment-induced segregation are responsible for the observed effects. They[(302)] also used AES and EELS to study the effects of 1- to 3-keV Ar ions on $ErSi_2$. The ion bombardment results in an Er-enriched surface with a stoichiometry close to ErSi. The authors attribute this steady-state surface composition to preferential sputtering of Si, the lighter component of the silicide.[(302)]

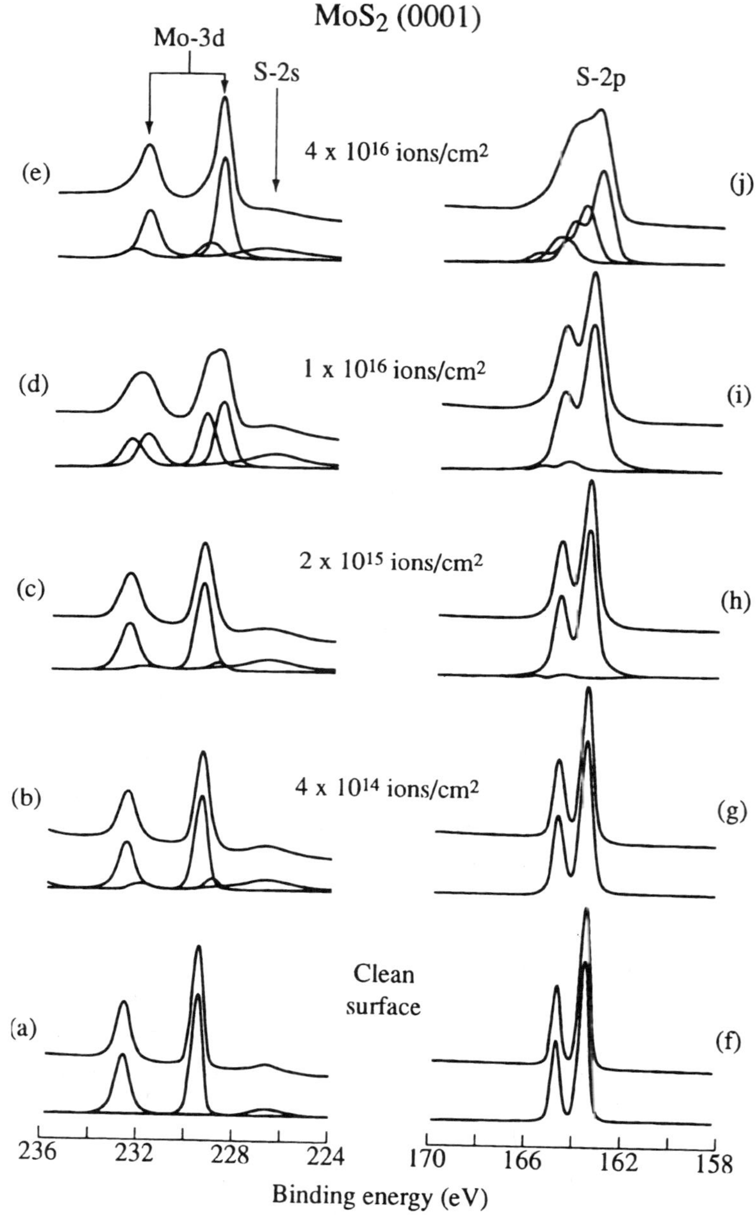

Figure 5.10. Mo 3d and S 2p photoemission spectra from a MoS_2 (0001) surface that had been bombarded with 1-keV Ne ions. Surface chemical changes as a function of the ion dose are evident in both the Mo and S spectra. Adapted and reprinted with permission from[299] J. R. Lince, et al., *Surf. Sci.* **210**, 387–405 (1989), and with kind permission of Elsevier Science - NL, Sara Burgerhartstraat 25, 1055 KV Amsterdam, The Netherlands.

Swart and Berning[303] studied the effects of Ar-ion bombardment on $Tb_{1.7}Si$ and FeSi and analyzed the surfaces with AES. For $Tb_{1.7}Si$, ion bombardment results in a Si-rich surface. In contrast, sputtering the FeSi produces an Fe-rich surface. The surface composition of FeSi changes when the ion energy is varied from 0.5 to 4 keV. They explain the Si surface enrichment of $TbSi_{1.7}$ by inward diffusion ("RED") and "radiation-induced segregation." These effects are also observed in FeSi, but the Si is also preferentially sputtered, resulting in an Fe-rich surface. Depth profiles were successfully obtained with 500- to 700-eV ions to profile the damaged layers in these systems. In addition to the experimental work, the data were compared to Monte Carlo simulations of ion–sample interactions.[303] Hofmann and Stepanova[304] solved equations for linear and binary collision cascades to explain the differential sputtering of Ni_3Al and $TaSi_2$ with 0.1 to 4-keV Ar ions. The effects on compositional depth profiles were discussed.

Harper *et al.*[305] reported that highly selective sputtering of Si from samples of $TiSi_2$ can be achieved with 300-eV Ar ions and by varying the temperature of the sample. Near room temperature, Si and Ti are sputtered from the sample, but between 500 and 700°C the silicon-depleted surface from sputtering is replenished from the bulk through diffusion processes that result in almost 100% selective sputtering of Si. Zaporozchenko *et al.*[306] studied the surface compositional changes at different ion beam incidence angles for Ar-ion-bombarded Co_2Si. They report that the Co concentration in the altered layer of 4-keV Ar-ion-bombarded Co_2Si is a maximum at a depth of 5–10 nm. The experimental AES data were also compared with model calculations for this system.

Brijs *et al.*[307] used RBS to analyze the subsurface effects of ion bombardment of Pt on Si or $CoSi_2$. They studied the effects of segregation, differential sputtering, and strong variations in the internal distribution of elements within samples during the sputtering process with a variety of ions. The *in situ* RBS data were also correlated with XPS, AES, and SIMS measurements. Silicides have been produced from ion-bombarded silicon–metal multilayers.[308] The formation of $TaSi_2$ from Ta and Si multilayers bombarded with 3-keV Ar ions has been confirmed with XPS and AES.[308]

Metal phosphates such as cubic ZrP_2O_7, $HZr_2(PO_4)_3$, and amorphous $Zr(HPO_4)_2$ are drastically modified by 2-keV Ar-ion doses of 10^{15} to 10^{17} ions/cm^2.[309] XPS reveals that the surfaces of the phosphates decompose yielding mixed metal oxides and polyphosphates as shown in Fig. 5.11. Chen and Rabalais[310] detected a stoichiometry change from LiF to Li_4F_3 with XPS after lithium fluoride was bombarded with 3-keV Ar ions at a dose of about 10^{17} ions/cm^2. The surface composition was metallic lithium and depleted in fluorine. The noble-gas-ion-induced decomposition of carbonates, sulfates, nitrates, and chromates has also been reported.[310,311] Asami *et al.*[312] have used AES and XPS to study the effect of 1- to 10-keV Ar-ion etching on iron and tungsten oxides and sodium chromate. In all cases, the ion bombardment resulted in the reduction of the metal oxides, producing complex XPS spectra with iron and tungsten being reduced to the metallic state.

One of the most widely used procedures in surface science is cleaning sample surfaces with ion beams; the most common contaminant is adventitious carbon. Care must be taken, as demonstrated by Selvam *et al.*,[313] who showed that carbon contamination present on samples of intermetallic compounds, such as $LaNi_5$,

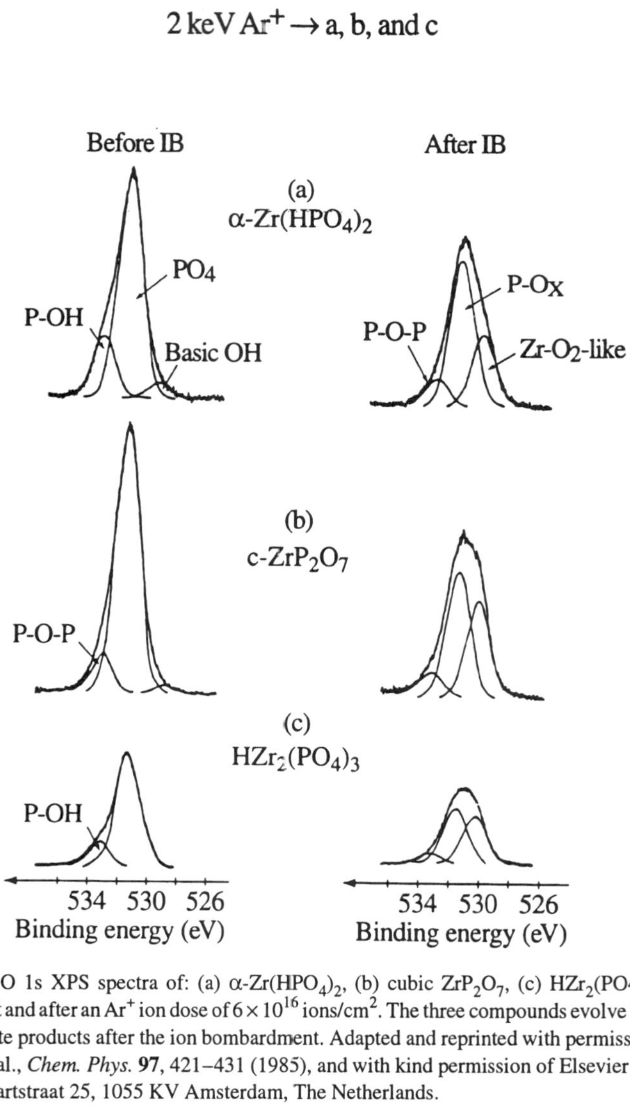

Figure 5.11. O 1s XPS spectra of: (a) α-$Zr(HPO_4)_2$, (b) cubic ZrP_2O_7, (c) $HZr_2(PO4)_3$ before ion bombardment and after an Ar^+ ion dose of 6×10^{16} ions/cm^2. The three compounds evolve toward similar final phosphate products after the ion bombardment. Adapted and reprinted with permission from[309] S. Pignataro, et al., *Chem. Phys.* **97**, 421–431 (1985), and with kind permission of Elsevier Science - NL, Sara Burgerhartstraat 25, 1055 KV Amsterdam, The Netherlands.

$CaNi_5$, TiFe, Mg_2Ni, and Mg_2NiH_4, can form implanted carbides on the surface of the samples with 4- to 8-keV Ar-ion bombardment. Figure 5.12 shows C 1*s* XPS spectra of TiFe as a function of sputter time. The growth of a carbide species is clearly evident from a peak shifted to a lower binding energy from the main carbon XPS peak at 284.8 eV. Possible ways to minimize carbide formation and identify the elements most susceptible to carbide formation were also discussed.(313,314)

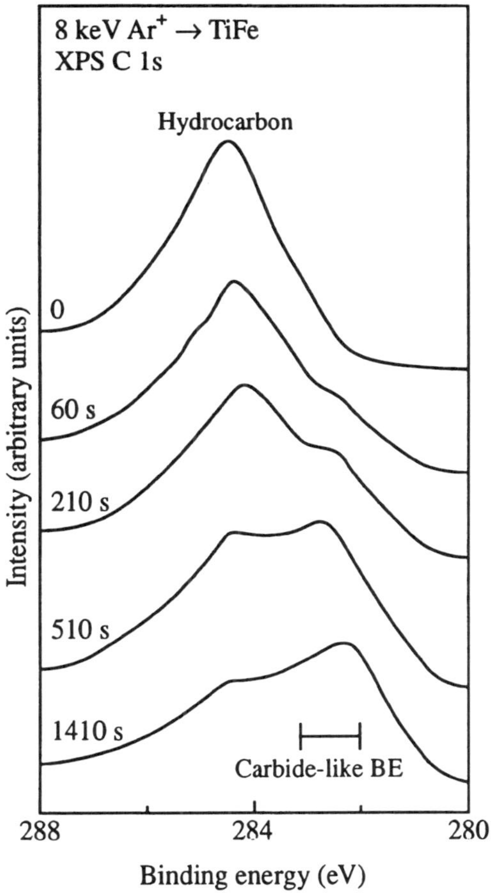

Figure 5.12. C 1s XPS spectra from a TiFe surface that was Ar ion bombarded for different times from zero to 1410 s. Note the formation of an XPS-detectable carbide-like species shifted to lower binding energy from the main hydrocarbon C 1s peak. Adapted and reprinted with permission from[314] E. Selvam, et al., *Journal of Less-Common Metals* **161**, 77–85 (1990), and with kind permission of Elsevier Science - NL, Sara Burgerhartstraat 25, 1055 KV Amsterdam, The Netherlands.

5.7. Calculations and Simulations

In addition to the material already discussed, researchers have used computer simulations and models to try to understand the mechanisms of low-energy ion beam damage in simple systems. Chakarov and Karpusov[(315)] have applied a computer model to the ion-beam-induced atomic mixing of thin films of Co on Si, using only the process kinematics. More general mathematical models have been developed by Sigmund[(316)] and Peinador *et al.*[(317)] to include various thermal and nonthermal effects to describe compositional changes under ion bombardment and sputtering of alloys, compounds, and isotropic mixtures. Krivobokov and Pashchenko[(318)] also developed a model to explain general atomic transport properties in ion-irradiated solids. Monte Carlo simulations for the ion-beam-induced separation of ^{10}B and ^{11}B isotopes have been employed in models by Zheng *et al.*[(319)] Their results indicate that the observed isotope separation is independent of the emission angle of the ejected beam from 0° to 60° at fluences of less than 10^{15} ions/cm^2. Carter and co-workers[(320,321)] developed an altered-layer model for sputter depth profiling consisting of an outer sputtering layer, from which atoms are ejected by the ion flux, and an adjacent layer in which the atomic constituents are spatially homogenized by bombardment-induced mixing.

Kelly and Miotello[(322,323)] addressed ion-beam-mixing phenomena for a variety of metallic bilayer systems, using thermodynamic considerations of the heat of mixing of the materials, i.e., the change in enthalpy of mixing, which they refer to as a chemical driving force. After studying many bilayer systems, they determined that the extent of ion beam mixing is sensitive to both the sign and magnitude of the enthalpy of mixing. Protsenko and Chaikovskii[(324)] extended models to explain how impurities diffuse in samples exposed to low-energy ion beam bombardment. Kang *et al.*[(325)] compared Monte Carlo computer simulations with experimental data in the bombardment of Au–Cu and TiC with 2-keV Ar ions. The results suggest that the altered-layer formation in TiC is caused by the preferential sputtering of the lighter atom, whereas in the Au–Cu system differential sputtering, "radiation-enhanced diffusion," and "radiation-induced" segregation all play significant roles in the formation of a Au-enriched top atomic layer. This altered layer is followed by a layer depleted in Au and then by alloys with bulk composition. Zheng *et al.*[(326,327)] used Monte Carlo methods to study differential sputtering of PtCu binary alloys. The effects of differential sputtering and enrichment transformation were discussed. Ackermans *et al.*[(328)] have studied with low-energy ion scattering the surface compositional changes of CuPd alloys bombarded with 3-keV Ne ions. Their results suggest that contaminant levels of S segregate to the top of surface Cu atoms, and that Cu is preferentially sputtered.

6. Depth Resolution: Sample, Beam, and Instrumental Effects

6.1. Introduction

In this section, we consider the effect of the initial sample condition, ion beam variables, and instrumental parameters on ion-beam-induced effects, especially roughening and depth resolution. Although the intrinsic nature of the sputtering process will induce changes in the sample and degrade the depth resolution, the initial condition and nature of the sample may further contribute to the changes. In many cases, the initial sample condition cannot be changed, but an understanding of the likely effects of the sample can aid the selection of more desirable experimental conditions and simplify data interpretation. Instrumental parameters can often be optimized to minimize the interface degradation; conversely, poorly chosen instrument settings can cause further degradation. Examples of optimizing the instrumental parameters and conditions are presented, along with a discussion about the ultimate limits of depth resolution in sputter profiling.

6.2. Nature and Condition of the Sample

The condition of the sample can have a profound effect on the type and extent of changes induced by ion beam bombardment. The cleanliness of the sample surface, the form of the sample (single crystal, polycrystalline, amorphous, etc.), the initial surface and interface roughnesses, sample composition, homogeneity, and even the method of preparation can all be important factors. The depth of an interface and the number of layers may also have an effect on the depth resolution.

Sputtering is often used to obtain a clean surface by removing adventitious carbon and other contaminants, particularly those resulting from sample exposure to air or from diffusion of impurities from the bulk to the surface during annealing. However, the contamination can be implanted deeper into the surface as a result of the ion beam impact, and may seed the formation of sputter-induced roughness, as discussed in Sec. 4. Contamination may also change the sputter yields and erosion rate. Some impurities, such as carbon or oxygen on reactive metal surfaces, or contaminants on a scratched or rough surface, may not be completely removed with sputtering, although the surface concentration can generally be reduced to the order of a few percent. Heating the sample, either during sputtering or alternating with sputtering, is sometimes helpful for reducing contamination levels.

The physical form of the sample affects the uniformity of the sputtering process. The uniformity of erosion generally decreases along the series of amorphous, crystalline, and polycrystalline material forms.[329] Although single-crystal surfaces may sputter relatively uniformly, channeling of the ion beam along low-atom-density directions may greatly increase the ion beam penetration and increase the depth of the damage zone while decreasing the erosion rate. Defects in single-crystal samples also tend to increase the roughness generated by sputtering.[183,329] The sputtering rate usually varies for different crystalline faces (Fig. 2.15); thus, the individual grains in a polycrystalline sample are likely to erode at different rates. By this mechanism, considerable roughness may be generated across the sample surface, accompanied by shadowing and redeposition effects. The roughness caused by sputtering depends on the particular crystal face. Dawson and Petrone report[330] that in sputtering polycrystalline Mo, high-index crystal planes become rougher than low-index faces. The variation in roughness affects the intensities of the Auger signals and results in errors of up to 20% in the compositional analysis. Powder samples represent a particular challenge. The outer surface material is continually exposed as the edges are uncovered with sputtering, and individual particles may be moved around or even blown off the sample with sputtering.[331] A slow erosion rate may remove small amounts of surface contamination and allow for a more representative analysis of the bulk powder material, but true profiling through the particles may not be possible. Smooth, homogeneous amorphous layers tend to have the most uniform erosion rates, because orientation and nonuniformity effects are not a factor.

Defects in the sample are still a problem; small hillocks may be smoothed, but pinholes become larger and shallower, with degradation of the depth resolution Δz until the pinholes coalesce.[183] Surface defects such as scratches and local stresses will roughen initially with sputtering and will gradually be smoothed under the proper conditions. However, a homogeneous sample containing a statistical distribution of defects in the bulk will continually roughen with sputtering.[6] With the use of sample rotation (Sec. 6.4); roughness may be reduced to 1–2 nm, at which point atomic mixing becomes the dominant factor in Δz degradation.[183]

Most samples will roughen with prolonged ion beam bombardment (Sec. 6.5), and any initial roughness of the sample surface will generally propagate with depth and increase as the sample is sputtered. In polycrystalline samples, different crystalline faces not only sputter at different rates, but on a rough surface planes inclined at different angles with respect to the ion beam also have different sputter rates. This is because the effective ion beam incidence angle is different for each incline.[6] Again, the situation is compounded by the shadowing of some areas of the surface by protrusions and redeposition and resputtering effects. The effect on the interface width is that the smallest values of Δz are obtained on smooth samples, and progressively larger values of Δz are found on rough samples. Zalar and Hofmann[332] studied the effects of different amounts of initial surface roughness on the values of Δz for sputtered Cr–Ni multilayer samples for which the initial surface consisted of smooth and rough areas. The smooth part of the surface had a roughness amplitude R_a, of 0.012 μm, and the rough part of the surface had an R_a of 0.920 μm. Figure 6.1 illustrates typical results for Δz as a function of sputtered depth, for samples with various proportions of smooth and rough surface areas.[332] The lower curve is for the relatively smooth surface and the top curve is for the relatively rough sample. The middle curve represents a sample in which half the analyzed area was smooth and half was rough. It is clear that Δz increases with an increasing fraction of rough surface in the sampled area. Also evident is the increase in Δz with eroded depth, even for the initially smooth surface. The magnitude of the initial roughness is not alone responsible for the increase in Δz; in fact, the angular distribution of the different microplanes is a more critical factor for causing different sputtering rates, shadowing, and redeposition;[(329)] again, powdered samples are a particular problem. The increase in Δz with depth saturates at large erosion depths and is sample specific.

The possibility that sputter-enhanced surface roughening might be minimized by increasing surface diffusion by sputtering at elevated sample temperatures has been examined in several studies.[(333–335)] For silver films on polished copper substrates, the sputter-induced surface roughness decreased substantially at elevated sample temperatures,[(333,334)] but there was no direct correlation between the interface width and the surface roughening. At 423 K, the interface width was significantly smaller than at room temperature, but Δz increased at 498 K even though the surface was smoother than it was for the sample sputtered at 423 K.[(333,334)] Studies performed on a Cr–Ni multilayer stack showed a large asymmetric increase in interface broadening at elevated temperatures, with much greater

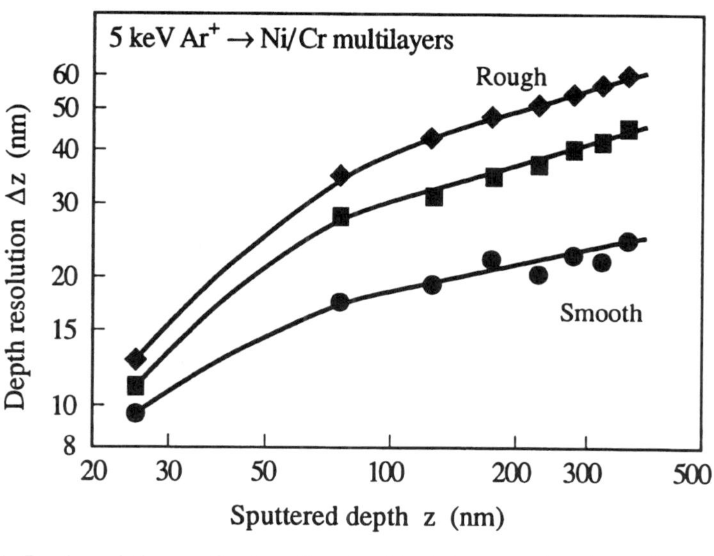

Figure 6.1. Depth resolution as a function of sputtered depth for a smooth and a rough sample, and a sample (middle curve) with both smooth and rough areas, illustrating degradation of the value of Δz with an increasing proportion of rough surface in the sample area. Adapted and reprinted with permission from[332] A. Zalar and S. Hofmann, *J. Vac. Sci. Technol., A* **5**(4), 1209–1212 (1987). Copyright 1987 American Vacuum Society.

broadening at the Cr-to-Ni interface than at the Ni-to-Cr interface.[(335)] Clearly, surface roughness is not the only factor contributing to interface broadening. Ion-beam-bombardment-enhanced interdiffusion of material across interfaces and film nucleation may also be important, and these effects may increase at higher temperatures. Thus, the effects of increased sample temperature on Δz are highly material dependent, and they also depend on temperature. For thin films, interdiffusion between layers may dominate and counteract any smoothing effect of surface diffusion, but for thicker films or for systems in which the interdiffusion is slow, increasing the sample temperature may significantly improve Δz by reducing topographical roughness.[(334)]

The chemical form of the sample also affects the depth resolution. Some idealized situations are shown in Fig. 6.2.[(336)] Obviously, planar, homogeneous layers are most likely to have optimal depth resolution. Any inhomogeneities, which include defects and different chemical or physical phases that are either in the plane of the surface or normal to the surface, can affect the sputter-induced effects (e.g., local erosion rates, differential sputtering, diffusion, and segregation), and can affect the composition profile. Roughness or diffusion, or both, at the interfaces of multilayered films may be difficult to distinguish from the sputter-induced broad-

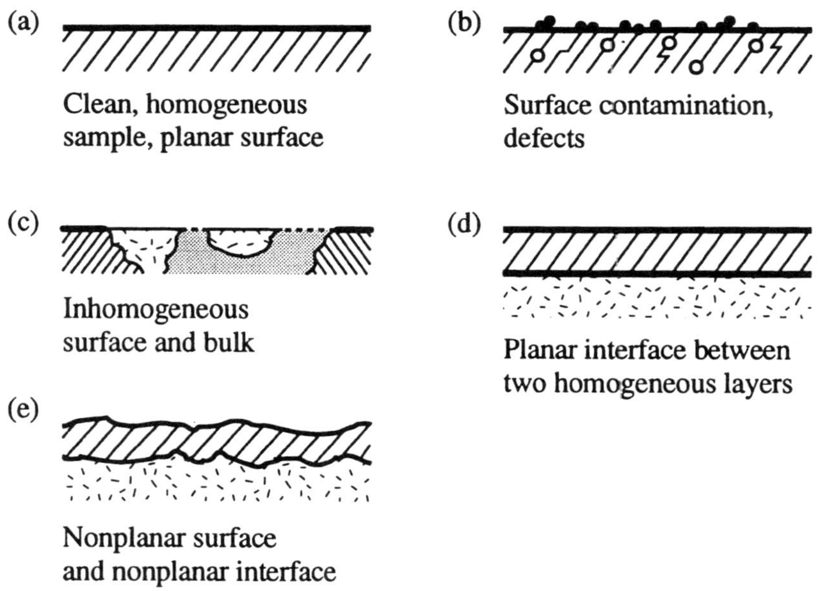

Figure 6.2. A variety of idealized sample, surface, and interface conditions that will affect the depth resolution. Adapted and reprinted with permission from[336] C. J. Powell, et al., in: *Ion Spectrotroscopies for Surface Analysis*, edited by A. W. Czanderna and D. M. Hercules (Plenum Press, New York, 1991), pp. 417–437. Copyright 1991 Plenum Press, New York.

ening of interfaces. Layers that are identical in chemical composition but fabricated in different ways may exhibit different sputter-induced characteristics, so sample preparation may affect the depth resolution.[(337)] Surface charge on insulating samples may deflect the ion beam, the analysis beam, or both, and result in nonuniformities in the erosion rate or affect the diffusion of charged species, often unpredictably.[(329)]

To summarize, degradation in depth resolution may result from the nature and condition of the sample because (1) any contamination may be implanted in the sample and may affect the erosion rate; (2) variations in the sample structure (crystalline, amorphous) have a strong effect on erosion rates and topography development; (3) defects in the sample initiate or enhance some bombardment-induced effects, or both; (4) an initially rough surface or interface is usually propagated and becomes rougher under ion beam bombardment; (5) sample inhomogeneity, including the presence of layers, affects the uniformity of erosion; (6) multilayers may exhibit asymmetric effects (i.e., an A/B edge may broaden differently from a B/A edge); (7) of the depth of the interface, (8) bombardment effects may be influenced by sample preparation and fabrication methods; and (9) changes in the sample conductivity may adversely affect the uniformity and alignment of the ion beam. Many of these sample parameters cannot be changed, but an understanding of the likely effects can aid in interpretation and understanding of the experimental results.

6.3. Ion Beam Bombardment Effects

The ion beam bombardment parameters (beam energy, current density, angle of incidence, etc.) play a large role in the depth resolution. Interface widths tend to increase with increased ion beam energy. Higher beam energies allow for deeper penetration into the sample, and hence a thicker altered layer. Furthermore, the depth at which differential sputtering reaches steady state also depends on ion beam energy, and roughly correlates with the range of the ion in the material [Sec. 2, Eq. (1)], a particularly important result for very thin films.[(338)] Figure 6.3 shows the dependence of Δz on ion beam energy for Ni–Cr interfaces in a multilayer thin film material[(339)] sputtered with 1-, 2-, 3-, and 4.5-keV Ar ions. There is a marked degradation in Δz with increasing energy, and the largest rate of change in Δz occurs between 1 and 2 keV. The curves have similar shapes, suggesting that similar mechanisms are contributing to the interface broadening at various eroded depths, independent of beam energy, with the beam energy primarily affecting the magnitude of the effect.[(339)] An ion gun has been developed that can be operated with energies as low as 300 eV and results in a further reduction in the interface widths observed with SIMS.[(340)] In another study,[(166)] AFM was used to examine the effect of beam energy of Ne, Ar, and Kr ions on the topography development in terms of the rms surface roughness of highly polished InP(100) samples. Energies of 0.5, 1,

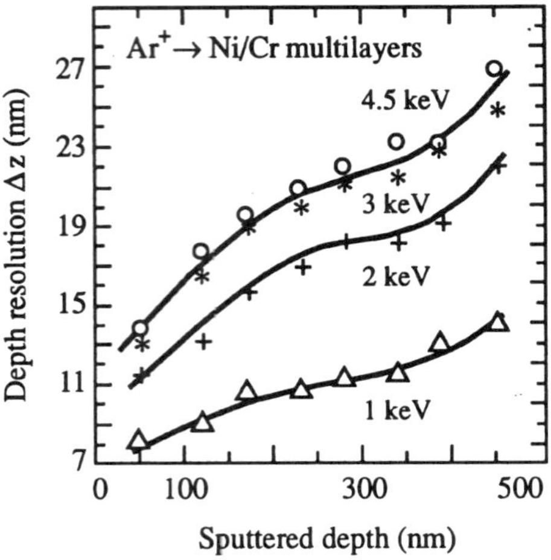

Figure 6.3. Depth resolution for Ni/Cr interfaces in a multilayered specimen as a function of sputtered depth and for different Ar ion beam energies. Depth resolution deteriorates at larger depths and for higher beam energies. Adapted and reprinted with permission from[339] J. Fine, et al., *J. Vac. Sci. Technol., A* **3**(3), 1413–1417 (1985). Copyright 1985 American Vacuum Society.

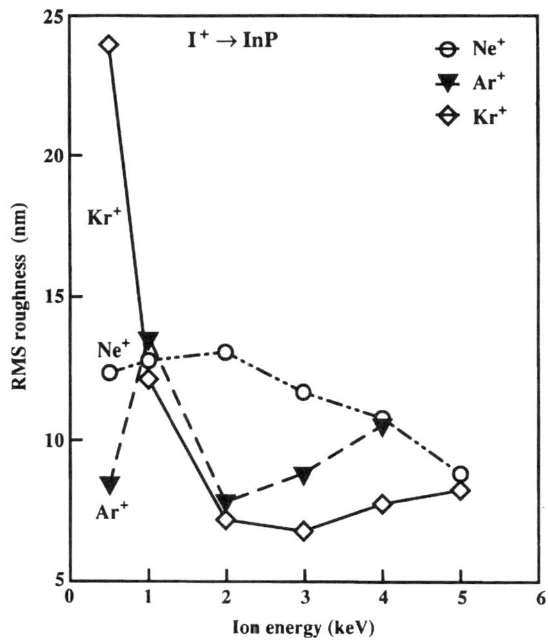

Figure 6.4. Rms surface roughness as a function of ion beam energy for Ne-, Ar-, and Kr-ion bombarded InP with an initial rms roughness of 0.19 nm. Adapted and reprinted with permission from[166] C. M. Demanet, et al., *Surf. Interface Anal.* **24**, 497–502 (1996). Copyright 1996 John Wiley and Sons.

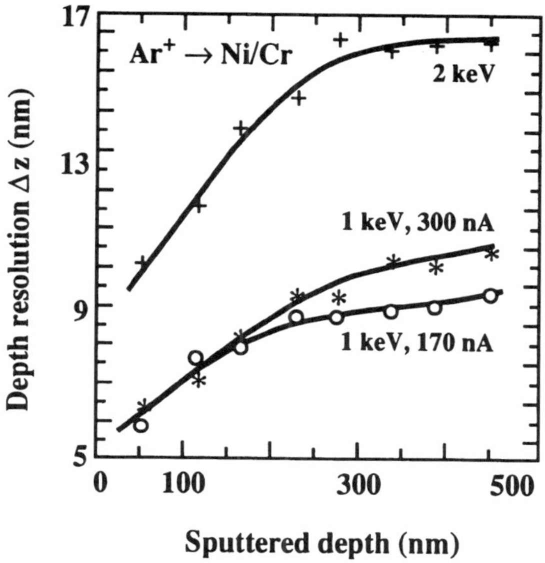

Figure 6.5. Depth resolution as a function of sputtered depth for Ni/Cr interfaces in a multilayered specimen for different ion beam currents and energies. Adapted and reprinted with permission from[339] J. Fine, et al., *J. Vac. Sci. Technol., A* **3**(3), 1413–1417 (1985). Copyright 1985 American Vacuum Society.

2, 3, 4 and 5 keV were used, with a constant current density for each ion species and an ion beam incidence angle incidence of 41°. The results are shown in Fig. 6.4. There are differences in the shapes of the curves for the three ion species, but in all cases there is a critical energy for which the surface roughness is maximized. This energy tends to be at the low range of the beam energies investigated, and is at a lower energy for the higher-ion-mass species. However, both the maximum and minimum roughnesses depend on the ion beam incidence angle and ion dose, as well as on the beam energy and species.

Both the ion beam current density and the total ion dose affect the depth resolution. The depth resolution deteriorates with increased ion beam current density and is thought to result from the greater number of defects produced and the increased rate of material diffusion. If this is the case, the change in Δz is related to the defect lifetime (Sec. 3). However, there is also a known dependence of topographical development on current density, and this may also affect the depth resolution.[339] Figure 6.5 shows the effect of increased ion beam current on the Δzs of Ni–Cr interfaces, with the two lower curves representing measured interface widths for 1-keV Ar ions at beam currents of 170 and 300 nA.[339] For eroded depths greater than 150 nm, the difference is measurable. The upper curve is for 2-keV Ar-ion bombardment, and it appears that the effect of increased beam energy is greater than that of increased current density. In another study, the effect of total Ar-ion dose on topography was examined for InP by using AFM.[(174)] Figure 6.6 shows the rms roughness as a function of ion dose for two conditions: 5-keV Ar ions at 71° incidence and 0.5-keV Ar ions at 41° incidence angle. The rms roughness is seen to be nearly constant in both cases at doses below 5×10^{15} ions/cm^2 and increases with the logarithm of the total ion dose thereafter. Ripple formation was noted at doses of 5×10^{15} ions/cm^2 for 71° bombardment and at $\sim 10^{18}$ ions/cm^2 for 41° ion beam incidence.

The ion beam incidence angle also has an effect on the interface width. For smooth Ni–Cr samples with interface depths ranging between 50 and 226 nm, Δz is a maximum at an incidence angle of about 30° and then decreases for more grazing angles, as seen in the data of Fig. 6.7.[(339)] This result is thought to arise from a combination of lower effective beam current densities and a decrease in the depth of the collision cascade at more oblique incidence angles.[(339)] In a similar study on Ni–Cr interfaces performed with both Ar and Xe ions, consistently smaller

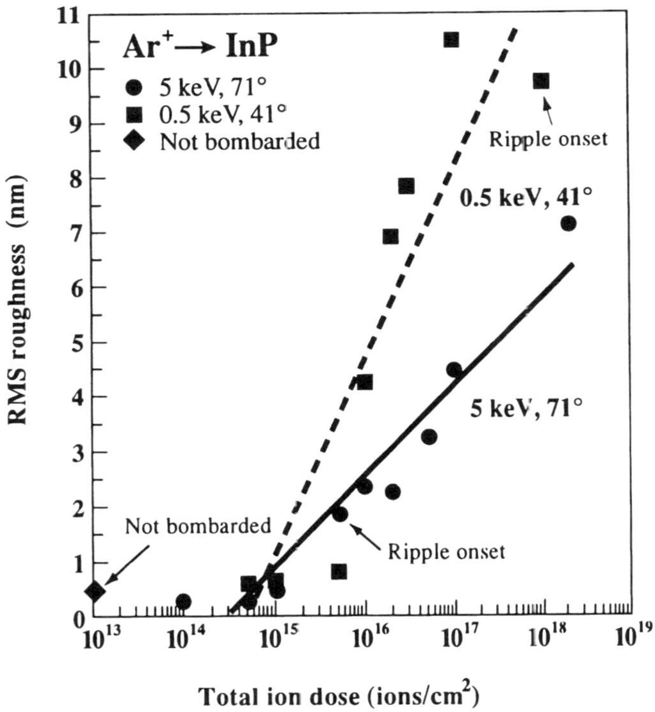

Figure 6.6. Rms surface roughness as a function of total ion dose for Ar-ion-bombarded InP at two different ion beam energies and incidence angles. Adapted and reprinted with permission from[174] C. M. Demanet, et al., *Surf. Interface Anal.* **23**, 433–439 (1995). Copyright 1995 John Wiley and Sons, Ltd.

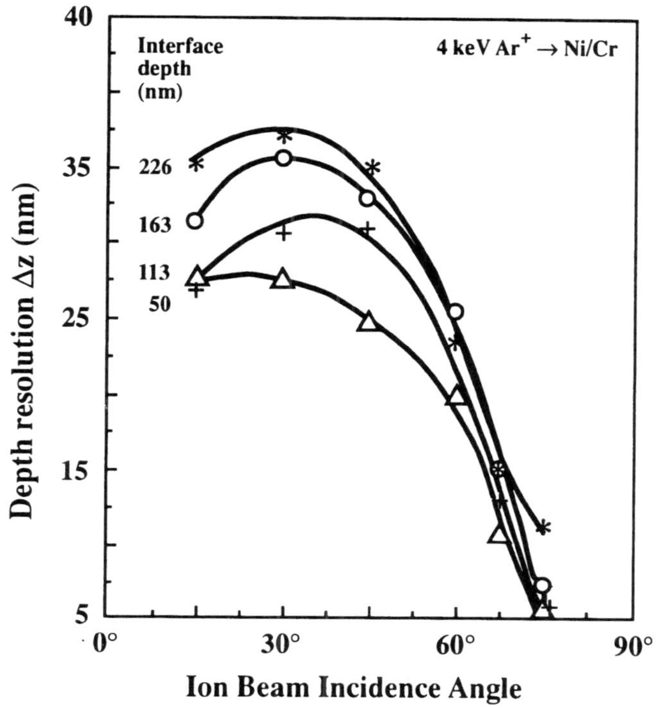

Figure 6.7. Depth resolution for Ni/Cr interfaces in a multilayered specimen as a function of ion beam incidence angle and for different interface depths for 4-keV Ar-ion bombardment. Adapted and reprinted with permission from[339] J. Fine, et al., *J. Vac. Sci. Technol., A* 3(3), 1413–1417 (1985). Copyright 1985 American Vacuum Society.

values of Δz were found for Xe sputtering, except at 80° incidence, where the Δz values were essentially the same.[(341)] However, the behavior of Δz with ion beam incidence angle was different for Xe: Δz showed a local maximum at 35°, but the absolute maximum occurred at 0° (normal incidence).

The effect of angle of incidence on topographical development on InP was also investigated with AFM.[(191)] Samples were bombarded with 0.5-keV Ar ions and 0.5- and 5-keV Kr ions at ion beam incidence angles of 1°, 11°, 21°, 31°, 41°, 51°, 61°, 71°, and 81°. The results for the rms roughness as a function of ion incidence angle are shown in Fig. 6.8. Both sputter cone and ripple formation were observed. With 0.5-keV Ar ions, the size of the sputter cones increased with increasing incidence angle to a maximum at 45°. Further increases in the incidence angle resulted in a decreasing cone size and the appearance of ripples. For 0.5-keV Kr ions, the maximum cone size appeared at 21° and ripples formed at larger angles. Note also that the samples actually became smoother with sputtering at glancing

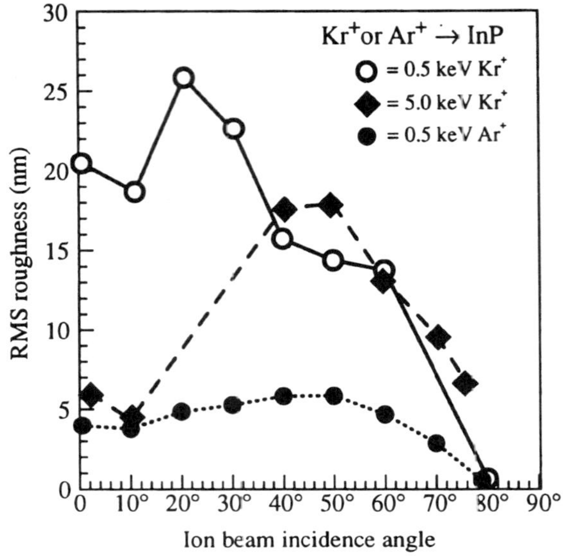

Figure 6.8. Rms surface roughness as a function of ion beam incidence angle for Ar- and Kr-ion bombarded InP. Adapted and reprinted with permission from[191] C. M. Demanet, et al., *Surf. Interface Anal.* **24**, 503–510 (1996). Copyright 1996 John Wiley and Sons, Ltd.

incidence. The authors noted a similarity between trends in the rms roughness and experimental sputter yield data.(191) For initially rough samples, using a grazing-ion-beam incidence angle severely degrades the depth resolution because the initial roughness propagates and becomes worse from the enhanced effects of shadowing and redeposition.(329,342) Normal-ion-beam incidence is often recommended for sputtering initially rough samples to minimize further interface degradation; however, Hofmann(329) reports that the effects of increasing ion beam incidence angle are not usually significant below 60°.

At grazing incidence, beam penetration is minimized and the collision cascade is restricted to the near-surface layers.(270) Profiles through 25 nm of Al deposited on InP substrates showed a considerable improvement in interface width with sputtering at glancing incidence angles. Measured interface widths using Ar and Xe ions at 3 and 5 keV and at 18° and 78° incidence angles are shown in Table 6.1. Mean interface widths of 13.5 nm were measured at an 18° incidence angle, and 7 nm at 78° for 5-keV Ar ions.(270) Lee *et al.*(343) used medium-energy ion scattering to examine the type and depth of the damage layer formed on Si(100) with 3-keV Ar-ion bombardment at different incidence angles. They found that an amorphous surface layer was formed with sputtering, and that the thickness of this layer at a saturation dose was approximately 14.2 nm thick at normal incidence but only 4.8

Table 6.1. Interface Width for Al–InP for Ar- and Xe-Ion Bombardment at 18° and 78° Relative to the Surface Normal

Ion	Ion energy (keV)	Incidence angle θ	Interface width (nm) In	P	Mean
Ar	31	18°	10.8	6.6	8.2
Ar	3	78°	6.2	4.9	5.5
Ar	5	18°	18.7	8.3	13.5
Ar	5	78°	8.7	5.4	7.0
Xe	5	18°	14.4	8.8	11.6
Xe	5	78°	8.5	7.7	8.2

From Ref. 270.

nm thick at 80° incidence. In another study the optimal combination of ion energy, total dose, angle of incidence, and substrate temperature was explored for cleaning GaAs substrates with minimum damage.(344)

The primary-ion species used for sputtering affects the depth resolution; in general, Δz is expected to decrease with higher-mass primary ions, although this is not always the case. Figure 6.9 shows measured interface widths for polycrystalline Al, Mo, Ag, and Ta films sputter-deposited onto Si wafers and for which Ar, Ne, and Xe ions were used.(345) In all cases the best value for Δz was obtained with Xe ions, but it can be seen that the differences are not always large, especially for films with high-atomic-weight components. The erosion rates also vary significantly with the mass of the incident ions; for Mo on Si, for example, an order-of-magnitude difference in sputtering time is obtained with Xe and Ne ions, i.e., ~30 min for Xe versus ~380 min for Ne. Extremely slow erosion rates may permit build up of detrimental contamination. In another study,(346) Δz values for Cr–Ni multilayers were found to decrease by about 50% when using an unrastered 0.9-keV Xe-ion beam as compared to an Ar-ion beam, and the difference was attributed to reduced surface roughness when using Xe ions. However, in Fig. 6.5 we note that for InP there is no simple dependence of sample roughness on ion mass, especially at low ion beam energies, and the roughness varies greatly at different ion beam incidence angles. Referring to Table 6.1, one can see that the interface width for P in an Al–InP system is the same, or even slightly larger with Xe ions compared to Ar ions. SIMS profiles(347) through alternating layers of isotopic Si (^{30}Si/^{28}Si) showed the expected decrease in Δz values with ion mass for Ne, Ar, Xe ions for the leading edges of each interface, but did not show a dependence on ion mass for the values of Δz for the trailing edge (note Fig. 1.1c that was done for Ta_2O_5).

Marker layers, consisting of a buried delta-like distribution of a dopant material in a sample, are very useful for theoretical and experimental studies of depth

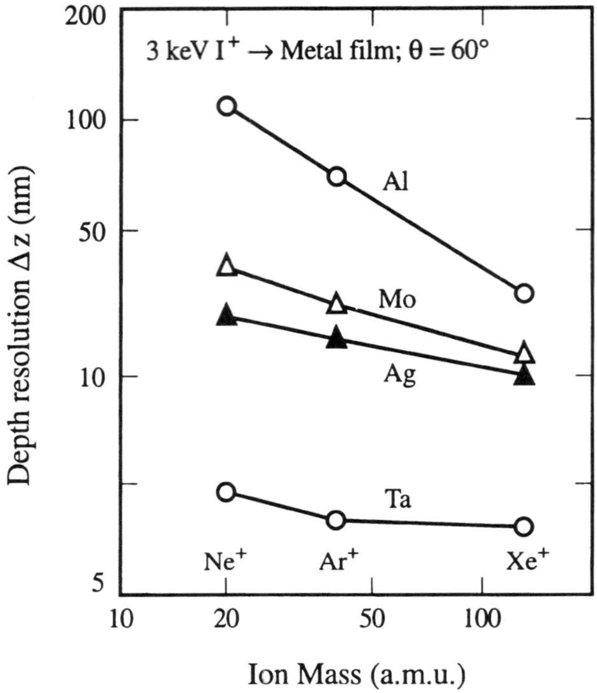

Figure 6.9. Depth resolution for Al, Mo, Ag, and Ta metal films on Si wafers as a function of ion mass (Ne, 20; Ar, 40; Xe, 131). Adapted and reprinted with permission from[345] M. Tanemura, et al., *Surf. Sci.* **230**, 283–289 (1990), and with kind permission of Elsevier Science - NL, Sara Burgerhartstraat 25, 1055 KV Amsterdam, The Netherlands.

resolution. For sufficiently thin (i.e., ~1 nm or less) atomically sharp marker layers, the experimentally measured depth profile identifies the sputter-induced resolution function.[(13,329)] The resolution function can be used to numerically deconvolute experimental depth profiles through thicker layers to obtain the "true" sample profile. Further details of this procedure can be found in Refs. 13 and 329 (see also Chap. 5). Experimental profiles through marker layers clearly show the asymmetry of the broadening, as evidenced by much sharper slopes in the profile at the leading edge and exponential-like curves at the trailing edge of the distribution. Experimental evidence indicates that the broadening of the leading edge depends on the mass and energy of the primary-ion beam (the broadening increases with increasing energy and decreases with increasing mass), while the trailing-edge broadening is independent of the ion beam mass.[(347,348)] By extension, trailing-edge broadening was found to be independent of the stopping power or ion beam range; and the experimentally measured broadening of B marker layers in Si or SiGe confirmed this.[(349)] Recent theoretical formulations have been derived that reproduce the experimentally observed asymmetry.[(348,350,351)]

From this discussion, it is clear that some general trends can be identified. However, the effects of ion beam parameters on the depth resolution are complex, interdependent, and highly sample dependent.

6.4. Instrumental Effects

Instrumental parameters and capabilities may have a pronounced effect on the depth dependence of Δz. Although the initial sample condition may not be altered unless permitted by the research being done, some instrumental parameters can be optimized as well as the beam parameters discussed in Sec. 6.3. First, it is assumed that an ultrahigh vacuum system is being used. At base pressures above 10^{-9} Torr, the arrival rates of contaminants from the vacuum onto the sample may be high enough to compete with and affect the rate of erosion.[329,352] High-purity (99.99+%) gases must be used for the ion beam to avoid implanting impurities. Differentially pumped ion guns, which allow system pressures $<5 \times 10^{-8}$ Torr during sputtering, are preferred over guns that require backfilling the entire chamber to pressures of ca. 10^{-5} to 10^{-6} Torr. If backfilling is required, a system maintaining some pumping or flow of gas reduces adsorption from the background compared to a static, unpumped system, particularly for prolonged sputtering times.[329]

The instrument used for sputtering must be designed properly and kept at performance specifications. The ion beam must have a narrow energy spread, constant-current density, and Gaussian-shaped beam to ensure uniform erosion. The proper alignment of the ion beam and analysis beam (when using AES or XPS), or appropriate gating of the signal (for ISS and SIMS), is critical to avoid contributions from the crater wall to the detected signal and serious loss of depth resolution.[353] For most applications and where optimal depth resolution is desired, rastering the ion beam is recommended, both to minimize the effect of any beam inhomogeneities and to provide a large enough area for analysis without interference from crater walls.[329] The flatness of the crater in the analyzed area (volume) is absolutely critical.[6] Figure 6.10 shows three crater profiles.[10] The crater in part (a) is nearly ideal, and was obtained on a Si single wafer with a 300-μm-diameter ion beam that was rastered over an area of 1.6 mm × 1.6 mm. The second profile (b) is from a Ni alloy and was obtained with a static, unrastered 600-μm-diameter ion beam. It would be difficult to find an area to analyze that would not span multiple depths in that sample, and the area of maximum eroded depth is not in the center of the sputter crater, thus making alignment of the analysis beam difficult. The last profile was taken on oxygen-free, high-conductivity (OFHC) Cu, and in this case a 300-μm beam was rastered over an area of 1 mm × 1 mm. While this crater does have a near-Gaussian "average" envelop, the local nonuniformities are likely to degrade the depth resolution. Well-characterized standards, e.g., Ta_2O_5 on Ta and Ni–Cr multilayers, are available for use in characterizing and optimizing the basic ion gun

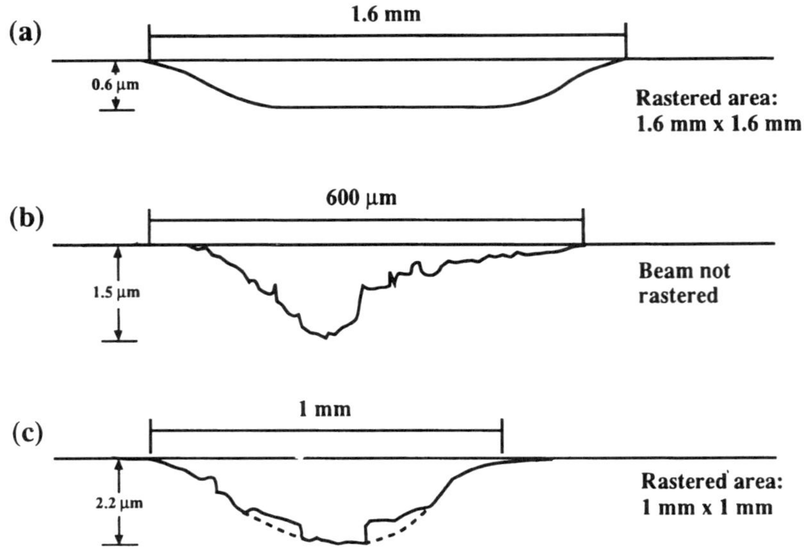

Figure 6.10. Crater profiles for rastered and unrastered samples with 0.3-mm FWHM ion beams; (a) rastered 5.5-keV O_2 at 45° on a Si single crystal, (b) unrastered 5.5-keV O_2 at 45° onto NiBe alloy with dz/dt at 260 nm/min, and (c) rastered 5.5-keV O_2 at 45° onto OFHC Cu with dz/dt at 160 nm/min. Adapted and reprinted with permission from H. W. Werner and H. A. M. de Grefte, Nat. Lab. Rep., 4723 (1972), unpublished. See also Fig. 5.49 in Ref. 10.

parameters and to determine nominal etch rates; their use is highly recommended.[6,339,353]

Rotation of the sample during sputter etching can significantly reduce sputter-induced topography that arises from variations in the local sputtering rate. Effects of shadowing and redeposition are also minimized.[200,342,354] Since topography development is one of the most significant factors causing degradation in Δz, sample rotation often results in a marked reduction in Δz. The best results are obtained on initially smooth surfaces and with grazing-incidence ion beams, but even on initially rough surfaces significant improvement may result and those surfaces may become smoother.[200,342,354] Figure 6.11[355] shows AFM images of two layers of Ti–TiN on a Si substrate after sputtering with and without rotation. The initial roughness of the polycrystalline surface is evident in the top image (a). The second image (b) shows the sample after bombardment with 4-keV Ar ions, rastered over a 3-mm × 3-mm area. The roughness is even more pronounced after sputtering, and larger features are evident. The last image (c) shows results for a similar amount of sputtering done while rotating the sample; it is seen that the roughness is greatly reduced, and appears to be even less than the initial roughness. Rotation helps counteract the effects of sample geometry and variations in the crystalline orienta-

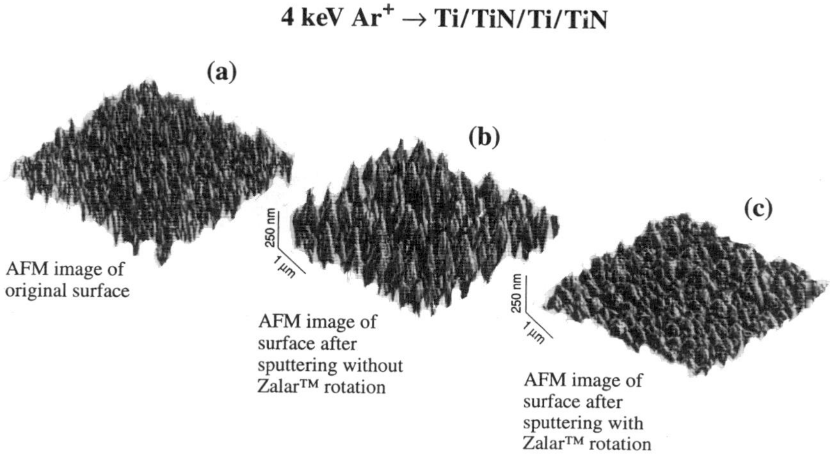

Figure 6.11. The effect of sample rotation on the ion-beam-induced roughness as shown by AFM images of Ti/TiN multilayer samples (a) before and (b,c) after 4-keV Ar-ion bombardment (b) without and (c) with Zalar™ rotation. Adapted and reprinted with permission from[355] Charles Evans and Associates®, Zalar rotation enhanced Auger depth profiles, Auger Application Note, Charles Evans and Associates, Redwood City, CA, (1996).

tion on the development and propagation of roughening. Figure 6.12, which is for the same sample used in Fig. 6.11, shows the effect of sample rotation on a sputter depth profile with analysis by AES. This sample was eroded to a depth in excess of 500 nm, so the depth scale is compressed, but the improvement in Δz is still evident, particularly for the deeper layers. The interface width for the rotated sample is less than half that for the nonrotated sample; furthermore, the degradation of Δz with total sputter depth is reduced. Some features apparent in the rotated sample are not detectable in the data from the unrotated sample because of the lack of depth resolution. This example illustrates how an increase in Δz values may smear out subtle features and make them undetectable.

Other studies illustrate that not only is Δz markedly improved with sample rotation, but that for some systems the value of Δz remains essentially constant for multilayer samples up to erosion depths of micrometers, even for polycrystalline samples.[(183,200,354,356)] Figures 6.13–6.15 show the interface width as a function of depth for stationary (S) and rotated (R) samples obtained from AES, XPS, or SIMS analyses.[(357)] Using grazing ion beam incidence results in marked improvement in Δz for smooth samples; grazing-ion beam incidence along with rotation gives some further improvement in Δz for smooth samples, but the incremental improvement is not large. However, for initially rough samples, grazing-ion beam incidence alone results in severe degradation of Δz. When used with sample rotation, the Δz value for grazing-ion-beam incidence for rough samples is greatly reduced, often to a

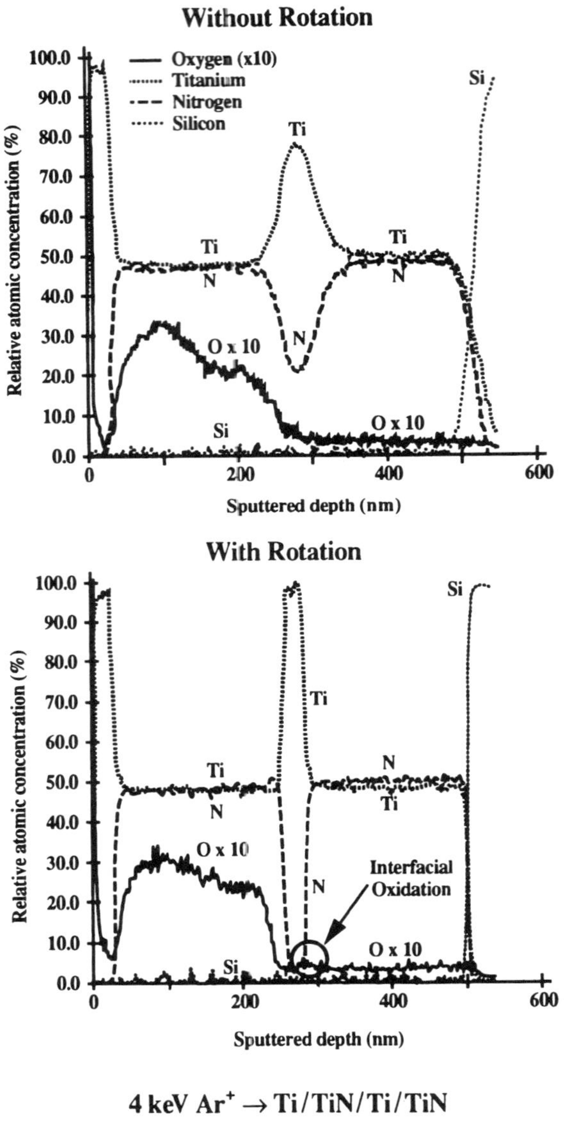

Figure 6.12. Auger compositional depth profiles for 4 keV Ar ion sputtered Ti/TiN multilayer samples obtained with and without Zalar™ rotation, showing improved depth resolution for the rotated sample. Adapted and reprinted with permission from[355] Charles Evans and Associates®, Zalar rotation enhanced Auger depth profiles, Auger Application Note, Charles Evans and Associates, Redwood City, CA, (1996).

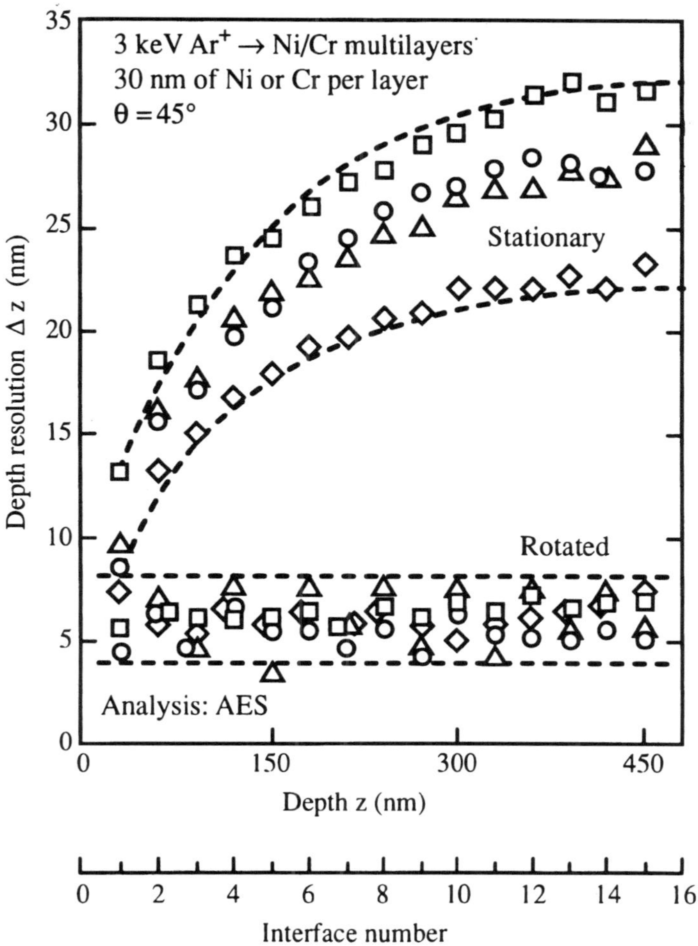

Figure 6.13. Depth resolution for Ni/Cr as a function of interface depth for Ni/Cr multilayer specimens measured from AES profiles with and without sample rotation during ion bombardment. The different symbols represent measurements obtained at four different laboratories. Adapted and reprinted with permission from[357] S. Hofmann, et al., *Surf. Interface Anal.* **20**, 621–626 (1993). Copyright 1993 John Wiley and Sons.

comparable or better value than that obtained with stationary sputtering at normal incidence.[(342)] In general, sample rotation greatly reduces the effect of ion beam incidence angle on Δz, in many cases to the point where Δz is nearly independent of the incidence angle.[(39)] Figure 6.16 shows the interface width as a function of incidence angle for stationary and rotated specimens for smooth Al films deposited on TiN.[(209,358)] Interface widths generally improve with sample rotation and are

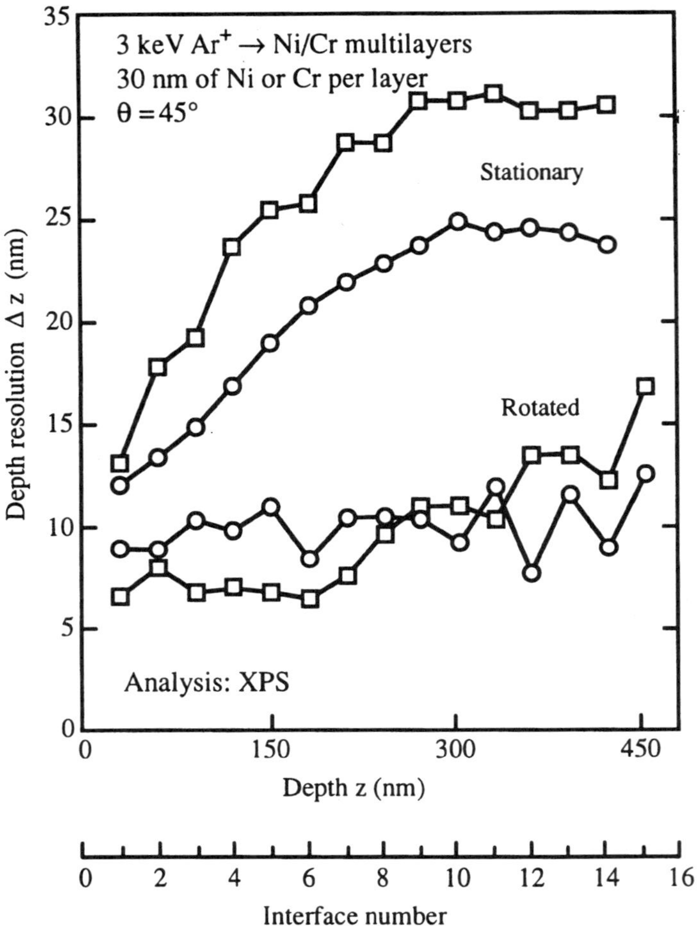

Figure 6.14. Depth resolution as a function of interface depth measured from XPS profiles for Ni/Cr multilayer samples with and without sample rotation during ion bombardment. The two symbols represent measurements obtained by two laboratories. Adapted and reprinted with permission from[357] S. Hofmann, et al., *Surf. Interface Anal.* **20**, 621–626 (1993). Copyright 1993 John Wiley and Sons.

largely independent of ion beam incidence angle. It is noteworthy that comparable interface widths are obtained with and without rotation for ion beam incident angles between 65° and 75°.

There is a critical value for the rotation speed that depends on the ion energy and mass, as well as the ion beam angle of incidence and erosion rate.[(342,359)] Values for Δz improve up to the critical value and then remain the same for faster speeds. For Cu–Fe multilayers, the critical rotation speed varied between 0.01 and 1 rpm,

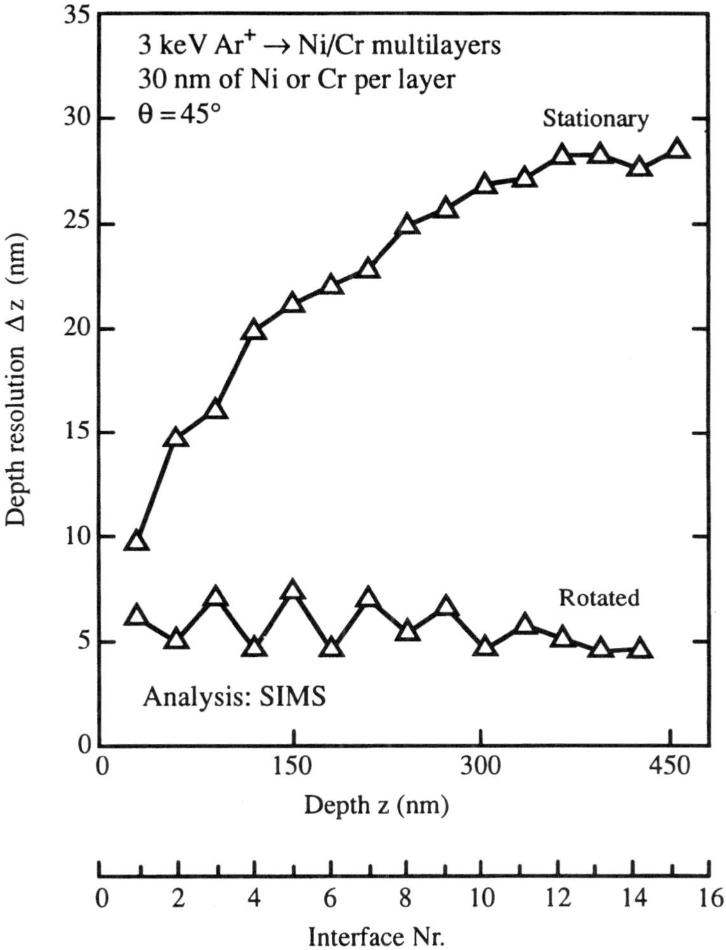

Figure 6.15. Depth resolution as a function of interface depth measured from SIMS profiles for Ni/Cr multilayer samples with and without sample rotation during ion bombardment. Adapted and reprinted with permission from[357] S. Hofmann, et al., *Surf. Interface Anal.* **20**, 621–626 (1993). Copyright 1993 John Wiley and Sons.

depending on the ion mass and energy, but under optimal conditions the same value of Δz was ultimately obtained in all cases. The critical speed is generally lower for higher-mass ions.(359) However, for sputtering Cu–Fe multilayer systems with 1-keV Xe ions, no improvement in Δz was obtained with variations in the rotation speed, and Δz is essentially the same (and independent of z) with and without sample rotation. SEM images have confirmed that there is an association between a decrease in sample roughness and a decrease in Δz; for 1-keV Xe ions incident on

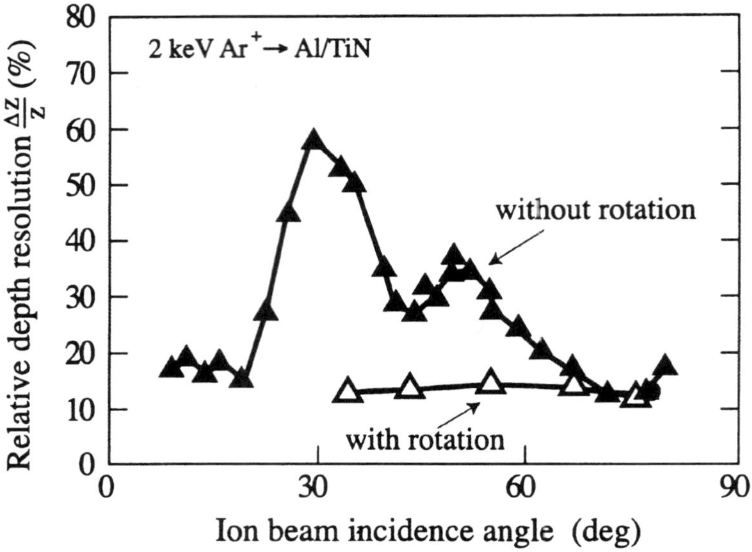

Figure 6.16. Depth resolution for Al films on TiN substrates sputtered by 2-keV Ar-ion beams as a function of incidence angle with and without sample rotation. Adapted and reprinted with permission from[209] W. Pamler and K. Wangemann, *Surf Interface Anal.* **18**, 52–58 (1992), and with kind permission of Elsevier Science - NL, Sara Burgerhartstraat 25, 1055 KV Amsterdam, The Netherlands.

the Cu–Fe multilayers, the roughness was the same with and without rotation.[(359)] For relatively smooth samples and when using an ion beam incidence angle of more than 50°, sample rotation may not give a significant improvement in Δz over nonrotated samples.[(342)]

The proper choice of the ratio of rotation speed and erosion rate may be more important than the rotation speed alone, especially for smooth samples.[(342)] In general, this ratio should be kept low because longer dwell times increase the adverse effect of beam inhomogeneities. Any eccentricity in the rotation axis about the center of the area analyzed can have serious consequences on the depth resolution that is greatly magnified by even small misalignments of the analysis beam. Reference 342 gives an example showing severe deterioration in Δz for a misalignment of only 400 μm; periodic spikes are apparent in the profile, and the Δz deterioration is significantly worse with increasing sputter depth. Impressive results have been obtained on multilayer and buried interface systems when sputtering with rotation and with grazing angles (80° to 87°) of ion beam incidence.[(183,359–361)] For these grazing-incidence angles, monotonic and significant decreases in the interface broadening were observed down to the lowest (0.2 keV) ion beam energies used. The depth resolution was constant to erosion depths of 500 nm. However, large and immediate changes in Δz were observed with changes in the ion beam energy; this result is attributed to changes in the atomic mixing depth.

For rough samples, Δz is still dependent on the initial roughness, even with rotation. However, the improvement in Δz with sample rotation may be even more dramatic than for smooth samples. Grazing-incidence ion beams are not recommended for stationary sputtering of rough samples, because shadowing effects are pronounced and may result in sharp increases in Δz. The situation is greatly improved, with rotation as is evident in Fig 6.11. The improvement gained depends on the distribution of angles of microplanes on the surface with respect to the ion beam. Rotation in combination with grazing-ion incidence also reduces the influence of ion mass on Δz.[354,359] For rough samples, higher beam energies at grazing incidence along with rotation may improve the "sputter smoothing" of the sample.[329]

It is not always necessary to rotate samples to obtain optimal depth resolution, especially for amorphous materials and single crystals. Using two or more ion beams that are incident onto the sample from different directions and at different angles give similar results to those obtained with sample rotation,[329] although the improvements may not be as dramatic. The previously mentioned study on Cu–Fe multilayers[359] showed no improvement in Δz with sample rotation when sputtered with 1-keV Xe ions. Pamler and Wangemann,[209] using oxygen ions at near-normal incidence, sputtered aluminum films and found interface widths better than those obtained with inert ions and sample rotation. However, significant improvement may be achieved on a wide variety of samples by using rotation. Although it is generally true, especially for large sputter depths, that optimal results are obtained with grazing-incidence angles, sample rotation, and low ion energy, the parameters are sample dependent and in most cases require individual determination of the optimal conditions.[183,329] We stress that determining the optimal sputtering conditions is worth the effort if improved Δz values are required. The difference in Δz obtained for optimized and nonoptimized conditions is often in excess of an order of magnitude.[209]

6.5. Ultimate and Practical Limits on Δz

The previous discussion indicates that considerable optimization of sputtering parameters can be accomplished, but, for a given material, it may be difficult to predict *a priori* the specific optimal parameters. Varying the different instrumental parameters is not straightforward because many parameters are coupled and may additionally depend on the sample parameters discussed previously. The effect of instrumental parameters on sputtering is complex because the various factors all interact, often in a complex manner, and are material dependent. Thus, experimentation is required to determine the optimal parameters for a given sample.

There is also a depth dependence to Δz; in general, Δz increases with increasing erosion depth, particularly when no sample rotation is used.[6,329,339] This result is illustrated in Fig. 1.2. The expected dependence on depth varies with the layer thickness.[329] For layers ≤ 5 nm, Δz increases approximately linearly with depth, as the sputter-induced mixing, statistical roughening, and differential sputtering reach a steady state (Fig. 1.2). The interface width may remain fairly constant for the next several hundred nanometers, especially for amorphous, homogeneous materials. However, for polycrystalline metals, Δz may continue to increase with depth in this region. Perhaps more subtle is evidence that for a given total erosion depth through a multilayer system, Δz may depend on the thicknesses of the individual layers.[362] For the most ideal materials, Δz increases after eroding through 200 to 300 nm, without sample rotation, as effects of beam inhomogeneities, even for the best-tuned instruments, become important factors. The extent and type of such roughening, as well as the depth dependence, are highly sample dependent, but polycrystalline metallic films tend to develop the most severe topography upon sputtering, while amorphous materials, particularly oxides, tend to develop much less roughening. This depth-dependent roughening may account for most of the degradation in Δz for samples sputtered to depths of more than 100 nm.[6,338,363] Sample rotation during sputtering may significantly reduce the degradation in Δz for large sputter depths.

Those who have devoted a great deal of effort to obtaining the lowest ultimate depth resolution when sputter-depth-profiling agree that under the most favorable conditions, and for an ideal sample, the practical limit for Δz is about 2 nm; e.g., see Hoffman,[329,364] Wittmaack,[13,358] Seah *et al.*,[6,365] and Barna and Menyhard.[183] The lower bound for Δz is limited primarily by the generation of an altered interface layer from atomic mixing. The anticipated lower limit for Δz obtainable by using sample rotation is shown by the dashed line in Fig. 1.2. For systems with atomically sharp interfaces, e.g., Ta_2O_5–Ta_1 SiO_2–Si, and GaAlAs–GaAs, with interfaces a few nanometers below the surface, Δz values of less than 1 nm have been obtained (Fig. 1.2), and these values are very close to the intrinsic limit imposed by the statistical nature of ion beam mixing in the sputtering process. For some samples and for very deep interfaces, Δz values may be lowered by bombarding with 200- to 300-eV Xe ions at grazing-incidence angles. However, the slow erosion rate under these idealized bombarding conditions has practical consequences that preclude using them as a routine procedure. For example, Iltgen *et al.*[366] showed explicitly from TOF–SIMS analyses the improvement of depth resolution for a delta layer of B implanted in Si at 100 eV for Ne, Ar, Kr, and Xe at 1 keV; Δz ranged from 2.0 to 1.1 nm, respectively. They also used a molecular ion, SF_6, for the same sample at the same energy and incidence angle to obtain a Δz of 0.65 nm. In their most recent work, Benninghoven and coworkers[367] have obtained a Δz of 0.4 nm, again using SF_6 but at 400 eV. They thus have nearly reached the theoretical limit as can be deduced from Eq. (1) and the discussion of it.[9] Finally, we emphasize that the values for the lower limits in Δz are for *ideal* specimens. For a typical practical specimen a realistic lowest value for Δz will be about 5 nm and may be much higher because of the synergistic interaction of the sample, beam, and instrumental parameters.

7. Combined Beam Effects

When two or more energetic beams are incident on the same region of a target material, the response may be quite different than the linear superposition of the independent responses. Such synergistic responses have been observed by the authors during Auger analysis, using a focused electron beam to analyze the etch pit of an Ar-ion beam incident on thin metal films on glass. Both the Ar-ion beam and the electron beam were operated continuously, and significant increases in the mixing of the layers were noted, as well as topographical anomalies. Using Auger analysis during sputter depth profiling is a common practice within the surface analysis community, but it may lead to spurious results, depending on the system being studied.

Only a few recent studies have treated artifacts arising from the use of multiple beams on a target. The most graphic presentation of combined beam effects on a target is presented by Gries.[168] Using a doped InP target, a 4.5-keV Ar-ion beam at 45°, and a scanning Auger beam, Gries noted spectacular differences in the topography of the target in the region subjected to the combined beams compared to the region of ion bombardment only, as shown in Fig. 7.1. In this case, the combined beams resulted in a raised plateau that was substantially smoother than the ion bombarded region. For a target that was annealed after implantation, the combined beams still produced smoothing, but the area formed a depression below the surrounding ion-bombarded area (Fig. 7.2). The plateau was raised by about 20% of the nominal sputtered depth, whereas the depression was lowered by about 30% of the nominal sputtered depth.

Synergistic effects have also been reported for direct synchrotron radiation etching of a resist with a low-energy (several tens of electron volts) oxygen-ion beam.[368] Increases in etch rate by a factor of 3 were observed, with a synergistic effect as large as the original etch rate. In another study, photoluminescent behavior in a GaAs multiple-quantum-well structure subject to 500-eV Ar-ion bombardment was significantly reduced by simultaneous illumination with a 4-mW HeNe laser beam.[162] Finally, a recent theoretical study[369] of simultaneous bombardment of a carbon substrate with C and Pt ions indicated synergistic sputtering effects for the two ion species. Changes in the sputter ratio varied by as much as a factor of 2 for incident-ion energies from 200 to 1000 eV. Although these Monte Carlo simulations are not good approximations of conditions used for surface analysis, they clearly illustrate the existence of synergistic effects for combined beams.

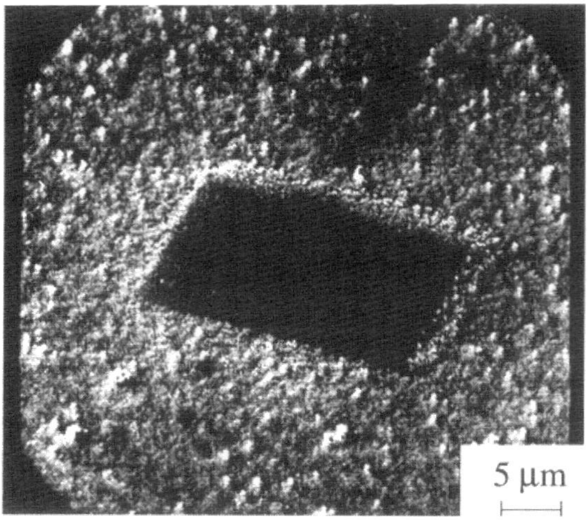

Figure 7.1. Topography of an InP wafer after sputtering to a depth of about 1 μm by 4-keV Ar ions at 45° incidence (in the direction of bottom to top). The much smoother, rectangular area is the result of simultaneous electron beam scanning and is about 0.2 μm *above* the rough structure. The InP wafer was pre-implanted with 350-keV Si (at a dose of 1×10^{15} ions/cm^2) and was not annealed before sputtering. Adapted and reprinted with permission from[168] W. H. Gries, *Surf. Interface Anal.* **14**, 611–618 (1989). Copyright 1989 John Wiley and Sons.

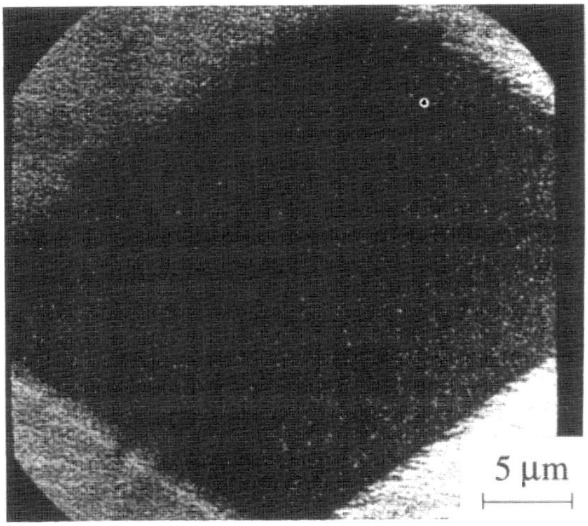

Figure 7.2. Topography on an InP wafer as in Fig. 7.1, with the exception that the wafer was annealed at 750°C for 30 min after implantation and before sputtering (to approximately the same depth as in Fig. 7.1). The direction of ion incidence was from right to left. The smoother area results from the simultaneous electron beam scanning and is about 0.3 μm *below* the rough surrounding area. Adapted and reprinted with permission from[168] W. H. Gries, *Surf. Interface Anal.* **14**, 611–618 (1989). Copyright 1989 John Wiley and Sons.

8. Applications

During our review of the literature, we encountered numerous applications of ion beams in addition to cleaning surfaces and sputter depth profiling. We list these in Table 8.1 without comment or referencing a source, because in most cases many sources exist. Many applications articles are published in the journal *Nuclear Instruments and Methods in Physics Research B*, such as Ref. 370. Specific applications we encountered more by chance than design are listed in Table 8.2 with reference citations.[66,78,203,371–396]

Table 8.1. Some Applications of Ion Beams and Ion Bombardment

Surface analysis with SIMS, SNMS, SALI, ISS, TOF-ISS, TOF-SIMS and MEIS
Sputter depth profiling and ion beam etching of surfaces
Controlled fabrication of coatings for microelectronics
Ion beam implantation for modifying materials
Formation of corrosion-resistant surfaces of materials
Preparation of films with tailored optical and electronic properties
Reactive ion beam etching of materials or reactive sputtering of materials
Ion-beam-assisted deposition of films
Patterning of surfaces and topography development
Forming damaged layers in near surface regions (structural changes)
Making compositional changes and altering chemical reactivity
Modifying the local electronic properties
Fabricating photovoltaic devices
Introducing surface defects to modify structural and other properties
Growing epitaxial layers by ion-assisted molecular beam epitaxy
Eliminating strain and suppressing island nucleation
Modifying materials by amorphizing the near-surface region
Modifying materials by recrystallization of amorphized layers
Modifying materials by differential or preferential sputtering yields
Enhancing deposition of films
Enhancing adhesion of surfaces
Enhancing reactivity to form new chemical compounds by ion beam mixing followed by rapid thermal annealing
Enhancing wear resistance and other tribological applications
Enhancing corrosion resistance
Enhancing interface diffusion rates
Modifying polymers, insulators, and ceramic materials to improve properties
Broadening of interfaces in superlattices
Deposition by radio-frequency sputtering and using electron–cyclotron resonance
Preparing TEM specimens by ion milling
Modifying near-surface properties with trapped beam atoms
Using plasma deposition processes

Table 8.2. Specific Applications of Ion Beams and Ion Beam Bombardment

Ref.	Application
371	Etching semiconductors with an inert-gas ion beam and using precursors
372	Preparing garnet waveguides by ion beam etching
373	Forming ridges in Al or Ga garnet crystals for waveguides
374	Sharpening Mo tips for use in focused ion and electron beam sources
375	Forming surface emitting lasers by forming ridges with focused ion beam etching
376	Minimizing the soft magnetism of Fe_xN films by bombardment and annealing
377	Growing and etching zirconium oxide films using ion beams
378	Preserving epitaxial growth of Si at temperatures of 300–350°C using ion beam bombardment
379	Depositing high-quality lanthanum-modified zirconate titanate by ion beam sputtering
380	Forming amorphous aluminum oxide, Ti–6A1–4V alloys, and iron films by sputtering or ion implantation
381	Lowering the limit of detectability of H in Al using ion beam etching
382	Patterning and modifying organosilane self-assembled monolayers (SAMs) on native silicon oxide using ion beams
383	Patterning SAMs on Ag by microfocus beam bombardment
384	Ion beam etching for pattern transfer or cleaning before deposition
385	Oxidizing Si or eroding silicon oxide during oxygen ion beam bombardment
386	Improving adhesion of polycarbonate and acrylonitrile butadiene styrene polymers by ion beam surface modification
387	Improving H permeation through V after ion bombardment cleaning
388	Improving the adhesion by Ar-ion etching microcavities in E24 carbon steel
389	Modifying surfaces by ion beam masked and maskless etching for lithographic applications
390	Modifying the optical properties and density of titanium oxide films
391	Elucidating the limiting resolution of focused ion beams prepared by sputtering
392	Studying reactions of ions with organic surfaces
393	Cleaning and etching Si with chlorine ion bombardment
203	Forming ion-etched grooves in $LiNbO_3$ for waveguide applications
394	Removing SiO_2 from Si with H-ion beam cleaning
395	Modifying glass surface properties using ion beams
78	Growing crystalline films at lower temperatures from near-surface damaged layers
66	Improving insulator performance in lithography, erosion, and coatings
396	Using ion beam sputtering for biomedical applications

9. Summary and Concluding Remarks

We conclude that sputtering, in general, is *not* a technique in which successive layers of the sample are removed in an ordered, layer-by-layer manner. Sputtering is a complex process that involves an often synergistic interplay between sample and ion beam parameters. Ion beam bombardment of a sample almost always introduces structural, topographical, physical, or chemical changes that significantly alter the properties of the solid. As yet, no model accurately predicts all the effects of ion beam bombardment on an arbitrarily chosen sample.

The ejection of material by sputtering is accompanied by implantation and mixing of the ion beam and sample material, which results in an altered layer that may extend for tens of nanometers into the sample. Most of the incident ions in a bombarding beam penetrate the sample, undergo a series of collisions with the substrate atoms, and transfer momentum and energy to the solid in the process; incident ions are implanted and neutralized, except for the small fraction that scatter as ions or escape as neutrals after multiple collisions in the near-surface region. The mean ion range in the sample generally increases with increasing ion energy, decreases as the angle of incidence is increased, and decreases with increasing ion mass; the ion range also depends on the substrate mass, elemental composition, electronic properties, and crystallographic structure. The interaction of the ion beam with the sample sets into motion a collision cascade that results in sputtering and mixing of atoms in the substrate material. Ejected particles with energies below a few electron volts are easily redeposited in or near the bombarded area. Diffusional mixing and segregation processes occur and their extent increases with increasing beam energies and fluences. The resulting altered layer may differ from the original material in composition, density, phase, and other properties. In layered samples, the interface between the layers is usually broadened considerably from the effects of ion beam bombardment.

The onset of surface and bulk defect production occurs at ion beam energies of less than 100 eV, i.e., at energies and for fluences considerably lower than those generally used for sample cleaning or depth profiling. The number and rate of defects produced depends on the ion beam mass, energy, current density, total fluence, angle of incidence, and bombardment time. Surface and bulk vacancies, dislocations, and interstitials are formed. At higher doses, point, complex, continuous, and interface defects may form and interact, and result in enhanced diffusion processes. Vacancies coalesce into extended defects at high concentrations, but the vacancy concentration can be reduced by annealing at elevated temperatures. At the highest beam energies and fluences, the bombarded zone is almost always amorphized, and damage layers much deeper than the ion beam range usually develop. Alteration of the chemical, electrical, optical, structural, mechanical, and other physical properties during ion beam bombardment should be expected.

The sample topography may be drastically altered by ion beam bombardment. Both thermodynamic factors and kinetic processes contribute. Many different types of topographical features have been observed, including etch pits and craters, pyramids, cones, whiskers, and a variety of corrugated and periodic features. Both random and periodic structures have been reported. Topography develops on both the micro- and macroscopic levels and may be considerably different on the two levels. Ion beam bombardment generally induces some roughness, even on initially smooth samples, but under certain circumstances smoothing or polishing occurs.

Topographical development is influenced by the ion beam parameters and sample characteristics. The angle of incidence of the ion beam is particularly critical, as is the distribution of microplanes with different inclines on the sample. Contamination and impurities are also implicated in some types of topography development. The erosion rate and changes in the apparent "surface" composition are altered as topography develops. Microtopography development resulting from the statistics of sputtering and mixing limits the ultimate depth resolution obtainable with sputter depth profiling, and macrotopography development may severely degrade the depth resolution.

As might be expected, compositional changes and chemical reactivity are also often affected by ion beam bombardment. Differential sputtering, ion-beam-enhanced diffusion and segregation processes, and structural changes affect the reactivity and chemical state of many samples. Beam-induced defects enhance the surface and bulk reactivity because of broken bonds, and annealing of defects in depleted surfaces also complicates the interpretation of data, especially that taken at different temperatures. The properties of the ion beam may influence the type of degradation processes that occur, especially for organic samples. Properties of the sample are also important in the type and extent of ion-beam-induced composition and reactivity changes, including surface chemical binding energy, mass, and diffusion properties. The surface concentration of alloys is often altered, and many oxide materials and other compounds are chemically reduced with ion beam bombardment. Highly mobile species can migrate for considerable distances, resulting in either enrichment or depletion at the surface region. Contaminants and impurities imbedded in the sample by the ion beam may also react to form new compounds, and the original composition may be difficult to deduce. In most cases, the effects of ion beam bombardment on composition and chemical reactivity enhancement cannot be predicted.

For depth profiling layered samples, the depth resolution is influenced by the initial sample characteristics, ion beam variables, and instrumental parameters. The cleanliness of the sample surface, form of the sample (amorphous or crystalline), initial surface and interface roughnesses, composition, and homogeneity are important. Smooth, homogeneous, amorphous, noninteracting layers tend to have the most uniform erosion rates and best ultimate depth resolutions. Important ion beam parameters include beam energy, current density, angle of incidence, mass, and total ion dose. Interface widths increase with increased beam energy because of deeper beam penetration into the sample and an increased zone of mixing. The interface widths in general decrease with the use of more massive ions, and the erosion rate is greater at higher beam energies, higher current densities, and with ions of larger mass. The depth resolution generally deteriorates with increased ion beam current density and dose. For smooth samples, grazing-ion-beam incidence generally yields the best depth resolution, while normal incidence is often better for initially rough samples. For optimal depth resolution, the instrument used for sputtering must be kept at performance specifications. Ultrahigh vacuum systems with base pressures of 10^{-9} Torr or better are preferred. The ion beam ideally will have a narrow energy spread, uniform current density, and Gaussian shape. The ion beam and analysis beam must be properly aligned. Simultaneous use of ion beams and other probe beams (e.g., electron or photon beams) may result in artifacts. Sample rotation during sputtering can greatly reduce topography development and often results in a significant improvement in the depth resolution. For the most ideal cases, depth-resolution values as small as 0.4 nm have been obtained; more typical values for initially sharp interfaces are 2–5 nm, and may be much larger.

Although sputtering is an invaluable technique for sample cleaning and for determining the composition in depth in solids, changes introduced in the sample by ion beam bombardment can be large, varied, and elusive to determine. Great care must be taken in interpreting data to allow for the effects of ion beam bombardment. Optimization of experimental parameters, while not always straightforward, can minimize ion-beam-bombardment effects; with optimization, improvements in excess of an order of magnitude in depth resolution from arbitrarily chosen parameters can be realized for sputter depth profiles. Whenever ion beams are used for obtaining composition-in-depth profiles, extreme caution must be used in interpreting the data.

Acknowledgments

The authors are deeply grateful to C. J. Powell and T. E. Madey for their careful review of the manuscript. The authors especially appreciate the following NREL staff members: Annette Berger, for her superb expertise and help with the literature search; Paula Robinson, for her outstanding assistance and patience in word-processing the manuscript and references; and Al Hicks, for his meticulous preparation of the illustrations. We thank the NREL-FIRST program for their financial support of this work and are grateful to the U.S. Department of Energy for their support of this work under Contract No. DE-AC36-83CH10093.

References

1. A. W. Czanderna, in: *Ion Spectroscopies for Surface Analysis* (A. W. Czanderna and D. M. Hercules, eds.), Plenum Press, New York (1991), pp. 1–44.
2. J. A. McHugh, in: *Methods of Surface Analysis* (A. W. Czanderna, ed.), Elsevier, Amsterdam (1975), pp. 223–278.
3. ASTM E673-194a Standard terminology relating to surface analysis, in: *1995 Annual Book of ASTM Standards*, Vol. 03.06, American Society for Testing and Materials, W. Philadelphia, PA (1989), pp. 789–798.
4. G. K. Wehner, in: *Methods of Surface Analysis* (A. W. Czanderna, ed.), Elsevier, Amsterdam (1975), pp. 1–37.

5. A. W. Czanderna, ed., *Methods of Surface Analysis*, Elsevier, Amsterdam (1975).
6. M. P. Seah and C. P. Hunt, Interface depth resolution for Ta_2O_5/Ta and SiO_2/Si interfaces, *Surf. Interface Anal.* **5**, 33–37 (1983).
7. H. F. Winters, in: *Radiation Effects on Solid Surfaces* (M. Kaminsky, ed.), American Chemical Society, Washington, DC (1976), pp.1–29.
8. H. W. Werner, in: *Electron and Ion Spectroscopy of Solids* (L. Fiermans, J. Vennik, and W. Dekeyser, eds.), Plenum Press, New York (1978), pp. 324–441.
9. L. C. Feldman and J. W. Mayer, *Fundamentals of Surface and Thin Film Analysis*, North-Holland, New York (1986).
10. A Benninghoven, F. G. Rüdenauer, and H. W. Werner, *Secondary Ion Mass Spectrometry*, Wiley, New York (1987).
11. E. Taglauer, in: *Ion Spectroscopies for Surface Analysis* (A. W. Czanderna and D. M. Hercules, eds.), Plenum Press, New York (1991), pp. 363–416.
12. J. B. Malherbe, Sputtering of compound semiconductor surfaces. I. Ion-solid interactions and sputtering yields. *Crit. Rev. Solid St. Mater. Sci.* **19**(2), 55–127 (1994).
13. K. Wittmaack, in: *Practical Surface Analysis*, Vol. 2-Ion and Neutral Spectroscopy (D. Briggs and M. P. Seah, eds.), Wiley, Chichester, UK (1992), pp. 105–175.
14. A. M. Stoneham, in: *Materials Modification by High-Fluence Ion Beams* (R. Kelly and M. Fernanda da Silva, eds.), Kluwer, Dordrecht, The Netherlands (1989), pp. 217–230.
15. W. Heiland, H. Derks, and T. Bremer, Slowing down and scattering of ions in solids, *Scan. Microsc. Suppl.* 4, **1990**, 257–263.
16. N. Q. Lam, Physical sputtering of metallic systems by charged-particle impact, *Scan. Microsc. Suppl.* 4, **1990**, 311–352.
17. E. V. Kornelsen, The ionic entrapment and thermal desorption of inert gases in tungsten for kinetic energies of 40 eV to 5 keV, *Can. J. Phys.* **42**, 364–381 (1964).
18. W. Choi, C. Kim, and H. Kang, Interactions of low energy (10-600 eV) noble gas ions with a graphite surface: surface penetration, trapping and self-sputtering behaviors, *Surf. Sci.* **281**, 323–335 (1993).
19. T. Michely and G. Comsa, The scanning tunneling microscope as a means for the investigation of ion bombardment effects on metal surfaces, *Nucl. Instrum. Meth. Phys. Res., Sect. B* **82**, 207–219 (1993).
20. G. Carter, I. V. Katardjiev, M. J. Nobes, and J. L. Whitton, Topographic evolution in the atomic scale growth and erosion continuum, *Scan. Microsc.* **2**(3), 1293–1327 (1988).
21. G. Carter, M. J. Nobes, I. V. Katardjiev, and J. L. Whitton, The role of bombardment induced defects in the initiation of topography on ion bombarded f.c.c. metals, *Defect Diffusion Forum* **57-58**, 97–126 (1988).
22. J. Lapujoulade, The roughening of metal surfaces, *Surf. Sci. Rep.* **20**, 191–249 (1994).
23. I. H. Wilson, in: *Materials Modification by High-Fluence Ion Beams* (R. Kelly and M. Fernanda Da Silva, eds.), Kluwer, Dordrecht, The Netherlands (1989), pp. 421–466.
24. J. B. Malherbe, Sputtering of compound semiconductor surfaces. II. Compositional changes and radiation-induced topography and damage, *Crit. Rev. Solid St. Mat. Sci.* **19**(3), 129–195 (1994).
25. R. Kelly and A. Miotello, in: *Materials and Processes for Surface and Interface Engineering* (Y. Pauleau, ed.), Kluwer, Dordrecht, The Netherlands (1995), pp. 67–109.
26. V. I. Zaporozchenko and M. G. Stepanova, Preferential sputtering in binary targets, *Prog. Surf. Sci.* **49**, 155–196 (1995).
27. S. Hofmann, in: *Practical Surface Analysis*, Vol. 1-*Auger and X-ray Photoelectron Spectroscopy* (D. Briggs and M. P. Seah, eds.), Wiley, New York (1991), pp. 143–200.
28. J. M. Millunchick, L. Hultman, and S. A. Barnett, Effect of 20-95 eV Ar ion bombardment of GaAs (001): in pursuit of damage-free ion-assisted growth and etching, *J. Vac. Sci. Technol. A* **13**(3), 1155–1159 (1995).

29. H. G. Hagstrum, in: *Electron and Ion Spectroscopy of Solids* (L. Fiermans, J. Vennik, and W. Dekeyser, eds.), Plenum Press, New York (1978), pp. 273–323.
30. P. Varga, Neutralization of multiply charged ions at a surface, *Appl. Phys. A* **44**, 31–41 (1987).
31. J. W. Rablais, Ion-induced surface alterations due to electronic charge exchange and chemical reactions, *Israel J. Chem.* **22**(3), 365–374 (1982).
32. P. Nordlander, Charge transfer processes in atom-surface collisions, *Scan. Microsc. Suppl.* 4, **1990**, 353–370.
33. U. Wille, Theoretical neutralization rates for Rydberg states in ion-metal-surface scattering, *Nucl. Instrum. Meth. Phys. Res., Sect. B* **67**, 132–137 (1992).
34. M. Alducin, A. Arnau, and P. M. Echenique, Auger capture rates for a slow ion close to a metal surface, *Nucl. Instrum. Meth. Phys. Res., Sect. B* **67**, 157–159 (1992).
35. V. A. Esaulov, L. Guillemot, and S. Lacombe, Autoionising state excitation in collisions of inert gas projectiles with an Al surface and effects of initial stages of oxidation, *Nucl. Instrum. Meth. Phys. Res., Sect. B* **90**, 305–309 (1994).
36. W. Heiland and A. Närmann, Excitation and loss processes of ions at surfaces, *Nucl. Instrum. Meth. Phys. Res., Sect. B* **78**, 20–28 (1993).
37. D. A. Thompson, High density cascade effects, *Rad. Eff.* **56**, 105–150 (1981).
38. E. K. Vasileva, Mechanism of formation of continuous emission bands in ion-photon emission spectra of rare-earth elements, *Opt. Spektrosk.* **66**(1), 26–27 (1989).
39. D. Ghose, U. Brinkmann, and R. Hippler, Light emission from a Si target during ion beam sputtering, *Surf. Sci.* **327**, 53–58 (1995).
40. P. G. Fournier, J. Fournier, and B. Bellaoui, Optical emissions from sputtering species formed by the fast ion bombardment of clean and oxygenated Mg surfaces, *Nucl. Instrum. Meth. Phys. Res., Sect. B* **67**, 604–609 (1992).
41. R. Hippler and S. Reinke, Oxygen dependence of light emission from sputtered atoms and ions following ion bombardment of polycrystalline aluminum targets, *Nucl. Instrum. Meth. Phys. Res., Sect. B* **68**, 413–416 (1992).
42. V. V. Gritsyna, A. G. Koval, and D. A. Kleopov, Photon emission on ion bombardment of single- and polycrystalline NaCl, *Bull. Russ. Acad. Sci. Phys.* **58**(3), 397–401 (1994).
43. F. Aumayr and H. Winter, Electron emission induced by slow highly charged ions on a clean metal surface, *Nucl. Instrum. Meth. Phys. Res., Sect. B* **90**, 523–532 (1994).
44. V. E. Yurasova, Secondary particle emission from metals under ion bombardment in the region of phase transitions-Part 2. Charged-particle and photon emission, *Vacuum* **36**(10), 609–630 (1986).
45. W. O. Hofer, Ion-induced electron emission from solids, *Scan. Microsc. Suppl.* 4, **1990**, 265–310.
46. K. Ohya and J. Kawata, Monte Carlo study of incident-angle dependence of ion-induced kinetic electron emission from solids, *Nucl. Instrum. Meth. Phys. Res., Sect. B* **90** , 552–555 (1994).
47. Z. Šroubek, Formation of localized states and electronic excitation in solids during low-energy particle bombardment, *Nucl. Instrum. Meth. Phys. Res., Sect. B* **78**, 140–143 (1993).
48. M. Soszka and W. Soszka, Surface electron emission induced by ion bombardment: methods of observation, *Nucl. Instrum. Meth. Phys. Res., Sect. B* **67**, 632–635 (1992).
49. K. Ohya and I. Mori, Statistics of kinetic secondary electron emission due to ion-atom collisions in solids, *Nucl. Instrum. Meth. Phys. Res., Sect. B* **67**, 628–631 (1992).
50. J. Kawata and K. Ohya, Monte Carlo simulation of ion-induced kinetic electron emission from a metal surface, *Appl. Surf. Sci.* **100/101**, 338–341 (1996).
51. J. Fine, M. Szymonski, J. Kolodziej, M. Yoshitake, and K. Franzreb, Collisional deexcitation at ion-bombarded surfaces, *Phys. Rev. Lett.* **71**(21), 3585–3588 (1993).
52. R. A. Baragiola, Comment on "Collisional deexcitation at ion-bombarded surfaces," *Phys. Rev. Lett.* **77**(2), 408 (1996).
53. Z. Sroubek and J. Fine, Electronic excitations produced by deep-level promotion during atomic collisions in solids, *Phys. Rev. B* **51**(9), 5635–5643 (1995).

54. E. A. Maydell and D. J. Fabian, Ion-induced Auger emission and sputtering processes in solids, *Nucl. Instrum. Meth. Phys. Res., Sect. B* **67**, 610–615 (1992).
55. K. O. Legg, W. A. Metz, and E. W. Thomas, Auger spectra induced by ion impact on metals, *Nucl. Instrum. Meth.* **170**, 561–564 (1980).
56. S. Valeri and R. Verucchi, Ion beam-stimulated Auger electron emission from Cr and Cr-silicides, *Phys. Scr., T* **41**, 246–250 (1992).
57. R. G. Albridge, A. V. Barnes, R. F. Hagulund, Jr., M. H. Mendenhall, H. Mogul, D. P. Russell, P. M. Savundararaj, N. H. Tolk, and P. Wang, Desorption induced by ion- and neutral-beam excitation, *Nucl. Instrum. Meth. Phys. Res., Sect. B* **33**, 848–850 (1988).
58. M. L. Lu, Bond breaking and electron tunneling mechanisms in secondary ion emission from surfaces, *Comments At. Mol. Phys.* **20**(2), 79–91 (1987).
59. L. Guillemot, S. Lacombe, M. Huels, Vu Ngoc Tuan, and V. A. Esaulov, Inelastic processes in Ne^+ collisions with a Mg surface; charge fractions and surface roughness, *Nucl. Instrum. Meth. Phys. Res., Sect. B* **90**, 270–273 (1994).
60. A. Benninghoven, in: *Chemistry and Physics of Solid Surfaces* (R. Vanselow and S.Y. Tong, eds.), CRC Press, Cleveland (1977), pp. 207–233.
61. N. Q. Lam, Ion bombardment effects on the near-surface composition during sputter profiling, *Surf. Interface Anal.* **12**(1-12), 65–77 (1988).
62. L. Guillemot, S. Lacombe, M. Maazouz, E. Sanchez, and V. A. Esaulov, Inelastic collisions of Ne atoms and ions with a Si surface and effects of initial stages of oxidation, *Surf. Sci.* **356**, 92–100 (1996).
63. E. Taglauer, Probing surfaces with low-energy ions, *Nucl. Instrum. Meth. Phys. Res., Sect. B* **98**, 392–399 (1995).
64. V. A. Esaulov, L. Guillemot, S. Lacombe, and Vu Ngoc Tuan, Inelastic inert gas ion collisions with metal targets. Rearrangement and adsorbate effects, *Nucl. Instrum. Meth. Phys. Res., Sect. B* **100**, 232–241 (1995).
65. M. Suchanska, Ion induced photon emission of metals, *Prog. Surf. Sci* **54**, 165–209 (1997).
66. A. M. Stoneham, Radiation effects in insulators, *Nucl. Instrum. Meth. Phys. Res., Sect. B* **91**, 1–11 (1994).
67. Z. Postawa, Desorption of alkali halides stimulated by ion bombardment, *Radiat. Eff. Defects Solids* **128**, 107–125 (1994).
68. A. Poradzisz, Z. Postawa, J. Rutkowski, and M. Szymonski, Electronic sputtering of halides by low energy ion and electron bombardment, *Nucl. Instrum. Meth. Phys. Res., Sect. B* **33**, 830–833 (1988).
69. Z. Šroubek and J. Fine, Electronic structure of collision cascades in semiconductors, *Nucl. Instrum. Meth. Phys. Res., Sect. B* **100**, 253–255 (1995).
70. D. Fink, M. Müller, R. Klett, L. T. Chadderton, L. Palmetshofer, J. Kastner, J. Vacik, V. Hnatowicz, and P. Popok, Sputtering of fullerene by noble gas ions at high fluences, *Nucl. Instrum. Meth. Phys. Res., Sect. B* **103**, 415–422 (1995).
71. M. Kitajima, T. Noda, and M. Okada, The effect of stoichiometry on sputtering, deuterium retention and surface damage in titanium carbide films, *Thin Solid Films* **156**, 231–238 (1988).
72. H. A. Filius, A. Van Veen, and K. R. Bijkerk, The retention of Ar in low energy high fluence Ar-irradiated Mo and Si, *Radiat. Eff. Defects Solids* **108**, 1–8 (1989).
73. E. K. Baranova, M. I. Guseva, Y. K. Kovneristyi, and S. A. Minaeva, Blistering and sputtering of a new vanadium alloy, *Fiz. Khim. Obrab. Mater.* **21**(1), 28–32 (1987).
74. D. Ghose and S. B. Karmohapatro, Topography of solid surfaces modified by fast ion bombardment, *Ad. Electron. Electron Phys.* **79**, 73–150 (1990).
75. B. A. Kalin, V. P. Gladkov, N. V. Volkov, and S. E. Sabo, Atom penetration from a thin film into the substrates during sputtering by polyenergetic Ar^+ ion beam with mean energy of 9.4 keV, *Mater. Res. Soc. Symp. Proc.* **354**, 237–242 (1995).

76. R. A. Langley, Helium-ion-induced release of hydrogen from graphite, *J. Nucl. Mater.* **162–164**, 1030–1034 (1989).
77. H. Kang, K. H. Park, C. Kim, B. S. Shim, S. Kim, and D. W. Moon, Point defect formation on graphite surface induced by ion impact at energies near penetration threshold, *Nucl. Instrum. Meth. Phys. Res., Sect. B* **67**, 312–315 (1992).
78. D. G. Armour and A. H. Al-Bayati, Surface radiation damage, *Nucl. Instrum. Meth. Phys. Res., Sect. B* **67**, 279–286 (1992).
79. B. A. Kalin, N. V. Volkov, S. E. Sabo, V. F. Rybalko, G. D. Tolstolutskaya, and I. E. Kopanets, Topography change of Be, Al, Ni, Cu, and Mo surfaces sputtered with polyenergetic H^+, He^-, and Ar^+ ion beams, *Bull. Russ. Acad. Sci. Phys.* **58**(3), 476–480 (1994).
80. E. V. Kornelsen and M. K. Sinha, Thermal release of inert gases from (110) and (211) tungsten surfaces, *J. Appl. Phys.* **40**(7), 2888–2894 (1969).
81. E. V. Kornelsen, Entrapment of helium ions at (100) and (110) tungsten surfaces, *Can. J. Phys.* **48**, 2812–2823 (1970).
82. S. S. Todorov and E. R. Fossum, Oxidation of silicon by a low-energy ion beam: experiment and model, *Appl. Phys. Lett.* **52**(1), 48–50 (1988).
83. P. Williams and J. E. Baker, Implantation and ion beam mixing in thin film analysis, *Nuclear Instrum. Meth.* **182/183**, 15–24 (1981).
84. R. P. Webb, in: *Practical Surface Analysis*, Volume 2-*Ion and Neutral Spectroscopy* (D. Briggs and M. P. Seah, eds.), Wiley, Chichester, England (1992), pp. 657–704.
85. R. Kelly and A. Miotello, Metal-ceramic ion-beam mixing: a quest for general principles, *Surf. Coat. Technol.* **83**, 134–145 (1996).
86. L. J. Huang, W. M. Lau, H. T. Tang, W. N. Lennard, I. V. Mitchell, and P. J. Schultz, Near-surface structure of low-energy-argon-bombarded Si (100), *Phys. Rev. B* **50**(24), 18453–18468 (1994).
87. P. C. Zalm, Quantitative sputtering, *Surf. Interface Anal.* **11**, 1–24 (1988).
88. G. Betz, G. K. Wehner, H. H. Andersen, and H. L. Bay, in: *Sputtering by Particle Bombardment II* (R. Behrisch, ed.), Springer-Verlag, Berlin (1983), pp. 11–90.
89. H. H. Andersen and H. L. Bay, in: *Sputtering by Particle Bombardment I* (R. Behrisch, ed.), Springer-Verlag, Berlin (1982), pp. 145–218.
90. K. W. Pierson, J. L. Reeves, T. D. Krueger, C. D. Hawes, and C. B. Cooper, Total sputtering yield of Ag/Cu alloys for low energy argon ions, *Nucl. Instrum. Meth. Phys. Res., Sect. B* **108**, 290–299 (1996).
91. T. Nenadović, B. Perraillon, Z. Bogdanov, Z. Djordjević, and M. Milić, Sputtering and surface topography of oxides, *Nucl. Instrum. Meth. Phys. Res., Sect. B* **48**, 538–543 (1990).
92. S. Priggemeyer and W. Heiland, Sputtering and crater formation in oxidic materials due to ion induced sputtering dependent on energy, *Nucl. Instrum. Meth. Phys. Res., Sect. B* **78**, 198–203 (1993).
93. A. Wucher, M. Watgen, C. Mössner, and H. Oechsner, Energy dependent studies of anisotropic atomic sputtering of Ni (111), *Nucl. Instrum. Meth. Phys. Res., Sect. B* **67**, 531–535 (1992).
94. B. J. Garrison, Preferential sputtering of binary compounds: a model study, *Surf. Sci.* **114**, 23–37 (1982).
95. B. Baretzky and E. Taglauer, Does preferential sputtering depend on the angle of incidence?, *Surf. Sci.* **162**, 996–1002 (1985).
96. J. Okano, T. Ochiai, and H. Nishimura, Ion-beam-induced isotope composition changes in metal surfaces and recoil implantation, *Appl. Surf. Sci.* **22/23**, 72–81 (1985).
97. P. Sigmund and M. W. Sckerl, Momentum asymmetry and the isotope puzzle in sputtering by ion bombardment, *Nucl. Instrum. Meth. Phys. Res., Sect. B* **82**, 242–254 (1993).
98. R. Kelly and D. E. Harrison, A summary of the theory of the preferential sputtering of alloys, *Mater. Sci. Eng.* **69**, 449–455 (1985).
99. H. H. Andersen, The depth resolution of sputter profiling, *Appl. Phys.* **18**, 131–140 (1979).

100. G. Bahir and E. Finkman, Ion beam milling effect on electrical properties of $Hg_{1-x}Cd_xTe$, *J. Vac. Sci. Technol., A* 7(2), 348–353 (1989).
101. W. Szyszko and J. Sielanko, Monte Carlo calculation of depth distribution of implanted ions and sputtering yield, *Phys. Status Solidi A* **112**, 315–318 (1989).
102. U. Conrad and H. M. Urbassek, Monte Carlo study of fluence dependent mixing and sputtering of isotopic targets under ion bombardment, *Surf. Sci.* **278**, 414–426 (1992).
103. R. P. Webb and D. E. Harrison, Jr., Computer simulation of pit formation in metals by ion bombardment, *Phys. Rev. Lett.* **50**(19), 1478–1481 (1983).
104. A. Azens, G. Romanovskis, and U. Kanders, A molecular dynamics study of the sputtering of the W (100) surface by low-energy He, Ar, and Kr ions, *J. Phys.: Condens. Matter* **4**, 5053–5060 (1992).
105. R. S. Taylor and B. J. Garrison, Molecular dynamics simulations of keV particle bombardment. Correlation of intact molecular ejection with adsorbate size, *Chem. Phys. Lett.* **230**, 495–500 (1994).
106. D. E. Sanders, K. B. S. Prasad, J. S. Burnham, and B. J. Garrison, Mechansitic analysis from molecular-dynamics simulations: keV-particle-induced desorption from Si{001}, *Phys. Rev. B* **50**(8), 5358–5370 (1994).
107. C.-C. Chang, H. H. Chen, and P. Wu, The effect of crystal structure on sputtering: molecular dynamics calculations, *Mater. Res. Soc. Symp. Proc.* **373**, 35–40 (1995).
108. M. G. Stepanova, Effect of ion beam mixing and compound formation on sputter depth profile of a Ta/Si multilayer thin film, *Surf. Interface Anal.* **24**, 416–418 (1996).
109. S. H. Han, G. L. Kulcinski, and J. R. Conrad, Computer simulation of ion beam mixing, *Nucl. Instrum. Meth. Phys. Res., Sect. B* **45**, 701–706 (1990).
110. I. R. Chakarov, S. S. Todorov, and D. S. Karpuzov, Dynamic Monte-Carlo simulation of compositional change and atomic redistribution in multicomponent targets under ion bombardment, *Nucl. Instrum. Meth. Phys. Res., Sect. B* **69**, 193–199 (1992).
111. M. G. Stepanova, Master equation for atomic collision cascades with fixed path length between collisions, *Nucl. Instrum. Meth. Phys. Res., Sect. B* **103**, 33–37 (1995).
112. R. Smith and D. E. Harrison, Jr., High-yield sputtering events for Ar^+-ion bombardment of Cu in the energy range 1–20 keV, *Phys. Rev. B* **40**(4), 2090–2096 (1989).
113. V. I. Shulga and P. Sigmund, Simulation of energy-dependent isotope sputtering, *Nucl. Instrum. Meth. Phys. Res., Sect. B* **103**, 383–386 (1995).
114. S. S. Todorov and I. R. Chakarov, On the mechanism of sputtering of SiO_2 by Ar at ion energies near the sputtering threshold, *Vacuum* **39**(11/12), 1101–1103 (1989).
115. A. Wucher and B. J. Garrison, Cluster formation in sputtering: a molecular dynamics study using the MD/MC-corrected effective medium potential, *J. Chem. Phys.* **105**(14), 5999–6007 (1996).
116. A. Wucher and B. J. Garrison, Sputtering of silver dimers: a molecular dynamics calculation using a many-body embedded-atom potential, *Surf. Sci.* **260**, 257–266 (1992).
117. J. W. Burnett, J. P. Biersack, D. M. Gruen, B. Jorgensen, A. R. Krauss, M. J. Pellin, E. L. Schweitzer, J. T. Yates, Jr., and C. E. Young, Depth of origin of sputtered atoms: experimental and theoretical study of Cu/Ru(0001), *J. Vac. Sci. Technol., A* **6**(3), 2064–2068 (1988).
118. D. Y. Lo, T. A. Tombrello, M. H. Shapiro, and D. E. Harrison Jr., Molecular dynamics simulation of sputtering with many-body interactions, *Mater. Res. Soc. Symp. Proc.* **100**, 145–150 (1988).
119. M. T. Robinson, Computer simulation of atomic collision processes in solids, *Mater. Res. Soc. Symp. Proc.* **279**, 3–16 (1993).
120. R. Collins and P. Sigmund, Pressure relaxation in atomic mixing and preferential sputtering, *J. Appl. Phys.* **72**(5), 1993–1996 (1992).
121. R. Smith and K. Beardmore, Molecular dynamics studies of particle impacts with carbon-based materials, *Thin Solid Films* **272**, 255–270 (1996).

122. M. Ghaly, R. S. Averback, and T. Diaz de la Rubia, Surface effects on damage production during ion bombardment: a molecular dynamics study, *Nucl. Instrum. Methods Phys. Res., Sect. B* **102**, 51–57 (1995).
123. M. W. Roberts and C. S. McKee, *Chemistry of the Metal-Gas Interface*, Oxford University Press, Oxford, UK (1978).
124. H. Gnaser, Processes in low-energy ion-surface collisons: preferential sputtering, defect and adatom formation, *Appl. Surf. Sci.* **100/101**, 316–328 (1996).
125. P. S. Mangat, P. Soukiassian, Y. Huttel, Z. Hurych, B. Gruzza, and A. Porte, Low-energy Ar^+ ion bombardment-induced modification of surface atomic bond lengths on InP (100) wafer, *Appl. Phys. Lett.* **63**(14), 1957–1959 (1993).
126. R. Ai, S. Savage, P. Xu, J. P. Zhang, and L. D. Marks, Microstructures of Si (111) on ion sputtering and electron annealing, *Mater. Res. Soc. Symp. Proc.* **236**, 259–264 (1992).
127. D. Muller, F. Ringeisen, D. Bolmont, and J. J. Koulmann, Hydrogenation of amorphized silicon by low energy H^+ ion bombardment studied by UPS, *Solid State Commun.* **72**(2), 219–222 (1989).
128. M. V. Ramana Murty and H. A. Atwater, Low energy ion irradiation of H-terminated Si (001): hydrogen sputtering, beam-induced (2 × 1) reconstruction, and Si epitaxy, *Nucl. Instrum. Meth. Phys. Res., Sect. B* **102**, 293–300 (1995).
129. T. Michely and G. Comsa, Sputtering damage in the Pt (111) surface by He^+ ion bombardment, *J. Vac. Sci. Technol., B* **9**(2), 862–866 (1991).
130. T. Michely, K. H. Besocke, and G. Comsa, Observation of sputtering damage on Au (111), *Surf. Sci. Lett.* **230**, L135–L139 (1990).
131. S. Gauthier, Y. Samson, J. C. Girard, S. Rousset, and J. Klein, Study of ion bombardment induced vacancy islands on Au (100) by scanning tunneling microscopy, *J. Vac. Sci. Technol., B* **12**(3), 1754–1757 (1994).
132. P. Bedrossian, Generation and healing of low-energy ion-induced defects on Si (100)-2×1, *Surf. Sci.* **301**, 223–232 (1994).
133. J. C. Girard, Y. Samson, S. Gauthier, S. Rousset, and J. Klein, STM study of the nucleation and annealing of ion bombardment induced defects on Cu (100), *Surf. Sci.* **302**, 73–80 (1994).
134. T. Michely and G. Comsa, Generation and nucleation of adatoms during ion bombardment of Pt (111), *Phys. Rev. B* **44**(15), 8411–8414 (1991).
135. P. Bedrossian and T. Klitsner, Surface reconstruction in layer-by-layer sputtering of Si (111), *Phys. Rev. B* **44**(24), 13783–13786 (1991).
136. T. Michely and C. Teichert, Adatom yields, sputtering yields, and damage patterns of single-ion impacts on Pt (111), *Phys. Rev. B* **50**(15), 11156–11166 (1994).
137. K. Yoneyama and K. Ogawa, Scanning tunneling microscopy studies on recovery processes of sputter-induced surface defects on Si(111)-7×7, *Jpn. J. Appl. Phys., Part 1* **35**, 3719–3723 (1996).
138. H. X. Liu, T. P. Schneider, J. Montgomery, Y. L. Chen, A. Buczkowski, F. Shimura, R. J. Nemanich, D. M. Maher, D. Korzec, and J. Engemann, A study of surface and subsurface properties of Si (100) after hydrogen ion-beam exposure, *Mater. Res. Soc. Symp. Proc.* **315**, 231–236 (1993).
139. J. A. Floro, B. K. Kellerman, E. Chason, S. T. Picraux, D. K. Brice, and K. M. Horn, Intra-cascade surface recombination of point defects during ion bombardment of Ge (001), *Mater. Res. Soc. Symp. Proc.* **316**, 993–998 (1994).
140. G. P. Malafsky, The effect of sputter temperature on vacancy island behavior on Ni (111) measured by photoemission of adsorbed xenon, *Surf. Sci. Lett.* **306**, L539–L544 (1994).
141. H. J. Steffen, D. Marton, and J. W. Rabalais, Defect formation in graphite during low energy Ne^+ bombardment, *Nucl. Instrum. Meth. Phys. Res., Sect. B* **67**, 308–311 (1992).
142. L. K. Verheij, E. Van Loenen, J. A. Van Den Berg, and D. G. Armour, Creation of surface damage on a nickel (110) surface by bombardment with 3-30 keV noble gas ions, *Nucl. Instrum. Meth.* **168**, 595–599 (1980).

143. H. Gnaser, B. Heinz, W. Bock, and H. Oechsner, Surface modification of GaAs (110) by low-energy ion irradiation, *Phys. Rev. B* **52**, 14086–14092 (1995).
144. W. D. Sawyer, J. Weber, G. Nabert, J. Schmalzlin, and H.-U. Habermeier, Implantation and diffusion of noble gas atoms during ion-beam etching of silicon, *J. Appl. Phys.* **68**(12), 6179–6186 (1990).
145. A. Vaseashta, A. Elshabini-Riad, and L. C. Burton, Effects of low energy Ar^+ ion bombardment on GaAs, *Mater. Sci. Eng. B* **9**, 489–500 (1991).
146. G. F. Feng, R. Zallen, J. M. Epp, and J. G. Dillard, Raman-scattering and optical studies of argon-etched GaAs surfaces, *Phys. Rev. B* **43**(12), 9678–9686 (1991).
147. M. Song, H. Hashimoto, Y. Yokota, and M. Awaji, Atomic scale morphology of the surface of the crystalline Si formed by ion etching, *Scanning* **12**, 29–32 (1990).
148. I. H. Wilson, Scanning tunnelling microscope studies of sputtering, *Surf. Interface Anal.* **20**, 637–644 (1993).
149. Mai Ghaly and R. S. Averback, Effect of viscous flow on ion damage near solid surfaces, *Phys. Rev. Lett.* **72**(3), 364–367 (1994).
150. M. H. Ludwig and R. E. Hummel, Differential reflectance studies of structural changes in AgAs caused by Ar^+ ion bombardment, *Mater. Res. Soc. Symp. Proc.* **300**, 345–350 (1993).
151. N. Seifert, W. Husinsky, G. Betz, Q. Yan, and N. H. Tolk, Sputtering of LiF films induced by low-energy Ar^+ impact, *Phys. Rev. B* **51**(18), 12202–12208 (1995).
152. Y. Z. Hu, M. Li, J. W. Andrews, K. A. Conrad, and E. A. Irene, in: *Defects in Silicon II* (W. Murray Bullis, U. Gösele, and F. Shimura, eds.), The Electrochemical Society, Pennington, NJ (1991), pp. 341–348.
153. Y. Z. Hu, J. W. Andrews, M. Li, and E. A. Irene, In situ spectroscopic ellipsometric investigation of argon ion bombardment of single-crystal silicon and silicon dioxide films, *Nucl. Instrum. Meth. Phys. Res., Sect. B* **59/60**, 76–79 (1991).
154. V. S. Belyakov and A. I. Titov, Formation of quasisteady states in semiconductors on irradiation by heavy ions, *Bull. Russ. Acad. Sci. Phys.* **58**(3), 430–433 (1994).
155. L. W. Hobbs, Topology and geometry in the irradiation-induced amorphization of insulators, *Nucl. Instrum. Meth. Phys. Res., Sect. B* **91**, 30–42 (1994).
156. A. Hoffman, S. Prawer, and R. Kalish, Structural transformation of diamond induced by 1-keV Ar-ion irradiation as studied by Auger and secondary-electron spectroscopies and total-secondary-electron-yield measurements, *Phys. Rev. B* **45**(22), 12736–12745 (1992).
157. T. Neffati, G. N. Lu, and C. Barret, Electrical characterization of Schottky diodes on very low energy ion-etched GaAs surfaces, *Solid-State Electron.* **31**(8), 1335–1342 (1988).
158. D. Marton, J. Fine, and G. P. Chambers, Temperature-dependent radiation-enhanced diffusion in ion-bombarded solids, *Phys. Rev. Lett.* **61**(23), 2697–2700 (1988).
159. Y. Yuba, T. Ishida, K. Gamo, and S. Namba, Characterization of ion beam etching induced defects in GaAs, *J. Vac. Sci. Technol., B* **6**(1), 253–256 (1988).
160. H. Neff, K. Y. Lay, M. S. Su, P. Lange, and K. J. Bachmann, Sputter induced near surface electronic defects in group II-VI compound semiconductor alloys, *Surf. Sci.* **189/190**, 661–668 (1987).
161. F. D. Auret, G. Myburg, S. A. Goodman, L. J. Bredell, and W. O. Barnard, Effect of sputter voltage on the electrical characteristics of argon ion sputtered n-type GaAs, *Nucl. Instrum. Meth. Phys. Res., Sect. B* **67**, 410–414 (1992).
162. C.-H. Chen, D. G. Yu, E. L. Hu, and P. M. Petroff, Photoluminescence studies on radiation enhanced diffusion of dry-etch damage in GaAs and InP materials, *J. Vac. Sci. Technol., B* **14**(6), 3684–3687 (1996).
163. G. Carter, I. V. Katardjiev, M. J. Nobes, and J. L. Whitton, Sputter-induced surface topography on f.c.c. metals, *Mater. Sci. Eng.* **90**, 21–32 (1987).

164. D. W. Moon and K. J. Kim, Surface topography development on ion-beam-sputtered surfaces: role of surface inhomogeneity induced by ion-beam bombardment, *J. Vac. Sci. Technol., A* **14**(5), 2744–2756 (1996).
165. V. Vishnyakov, S. E. Donnelly, and G. Carter, Scanning tunnelling microscopy observations of the evolution of small-scale topography on gold surfaces following irradiation with low-energy argon ions, *Phil. Mag. B* **70**(1), 151–157 (1994).
166. C. M. Demanet, K. Vijaya Sankar, J. B. Malherbe, N. G. van der Berg, and R. Q. Odendaal, Atomic force microscopy investigation of noble gas ion bombardment on InP: effect of ion energy, *Surf. Interface Anal.* **24**, 497–502 (1996).
167. G. Carter, I. V. Katardjiev, and M. J. Nobes, Reproducibility and stability in surface morphological evolution, *Phil. Mag. B* **63**(4), 849–866 (1991).
168. W. H. Gries, Radiation-induced sample modification in surface analysis: InP as an extreme example, *Surf. Interface Anal.* **14**, 611–618 (1989).
169. G. K. Wehner, Cone formation as a result of whisker growth on ion bombarded metal surfaces, *J. Vac. Sci. Technol., A* **3**(4), 1821–1835 (1985).
170. A. Žalar and S. Hofmann, Redeposition in AES sputter depth profiling of multilayer Cr/Ni thin films, *Surf. Interface Anal.* **12**, 83–86 (1988).
171. A. C. Miller and A. W. Czanderna, Redeposition of sputtered species during ion etching of Cu, Ag, and Au, *J. Vac Sci. Technol.* **12**, 1086–1087 (1975).
172. H. F. Helbig, P. J. Adelman, A. C. Miller, and A. W. Czanderna, Redeposition of sputtered species by the electric fields of an incident ion beam and sputtered ions, *J. Vac. Sci. Technol.* **13**, 379–383 (1976).
173. I. V. Katardjiev, G. Carter, M. J. Nobes, S. Berg, and H.-O. Blom, Three-dimensional simulation of surface evolution during growth and erosion, *J. Vac. Sci. Technol., A* **12**(1), 61–68 (1994).
174. C. M. Demanet, J. B. Malherbe, N. G. van der Berg, and Vijaya Sankar, Atomic force microscopy investigation of argon-bombarded InP: effect of ion dose density, *Surf. Interface Anal.* **23**, 433–439 (1995).
175. V. Vishnyakov, G. Carter, D. T. Goddard, and M. J. Nobes, Topography development on selected inert gas and self-ion bombarded Si, *Vacuum* **46**(7), 637–643 (1995).
176. Y. S. Toh, M. J. Nobes, and G. Carter, A cellular automata simulation study of surface roughening resulting from multi-atom etch pit generation during sputtering, *Nucl. Instrum. Meth. Phys. Res., Sect. B* **67**, 586–589 (1992).
177. G. Carter and M. J. Nobes, Probabilistic and deterministic approaches to surface contour evolution during sputtering, *Vacuum* **45**(5), 539–546 (1994).
178. L. Pranevicius, S. Raila, and J. Vosylius, Initial stages of development of the surface topography induced by ion bombardment, *Surf. Coat. Technol.* **63**, 57–63 (1994).
179. I. A. Abroyan, V. Ya. Velichko, M. Ya. Litvak, V. I. Malyugin, O. A. Merkulova, and A. B. Fadeev, Statistical properties of surface microtexture arising from ion bombardment, *Bull. Russ. Acad. Sci. Phys.* **58**(3), 465–471 (1994).
180. S. Zekorn, Th. Fries, C. Becker, and K. Wandelt, Topography of sputtered semiconductor surfaces by STM in air, *Phys. Status Solidi A* **131**, 99 (1992).
181. V. Vishnyakov and G. Carter, Ne^+, Ar^+ and Xe^+ ion bombardment induced and suppressed topography on Si, *Nucl. Instrum. Meth. Phys. Res., Sect. B* **106**, 174–178 (1995).
182. G. Carter and V. Vishnyakov, Ne^+ and Ar^+ ion bombardment-induced topography on Si, *Surf. Interface Anal.* **23**, 514–520 (1995).
183. A. Barna and M. Menyhard, Auger depth profile analysis of deeply buried interfaces, *Phys. Stat. Sol. A* **145**, 263–274 (1994).
184. T. Motohiro and Y. Taga, Characteristic erosion of silica by oblique argon ion beam bombardment, *Thin Solid Films* **147**, 153–165 (1987).

185. G. W. Lewis, G. Kiriakides, G. Carter, and M. J. Nobes, Ion bombardment induced surface topography modification of clean and contaminated single crystal Cu and Si, *Surf. Interface Anal.* **4**(4), 141–150 (1982).
186. L. Clapham, J. L. Whitton, A. Mashayekhi, G. Carter, I. Katardjiev, and M. J. Nobes, Effect of alloying on ion-bombardment-induced surface topography on face-centred cubic metals, *Nucl. Instrum. Meth. Phys. Res., Sect. B* **48**, 571–575 (1990).
187. M. Tanemura and F. Okuyama, Topographical features of ion bombardment-induced microcones on Ag thin films, *Nucl. Instrum. Meth. Phys. Res., Sect. B* **47**, 126–132 (1990).
188. J. K. G. Panitz, Conical surface textures formed by ion bombarding 2% Be-Cu alloy, *J. Micromech. Microeng.* **1**, 52–59 (1991).
189. T. K. Chini, S. R. Bhattacharyya, D. Ghose, and D. Basu, Topography of the Ar^+-sputtered Si surface, *Jpn. J. Appl. Phys.* **30**(11A), 2895–2896 (1991).
190. J. B. Malherbe, H. Lakner, and W. H. Gries, Composition and structure of ion-bombardment-induced growth cones on InP, *Surf. Interface Anal.* **17**, 719–725 (1991).
191. C. M. Demanet, K. Vijaya Sankar, and J. B. Malherbe, Atomic force microscopy investigation of ion-bombarded InP: effect of angle of ion bombardment, *Surf. Interface Anal.* **24**, 503–510 (1996).
192. S. W. MacLaren, J. E. Baker, N. L. Finnegan, and C. M. Loxton, Surface roughness development during sputtering of GaAs and InP: evidence for the role of surface diffusion in ripple formation and sputter cone development, *J. Vac. Sci. Technol., A* **10**(3), 468–476 (1992).
193. T. K. Chini, J. Kato, M. Tanemura, and F. Okuyama, Monocrystalline InP surfaces normally sputtered with Ar^+: SEM and TEM observations, *Nucl. Instrum. Meth. Phys. Res., Sect. B* **95**, 313–318 (1995).
194. Y. Q. Zhang, F. Fujimoto, and F. Okuyama, Unusual texturing of Ar^+-sputtered InP surfaces associated with Ni-seeding, *Nucl. Instrum. Meth. Phys. Res., Sect. B* **101**, 306–310 (1995).
195. R. M. Bradley and J. M. E. Harper, Theory of ripple topography induced by ion bombardment, *J. Vac. Sci. Technol., A* **6**(4), 2390–2395 (1988).
196. J. J. Vajo, R. E. Doty, and E.-H. Cirlin, Influence of O_2^+ energy, flux, and fluence on the formation and growth of sputtering-induced ripple topography on silicon, *J. Vac. Sci. Technol., A* **14**(5), 2709–2720 (1996).
197. A. N. Protsenko, A mechanism of ion bombardment induced ripple topography, *Nucl. Instrum. Meth. Phys. Res., Sect. B* **82**, 417–420 (1993).
198. G. Carter, M. J. Nobes, G. W. Lewis, and C. R. Brown, The effect of surface topography evolution on sputter profiling depth resolution in Si, *Surf. Interface Anal.* **7**(1), 35–41 (1985).
199. G. Carter, M. J. Nobes, H. Stoere, and I. V. Katardjiev, The evolution of atomic scale topography by sputtering erosion, *Surf. Interface Anal.* **20**, 90–94 (1993).
200. A. Barna, P. B. Barna, and A. Zalar, Ion beam induced roughness and its effect in AES depth profiling of multilayer Ni/Cr thin films, *Surf. Interface Anal.* **12**, 144–150 (1988).
201. G. Carter, M. J. Nobes, and I. V. Katardjiev, The theory of ion beam polishing and machining, *Vacuum* **44**(3/4), 303–309 (1993).
202. A. Barna, P. B. Barna, and A. Zalar, Analysis of the development of large area surface topography during ion etching, *Vacuum* **40**(1/2), 115–120 (1990).
203. M. A. Venables and S. S. Makh, Polishing of lithium niobate by ion-beam etching, *J. Mater. Sci. Lett.* **6**, 1254–1256 (1987).
204. E. Chason and T. M. Mayer, Low energy ion bombardment induced roughening and smoothing of SiO_2 surfaces, *Appl. Phys. Lett.* **62**(4), 363–365 (1993).
205. Z. Csahok, Z. Farkas, M. Menyhard, G. Gergely, and Cs. S. Daroczi, Surface morphology development during ion sputtering: roughening or smoothing?, *Surf. Sci. Lett.* **364**, L600–L604 (1996).

206. R. Michael and D. Stulik, A semidynamic study of polymer surface morphology development in Kapton-H™ sputter etched by 6 keV xenon atoms, *Nucl. Instrum. Meth. Phys. Res., Sect. B* **28**, 259–263 (1987).
207. A. Karen, Y. Nakagawa, M. Hatada, K. Okuno, F. Soeda, and A. Ishitani, Quantitative investigation of O_2^+ -induced topography of GaAs and other III-V semiconductors: an STM study of the ripple formation and suppression of the secondary ion yield change by sample rotation, *Surf. Interface Anal.* **23**, 506–513 (1995).
208. A. Zalar, E. W. Seibt, and P. Panjan, Oxide thin films formed during rotational AES sputter depth profiling of Ni/Cr multilayers using oxygen ions, *Appl. Surf. Sci.* **100/101**, 92–96 (1996).
209. W. Pamler and K. Wangemann, Depth profile analysis of aluminium metallization in microelectronics: optimization of depth resolution, *Surf. Interface Anal.* **18**, 52–58 (1992).
210. T. Chassé, W. Heichler, J. Langhammer, and W. Zwanzig, Sputter depth profiling of GaAlAs double heterostructures using Auger electron spectroscopy, *Cryst. Res. Technol.* **22**(11), 1425–1429 (1987).
211. T. D. Bestwick, M. E. Welland, and R. B. Leane, Scanning tunnelling microscopy of ion etched silicon, *Vacuum* **38**(8-10), 823–826 (1988).
212. S. Hofmann, Quantitative depth profiling in surface analysis: a review, *Surf. Interface Anal.* **2**(4), 148–160 (1980).
213. R. Kelly, in: *Materials Modification by High-Fluence Ion Beams* (R. Kelly and M. Fernando da Silva, eds.), Kluwer, Dordrecht, The Netherlands (1989), pp. 305–327.
214. Z. W. Kowalski, A review of compositional changes of multicomponent materials (especially biomaterials) induced by ion sputtering, *J. Vac. Sci. Technol., A* **1**(2), 494–496 (1983).
215. A. Delcorte, L. T. Weng, and P. Bertrand, Secondary molecular ion emission from aliphatic polymers bombarded with low energy ions: effects of the molecular structure and the ion beam induced surface degradation, *Nucl. Instrum. Meth. Phys. Res., Sect. B* **100**, 213–216 (1995).
216. G. J. Leggett and J. C. Vickerman, Role of electronic particle-surface interactions during the sputter degradation of polymers, *Anal. Chem.* **63**, 561–568 (1991).
217. R. Schlögl, Modification of the electronic structure of graphite by intercalation, chlorination and ion etching, *Surf. Sci.* **189/190**, 861–872 (1987).
218. R. Kelly, On the role of Gibbsian segregation in causing preferential sputtering, *Surf. Interface Anal.* **7**(1), 1–6 (1985).
219. A. Oliva, R. Kelly, and G. Falcone, Sputtering of multicomponent materials: the diffusion limit, *Surf. Sci.* **166**, 403–418 (1986).
220. R. Kelly, On the problem of whether mass or chemical bonding is more important to bombardment-induced compositional changes in alloys and oxides, *Surf. Sci.* **100**, 85–107 (1980).
221. R. Kelly, Bombardment-induced compositional change with alloys, oxides, oxysalts and halides, III. The role of chemical driving forces, *Mater. Sci. Eng., A* **115**, 11–24 (1989).
222. N. Q. Lam, Mechanisms and kinetics of ion beam-induced compositional modifications, *Mater. Res. Soc. Symp. Proc.* **100**, 29–42 (1988).
223. A. Galdikas and L. Pranevičius, Redistribution of components in the altered layer formed by preferential sputtering, *Surf. Coat. Technol.* **64**, 167–172 (1994).
224. B. Baretsky, in: *Materials Modification by High-Fluence Ion Beams* (R. Kelly and M. Fernando da Silva, eds.), Kluwer, Dordrecht, The Netherlands (1989), pp. 329–337.
225. V. Miteva, D. Karpuzov, P. Ivanov, and St. Angelova, Stoichiometry effects at Cu-Ni alloy surfaces during 5 keV Ar ion sputtering at room temperatures, *Nucl. Instrum. Meth. Phys. Res., Sect. B* **85**, 340–343 (1994).
226. N. Q. Lam, H. A. Hoff, H. Wiedersich, and L. E. Rehn, Subsurface compositional modifications in Ni-Cu alloys during high-temperature ion sputtering, *Surf. Sci.* **149**, 517–536 (1985).
227. S. Sekine, H. Shimizu, and S. Ichimura, Composition of sputtered material from Cu-Ni alloy during Xe^+ ion sputtering at elevated temperatures, *Appl. Surf. Sci.* **84**, 401–406 (1995).

228. S. L. Jiang, A. Oliva, A. Amoddeo, and R. Kelly, Composition profiles due to bombardment-induced segregation in Cu-Ni samples as deduced by Auger spectroscopy, *Nucl. Instrum. Meth. Phys. Res., Sect. B* **45**, 646–650 (1990).
229. A. Kurokawa, H. J. Kang, and R. Shimizu, Computer simulation of the surface composition change of a Cu-Ni alloy under ion bombardment at different sample temperatures, *Surf. Sci.* **255**, 120–126 (1991).
230. R.-S. Li, Temperature dependence of the surface composition changes under Ar ion bombardment of an $Au_{0.5}Cu_{0.5}$ alloy between 30°C and 700°C, *Surf. Sci.* **193**, 373–386 (1988).
231. R.-S. Li and Y.-Z. Sun, Quantitative Auger electron spectroscopy analysis of the Au-Cu alloys, *Surf. Sci.* **191**, 339–352 (1987).
232. T. Koshikawa and K. Goto, in: *Secondary Ion Mass Spectrometry SIMS V* (A. Benninghoven, R. J. Colton, D. S. Simmons, and H. W. Werner, eds.), Springer Verlag, Berlin (1985) pp. 394–396.
233. R.-S. Li, L.-X. Tu, and Y.-Z. Sun, Observation of post-bombardment segregation in an Au-Cu alloy at room temperature, *Appl. Surf. Sci.* **26**, 77–92 (1986).
234. H. J. Kang, R. Shimizu, and T. Okutani, ISS measurement of surface composition of Au-Cu alloys by simultaneous ion bombardments with Ar^+ and He^+ ions, *Surf. Sci.* **116**, L173–L178 (1982).
235. D. Marton, J. Fine, and G. P. Chambers, Ion-induced radiation-enhanced diffusion of silver in nickel, *Mater. Sci. Eng., A* **115**, 223–227 (1989).
236. N. Q. Lam and H. A. Hoff, Surface and subsurface composition changes in Ni-Si alloys during elevated-temperature sputtering, *Surf. Sci.* **193**, 353–372 (1988).
237. G. N. Van Wyk, J. du Plessis, and E. Taglauer, Sputtering and radiation-enhanced segregation in amorphous Cu-Ti alloys, *Surf. Sci.* **254**, 73–80 (1991).
238. C. G. Bondarenko, L. I. Ivanov, and S. I. Kucheryavyi, Examination of sputtering of aluminum alloys in ion bombardment, *Fiz. Khim. Obrab. Mater.* **20**(5), 24–25 (1986).
239. A. R. Krauss, D. M. Gruen, and A. B. DeWald, Modification of alloy surface composition by segregation processes as a means of impurity control in fusion devices, *J. Nucl. Mater.* **121**, 398–403 (1984).
240. H. J. Kang, R. H. Roberts, S. He, D. J. O'Connor, and R. J. MacDonald, Surface composition changes of Ni-Zr alloy under Ar^+ ion bombardment by ISS, AES and XPS, *Surf. Sci.* **302**, 363–370 (1994).
241. H. J. Kang, C. H. Kim, N. S. Park, D. J. O'Connor, and R. MacDonald, Preferential sputtering and surface segregation of CuZr alloys, *Appl. Surf. Sci.* **100/101**, 329–332 (1996).
242. C. Fan and D. Yang, Changes in alloy surface composition induced by low energy ion bombardment, *Vacuum* **43**(3), 231–234 (1992).
243. R. P. Frankenthal and D. J. Siconolfi, Effect of ion mass and ion energy on the surface composition of two sputtered tin-lead alloys, *J. Vac. Sci. Technol., A* **2**(2), 1089–1092 (1984).
244. M. Kudo, A. Ishijima, and T. Morohashi, Ion induced alteration at Pb-Sn alloy surface investigated by Auger electron spectroscopy and X-ray photoelectron spectroscopy, *Appl. Surf. Sci.* **100/101**, 134–137 (1996).
245. G. E. Hammer and R. M. Shemenski, Ion bombardment effects in the sputtering of single- and two-phase brass alloys; Cu: 18–48% Zn, *Surf. Interface Anal.* **10**, 355–359 (1987).
246. E. Darque-Ceretti, F. Delamare, and G. Blaise, Ion induced surface compositional changes of α brass; a comparative study by AES and sputtered thermal ion mass spectrometry. Application to quantitative AES, *Surf. Interface Anal.* **7**(3), 141 (1985).
247. J. Du Plessis, G. N. Van Wyk, and E. Taglauer, Preferential sputtering and radiation enhanced segregation in palladium-platinum alloys, *Surf. Sci.* **220**, 381–390 (1989).
248. S.-P. Jeng and P. H. Holloway, Preferential sputtering of Ni/Cr (110), *Surf. Sci.* **227**, 273–277 (1990).
249. Th. Lill, W. F. Calaway, and M. J. Pellin, Sputtering of tin and gallium-tin clusters, *Nucl. Instrum. Meth. Phys. Res., Sect. B* **100**, 361–365 (1995).

250. S. K. Sharma and S. Hofmann, Study of amorphous and crystallized $Fe_{75}B_{25}$ by Auger electron spectroscopy and X-ray photoelectron spectroscopy, *Appl. Surf. Sci.* **51**, 139–155 (1991).
251. Qu Zhe, A general expression of the steady state composition for multicomponent system during ion sputtering, *Surf. Sci.* **161**, L549–L553 (1985).
252. G. Betz, J. Dudonis, and P. Braun, Surface enrichment in Ag-Au-Pd alloys under ion bombardment, *Surf. Sci.* **104**, L185–L188 (1981).
253. J. Dudonis and L. Pranevicius, Changes in surface composition induced by ion bombardment of multicomponent alloys, *Surf. Coat. Technol.* **56**, 233–242 (1993).
254. S. Tang, N. Q. Lam, and A. Yacout, Near-surface composition modifications in Ag-Au-Cu alloys by ion sputtering at elevated temperatures, *Surf. Sci.* **231**, 253–262 (1990).
255. N. Q. Lam, S. Tang, A. M. Yacout, and L. E. Rhen, Ion beam modifications of near-surface compositions in ternary alloys, *Nucl. Instrum. Meth. Phys. Res., Sect. B* **59/60**, 889–892 (1991).
256. A. M. Yacout, N. Q. Lam, and J. F. Stubbins, Near-surface compositional modifications of ternary alloys during sputtering at elevated temperatures, *Nucl. Instrum. Meth. Phys. Res., Sect. B* **42**, 49–60 (1989).
257. Qu Zhe, Determination of the steady state composition in a ternary system from the corresponding binary systems under ion sputtering, *Surf. Sci.* **176**, L873–L878 (1986).
258. G. C. Tyrrell, I. W. Boyd, and R. B. Jackman, Ion beam assisted etching of silicon with bromine, the role of the adsorbed state, *Appl. Surf. Sci.* **43**, 439–446 (1989).
259. S. S. Todorov, I. R. Chakarov, and D. S. Karpuzov, Low energy ion beam oxidation of silicon surfaces: ballistics, diffusion and chemistry, *Nucl. Instrum. Meth. Phys. Res., Sect. B* **65**, 79–83 (1992).
260. F. R. McFeely and J. A. Yarmoff, in: *The Physics of Semiconductors, Vol. 1* (O. Engström, ed.), World Scientific, Singapore (1987) pp. 137–140.
261. N. Menzel and K. Wittmaack, RBS study on bombardment-induced redistribution of copper impurities in silicon using neon, oxygen and nitrogen ion beams at different impact angles, *Nucl. Instrum. Meth. Phys. Res., Sect. B* **45**, 219–222 (1990).
262. E. Collart and R. J. Visser, Deep surface damage of SiO_2 by sputtering with low energy Ar-ions, *Surf. Sci.* **218**, L497–L504 (1989).
263. E. Paparazzo, M. Fanfoni, C. Quaresima, and P. Perfetti, Evidence of SiO_x suboxides at Ar ion etched silica surfaces, *J. Vac. Sci. Technol., A* **8**(3), 2231–2234 (1990).
264. T. Mizutani, Preferential sputtering of oxygen from SiO_2 by low-energy ion beam and neutral beam bombardment, *Jpn. J. Appl. Phys., Part 2* **30**(4A), L628–L631 (1991).
265. K. Miyake, Removal of a thin SiO_2 layer by low-energy hydrogen ion bombardment at elevated temperatures, in: *Proceedings of the 1989 International Symposium on Microprocess Conference, JJAP Series 3*, 277–282 (1989).
266. J. Morais, T. A. Fazan, and R. Landers, Ion induced composition changes on (110) InP cleaved surface, *Phys. Status Solidi A* **141**, K19 (1994).
267. C. Jardin, D. Robert, B. Achard, B. Gruzza, and C. Pariset, An AES and ELS study of InP (100) surface subjected to argon ion bombardment, *Surf. Interface Anal.* **10**(6) 301–305 (1987).
268. J. B. Malherbe, Ion angle dependence in the preferential sputtering of InP, *Appl. Surf. Sci.* **70/71**, 322–326 (1993).
269. M. Bouslama, C. Jardin, and M. Ghamnia, The InSb (100) surface change during the argon ion bombardment and the electron stimulated oxidation, *Vacuum* **46**(2), 143–146 (1995).
270. D. K. Skinner, J. G. Swanson, and C. V. Haynes, Ion beam induced composition changes during Auger sputter profiling of thin Al films on InP, *Surf. Interface Anal.* **5**(1), 38–42 (1983).
271. H. J. Kang, Y. M. Moon, T. W. Kang, J. Y. Leem, J. J. Lee, and D. S. Ma, Surface composition and structure changes in GaAs compounds due to low-energy Ar^+ ion bombardment, *J. Vac. Sci. Technol., A* **7**(6), 3251–3255 (1989).

272. N. A. Bert, S. G. Konnikov, A. V. Korol'kov, and K. Yu. Pogrebitskii, Change in the composition of gallium arsenide during bombardment of its surface by Ar^+ ions, *Sov. Tech. Phys. Lett.* **14**(4), 298–299 (1988).
273. J. B. Malherbe, W. O. Barnard, I. Le R. Strydom, and C. W. Louw, Preferential sputtering of GaAs, *Surf. Interface Anal.* **18**, 491–495 (1992).
274. A. El. Khalki, F. Proix, and C. A. Sébenne, Room temperature interaction of ionized nitrogen with cleaved GaAs, *Vacuum* **39**(11/12), 1131–1133 (1989).
275. F. Xu, Z. Lin, D. M. Hill, and J. H. Weaver, Temperature-dependent interface evolution for Ti/GaAs (100) and Cr/GaAs (100), *Phys. Rev. B* **36**(12), 6624–6630 (1987).
276. M. Inoue, T. Tsutsui, S. Nishigaki, and T. Noda, Accommodation of argon and oxygen gases on a GaAs (110) surface under Ar^+ ion bombardment, *Surf. Sci. Lett.* **273**, L457–L460 (1992).
277. I. N. Eydokimov, R. Valizadeh, D. G. Armour, N. V. Richardson, and C. F. McConville, Oscillatory evolution of an ion bombarded InSb(100) surface, *Surf. Sci.* **318**, 281–288 (1994).
278. L. L. Kazmerski, N. A. Burnham, A. B. Swartzlander, A. J. Nelson, and S. E. Asher in: *Conference Record of the 19th IEEE Photovoltaic Specialists Conference*, IEEE, New York (1987), pp. 1315–1321.
279. E. Niemi and L. Stolt, Characterization of $CuInSe_2$ thin films by XPS, *Surf. Interface Anal.* **15**, 422–426 (1990).
280. J. B. Malherbe, S. Hofmann, and J. M. Sanz, Preferential sputtering of oxides: a comparison of model predictions with experimental data, *Appl. Surf. Sci.* **27**, 355–365 (1986).
281. D. F. Mitchell, G. I. Sproule, and M. J. Graham, Sputter reduction of oxides by ion bombardment during Auger depth profile analysis, *Surf. Interface Anal.* **15**, 487–497 (1990).
282. I. Bertóti, A. Tóth, M. Mohai, R. Kelly, and G. Marletta, Effect of ion bombardment on Cr-Si-O layers: an X-ray photoelectron spectroscopic study, *Thin Solid Films* **241**, 211–217 (1994).
283. P. Varga and E. Taglauer, Depth profiling of the altered layer in Ta_2O_5 produced by sputtering with He ions, *Nucl. Instrum. Meth. Phys. Res., Sect. B* **2**, 800–803 (1984).
284. J. C. Hamilton, B. E. Mills, and R. E. Benner, Depth profiling of metal oxides using Raman spectroscopy with ion bombardment, *Appl. Phys. Lett.* **40**(6), 499–502 (1982).
285. P. C. Karulkar, Effects of sputtering on the surface composition of niobium oxides, *J. Vac. Sci. Technol., A* **18**(2), 169–174 (1981).
286. C. S. Fang, F. M. Pan, W. S. Tse, and S. R. Horng, Photoemission studies of ion bombardment effect on SnO_2 surfaces, *Surf. Sci.* **211/212**, 279–288 (1989).
287. C. Morant, J. M. Sanz, and L. Galán, Ar-ion bombardment effects on ZrO_2 surfaces, *Phys. Rev. B* **45**(3), 1391–1398 (1992).
288. A. R. González-Elipe, A. Fernández, J. P. Holgado, A. Caballero, and G. Munuera, Mixing effects in CeO_2/TiO_2 and CeO_2/SiO_2 systems submitted to Ar^+ sputtering, *J. Vac. Sci. Technol., A* **11**(1), 58–65 (1993).
289. D. W. Moon and K. J. Kim, Sputter depth profiling analysis of Ta_2O_5 on Si without preferential sputtering by energetic oxygen ion beams, *Appl. Phys. Lett.* **62**(24), 3094–3096 (1993).
290. K. J. Kim, D. W. Moon, S. H. Nam, W. J. Lee, and H. G. Kim, XPS sputter depth profiling of the chemical states for $SrTiO_3$/Si interface by O_2^+ ion beams, *Surf. Interface Anal.* **23**, 851–857 (1995).
291. M. Saidoh and R. Yamada, Study of modification processes in sputtering during oxygen exposure, *Nucl. Instrum. Meth. Phys. Res., Sect. B* **39**, 599–602 (1989).
292. C. N. Christy, *ESCA Examination of Metal Oxides and Electronic Ceramic Materials: The Effect of a Low Energy Argon Ion Beam*, Ph.D. Thesis, University of South Florida, Tampa (1989).
293. V. I. Nefedov, A. N. Sokolov, M. A. Tyzykhov, N. N. Oleinikov, Ye. A. Yeremina, and M. A. Kolotyrkina, XPS study of high-tc superconductor surface degradation, *J. Electron Spectrosc. Relat. Phenom.* **49**, 47–60 (1989).
294. T. Matsui, D. Yamaguchi, and H. Kamijo, Damage to $YBa_2Cu_3O_y$ surface caused by Ar ion beam etching, *Jpn. J. Appl. Phys., Part 2* **35**, L97–L99 (1996).

295. S. M. Mukhopadhyay and T. C. S. Chen, Surface properties of perovskites and their response to ion bombardment, *J. Appl. Phys.* **74**(2), 872–876 (1993).
296. R. K. Brow, Glass surface modifications during ion beam sputtering, *J. Non-Cryst. Solids* **107**, 1–10 (1988).
297. S. M. Mukhopadhyay and S. H. Garofalini, Surface studies of TiO_2-SiO_2 glasses by X-ray photoelectron spectroscopy, *J. Non-Cryst. Solids* **126**, 202–208 (1990).
298. K. Hara and T. Itoh, in: *Secondary Ion Mass Spectrometry SIMS IV* (A. Benninghoven, J. Okano, R. Shimizu, and H. W. Werner, eds.), Springer-Verlag, Berlin (1984) pp. 248–251.
299. J. R. Lince, T. B. Stewart., M. M. Hills, P. D. Fleischauer, J. A. Yarnoff and A. Taleb-Ibrahimi, Chemical effects of Ne^+ bombardment on the $MoS_2(0001)$ surface studied by high-resolution photoelectron spectroscopy, *Surf. Sci.* **210**, 387–405 (1989).
300. S. Valeri and G. Ottaviani, Ion beam effects on surface and subsurface composition of near-noble silicides, *J. Vac. Sci. Technol., A* **5**(4), 1358–1361 (1987).
301. S. Valeri, A. di Bona, G. Ottaviani, and M. Procop, Ion beam effects on the surface and near-surface composition of $TaSi_2$, *Nucl. Instrum. Meth. Phys. Res., Sect. B* **59/60**, 98–101 (1991).
302. S. Valeri, U. del Pennino, G. Ottaviani, P. Sassaroli, and K. N. Tu, AES and EELS study of $ErSi_2$ and its behavior under ion bombardment and oxygen exposure, *Solid St. Commun.* **60**(7), 569–573 (1986).
303. H. C. Swart and G. L. P. Berning, The influence of sputtering on the surface composition of FeSi and $TbSi_{1.7}$, *Appl. Surf. Sci.* **73**, 260–263 (1993).
304. S. Hofmann and M. G. Stepanova, Preferential sputtering of argon ion bombarded Ni_3Al and $TaSi_2$, *Appl. Surf. Sci.* **90**, 227–233 (1995).
305. J. M. E. Harper, S. Motakef, and D. Moy, Highly selective sputtering of silicon from $TiSi_2$ at elevated temperature, *Appl. Phys. Lett.* **60**(10), 1196–1198 (1992).
306. V. I. Zaporozchenko, M. G. Stepanova, and S. S. Vojtusik, Effect of ion incidence angle on surface concentrations in Ar^+-bombarded Co_2Si, studied by AES, *Surf. Interface Anal.* **23**, 185–187 (1995).
307. B. Brijs, H. Bender, W. De Coster, R. Moons, and W. Vandervorst, In situ RBS analysis of ion beam mixing during low energy sputtering, *Nucl. Instrum. Meth. Phys. Res., Sect. B* **79**, 446–449 (1993).
308. B. R. Chakraborty and S. Hofmann, Ion bombardment induced silicide formation during sputter depth profiling of Ta/Si multilayer thin film structure as studied by X-ray photoelectron spectroscopy and Auger electron spectroscopy, *Thin Solid Films* **204** , 163–174 (1991).
309. S. Pignataro, G. Marletta, and O. Puglisi, Binary collisions inducing chemical reactions in 2 keV Ar^+ sputtering of Zr and Ti phosphates, *Chem. Phys.* **97**, 421–431 (1985).
310. J.-N. Chen and J. W. Rabalais, Preferential sputtering of LiF surfaces monitored by photoelectron spectroscopy and direct recoil spectrometry, *Surf. Sci.* **176**, L879–L885 (1986).
311. J. T. Rickman and R. W. Linton, Surface and depth profiling techniques using XPS applied to the study of nickel-containing environmental particles, *Appl. Surf. Sci.* **68**, 375–393 (1993).
312. K. Asami, M. A. De Sá, and V. Ashworth, The effect of ion etching on surface composition and chemical state, *Corros. Sci.* **26**(1), 15–25 (1986).
313. P. Selvam, B. Viswanathan, and V. Srinivasan, Note on the formation of surface carbides, *J. Electron Spectrosc. Relat. Phenom.* **50**, 277–287 (1990).
314. P. Selvam, B. Viswanathan, and V. Srinivasan, Ion-induced carbide formation of TiFe: evidence from XPS and AES studies, *J. Less-Common Met.* **161**, 77–85 (1990).
315. I. Chakarov and D.S. Karpusov, in: *Materials Modification by High-Fluence Ion Beams* (R. Kelly and M. Fernando da Silva, eds.), Kluwer, Dordrecht, The Netherlands (1989), pp. 185–190.
316. P. Sigmund, Exact solutions of balance equation governing ion-beam-induced composition changes and sputtering, *Appl. Phys. Lett.* **66**(4), 433–1995).

317. J. A. Peinador, I. Abril, J. J. Jiménez-Rodriĝez, and A. Gras-Marti, Atomic mixing of multi-component materials: the dilute limit, *Vacuum* **39**(7/8), 695–699 (1989).
318. V. P. Krivobokov and O. V. Pashchenko, Mathematical model for atomic transport processes in ion-irradiated metals, *Sov. Phys. Tech. Phys.* **36**(3), 367–369 (1991).
319. L. Zheng, R. Li, M. Li, and Y. Cheng, Boron isotope fractionation during prolonged ion bombardment, *Chin. Phys. Lett.* **8**(6), 325–328 (1991).
320. G. Carter, I. V. Katardjiev, and M. J. Nobes, An altered layer model for sputter-profiling, *Surf. Interface Anal.* **14**, 194–208 (1989).
321. G. Carter and M. J. Nobes, Ion-beam-induced topography and compositional changes in depth profiling, *Surf. Interface Anal.* **19**, 39–49 (1992).
322. A. Miotello and R. Kelly, Ion-beam mixing with chemical guidance, I. The bilayer problem, *Surf. Sci.* **268**, 340–350 (1992).
323. R. Kelly and A. Miotello, Thermodynamic effects in depth profiling and ion-beam mixing without invoking thermal spikes, *Appl. Phys. Lett.* **64**(20), 2649–2651 (1994).
324. A. N. Protsenko and E. F. Chaikovskii, Radiation-stimulated diffusion during ion etching, *Sov. Phys. Tech. Phys.* **31**(2), 242–244 (1986).
325. H. J. Kang, J. H. Kim, Y. S. Kim, D. W. Moon, and R. Shimizu, Dynamic Monte Carlo simulation of surface composition changes on TiC and Au-Cu under Ar^+ ion bombardment, *Surf. Sci.* **226**, 93–100 (1990).
326. L.-P. Zheng, R.-S. Li, M.-Y. Li, M. Chen, and X.-Q. Xia, Enrichment transformation between individual elements in the near-normal direction during binary alloy sputtering, *Nucl. Instrum. Meth. Phys. Res., Sect. B* **71**, 39–43 (1992).
327. L.-P. Zheng, R.-S. Li, M.-Y. Li, J.-Y. Lin, and Z.-Y. Zou, Correlation between isotope fractionation and surface composition profile during prolonged bombardment, *Nucl. Instrum. Meth. Phys. Res., Sect. B* **62**, 61–64 (1991).
328. P. A. J. Ackermans, G. C. R. Krutzen, and H. H. Brongersma, The use of a calibration in low-energy ion scattering: preferential sputtering and S segregation in CuPd alloys, *Nucl. Instrum. Meth. Phys. Res., Sect. B* **45**, 384–389 (1990).
329. S. Hofmann, in: *Practical Surface Analysis*, Vol. 1-*Auger and X-ray Photoelectron Spectroscopy* (D. Briggs and M. P. Seah, eds.), Wiley, New York (1991), pp. 143–200.
330. P. T. Dawson and S. A. Petrone, Crystal orientation effects on the surface morphology produced by ion bombardment of a pure element: implications for quantitative surface analysis, *Surf. Interface Anal.* **17**, 273–281 (1991).
331. H. Tsuji, J. Ishikawa, H. Itoh, Y. Toyota, and Y. Gotoh, Fundamental study on powder-scattering in positive- and negative-ion implantation into powder materials, *Appl. Surf. Sci.* **100/101**, 342–346 (1996).
332. A. Zalar and S. Hofmann, Superposition of Auger electron spectroscopy depth profiles obtained on Cr/Ni multilayer samples with different roughnesses, *J. Vac. Sci. Technol., A* **5**(4), 1209–1212 (1987).
333. M. Kühlein and I. Jäger, On the influence of sample temperature on depth resolution in sputtering, *Surf. Interface Anal.* **6**(3),129–134 (1984).
334. M. P. Seah and M. Kühlein, Temperature, roughness and depth resolution in ion sputter profiles, *Surf. Sci.* **150**, 273–288 (1985).
335. N. Tanović, L. Tanović, and J. Fine, Interface width dependence on sample temperature during Auger sputter depth profiling of Cr/Ni multilayered thin films, *Nucl. Instrum. Methods Phys. Res., Sect. B* **67**, 491–494 (1992).
336. C. J. Powell, D. M. Hercules, and A. W. Czanderna, in: *Ion Spectrotroscopies for Surface Analysis* (A. W. Czanderna and D. M. Hercules, eds.), Plenum Press, New York (1991), pp. 417–437.

337. N. Tanović, L. Tanović, J. Fine, B. Gaković, P. Panjan, and N. Popovi, Microtopography and interface width of sputter profiled Cr/Ni multilayered thin-film materials, *Nucl. Instrum. Methods Phys. Res., Sect. B* **67**, 590–594 (1992).
338. C. P. Hunt and M. P. Seah, Characterization of a high depth-resolution tantalum pentoxide sputter profiling reference material, *Surf. Interface Anal.* **5**(5), 199–209 (1983).
339. J. Fine, P. A. Lindfors, M. E. Gorman, R. L. Gerlach, B. Navinšek, D. F. Mitchell, and G. P. Chambers, Interface depth resolution of Auger sputter profiled Ni/Cr interfaces: dependence on ion bombardment parameters, *J. Vac. Sci. Technol. A* **3**(3), 1413–1417 (1985).
340. J. B. Clegg, N. S. Smith, M. G. Dowsett, M. J. J. Theunissen, and W. B. de Boer, Secondary ion mass spectroscopy resolution with ultra-low beam energies, *J. Vac. Sci. Technol. A* **14**(4), 2645–2750 (1996).
341. J. Liday, V. Kliment, P. Vogrinčič, and S. Tomek, The influence of Ar^+ and Xe^+ ion beam incidence angle on the depth resolution in AES, *Vacuum* **46**(1), 53–55 (1995).
342. S. Hofmann and A. Zalar, Depth profiling with sample rotation: capabilities and limitations, *Surf. Interface Anal.* **21**, 304–309 (1994).
343. J. C. Lee, C. S. Jeong, H. J. Kang, H. K. Kim, and D. W. Moon, Damage profiling of Ar^+ sputtered Si (100) surface by medium energy ion scattering spectroscopy, *Appl. Surf. Sci.* **100/101**, 97–101 (1996).
344. J. G. C. Labanda, S. A. Barnett, and L. Hultman, Effects of glancing-angle ion bombardment on GaAs (001), *J. Vac. Sci. Technol., B* **13**(6), 2260–2268 (1995).
345. M. Tanemura, S. Fujimoto, and F. Okuyama, Dependence of Auger depth resolution and surface texturing on primary ion species, *Surf. Sci.* **230**, 283–289 (1990).
346. J. Liday, R. Harman, G. Badin, J. Breza, and A. Zalar, Auger electron spectroscopy composition depth profiling of Cr/Ni multilayer structures using Ar^+ and Xe^+ ions, *Thin Solid Films* **208**, 290–294 (1992).
347. K. Wittmaack and D. B. Poker, Interface broadening in sputter depth profiling through alternating layers of isotopically purified silicon, *Nucl. Instrum. Meth. Phys. Res., Sect. B* **47**, 224–235 (1990).
348. M. Petravić, B. G. Svensson, and J. S. Williams, On the estimation of depth resolution during sputter profiling, *Appl. Phys. Lett.* **62**(3), 278–281 (1993).
349. P. A. Ronsheim and M. Tejwani, Effect of matrix stopping power on sputter depth profile broadening, *J. Vac. Sci. Technol., B* **12**(1), 254–257 (1994).
350. P. C. Zalm and C. J. Vriezema, On some factors limiting depth resolution during SIMS profiling, *Nucl. Instrum. Meth. Phys. Res., Sect. B* **67**, 495–499 (1992).
351. M. J. Caturla, I. Abril, and J. J. Jiménez-Rodriguez, Depth resolution in depth profiling of marker layers by energetic ion bombardment, *Nucl. Instrum. Meth. Phys. Res., Sect. B* **95**, 91–96 (1995).
352. J. F. O'Hanlon, *A User's Guide to Vacuum Technology*, Wiley, New York (1989).
353. C. P. Hunt and M. P. Seah, Method for the alignment of samples and the attainment of ultra-high-resolution depth profiles in Auger electron spectroscopy, *Surf. Interface Anal.* **15**, 254–258 (1990).
354. A. Zalar, Significance of sample rotation in Auger electron spectroscopy sputter depth profiling of thin films, *Thin Solid Films* **193/194**, 258–269 (1990).
355. Charles Evans and Associates, Zalar rotation enhanced Auger depth profiles, Auger Application Note, Charles Evans and Associates, Redwood City, CA (1996).
356. D. E. Sykes, A. Chew, J. Hems, and K. Stribley, High depth resolution depth profiling of metal films using SIMS and sample rotation, *Appl. Surf. Sci.* **100/101**, 77–83 (1996).
357. S. Hofmann, A. Zalar, E.-H. Cirlin, J.J. Vajo, H.J. Mathieu, and P. Panjan, Interlaboratory comparison of the depth resolution in sputter depth profiling of Ni/Cr multilayers with and without sample rotation using AES, XPS and SIMS, *Surf. Interface Anal.* **20**, 621–626 (1993).
358. K. Wittmaack, Towards the ultimate limits of depth resolution in sputter profiling: beam-induced chemical changes and the importance of sample quality, *Surf. Interface Anal.* **21**, 323–335 (1994).

359. M. Tanemura, S. Aoyama, and F. Okuyama, Optimization of rotational Auger depth profiling, *Surf. Interface Anal.* **18**, 475–480 (1992).
360. M. Menyhard, A. Barna, J. P. Biersack, K. Järrendahl, and J.-E. Sundgren, Study of ion mixing during Auger depth profiling of Ge-Si multilayer system. II. Low ion energy (0.2–2 keV) range, *J. Vac. Sci. Technol. A* **13**(4), 1999–1994 (1995).
361. A. Konkol, A. Sulyok, M. Menyhard, and A. Barna, Auger in-depth profiling of Mo-Si multilayers, *J. Vac. Sci. Technol. A* **12**(2), 436–442 (1994).
362. D. Marton, Influence of the individual layer thickness on the depth resolution determined in sputter depth profiling of multilayers, *Thin Solid Films* **146**, L11–L14 (1987).
363. I. Kojima, N. Fukumoto, and M. Kurahashi, STM and AES study of the relation between ion sputter induced roughness and depth resolution, *J. Electron Spectrosc. Relat. Phenom.* **50**, c9–c13 (1990).
364. S. Hofmann, Approaching the limits of high resolution depth profiling, *Appl. Surf. Sci.* **70/71**, 9–19 (1993).
365. M. P. Seah, H. J. Mathieu, and C. P. Hunt, The ultra-high resolution depth profiling reference material - Ta_2O_5 anodically grown on Ta, *Surf. Sci.* **139**, 549–557 (1984).
366. K. Iltgen, C. Bendel, A. Benninghoven, and E. Niehuis, Optimized time-of-flight secondary ion mass spectroscopy depth profiling with a dual beam technique, *J. Vac. Sci. Technol. A* **15**(3), 460–464 (1997).
367. A. Benninghoven, private communication to A. W. Czanderna (1997).
368. H. Murakami, S. Ichimura, H. Shimizu, I. Kudo, and N. Atoda, Synergistic enhancement of direct synchrotron radiation etching of a resist by a low-energy oxygen beam, *Jpn. J. Appl. Phys.* **26**(11), L1774–L1776 (1987).
369. S. Berg and I. V. Katardjiev, Synergistic sputtering effects during ion bombardment with two ion species, *J. Vac. Sci. Technol. A* **13**(3), 831–833 (1995).
370. S. P. Withrow and D. B. Poker, eds., Ion beam modification of materials, *Nucl. Instrum. Meth. Phys. Res. Sect. B* **59/60**, 1–1476 (1991).
371. R. B. Jackman, Ion beam-assisted etching of semiconductors: surface chemistry vs. surface physics, *Vacuum* **44**(3/4), 239–243 (1993).
372. Y. Katoh, N. Sugimoto, A. Tate, and A. Shibukawa, Ion incident angle dependence of ridge shapes formed on [111] oriented $Gd_3Ga_5O_{12}$ by ion beam etching, *Jpn. J. Appl. Phys. Part 2* **31**, L591–L593 (1992).
373. Y. Katoh, N. Sugimoto, and A. Tate, Ridge structures formed on $Nd_3Ga_5O_{12}$ and $Su_3Ga_5O_{12}$ single crystal substrates, *Jpn. J. Appl. Phys. Part 2* **33**, L726–L729 (1994).
374. S. Morishita and F. Okuyama, Sharpening of monocrystalline molybdenum tips by means of inert-gas ion sputtering, *J. Vac. Sci. Technol. A* **9**(1), 167–169 (1991).
375. H. P. Lee, A. Scherer, E. D. Beebe, W. P. Hong, R. Bhat, and M. A. Koza, 1.57 μm InGaAsP/InP surface emitting lasers by angled focus ion beam etching, *Electron. Lett.* **28**(6), 580–582 (1992).
376. M. Naoe, M. Nagakubo, and T. Yamamoto, Effects of annealing and periodic ion bombardment on the soft magnetism of Fe_xN films by a double ion beam sputtering system, *J. Appl. Phys.* **64**(10), 5449–5451 (1988).
377. K. H. Müller, R. P. Netterfield, and P. J. Martin, Dynamics of zirconium oxide thin-film growth and ion-beam etching, *Phys. Rev. B* **35**(6), 2934–2940 (1987).
378. W. Shindo and T. Ohmi, Ion energy, ion flux, and ion mass effects on low-temperature silicon epitaxy using low-energy ion bombardment process, *J. Appl. Phys.* **79**(5), 2347–2351 (1996).
379. L. L. Boyer, A. Y. Wu, and J. R. McNeil, Physical and electro-optical properties of ion beam sputtered thin film PLZT ceramic, *Mater. Res. Soc. Symp. Proc.* **200**, 97–102 (1990).
380. W. S. Sampath, F. M. Kustas, and P. J. Wilbur, in: *Proceedings of Industry–University Advanced Materials Conference* (J. G. Morse, ed.), Metallurgical Society, Warrendale, PA (1987), pp. 359–364.

381. Y. Kato, T. Kitamura, E. Isoyama, M. Hasegawa, T. Momose, and H. Ishimaru, A method of determining hydrogen in aluminum by removing surface gases with ion beam etching, *J. Vac. Sci. Technol. A* **6**(6), 3111–3114 (1988).
382. E. T. Ada, L. Hanley, S. Etchin, J. Melngailis, W. J. Dressick, M.-S. Chen, and J. M. Calvert, Ion beam modification and patterning of organosilane self-assembled monolayers, *J. Vac. Sci. Technol. B* **13**(6), 2189–2196 (1995).
383. G. Gillen, S. Wight, J. Bennett, and M. J. Tarlov, Patterning of self-assembled alkanethiol monolayers on silver by microfocus ion and electron beam bombardment, *Appl. Phys. Lett.* **65**(5), 534–536 (1994).
384. G. van Hassel, H. C. Heyker, and J. J. M. Kwaspen, Influence of in situ argon cleaning of GaAs on Schottky diodes and metal-semiconductor field-effect transistors, *J. Vac. Sci. Technol. B* **13**(6), 2245–2259 (1995).
385. B. G. Sevensson, B. Mohadjeri, and M. Petravić, Surface recession and oxidation of silicon during bombardment by low energy oxygen ions, *J. Appl. Phys.* **76**(6), 3831–3834 (1994).
386. B. A. Ratchev, G. S. Was, and J. H. Booske, Ion beam modification of metal-polymer interfaces for improved adhesion, *Nucl. Instrum. Meth. Phys. Res., Sect. B* **106**, 68–73 (1995).
387. M. Yamawaki, T. Namba, T. Yoneoka, K. Shida, and M. Kanno, Effect of argon ion sputtering of surface on hydrogen permeation through vanadium, *J. Nucl. Sci. Technol.* **20**(5), 405–413 (1983).
388. M. Cailler, A. Ouis, P. J. Schultz, and P. J. Simpson, Adhesion strength of copper thin-films to a E24 carbon steel: effects of substrate surface ion bombardment etching, *J. Adhesion Sci. Technol.* **7**(2), 141–157 (1993).
389. Z. W. Kowalski, Ion-bombardment modification of surface morphology of solids, *J. Mater. Sci.* **29**(13), 3542–3552 (1994).
390. M. Cevro and G. Carter, Ion beam sputtering and dual ion beam sputtering of titanium oxide films, *J. Phys. D: Appl. Phys.* **28**, 1962–1976 (1995).
391. J. Orloff, L. W. Swanson, and M. Utlaut, Fundamental limits to imaging resolution for focused ion beams, *J. Vac. Sci. Technol. B* **14**(6), 3759–3763 (1996).
392. R. G. Cooks, T. Ast, and T. Pradeep, Reactions of ions with organic surfaces, *Acc. Chem. Res.* **27**, 316–323 (1994).
393. I. Bello, W. H. Chang, and W. M. Lau, Mechanism of cleaning and etching Si surfaces with low energy chlorine ion bombardment, *J. Appl. Phys.* **75**(6), 3092–3097 (1994).
394. Y. Z. Hu, K. A. Conrad, M. Li, J. W. Andrews, J. P. Simko, and E. A. Irene, Studies of hydrogen ion beam cleaning of silicon dioxide from silicon using in situ spectroscopic ellipsometry and X-ray photoelectron spectroscopy, *Appl. Phys. Lett.* **58**(6), 589–591 (1991).
395. P. Mazzoldi, Ion beam modification of glass surface properties, *J. Non-Cryst. Solids* **120**, 223–233 (1990).
396. Z. W. Kowalski, Ion sputtering and its biomedical applications. Theoretical concepts and practical consequences. Clinical implications and potential use, *J. Mater. Sci.* **20**, 1521–1555 (1985).

Appendix 1. Acronyms and Abbreviations

AES	Auger electron spectroscopy
AFM	Atomic force microscopy
ARXPS	Angular resolved X-ray photoelectron spectroscopy
CB	Conduction band
EELS	Electron energy loss spectroscopy
FWHM	Full width at half maximum

HOPG	Highly oriented pyrolytic graphite
HREM	High resolution electron microscopy
HRTEM	High resolution transmission electron microscopy
IBB	Ion beam bombardment
ISS	Ion scattering spectroscopy
LEED	Low energy electron diffraction
LEIS	Low energy ion scattering (spectroscopy)
MBE	Molecular beam epitaxy
MEIS	Medium energy ion spectroscopy
ML	Monolayer
PEXAFS	Photoelectron extended X-ray analysis of fine structure
PLS	Photoluminescence spectroscopy
RBS	Rutherford backscattering spectroscopy
RED	Radiation enhanced diffusion
RGA	Residual gas analysis
RHEED	Reflected high energy electron diffraction
rms	Root mean square
S	Sputtering yield (see Y)
SALI	Surface analysis laser ionization (mass spectrometry)
SE	Spectroscopic ellipsometry
SEM	Scanning electron microscopy
SIMS	Secondary ion mass spectrometry
SNMS	Secondary neutral mass spectrometry
STM	Scanning tunneling microscopy
TDS	Thermal desorption spectroscopy
TEM	Transmission electron microscopy
TOF–ISS	Time-of-flight ion scattering spectroscopy
TOF–SIMS	Time-of-flight secondary ion mass spectrometry
VB	Valence band
XPS	X-ray photoelectron spectroscopy
XTEM	Cross-sectional transmission electron microscopy
Y	Sputtering yield (see S)
Δz	Interface thickness between 15.87% and 84.13% of a compositional surface analysis signal that is 100% for the elemental intensity before or after sputtering through the interface

4

Characterization of Surface Topography*†

T. V. Vorburger, J. A. Dagata, G. Wilkening, and K. Iizuka

1. Introduction

Surface topography plays an important role in the function of many kinds of components. Examples in engineering include the importance of cylinder-bore surface finish in automobile engines,[1–3] the disadvantage of increased drag of ships due to hull roughness,[4] and the advantage of making optical components as smooth as possible.[5,6] Examples in surface science include the sensitivity of certain surface chemical reactions to surface steps[7] and the role of roughness in surface-enhanced Raman scattering.[8] Hence, the techniques for studying surface topography form an important repertoire for scientists and engineers.

Several excellent reviews of the field have been published previously.[9–11] In this chapter we briefly describe the range of techniques available to characterize

*Contribution of National Institute of Standards and Technology, not subject to copyright.
†Sections of this chapter have appeared previously in two articles: T. V. Vorburger and G. G. Hembree, in: *Navy Metrology Research & Development Program Conference Report*, Naval Weapons Station, Corona, CA, April (1989); and T. V. Vorburger, J. A. Dagata, G. Wilkening, and K. Iizuka, *CIRP Ann.* **46/2**, 597 (1997).

T. V. Vorburger and J. A. Dagata • National Institute of Standards and Technology, Gaithersburg, MD 20899; G. Wilkening • Physikalisch–Technische Bundesanstalt, 38023 Braunschweig, Germany; K. Iizuka • Nikon Corporation, Tsukuba Research Laboratory, Tsukuba, Ibaraki 300-26, Japan.

Beam Effects, Surface Topography, and Depth Profiling in Surface Analysis, edited by Czanderna *et al.* Plenum Press, New York, 1998

surface topography. Then we discuss several commercially available profiling techniques in detail, because their description should be particularly useful for surface scientists and others who need to choose an instrument for measuring a particular surface topography.

We first discuss some common terms in surface topography measurement because the terminology, which differs from one technical field to another, could be a source of confusion. *Topography* is defined in the American National Standard (ASME B46.1),(12) intended largely for mechanical engineers, as "the three dimensional representation of geometric surface irregularities." This meaning is straightforward, but the terms describing components of surface topography differ among technical fields. To mechanical engineers, topography consists of three components: short spatial wavelengths called *roughness*, medium spatial wavelengths called *waviness*, and longer spatial wavelengths called *error of form*. By convention, the roughness and waviness comprise the *surface texture*.(12) Error of form is generally regarded as a dimensional quality outside the realm of surface texture or surface finish metrology. Discrete irregularities are called *flaws* and are regarded as components of surface topography, but because of its complexity the characterization of flaws is carefully avoided in ASME B46.1.(12)

For optical engineers the analog of error of form is *figure*, the analog of roughness is *surface finish* or *surface roughness*, and the analog of flaws is *imperfections*, which are described and classified in optics standards.(13,14) Perhaps the most carefully studied imperfection classes are *digs* and *scratches*.(13) Even waviness has an analog, sometimes called *mid-wavelength slope error*, which has been specified for optical surfaces from time to time.

In semiconductor manufacturing the analog of error of form is *flatness*, the analog of roughness is *microroughness*,(15) and flaws are categorized into several classes including *scratches*, *pits*, and other types of defects.(15) The term "orange peel" has also been used to describe a component of semiconductor surface topography somewhat analogous to the mechanical engineer's waviness.

In this chapter we generally use the terms "topography" and "roughness" or "surface roughness," the former to denote the overall geometrical structure of a surface and the latter two to denote the fine-scale irregularities whose characterization we will be emphasizing.

The vast majority of techniques used to quantify surface topography may be classified into three categories according to the ASME B46.1 standard.(12) These are *profiling*,(9–12,16–26) *area profiling*,(27–43) including those scanned probe microscopies (SPMs) that quantitatively assess topography, and *area averaging*.(5,6,24,44–67) To these categories we add those types of microscopy techniques(23,49,68–90) used for qualitative inspection rather than for quantitative characterization of the vertical dimension of surface topography. We focus on profiling and area-profiling techniques. We only briefly describe area-averaging techniques and microscopy.

Profiling techniques provide quantitative assessment of surface peaks and valleys through point-by-point measurement with a high-resolution probe. The profile may be represented, say, by interferometric fringes or by the trace produced by a contacting or noncontacting probe that traverses the surface. We review three types of profiling techniques—stylus, optical profiling, and scanned probe microscopy—and compare certain results from these methods.

Three-dimensional representations of the surface topography can be produced by using a procedure known as area profiling. The usual approach is to scan and step successively over the surface in a raster fashion to obtain a series of parallel profiles. The procedure can be used with most profiling techniques and is different from single-scan profiling and from area-averaging approaches to be discussed. Area profiling gives a much more complete representation of the surface topography than single-scan profiling, but the data gathering and analysis consume more time and computer memory and are thus more costly. Contact stylus instruments have lower-frequency mechanical limits than either optical or scanning probe instruments; the raster scanning process is more time consuming for stylus instruments. However, area-profile maps can be acquired in less than a minute with recently developed instruments.[37] Phase-measuring interferometers have array cameras and parallel processing, so the data acquisition process for them requires no lateral scanning motion and takes only a few seconds. Confocal scanning optical microscopes[91] can have nearly comparable data acquisition times, on the order of 10 s, for area profiling of surfaces. Scanned probe microscopes (SPMs) have data acquisition times on the order of a minute per frame, but under special conditions can be made to operate with frame acquisition times[92,93] of 6 s or less.

In contrast, an area-averaging technique probes the surface over a finite area all at once to produce a measured property that represents a statistical average of the peaks and valleys. Typical examples are pneumatic and hydraulic techniques,[9,44] the capacitance probe,[45,46] and angle-resolved[24,47–55] and angle-integrated[5,6,56,57] scattering of light as well as scattering of X-rays.[58] Light scattering has been used extensively for measurement of roughness of optical surfaces[50] and for scanned inspection of semiconductor wafers[52] to provide roughness measurement and defect or particle detection. Other instruments have been developed for roughness measurement of machined metal surfaces[51,55] and of specimens in vacuum.[57,58]

In addition, the spatial intensity distribution resulting from the scattering of laser light has a grainy fine structure known as speckle.[59–62] The spatial contrast and other properties of the speckle depend on the surface topography and hence can be used as indicators of surface roughness.

Area-averaging techniques offer the potential for high-speed inspection in quality control situations, but most have limited effectiveness for probing at atomic and near-atomic scales (or dimensions) typical of surface science applications. One area-averaging technique practiced in surface science is electron diffraction.[63,64]

It has been mostly used to study the crystal structure of surfaces and absorbed species, but may also be adapted for estimating topographic quantities such as the heights of surface steps and the widths of plateaus. Another area-averaging technique involves measurement of the quantity of gas adsorbed by a surface to estimate the total surface area.[65–67] Both physical and chemical adsorption processes have been employed in this way for estimating the surface areas of catalysts.[66,67]

Microscopy produces images that contain contrast due to topography as well as to other surface properties. These images can be used to measure accurately the lateral distance between surface features. However, the relative heights of the features cannot be measured so easily. The difficulty lies in the multiple origins of the image contrast in both optical and electron microscopy. Optical microscopy also has a limited depth of focus that makes even qualitative inspection of rough surfaces difficult.[68]

Scanning electron microscopy (SEM) has, on the other hand, a large depth of field along with several sources of image contrast. The SEM yields an excellent qualitative impression of surface topography.[75] The resolution of SEM for small surface features is material dependent, but for objects with large height-to-width ratios a resolution of 10 nm or less is achievable without difficulty. Quantitative measurement of topography at low magnifications is possible by use of SEM stereography.[39–42] However, measurement of surface feature heights with low uncertainty is only possible by cross-sectioning the object and imaging its profile.[79,80] Meaningful measurements from surface profile images require skill in sample preparation.

Transmission electron microscopy (TEM) of surface topography is possible after special sample preparation. In one technique a replica of the surface is made with a thin film of carbon or plastic. This replica is then shadowed by evaporating a heavy metal onto it at a shallow angle. Microdensitometry of TEM images of shadowed replicas has been developed into a quantitative surface roughness measuring method by Rasigni *et al.*[38] A more direct method of observing surface topography is by profile-imaging thin cross sections of crystalline specimens. High-resolution images of atomic-sized surface features can be obtained by profile-imaging with the TEM.[84] TEM has also been used effectively to inspect the thin cross sections of multilayer coatings.[85]

Reflection electron microscopy (REM) is a diffraction-imaging technique[86] in which high-energy electrons are incident at a glancing angle to the surface so that a profile image is formed directly. REM can be accomplished on either TEM or SEM instruments. The localized diffraction contrast in the resulting images is sensitive to surface topography as small as monoatomic steps.[87–89] The limitations of the technique are its requirements for small, relatively smooth, conductive crystalline specimens.

The remainder of this chapter is organized as follows. In Sec. 2 we discuss the statistical methods to quantify the results of profiling measurements. In Sec. 3 we review the stylus technique, and, in particular, we discuss its salient features and limitations as well as applications of the technique to the study of optical surfaces and area profiling. In Sec. 4 we discuss optical profiling, and in Sec. 5 we review the rapidly expanding field of SPM. In Sec. 6 we review a few experiments that have intercompared different profiling and area-profiling techniques. Finally, in Sec. 7 we compare the range and resolution, both vertical and lateral of three common techniques, stylus, interferometric microscopy, and one type of SPM known as atomic force microscopy (AFM); other features of these instruments are also tabulated.

2. Results Obtainable with Profiling Instruments

The results obtainable from profiling measurements may be represented by the profile recordings themselves. The profile recordings provide an immediate visual impression of the surface topography or of a feature on it. The stored profile data may also be statistically analyzed or used to measure individual surface features. The statistical tools provide quantitative measures for surface geometrical properties. The following sections are examples of profile records and statistical analysis.

2.1. Profile Recordings and Dimensional Measurement

Figures 1, 2, and 3 show some typical profiling results. Figures 1a and 1b are stylus profiles of a sputtered Si_3N_4 film and the adjacent polished fused-silica substrate measured by Bennett *et al.*[35] with a stylus instrument of the type described in Sec. 3. Aside from the statistical parameters that may be calculated from it, the measured profile is a combination of the noise of the instrument and the roughness of the surface and therefore the measured root-mean-square (rms) roughness of 0.06 nm for Figs. 1a and 1b is an upper limit on the vertical noise resolution of the stylus instrument and the rms roughness of the surface itself. For comparison, Fig. 1c shows the noise profile measured with the stylus held stationary. The measured rms roughness of 0.05 nm is a lower limit for the noise resolution of the instrument. Figure 2 shows a profile measured by a stylus instrument[94] of a pitch–height specimen[95]* used for calibration of the lateral and vertical scales of scanned probe microscopes. Figure 3 is a profile measured with a stylus instrument of a gold-coated diffraction-grating replica with a periodicity of 6000 lines/mm.[96] The profile demonstrates the instrument's lateral-resolution capabil-

*Certain commercial equipment or materials are identified in this paper to specify adequately an experimental procedure. Such identification does not imply recommendation or endorsement by NIST, nor does it imply that the equipment or materials are necessarily the best available for the purpose.

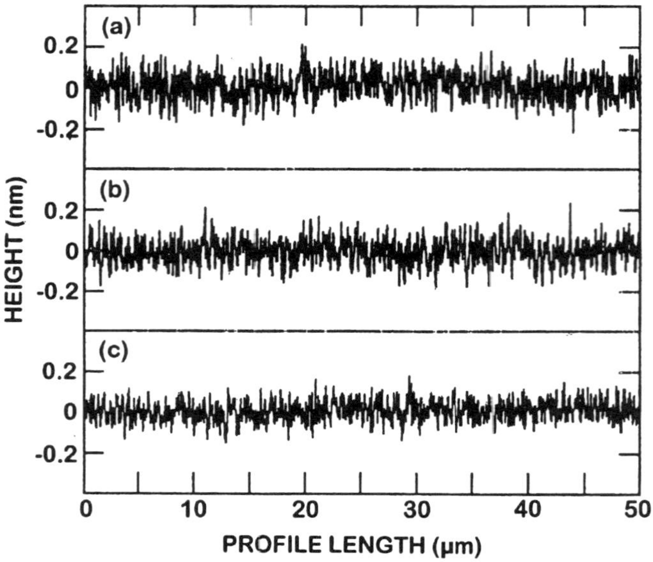

Figure 1. Surface profiles measured with a stylus instrument of (a) a Si_3N_4 film of 0.06-nm rms roughness, (b) fused-silica substrate adjacent to the Si_3N_4 film, also of 0.06 nm rms roughness, and (c) profile equivalent of the instrumental noise with the stylus stationary, 0.05-nm rms roughness (Adapted from Bennett *et al.*[35]).

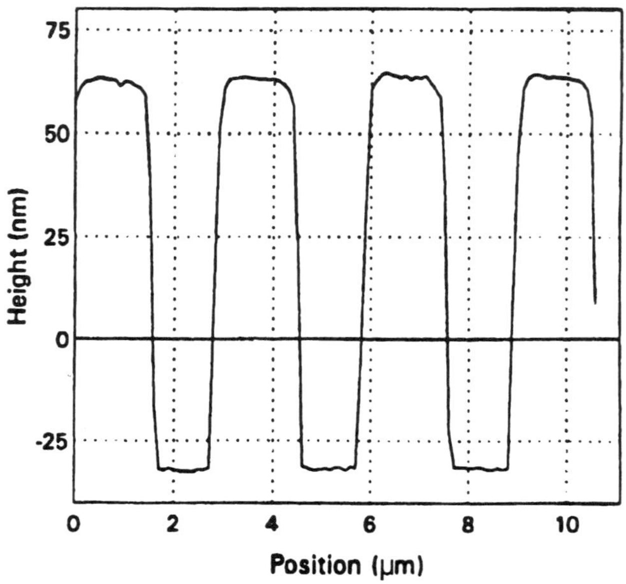

Figure 2. Surface profile of a pitch–height calibration standard measured with a stylus instrument by McWaid *et al.*[94]

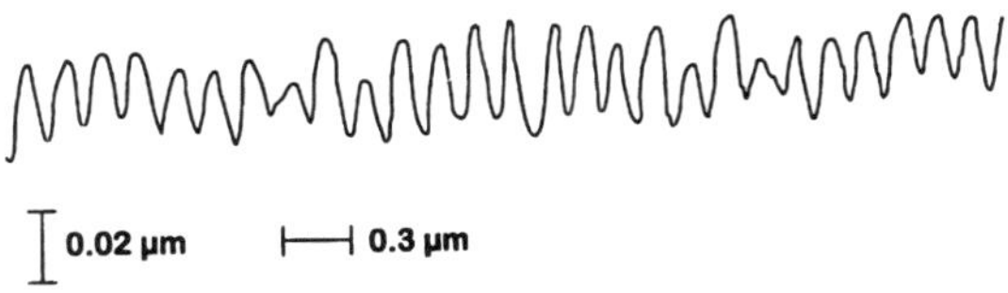

Figure 3. Surface profile of a 6000-lines/mm grating measured with a stylus instrument by Song and Vorburger.(97) The stylus tip had a nominal width of 50 nm.

ity, a quantity that depends largely on the stylus tip width. In this case the stylus is capable of at least partially resolving features about 0.17 µm apart.

As shown in Figs. 2 and 3, the surface profiles may be used to characterize the dimensions in the x (lateral) or z (vertical) directions of individual features on surfaces, such as the heights and linewidths of individual circuit elements on semiconductor wafers or photomasks, the dimensions of vias on semiconductor wafers, or the pitch and groove shape of a diffraction grating. The size and shape of the probe directly affect the accuracy of the linewidth and hole-width measurements. In particular, the probe response must be modeled to obtain an accurate assessment of linewidth. The probe size is also a factor for feature height measurements when the valley is a vee groove. For example, we cannot be certain that the stylus tip actually contacted the bottoms of the grating grooves shown in Fig. 3.

A potential source of confusion exists in displayed profile records because the horizontal and vertical magnifications of surface profile records are usually different. This aspect ratio, which is the vertical magnification divided by the horizontal magnification, can be 100:1 or higher in many applications that have profile records with highly sloped and sharply peaked structures. The aspect ratio can result in misperceptions about the appearance of surface topography and even miscalculations if one uses the recordings to determine certain parameters. Figure 4, taken from Reason,(97) shows a comparison between a profile displayed with a 1:1 aspect ratio and the same profile displayed with a 25:1 ratio. The qualitative impressions derived from the two pictures are quite different. In reality then, surfaces are much less jagged than they appear from conventional profile records.

2.2. Surface Statistics

In general, surface topographies are highly complex because many surface finishing processes such as polishing and shot blasting are statistical by nature. To characterize such surfaces, statistical descriptors are needed. There are basically two kinds of descriptors: *parameters*, such as rms roughness, which attempt to quantify some aspect of the surface statistics with a single number, and *surface statistical functions*, such as the power spectral density, which yield arrays of

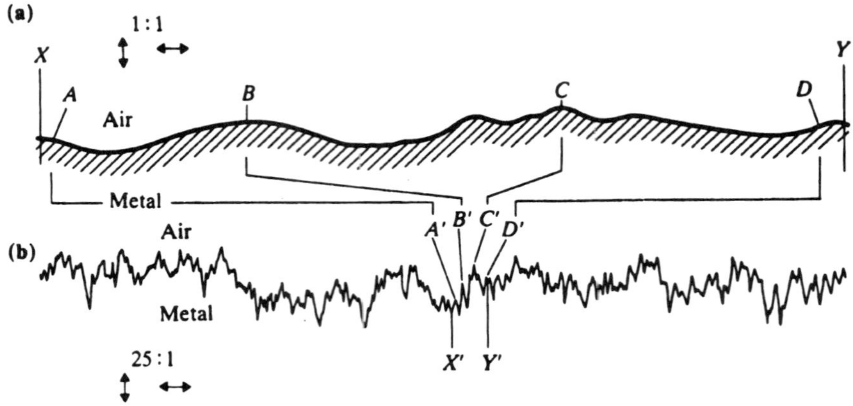

Figure 4. Stylus profiles displayed with two aspect ratios.[98] Profile (a) is an undistorted 1:1 representation. In (b) the horizontal scale has been compressed by a factor of 25.

numerical information about the surface. We will describe a few of these statistical descriptors.

2.2.1. Surface Parameters

A variety of surface parameters have been developed[9,12] to characterize the functions of surfaces for various applications.[1–4,19,98–100,101] Scores of parameters have been defined for industrial use, and many of these appear in national standards as well. Nevertheless, the confusing field of surface parameters may be classified adequately into four categories: height parameters, shape parameters, wavelength parameters, and combinations of these, known as hybrid parameters.

2.2.1a. Height Parameters. The most common statistical descriptors of surface height are the roughness average Ra and the rms roughness (also called Rq or σ). These two are closely related and are given by the following formulas in analytical and digitized form:

$$Ra = \frac{1}{L}\int_0^L |z(x)|\, dx \cong \frac{1}{N}\sum_{i=1}^{N} |z_i|$$

$$Rq = \left(\frac{1}{L}\int_0^L z^2(x)\, dx\right)^{1/2} \cong \left(\frac{1}{N}\sum_{i=1}^{N} z_i^2\right)^{1/2} \tag{1}$$

The various terms in these formulas are shown in Fig. 5. Both *Ra* and *Rq* are useful descriptors for characterizing the average heights and depths of surface profiles. The rms roughness is commonly specified for the surfaces of optical components. In general, lower values for the rms roughness of an optical component result in less stray light, so the quality of the component is higher. The *Ra* descriptor has been used in the automotive and other metalworking industries to specify the surfaces of many types of components ranging from cylinder bores to brake drums.(102) In addition to these two averaging height parameters, an assortment of other height parameters has been defined for various applications, including several for characterizing peak-to-valley height.(12,101)

2.2.1b. Spatial Wavelength Parameters. Spatial wavelength parameters are used to characterize the spacings of the peaks and valleys of the surface. The spatial wavelengths are related to the process variables that formed the surface, such as the shot size for a blasted surface, the grit size of a grinding wheel, or the feed of a tool. A typical wavelength parameter, included in a standard of the International Organization for Standardization (ISO),(101) is the mean width of profile elements *RSm*. Each profile element (Fig. 6) consists of a profile peak and the adjacent profile valley. As shown in Fig. 6, *RSm* has been calculated for a surface profile as the

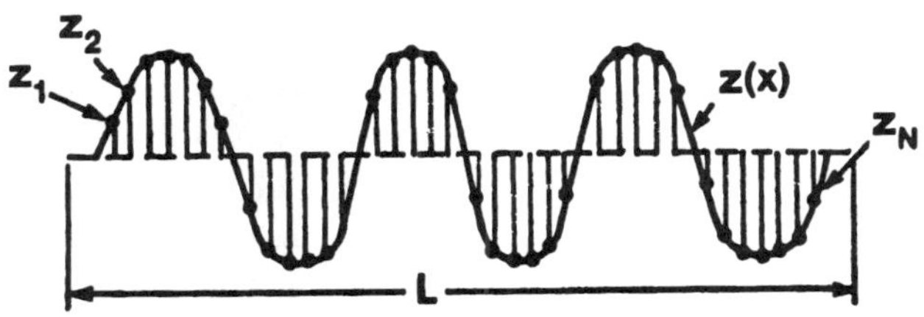

• R_a = **Average Deviation of Profile z(x) from the mean Line**

= **Total Shaded Area /L**

$$= \frac{1}{L}\int_0^L |z(x)|\,dx \cong \frac{1}{N}\sum_{i=1}^{N} |z_i|$$

• R_q = **rms Deviation. . .** $\cong \sqrt{\frac{1}{N}\sum_{i=1}^{N} z_i^2}$

Figure 5. Schematic surface profile showing the mean line, the evaluation length *L*, the digitized points z_i, and the definitions of the parameters *Ra* and *Rq*.

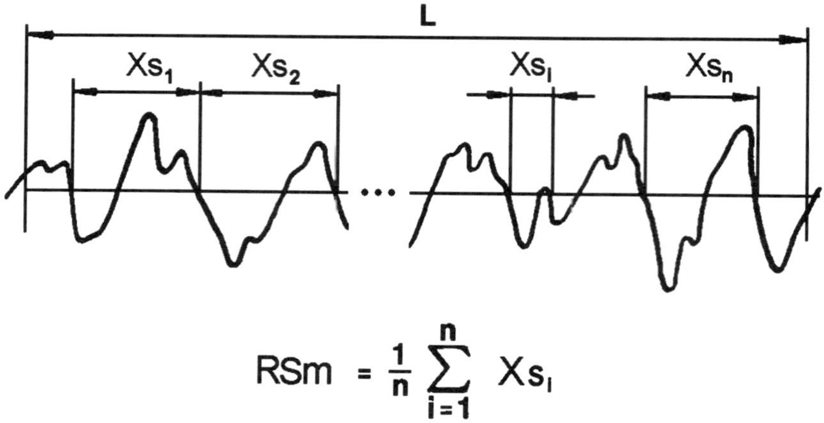

Figure 6. Surface profile showing a method of calculation for the peak spacing parameter, *RSm*, the mean width of profile elements (Adapted from the International Organization for Standardization (ISO) DIS 4287[(103)]).

average spacing between successive negative-slope crossings of the profile with the mean line, that is, the Xs_i's in Fig. 6.

2.2.1c. Shape Parameters. The periodic profiles in Fig. 7 all have the same *Ra* and spatial wavelength but different shapes and, hence, may perform differently for different applications. In particular, profile (b) represents a good bearing surface, profile (a) a poor one. Surfaces with facets like profile (c) are often used as diffraction gratings, whereas a surface with a sinusoidal profile like (d) has good properties as a standard. Other types of surfaces may be more suitable for lubricated sliding. Shape parameters help to quantify the functional differences between surfaces. The skewness (*Rsk*), for example, is a measure of the symmetry of the profile about the mean line. In digital form it is defined as

$$Rsk = \frac{1}{Rq^3} \frac{1}{N} \sum_{i=1}^{N} z_i^3 \tag{2}$$

According to this definition, the skewness is positive for profile (a) in Fig. 7 and negative for profile (b). The other two profiles are symmetrical about the mean line and have zero skewness.

Stout and Davis[(1)] studied the evolution of skewness and a similar shape parameter known as kurtosis for cylinder-bore surfaces during a laboratory simulation of the initial operating period of automobiles and demonstrated the usefulness of those parameters for describing the surface roughness during the initial opera-

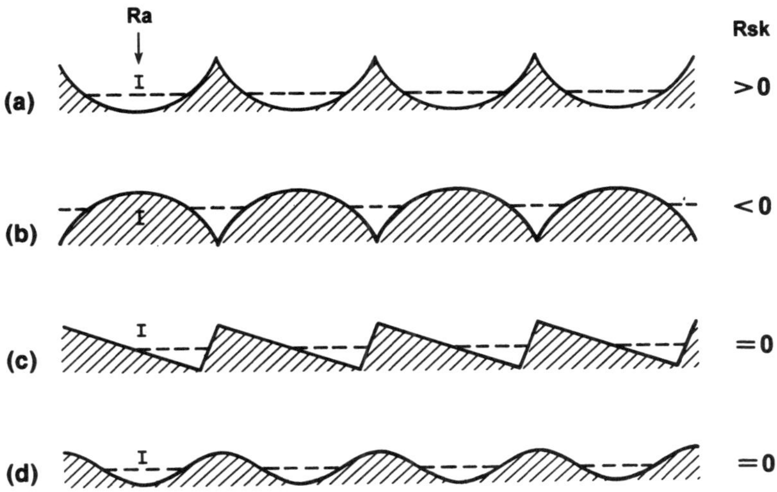

Figure 7. Profiles for four surfaces with the same *Ra* and wavelength. The dashed line shows the position of the mean line. *Rsk* is the skewness parameter.

tion. Even wider attention has been given to the use of a set of shape-related parameters, known as *Rk* parameters, to correlate with wear and oil retention in engines. The *Rk* parameters are derived from a statistical function called the material ratio curve and were formulated to characterize stratified surfaces such as those produced by plateau honing. An ISO standard(100) has been developed to describe the procedure for calculating *Rk* parameters.

2.2.1d. Hybrid Parameters. Slope and curvature are two examples of quantities that combine the concepts of height deviation and lateral displacement and hence are termed hybrid parameters. They may be defined analytically or digitally in many ways.(9) Hybrid parameters have found usefulness in tribology, such as theories dealing with elastic contact(103) and thermal conductance.(104)

2.2.2. Statistical Functions

More complete statistical descriptions of the properties of surface profiles may be obtained from statistical functions, such as those used in connection with random process theory and time-series analysis.(105,106) Four important ones are the (1) amplitude density function or height distribution, (2) the material ratio curve, also known as the bearing-area curve, (3) the autocorrelation function, and (4) the power spectral density. The definitions and applications of these functions are described in several works.(12,100,104–112) As examples, we discuss the power spectral density and the autocorrelation function in detail.

2.2.2a. Power Spectral Density. The power spectral density (PSD) decomposes the surface profile into its spatial Fourier component wavelengths. It may be defined analytically by

$$\mathrm{PSD}(F) = \lim_{L\to\infty} \frac{1}{L} \left| \int_{-L/2}^{L/2} z(x) e^{-i2\pi Fx}\, dx \right|^2 \tag{3}$$

or in digitized form by

$$\mathrm{PSD}(F) = \frac{\Delta}{N} \left| \sum_{j=1}^{N} z(j) e^{-2\pi F(j-1)\Delta} \right|^2 \tag{4}$$

where | | indicates the absolute value operation, Δ is the *sampling interval*, the lateral point spacing of the digitized data points, the total length of the profile L is equal to $N\Delta$, j is the running index representing individual data points in the profile, and F is the spatial frequency variable, which is calculated for discrete values over 0, $1/L$, $2/L$, . . . , $N/2L$ (= $1/2\Delta$). The calculation of the digital Fourier transform in Eq. (4) may be speeded up enormously by using fast Fourier transform (FFT) algorithms.(113,114)

In Fig. 8 we illustrate how the power spectral density is sensitive to the different characteristics of surfaces produced by different processes. Curve (a) shows the PSD plotted for a nearly sinusoidal surface,(115) a prototype of ones available from The National Institute of Standards and Technology (NIST) as standard reference materials.(116) The Fourier amplitude at the fundamental spatial frequency of 0.01 μm^{-1} (spatial wavelength = 100 μm) is the dominant feature in the curve, but imperfections in the sinusoidal nature of the surface are also evident from the presence of higher harmonics in the spectrum. In contrast, the PSD (curve 8c) for a more random surface (curve 8b) produced by grinding has a fairly monotonic, albeit somewhat randomized, distribution with little evidence for periodic components.

The PSD for the ground surface illustrates a ubiquitous phenomenon for all surface finishing processes: the power spectral density function generally increases with spatial wavelength up to wavelengths on the scale of the dimensions of the surface itself. This phenomenon results from random effects present in nearly all surface finishing processes(109) and may be described by statistical arguments associated with the wandering of particles in Brownian motion.(117) The phenomenon was dramatically illustrated by Sayles and Thomas,(117) who plotted the PSDs for different surfaces, from lapped steel to the surface of the moon. In all cases a generally monotonic increase of PSD with spatial wavelength was observed. In Fig.

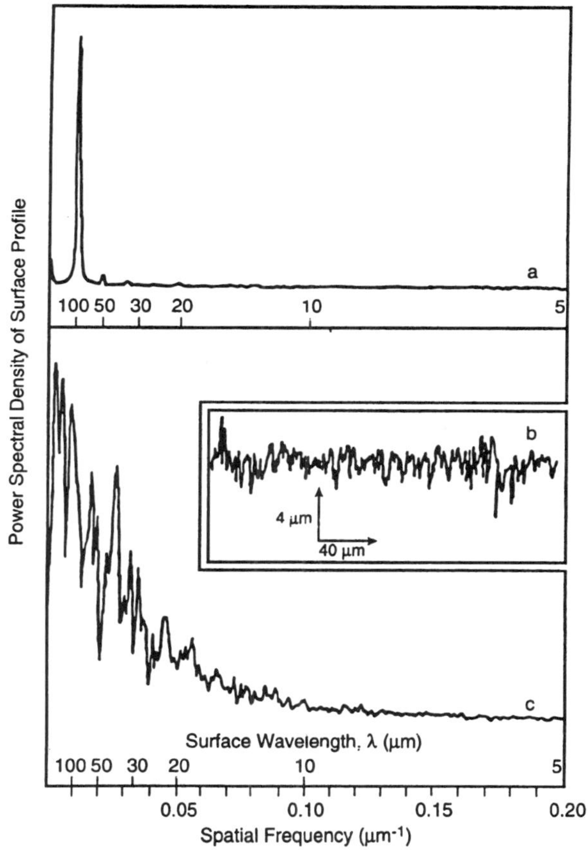

Figure 8. Power spectral density functions for profiles of two roughness specimens: (a) sinusoidal roughness specimen machined by diamond turning with $Ra = 3.0$ µm and spatial wavelength = 100 µm, and (c) commercially available specimen of a ground surface with $Ra = 0.60$ µm, whose profile is shown as curve b. Both power spectra are calculated for profiles having trace lengths of 3.8 mm that were measured with a stylus with an effective radius of ~7.5 µm. The ordinate values are proportional to the square root of the power spectral density values and are plotted in arbitrary units.

8c the PSD for the ground surface decreases suddenly near zero spatial frequency because the bandwidth of the measurement is limited by the profile evaluation length.

2.2.2b. Autocovariance and Autocorrelation.[9,12,114] Another way to process the information presented by the PSD is to take its Fourier transform to arrive at the autocovariance function (ACV):

$$\mathrm{ACV}(\tau) = \int_{-\infty}^{\infty} \mathrm{PSD}(F) e^{i2\pi F\tau}\, dF \tag{5}$$

where τ is a shift distance discussed later and the other terms are the same as in Eqs. (1) and (3). Alternatively, the autocovariance function may be calculated directly from the profile itself:

$$\mathrm{ACV}(\tau) = \frac{1}{L}\int_{0}^{L} z(x)z(x+\tau)\, dx \tag{6}$$

Equation (6) shows that the ACV function represents the overlap or correlation between the shifted and unshifted surface profiles with τ equal to the shift distance. The value of the autocovariance function at zero shift ($\tau = 0$) is, by definition, equal to the rms roughness of the profile, provided that an appropriate mean line has been subtracted out of the profile to arrive at $z(x)$. When the autocovariance function is normalized by dividing by the zero-shift value, the result is known as the autocorrelation function, ACF(τ):

$$\mathrm{ACF}(\tau) = \frac{\mathrm{ACV}(\tau)}{Rq^2} = \frac{\int_{0}^{L} z(x)z(x+\tau)\, dx}{\int_{0}^{L} z^2(x)\, dx} \tag{7}$$

Because the length of overlap between the shifted and unshifted profiles decreases as the shift distance increases, for a finite-length profile, and if we simultaneously use a digital formula, the autocorrelation function may be described by

$$\mathrm{ACF}(\tau) = (N-j)^{-1} \sum_{i=1}^{N-j} z(i)z(i+j) \Big/ \sum_{i=1}^{N} z^2(i) \tag{8}$$

where $\tau = j\Delta$.

Autocovariance and autocorrelation functions are useful for visualizing the relative degrees of periodicity and randomness in surface profiles. For example, Fig. 9 shows autocorrelation functions for Ge and Si surfaces calculated from surface profiles measured by a stylus instrument.[22] Both surfaces were machined by single-point diamond fly cutting under similar conditions with a feed of 3 μm, but the amplitude of the periodicity imposed by the feed of the tool is much larger on the Si surface. Hence, its autocorrelation function is highly periodic. Because its surface topography is dominated by random features, the Ge surface exhibits a strongly decaying autocorrelation function with a small amount of periodicity shown as a barely visible oscillation. Perhaps the differences between the two

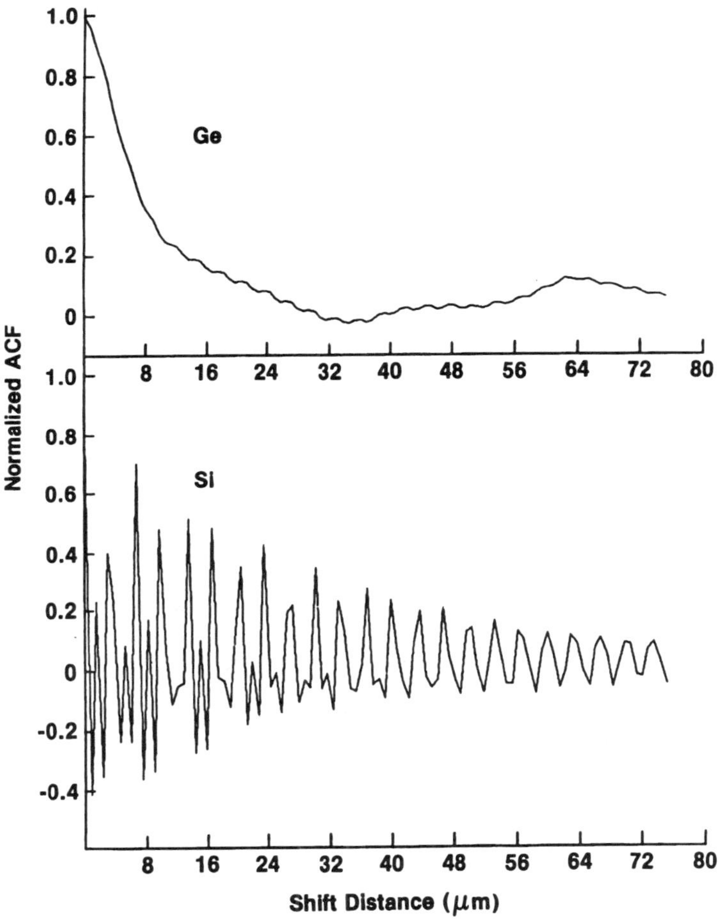

Figure 9. Autocorrelation functions (ACFs) for Ge and Si single-crystal surfaces machined by diamond turning. The curve for Ge is an average of three ACFs measured at different positions on the surface. For Si the curve is an average of four ACFs.

materials arise because Ge is softer than Si and a significant amount of Ge material is displaced across the surface by the tool rather than being removed from the surface.

Therefore, as measures of the lateral structure of surfaces, the PSD and the ACF seem to be useful in different ways. The PSD is useful for studying the strengths of various periodic components in the surface profile and for comparing these with the strength of the broad spectrum due to the random components. The autocorrelation function is useful for describing the lateral sizes of the random features on a surface by studying the decay in the function near-zero shift.

2.2.3. *Other Statistical Descriptors*

As stated, scores of parameters and functions have been developed to quantify profiles of surfaces, many of these for tribology and engineering. In surface science, also, certain topographic parameters correlate well with performance. For example, Somorjai and others[(118)] have studied the bonding and chemical activity of certain species on surfaces and correlated observations of reactivity with the density of lattice steps and associated step kinks on the surfaces. Other surface science phenomena that depend on roughness include surface plasmon coupling[(119)] and surface-enhanced Raman scattering.

2.3. Bandwidth Limits

We now consider the important concept of bandwidth limits[(120)] for surface metrology. Surface measuring instruments are sensitive to a finite bandwidth of surface spatial frequencies. This bandwidth is limited at the high spatial frequency (low spatial wavelength) end by the lateral resolution, which is often determined by the probe width, by the high-frequency response of the sensor, or by the sampling interval. At the low-spatial frequency (high spatial wavelength) end the limit may be the traverse length of the instrument or a long-wavelength cutoff filter. One recently developed type of stylus instrument 120+ has a traverse length of 50 mm and claims a lower spatial wavelength limit of 30 nm, likely due to the finite size of the stylus tip. Therefore, the effective bandwidth of spatial wavelengths that can be sensed by this instrument extends approximately from 30 nm to 50 mm. We have developed a scanning tunneling microscope system with a bandwidth limited at the long-wavelength end by a traverse length of approximately 600 μm and at the short-wavelength end by lateral-stage vibration of approximately 1 nm.[(121)] Spatial wavelengths larger or smaller than these ranges are detected with reduced or zero sensitivity by the respective instruments.

The bandwidth also affects the rms noise of the measurement. When the bandwidth is wider, more noise enters into the output profile. The noise contribution to a roughness measurement can be checked for many instruments by measuring an optical flat, because such a specimen ordinarily has a roughness less than the noise resolution of the instrument. If so, then the resulting value of measured *Ra* or *Rq* represents the noise resolution of the instrument, including the mechanical noise arising from the scanning motion and the electrical noise over the frequency bandwidth of the instrument.

3. Stylus Instruments

The stylus instrument (Fig. 10) is a widely used technique for measuring a surface profile. This technique uses a fine-diamond stylus with tip size approximately 0.1 to 10 μm to traverse the surface. As the stylus tracks the surface peaks and valleys, its vertical motion is converted to a time-varying electrical signal that represents the surface profile. Modern instruments digitize the electrical signal and store the resulting data set in a microprocessor or computer for statistical analyses, such as those described in Sec. 2. Stylus instrument technology has a 65-year history[121] and is highly developed.

3.1. Height Resolution and Range

The height resolution and range are strong advantages of stylus instruments. They depend largely on the (1) transducer design and (2) mechanical noise and straightness arising from the traverse mechanism.

In laboratory-level stylus instruments the transducer is often a linear variable differential transformer (LVDT)[9,97] that produces an output directly proportional to surface height. However, certain optics-based transducers are also sensitive to height.[123] Inductive and piezoelectric transducers may also be used. These latter types of transducers are sensitive to the stylus vertical motion—that is, to the time derivatives of the height variations rather than the height itself.[97,124] They generally cost less than LVDTs and have been widely used in manufacturing shops but less widely in research or metrology laboratories.

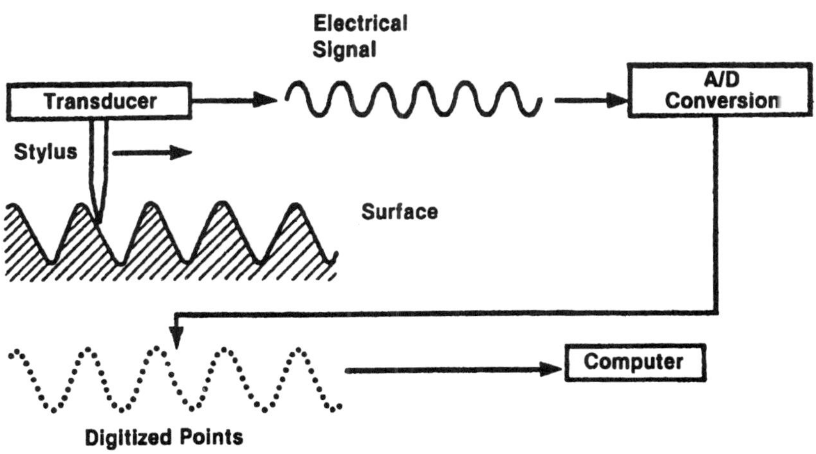

Figure 10. Schematic diagram of a stylus-profiling instrument. The transducer in this type of instrument produces an electrical signal that represents the surface profile. This signal may then be digitized, stored, and analyzed with statistical techniques.

One example of a LVDT is shown in Fig. 11. As the stylus moves vertically, the ferrite core is displaced from the balance position between the electromagnetic coils. This action results in an imbalance of an AC bridge, to which the coils are connected and which is driven at a high carrier frequency. The out-of-balance voltage signal, which is demodulated and amplified, is proportional to the vertical displacement of the stylus. As shown in Fig. 1, some transducers have noise levels of less than 0.06 nm[35] over fairly wide bandwidths of spatial wavelengths between about 0.7 and 50 μm. Thus, surface nanotopography approximately one atom high can be detected. Other LVDT transducers have a vertical range up to 1 mm, and, in general, the ratio of range to resolution is about 10^4–10^5.

Three types of designs for achieving low scanning noise and a good straightness of motion are shown in Fig. 12. A low-cost method (upper left) involves the use of a skid, which serves as a point of support and provides a nearby reference position with respect to which the vertical motion of the stylus is measured. The skid also serves as a high-pass mechanical filter for the instrument because its position is normally determined by contact with widely spaced surface peaks. Mechanical noise in the drive mechanism can be significantly reduced in the output profile through use of the skid. On the other hand, use of the skid can lead to unwanted distortions in the profile,[125] particularly where the skid trails the stylus (Fig. 13) rather than flanks or surrounds it.

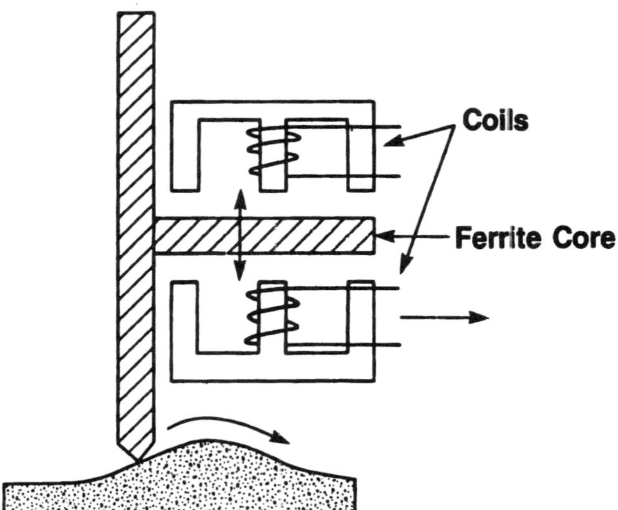

Figure 11. Schematic diagram of a high-sensitivity linear variable differential transformer (LVDT). As the stylus traverses the peaks and valleys of the surface, the ferrite core moves vertically with respect to the coils.

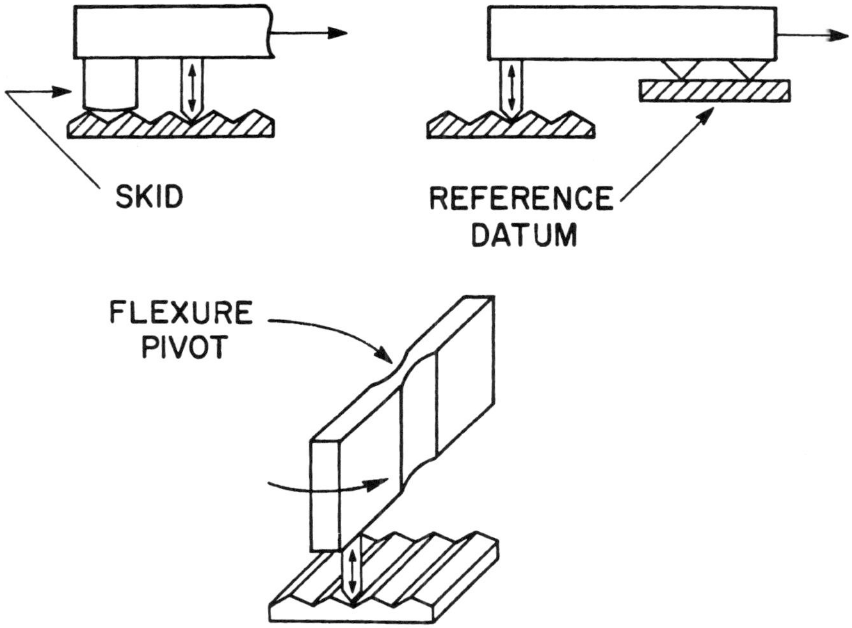

Figure 12. Schematic of three types of designs to constrain unwanted motion of the stylus.

Figure 12 (upper right) shows a schematic of an external reference surface datum that guides the motion of the tracing stylus. Well-engineered tracing systems may minimize the offset between the reference and measured surfaces by placing the stylus contact and the set of reference contacts as close to the same vertical line as possible or by minimizing the amount of angular motion and vibration in the carriage. The concept of designing a measuring machine with the displacement-measuring system in line with the displacement to be measured is called the Abbe principle.[126]

The flexure-pivot design in Fig. 12 contains no sliding components at all; a set of flexure pivots constrains the motion in a plane that is perpendicular to the axis of the pivots. Such monolithic designs provide highly-noise-free motion. A commercial system that exploits this mechanical design yielded the rms noise level of <0.06 nm,[35] as discussed. Another variation of the flexure-pivot approach[127] involved using multiple sets of flexures to allow for motion in two directions.

3.2. Lateral Resolution and Range

A major disadvantage of the flexure-pivot design is that the trace length is limited to the displacement at which the elastic bending forces of the flexure

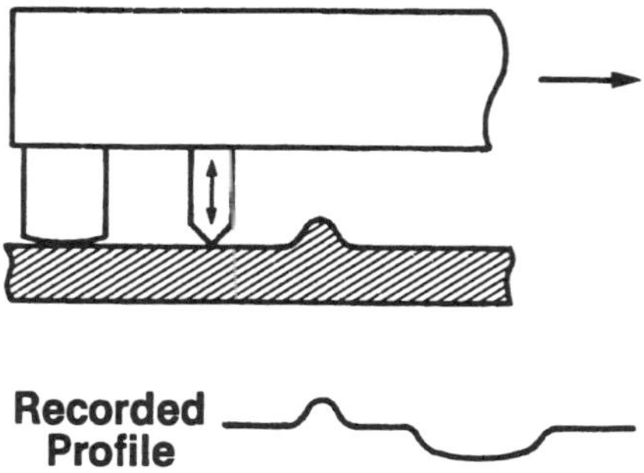

Figure 13. How an isolated peak in the surface can produce a false valley in the profile measured by a stylus instrument with a skid.[126]

become too large. In one example of the design discussed, the trace length limit is about 2 mm. However, lateral range is not a problem for other types of stylus instruments. The lead-screw drive mechanisms of some instruments permit trace lengths of about 300 mm or more.

Because the measured profile represents a convolution of the actual surface profile with the end form of the stylus, the lateral resolution depends critically on the size of the stylus tip and to a lesser degree on the flank angle.[128] Styli are often flattened at the end by production or by wear; the width across the flat essentially determines the lateral resolution. A commonly available width is 2.5 μm, but

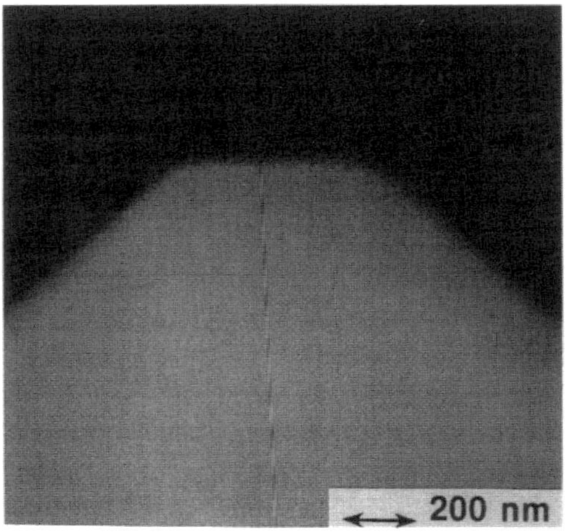

Figure 14. Scanning electron micrograph of the profile of a nominal 0.1 × 1 μm chisel-shaped stylus. The profile indicates that the nominal 0.1-μm edge is worn to a flat of width about 0.5 μm following 10 months of service.

pyramidal styli with tips as small as a nominal 0.1 μm × 0.1 μm may be readily obtained. Figure 14 shows a scanning electron micrograph of a nominal 0.1 μm × 1 μm chisel-shaped tip that had been in service for approximately 10 months. The nominal 0.1-μm edge was worn to a flat about 0.5 μm across.

In some instances highly spherical styli are also used, and Bennett and Dancy[17] and Vorburger *et al.*[128] show micrographs for these. Elson and Bennett[111] have calculated that the lateral resolution of such styli can be smaller than the spherical radius. Their calculations are based on a geometrical model of a stylus tracing over a sinusoidal surface profile with small slopes. We present here a related argument for a spherical stylus of radius r tracing over a surface step of height h. Figure 15 shows the geometrical construction when the stylus first contacts the step, which is assumed to be infinitely sharp. If $h \ll r$, the point of initial contact occurs at a position $\sqrt{2hr}$ to the left of the step edge; hence, $\sqrt{2hr}$ represents the width of the transition and is an estimator of the effective resolution of the stylus. The calculation assumes that no deformation occurs at the single point of contact. For a 1-μm radius stylus and a step of height 10 nm, the calculated resolution is 0.14 μm.

The stylus flank angle (Fig. 15) also affects its resolution. The stylus cannot accurately profile over slopes greater than 90° minus the flank angle. Flank angles of 45° and 30° are commonly available. The slope limitation is not a problem for most types of smooth surfaces where the average surface slopes are only a few

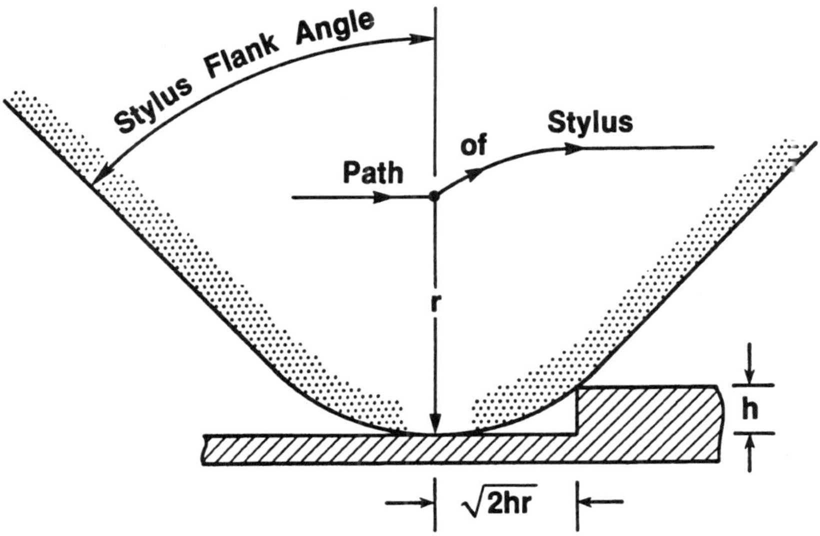

Figure 15. Diagram showing the path traced by a spherical stylus tip traversing over a small, sharp step.

degrees, but it poses problems for measurements of surfaces with steep cracks or holes, such as those on ceramics or cast materials.

An example of the lateral-resolution capability of the stylus instrument is given in Fig. 3, which shows the stylus profile for a diffraction grating with a periodicity of 6000 lines/mm. The stylus tip was nominally 0.1 μm × 0.1 μm, and its tip dimension along the profiling direction is comparable to or less than 0.1 μm, as measured by the razor-blade tracing technique[96,129] we normally use to check stylus quality *in situ*. This technique consists of traversing the stylus over the edge of a fine razor blade held vertically. If the razor-blade dimension is significantly smaller than the stylus tip and no deformation occurs during the measurement, the resulting profile represents fairly accurately the shape of the stylus in the direction of travel. Another artifact for this type of measurement is a vertically etched edge in silicon.[130]

Researchers have studied the effects of stylus shape or scanned probe tip shape on measured surface profiles. They have also studied the conditions under which the tip shape may be determined and its effects unfolded from the measured surface profile to produce one with improved accuracy. The work of Villarrubia[131,132] on scanned-probe-tip geometry is an excellent recent example.

To summarize the subject of lateral resolution, the finite size of the stylus as it follows the surface profile causes a reduction of the widths of valleys and an increase of the widths of peaks in the measured output profile. The minimum detectable valley width is roughly equal to the stylus tip width itself but may be smaller for spherical styli tracing over very flat valleys. From geometrical arguments the minimum detectable peak width can be quite small in principle, but if it becomes considerably smaller than the stylus tip width, the area of contact is decreased, the stylus pressure is increased, and the stylus might tend to plow through such an asperity.

3.3. Stylus Load and Surface Deformation

We now consider the question of damage to smooth surfaces by stylus loading. The logical parameters that determine whether any damage occurs are the surface hardness, the stylus force, the stylus tip width, and to a lesser extent the stylus speed. In our experience a stylus tip of width 2.5 μm should not produce detectable damage on reasonably hard metal surfaces, such as aluminum, as long as the stylus force is smaller than about 0.01 mN (1 mg).

Bennett and Dancy[17] studied the effects of stylus damage to KCl, a soft material with Mohs hardness less than 2, which is equivalent to that of gypsum. They used five different styli and systematically varied the stylus force of each; they then observed the damage by electron microscopy of shadow replicas of the surface. Figure 16 shows one set of micrographs for a series of traverses by a 1.2-μm radius stylus. For a stylus force of about 5 μN (0.5 mg), no damage was observed. Many

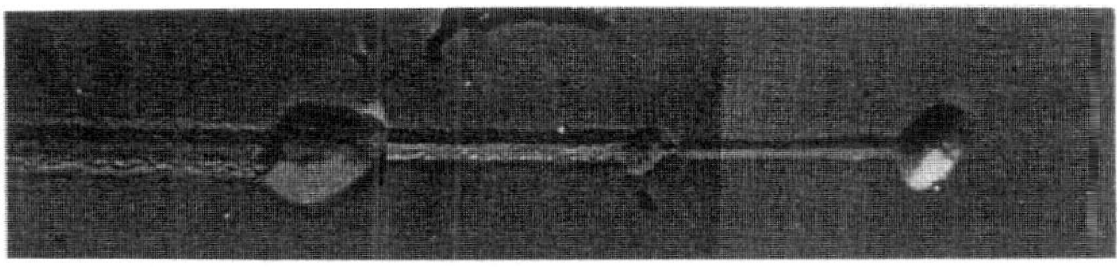

Figure 16. Electron micrographs of stylus tracks in KCl made with different stylus forces (Adapted from Bennett and Dancy[17]).

types of stylus instruments use stylus forces of about 500 μN (50 mg) or higher, but these are often used with stylus-tip radii of about 5 to 10 μm. Because the pressure is inversely proportional to the area of contact, the pressure on the surface due to stylus loading for a 10-μm radius stylus with a 500-μN force is about the same as that for a 1-μm radius stylus with a 5-μN force, assuming that the materials response between the two pressures is the same. Even if the stylus leaves a visible track, the resulting profile is likely to be accurate because the variation in the depth of the track over the surface should be significantly smaller than the depth itself.

So far we have discussed only plastic or irreversible deformation of the surface. It is also possible to estimate the amount of elastic or reversible deformation by varying the stylus force in a cyclical fashion and separating the reversible displacements from hysteresis effects in the vertical motion. Whitehouse[16] described preliminary studies of this kind, but noted that elastic deformation in the instrument would have to be isolated from that of the surface to estimate the latter. In any case the elastic deformation of metal surfaces by styli is expected to be very small.[133,134]

3.4. Other Distortions

We note several potential sources of profile distortion in addition to stylus forces and the finite stylus width. These include the nonlinearity of the transducer, the variation of the stylus speed, stylus flight, and digitization.

Linearity[135] of the transducer and the associated electronics are normally not a significant source of error in stylus instruments, although they can affect calibrations of step height or precision–roughness specimens. Figure 17 contains data on the nonlinearity of the combined LVDT and electronic circuitry for one of our stylus instruments at NIST. This figure shows measurements of the height of a step specimen at four places in the range of the transducer. These data were taken at the lowest magnification setting on the instrument, so the chart limits represent the maximum recommended range of the instrument. One point was measured just above the upper chart limit to test the linearity in this part of the range. The other three results indicate a variation in the calibration of the transducer of less than 0.5% over the full-scale range of 10 μm.

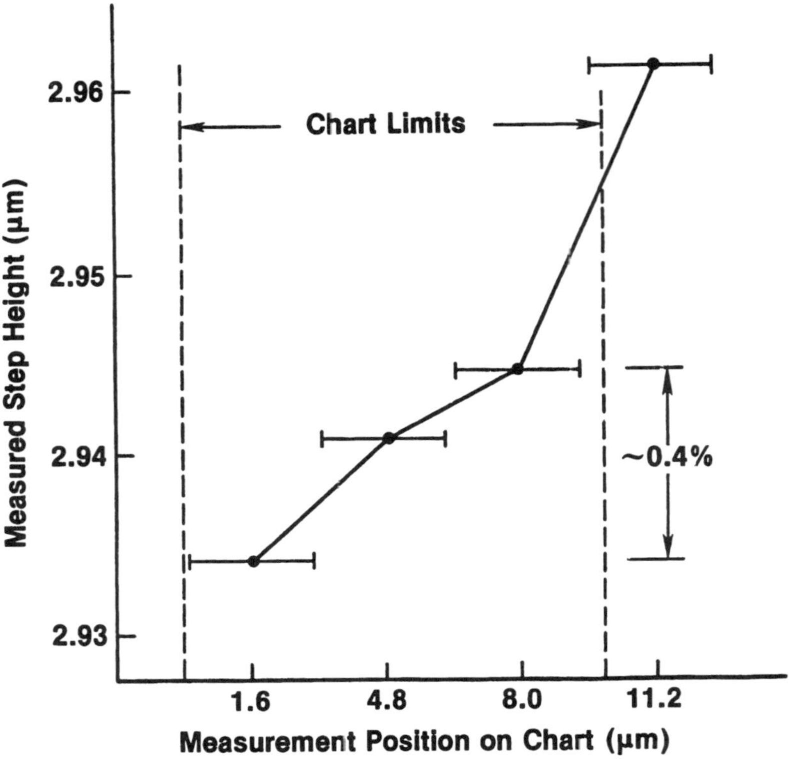

Figure 17. Nonlinearity data for a stylus-profiling instrument at NIST showing the measured height of a step versus vertical position on the chart recorder. The horizontal bars denote the span of the step on the chart recorder.

The relative nonlinearity is expected to be even less at smaller transducer displacements, i.e., higher magnifications. Teague[136] showed nonlinearity data that were taken at two magnifications for a specific LVDT that supported this expectation.

One potential source of nonlinearity, which has not been studied systematically, is the sticking of the transducer at subnanometer displacements. However, this type of misbehavior does not seem to appear on profiles showing such displacements.

A constant stylus speed usually is assumed for the horizontal scaling of the instrument. As with the transducer nonlinearities, typical variation of this speed produces uncertainty of about 1% in the lateral measurements.[136] Traversing errors due to variations in stylus speed can be reduced considerably by using laser interferometry to measure the lateral displacement of the stylus directly. This has been done for a series of precise measurements of surface wavelength in the calibration of NIST sinusoidal standard reference materials for surface roughness.[116]

Another problem is stylus flight—that is, the potential for the stylus to lose contact with the surface because of a rapid impulse, such as encountering a steeply rising surface step. The parameters that affect this phenomenon are stylus speed, stylus force on the surface, damping constant in the vertical direction, and rate of change of the surface slope itself—in other words, the amplitude and spatial wavelength of the surface. The interdependence between stylus force and speed for producing stylus flight is an important consideration. A magnetic cartridge for a phonograph with a force of 20 mN (2 g) can traverse a long-playing record at a tangential speed of 500 mm/s without losing contact, but a stylus with a force of 0.5 mN (50 mg) must travel as slow as about 1 mm/s to maintain contact. This problem has been studied,[137–141] and criteria for the onset of stylus flight have been developed. Typically, the stylus maintains contact with the surface for practical stylus forces (≥ 0.3 mN) and speeds (< 2.5 mm/s) even when the stylus traverses surfaces with average absolute slopes up to 15°.[140] However, occasional spikes in surface profiles due to stylus flight may take place after sharp peaks are encountered. The accuracy of profile data showing sharp, high peaks may be verified by remeasurement of the same profile at a slower speed.

As with other profiling techniques, the surface profiles obtained with stylus instruments are routinely digitized for computer storage or mass storage and subsequent quantitative analysis. Both the vertical quantization interval (i.e., the minimum recordable height difference in the digitized profile) and the lateral sampling interval are relevant to determining a realistic digital representation of the profile. The peak-to-valley height of the profile should be much larger than the vertical quantization interval, and the surface structures to be studied should be much wider than the lateral sampling interval. For studying a distribution of surface topographies, there should be enough points in the profile to give an adequate sampling of the statistical variability of these surfaces. For roughness profiles our system at NIST has used 4000 digitized points and a minimum of 2400 quantization levels over the vertical range, which is usually chosen so that it is three or four times the maximum peak-to-valley height of the profile. New standards now commonly recommend 8000 digitized points or more over a profile evaluation length.[12] Some high-resolution profilers now have 16-bit vertical digitization (about 65,535 quantization levels) and can acquire about 65,000 data points, features that facilitate zooming over several scales of observation in the display.[142]

3.5. Calibration

The lateral and vertical magnifications of stylus instruments may be calibrated to about 1% or better using roughness. The vertical displacement of the stylus may also be calibrated by traversing a step whose height has been calibrated by interferometric measurement. Vertical calibration becomes more complex at very high magnifications where the desired resolution may be at the nanometer or subnanometer level, at or beyond the resolution capabilities of conventional interferometric techniques. To verify the calibration of the instrument over subnanometer height changes, the linearity of the LVDT output with respect to vertical displacement must be verified, as discussed for Fig. 17.

In the lateral direction the relative displacement of the stylus over the surface may be measured directly by a laser interferometer.[116,127] Alternatively, calibrated grids or other types of periodic surface specimens may be used as secondary displacement standards.

3.6. Applications

Stylus instruments have been used to assess the topography of many types of surfaces, including mechanical parts, semiconductors, and optical surfaces. The measurements on semiconductors and optical components have demonstrated the resolution limits of stylus instruments and the high quality of the surfaces themselves. Although optical scattering and optical profiling are closely related to the functional properties of an optical surface, advantages of the stylus for measurement of optical surfaces include a large lateral range and a lateral resolution approaching 0.1 μm. Two disadvantages are the potential for surface damage and the extremely small fraction of the surface sampled as a profile.

A stylus instrument has also been used to profile steps with heights having molecular dimensions. Whitehouse[143] measured cleaved, single-crystal topaz with dislocation edges (Fig. 18). He concluded that the two left-hand edges were probably single-lattice steps of height 0.84 nm, whereas the third was likely a double-lattice step. Interesting results have also been obtained on optical gratings. Verrill[143] used the results of stylus traces of ruled gratings to adjust the alignment of the ruling tool to optimize the groove profile.

3.7. Area Profiling with Stylus Instruments

The stylus traverse may be repeated many times in a raster fashion over the surface to produce three-dimensional maps consisting of parallel stylus profiles.

Figure 18. Stylus trace of topaz showing lattice steps (Adapted from Whitehouse[144]).

Although this procedure requires a lot of effort, the measurements provide information about surfaces that might not be recognized from individual profiles alone.(144–148)

Williamson(144) was apparently the first to obtain three-dimensional maps with a stylus instrument. He displayed topography results as contour plots. The local surface height was also coded in terms of brightness like a relief map in cartography. This display enabled him to visualize the high spots in the surface and to study in detail the geometry of contact between rough surfaces.

Sayles *et al.*(145,146) represented human articular cartilage topography as isometric plots of parallel traces (Fig. 19). These plots show damaged areas in the tissue.

Stout *et al.*(148) provided a comprehensive review of their own and previous work and included discussions of mechanical requirements, digitization intervals, and filtering techniques for three-dimensional profile areas. They also showed algorithms and examples of area PSD and autocorrelation functions to characterize three-dimensional surface maps (Fig. 20).

We anticipate that an improved mathematical description of surface topography will result from three-dimensional functions and parameters and through the use of area-profiling instruments to be discussed. This improved mathematical description will yield a better understanding about physical processes, such as sliding and fluid flow, that take place on surfaces. However, many data points and surface profiles must be measured. Useful topography maps contain at least 200 profiles with at least 200 data points each. Stylus instruments have had the disadvantage of relatively long data-acquisition times because the mechanical response limits the maximum scan speed to about 1 mm/s. Profiling a 3 mm × 3 mm surface area thus

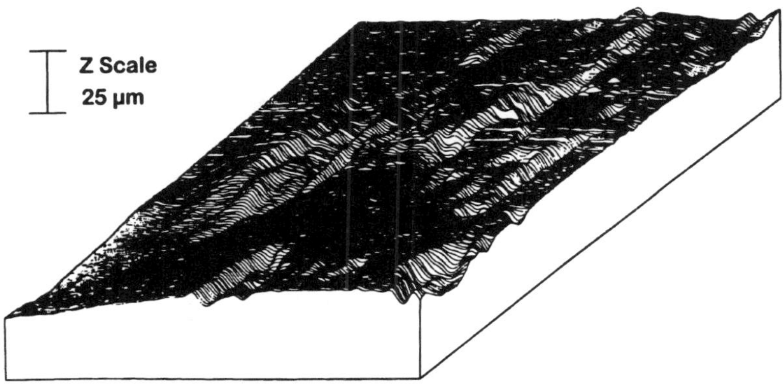

Figure 19. Perspective view of a 3.7 mm × 1.6 mm area of diseased human articular cartilage (arthritic knee joint). The long grooves (~5–10 μm in height) are parallel to the sliding direction and give evidence for wear damage of the tissue (Adapted from Sayles *et al.*(147,148)).

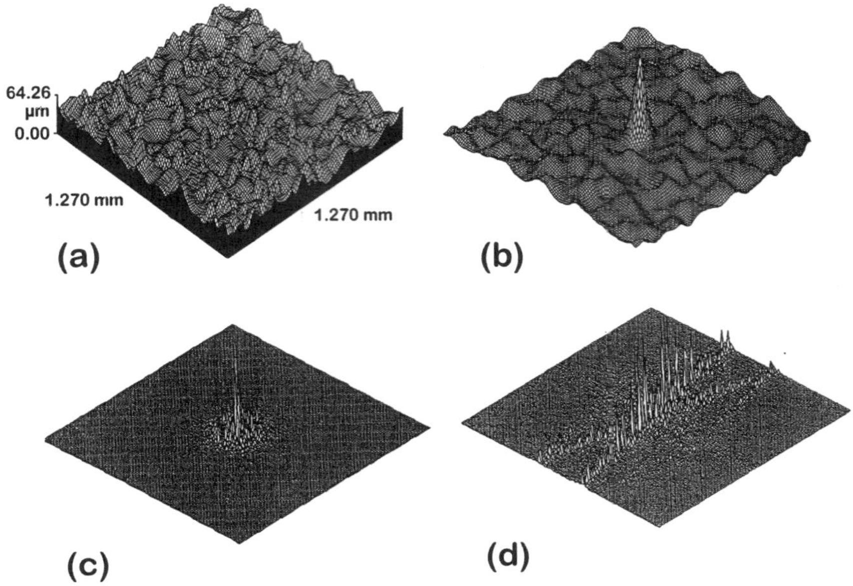

Figure 20. Surface topography measured with a stylus instrument and related statistical functions: (a) measured surface topography of an electrodischarge machined (EDM) surface; (b) autocorrelation function calculated from the EDM surface; (c) PSD function for the EDM surface; (d) PSD calculated for an engine-bore surface (Adapted from Stout *et al.*[150]).

might take close to an hour. However, a much faster instrument,[37] recently developed can measure a 0.2 mm × 0.2 mm area within 20 s.

4. Optical Profiling Techniques

Optical techniques offer the advantage of fast, noncontact area profiling of surface topography without mechanical scanning. Optical techniques are mostly of two types: focus sensing[43,91,149–151] and interferometry.[21,22,26,36,152–157] We emphasize interferometry next.

Interferometry is an important technique for measuring the roughness and figure of high-quality optical surfaces. The Twyman–Green two-beam interferometer (Fig. 21)[152] illustrates the basic principles. An optical wavefront incident at A is collimated by L_1 and split into two coherent beams by a partially transmitting mirror M. One beam is reflected from a smooth flat reference surface R, while the other is reflected from the surface S being studied. The reflected beams are then recombined at M. Lens L_2 produces an image of the surface S at O. With a small amount of tilt between S and R, the interferometer produces a set of generally

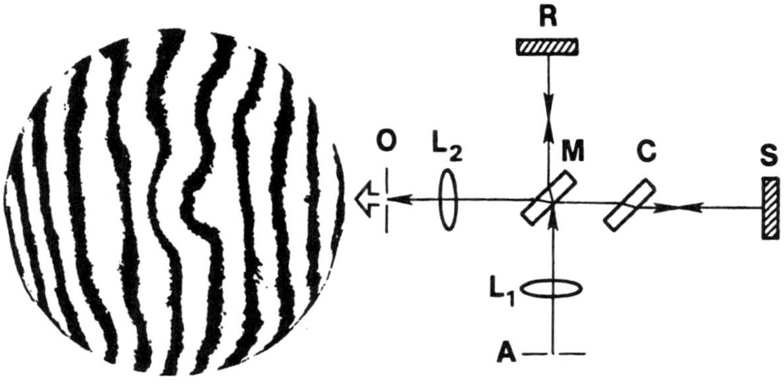

Figure 21. Twyman–Green two-beam interferometer. A is a source of monochromatic light. The fringes are observed at O. C is a compensator positioned to keep the optical path lengths of the two beams the same. The photo[154] (copyright 1979 AAAS) on the left illustrates interference fringes for a test surface.

parallel fringes, as shown, whose undulations reveal the peaks and valleys of the surface S in a manner similar to that of a stylus instrument. With interferometry, surface heights are measured directly in terms of optical wavelengths. However, conventional use of the technique involves image processing to convert the optical fringe data into stored profile data.

The speed and sensitivity of interferometry was vastly improved with the development of several types of phase-measuring techniques. If the light-reflecting properties of the specimen are uniform across the measured surface, the variation in phase ϕ at each point of the reflected wavefront (with wavelength λ) is proportional to the variation in the path length traveled from the surface and hence to the surface height z itself; i.e.,

$$z(x, y) = (\lambda/4\pi)\phi(x, y) \tag{9}$$

where (x, y) represents position on the surface.

Several types of interferometers[21,22,153–156] have been developed for use in both profiling and area-profiling modes of operation. One example is the phase-shifting interferometric (PSI) microscope.[21,22,36,153] If a linear array is used as a detector, as in the earliest versions, we call the instrument a *PSI profiler*, and if an area array camera is used, we call it a *PSI microscope* (PSIM). Depending on the desired magnification, PSIs may employ one of several optical configurations to produce interferometric fringes in the camera. The configuration in Fig. 22 is a Mirau interferometer. The beamsplitter plate and the reference surface are placed very close to the test surface in the converging wavefront of the microscope objective. The beamsplitter reflects part of the incident wave to the reference surface and transmits a portion to the test surface. The light reflected from both surfaces is

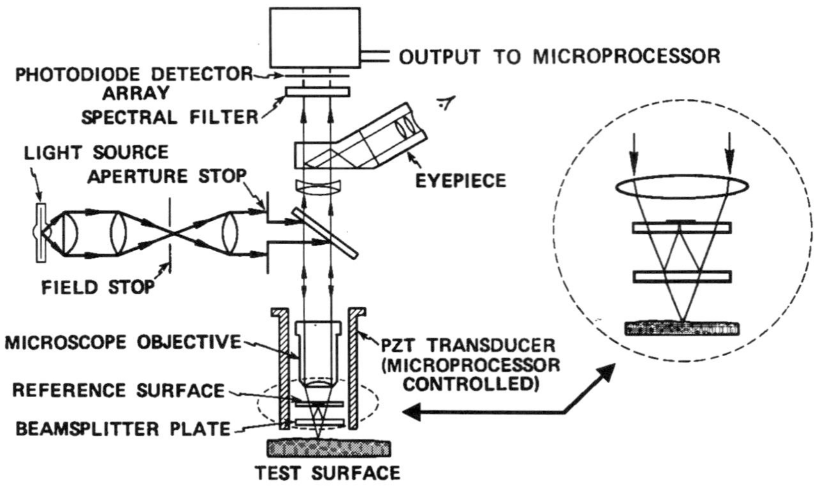

Figure 22. Schematic of the PSI developed by Wyant *et al.*[21] (illustration adapted from Bennett[24]). The dashed envelope shows details of the Mirau interferometer. PZT refers to a piezoelectric transducer.

recombined again at the beamsplitter and transmitted to the camera (or linear array). The interference fringes are analyzed in the camera by translating the reference surface in the z direction and by processing the resulting time-varying signals to produce the relative phase for each pixel in the detector array. Because the instrument does not have mechanical scanning in the lateral direction, there is no uncertainty in the measured surface maps or profiles associated with straightness of travel or velocity stability of a scanning probe or specimen stage as found in stylus instruments. However, the interferometric profiling accuracy does depend on several factors, including the quality of the reference mirror and the linearity of the phase-shifting piezoelectric z-drive. The length or area of the surface in the field of view is limited by the magnifications of the objective lenses available for the microscope. The instrument can provide a vertical resolution of about 0.1 nm and a lateral resolution as small as 0.5 μm for high-magnification objectives.

PSI profilers and microscopes have been used for practical surface applications, including measurements of the roughness of magnetic tapes[21] and of precision-finished optical components and measurements of the pole-tip recession of magnetic read and write heads.[157] Figure 23, for example, shows a topographic map of a stainless steel surface measured by Evans[158] with a PSI microscope to verify the successful machining of stainless steel to a smooth finish by cryogenic diamond turning.

When the measured rms roughness is less than about 0.4 nm, it is important to separate the topography of the reference surface from that of the test part by

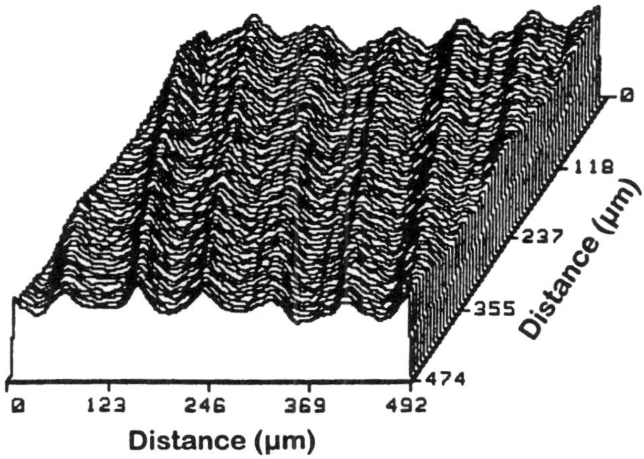

Figure 23. Surface roughness of a grade of 440 stainless steel machined by cryogenic diamond turning and measured with a PSI microscope by Evans.[160] The measured image was analyzed to give an rms roughness of 20.9 nm and an *Ra* value of 16.9 nm.

subtracting the reference-surface topography from the measured map. A technique known as reference-surface averaging aids in determining the reference-surface topography.[36] If a smooth, flat test part is measured at various well-separated positions, the average of the measured maps is a close approximation to the reference-surface topography, which is common to the otherwise statistically independent maps. This average reference-surface map may then be subtracted from the results of subsequent measurements. Reference-surface averaging has been used routinely for measurements of the topography of surfaces with rms roughnesses of about 0.1 nm. A complementary technique[159] to reference-surface averaging is that of taking the difference between two maps measured at different positions on the surface being studied. This procedure effectively eliminates the systematic effects of the reference surface and enables an estimate to be made of the rms roughness of the test surface.

The optical homogeneity of the reflecting surface is a critical factor for interferometers. Local variation in the surface material, possibly due to residual contamination layers, could produce variation in the optical phase shift upon reflection, which in turn would cause apparent variations in the surface height measurements.

Other interferometric profiling instruments[154,156] operate essentially as slope-measuring devices. The basic principles are similar to the Nomarski microscope.[160] Two closely spaced laser beams are reflected from different points on a surface, and the ensuing measurements of phase differences between the two beams reveal height variations on the surface. In Nomarski microscopy the phase differences between the beams due to surface topography result in interference contrast in the image. These instruments detect the phase differences electronically and then integrate the differential height measurements to obtain surface profiles.

The lateral resolution of the Nomarski profiler[154] instrument is about 1 μm. Noise in the differential measurements can accumulate if these are integrated to produce surface profiles or topographic maps, but the instrument has demonstrated the capability for rms roughness measurements of about 0.1 nm.

The phase-shifting interferometer is not useful for measurements of rough surfaces with roughness greater than about 150 nm because the patterns of light and dark fringes become too dense to analyze accurately. To overcome this problem and achieve measurements of rough surfaces, researchers have developed a modification of the instrument, called the vertical-scanning white-light interferometer.[161,162] Instead of a narrow bandwidth, white-light illumination is used in the instrument, which also features a large vertical-scanning range. As the reference surface is scanned vertically, the vertical position of maximum fringe contrast is sensed and recorded for each point in the image by a fast processor. The maximum fringe contrast occurs when the light path between the beamsplitter and the surface equals the light path between the beamsplitter and the reference. By this principle the vertical measurement range is only limited by the range of the vertical-scan stage. Measurements of topography with a peak-to-valley range of 500 μm are obtainable. Alternatively, focus-sensing techniques,[43,91,149–151] including the confocal scanning optical microscope, may be used to measure surfaces with rms roughness of several micrometers or more.

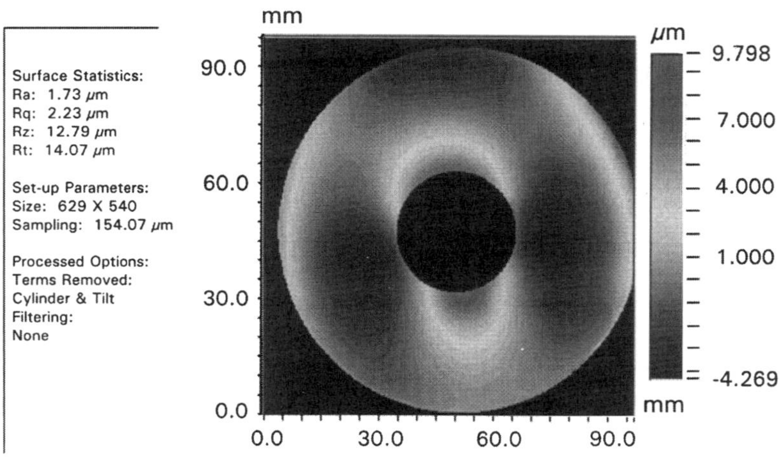

Figure 24. Measured topography of a magnetic hard disk consisting of about 285 stitched (combined) measurements. Measured by Wyant and Schmit[165] with a scanning white-light interferometric microscope.

The measurement area for interferometric microscopes is limited by the field of view of the optical system. With the increasing speed of computers, it is possible to measure large surface areas by stepping from one area to another and acquiring frames of information at each area. The individual frames may then be combined into a single large one by stitching (combining) them at the overlapping regions by using statistical techniques. Figure 24, for example, shows a topographic map[(163)] of a magnetic hard disk approximately 100 mm in diameter measured with a vertical-scanning white-light interferometer. The map is a combination of 285 overlapping regions.

5. Scanned Probe Microscopy

Scanned probe microscopy (SPM) offers researchers and technologists the capability to image surfaces at near-atomic and even atomic resolution. The field of SPM has expanded enormously because of the practical and scientific uses that atomic scale resolution provides and because of the subsequent proliferation of many types of high-resolution imaging mechanisms. We know of at least 20 types of probe-based techniques that have been used successfully in research on microscopic surface phenomena. The most important technique has been the atomic force microscope (AFM),[(164)] which is used to measure directly the surface topography. Three-dimensional topographic measurements with AFM and scanning tunneling microscopy (STM) have had many practical applications in industry and hold great promise for future developments.

In this section we emphasize STM and AFM. Their scientific merits and some results are discussed, and the capabilities of these instruments are highlighted. In Sec. 5.1 we briefly discuss the history of STM and AFM. In Sec. 5.2 we discuss the characterization and calibration of these instruments. In Sec. 5.3 we discuss other types of SPMs. In Sec. 5.4 we discuss current applications of these instruments. In Sec. 5.5 we discuss future applications.

5.1. Short History of Scanned Probe Microscopy

Scanning microscopy essentially began with the invention of the SEM.[(165)] This instrument produces a magnified image of a surface, not by direct imaging with lenses, by synchronizing the position of the measured output signal on the monitor screen with the position of the probe beam on the surface. SPM was invented by Young *et al.*[(27)] as the *topografiner*, a scanning method that sensed the surface through the field emission current that resulted when the probe was positioned within a small distance of the surface and a voltage was applied (Fig. 25). The magnitude of this current (J_T) was extremely sensitive to the distance between the sample and the emitter. Hence, as the emitter was scanned over the surface, its height

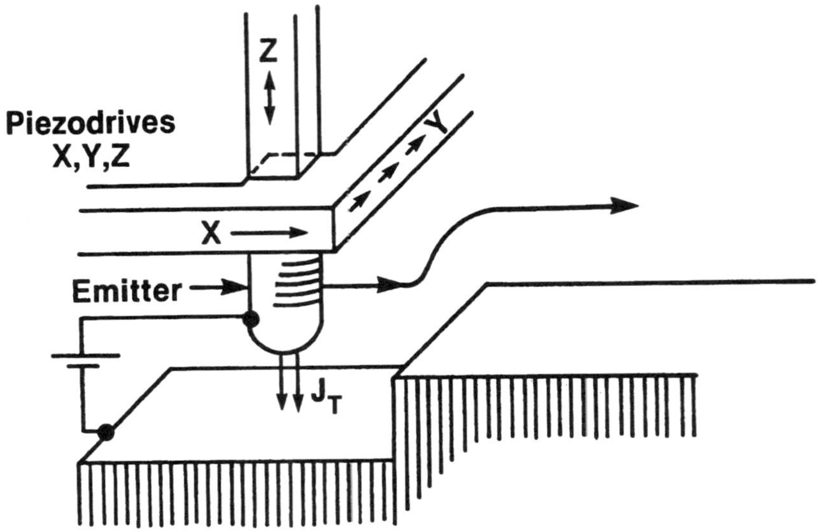

Figure 25. Schematic diagram of the principle of the topografiner and the STM.

z was servocontrolled to maintain a constant current across the gap between them. The x, y, and z motions were generated by piezoelectric elements controlled by signals that were also used to make $z(x, y)$ maps of the surface. The topografiner attracted little follow-on research until the observation of atoms on a Si(111) surface (Fig. 26) by Binnig *et al.*[32] in 1982, using an instrument they called the STM. The STM employed a conducting probe and sensed current and z-position feedback in the same way as the topografiner but did so by positioning the probe within about a nanometer of the surface and controlling that distance to a small fraction of a nanometer.

Largely because of its capability for imaging surfaces with atomic resolution, the development of the STM proceeded rapidly and its usage grew enormously. The techniques and applications of STM have been reviewed elsewhere.[166,167] Two examples of its capability for imaging surfaces with atomic resolution are shown in Figs. 27 and 28. Figure 27 is an interesting STM image of a stepped (100) face of a Si single-crystal surface.[167,168] The Si atoms on the surface pair to form rows of dimer pairs. These rows of dimers are resolved in the image, and one can see alternating terraces of dimers separated by single-atom steps with the rows running first parallel then perpendicular to the step edges. The step edges with the dimer rows running parallel are much straighter than the edges with the dimer rows running perpendicular. The image also shows a low density of randomly distributed atoms or dimers on top of the basic crystal structure. Figure 28 shows a remarkable differentiation[167,169] of Ga atoms and As atoms side by side in rows on the surface of GaAs. The Ga atoms (darker) were imaged with the applied tunneling voltage

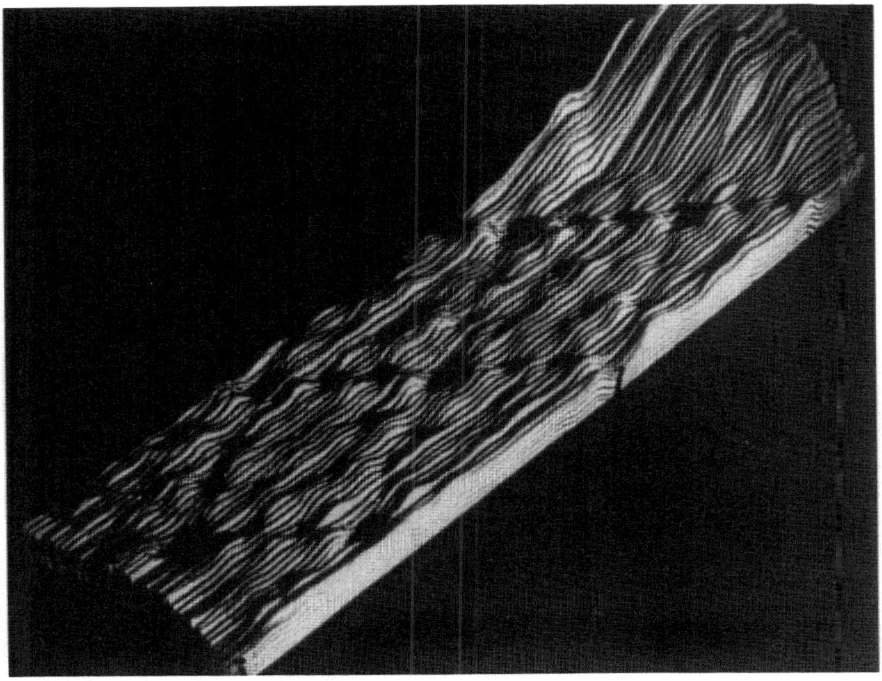

Figure 26. Surface topographic map of the 7×7 reconstruction of the (111) surface of clean single-crystalline Si obtained by Binnig *et al.*[32] using an STM. The map consists of parallel profiles. The clearly visible atomic structure is the most important early evidence of the capability of the STM for obtaining atomic spatial resolution.

in one polarity and the As atoms (paler) were imaged with the applied voltage having the reverse polarity.

In 1985, Binnig *et al.*[164] extended the scanned probe approach to a different type of sensor called the AFM. This instrument also used a fine tip, positioned close to the surface, mounted on a cantilever. The surface forces causing the cantilever to flex were sensed by an STM probe operating behind the cantilever. It was soon realized that the tip–surface interactions could be sensed by various techniques, and force sensors soon included optical levers,[170] null-sensing interferometers,[171] piezoresistive cantilevers,[172] and tuning forks.[173] The AFM has become the workhorse of scanned probe technology because it can be used to measure the topography of insulating surfaces as well as conducting ones at near-atomic resolution. However, many other sensing techniques have been developed to detect other physical properties of surfaces besides topography, notably magnetic force probes, frictional probes that detect sideways motion of the probe tip or cantilever, and the near-field scanning optical microscope. These techniques are discussed briefly in Sec. 5.3.

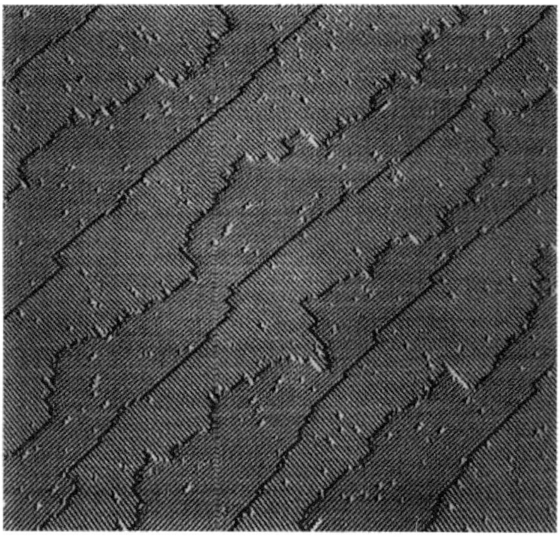

Figure 27. STM image of a stepped (100) surface of Si showing single-atom steps and terraces with alternating rows of dimer pairs of Si atoms (reprinted from Swartzentruber *et al.*[170]).

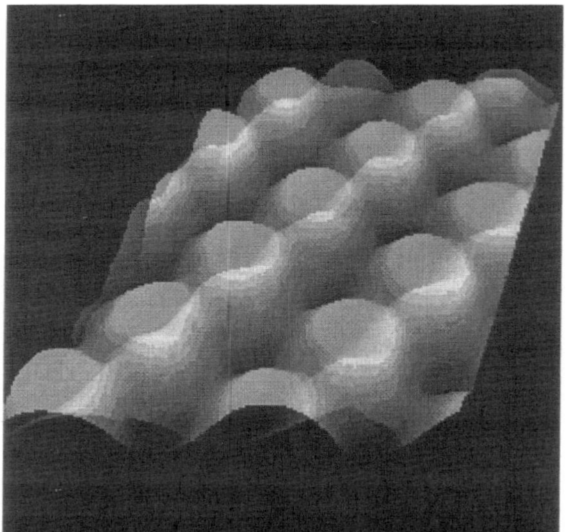

Figure 28. STM image of rows of individual Ga (darker) and As (paler) atoms on the surface of GaAs,[171] achieved by sensing the Ga atoms with opposite voltage polarity from that used to sense the As atoms. The original plate (in Stroscio and Kaiser[169] was in two colors.

5.2. Calibration and Characterization

Calibration is an important issue with SPMs. On one hand, the motion systems often consist of voltage-driven piezoelectric materials because of the need for high resolution. However, these materials have a high degree of nonlinearity. On the other hand, it is difficult to develop standards and displacement-measuring systems to calibrate the SPMs at the high resolution at which they are used.

Several properties of these instruments must be analyzed carefully. Overall displacements in the *x*, *y*, and *z* directions must be calibrated; hysteresis, drift, displacement nonlinearities, other motion errors, and noise effects must be understood and minimized. The size of the probe and its interaction area with the surface must also be minimized and understood, particularly to achieve accuracy in measurements of linewidths and hole widths. The smallest linewidths and hole widths designed on a semiconductor chip are called critical dimensions.[(174)] We discuss, in turn, instruments and standards for displacement calibration and for measuring critical dimensions (CDs).

5.2.1. *Instruments for Displacement Calibration*

Instruments for displacement calibration are divided into two principal categories. First, national laboratories have concentrated on developing STM and AFM instruments with traceability to the wavelength of light via displacement interferometry. These instruments are used primarily for calibrations of artifacts to be used as standards for calibrations of other SPMs. Commercial instruments then become traceable to the national laboratories primarily by means of calibration artifacts. The cost of incorporating laser interferometers into SPMs to provide traceability to the wavelength of light is significant relative to the cost of the instrument itself. As a result, only a few such instruments have been needed to date, and national laboratories are the primary manufacturers and users. Examples of metrology SPMs include the calibrated AFM (C-AFM)[(175)] and the molecular measuring machine (MMM)[(176)] at NIST, the probe-scan metrology SPM[(177)] and the laser interferometric SPM at the Physikalisch–Technische Bundesanstalt (PTB),[(178)] and the calibrated SPM at the Istituto di Metrologia G. Colonnetti (IMGC).[(179)] The MMM and the laser interferometric SPM are laser-based instruments. The probe-scan and IMGC instruments are capacitance based, but the capacitance sensors are calibrated off-line by laser interferometers. The C-AFM is a hybrid. Laser interferometers are integrated into the instrument for measuring displacement in *x* and *y*, and a capacitance gauge measures displacement in *z*. This capacitance gauge is calibrated against a laser interferometer installed periodically in place of the AFM sensor with a special fixture. Figures 29–32 show four of these instruments.

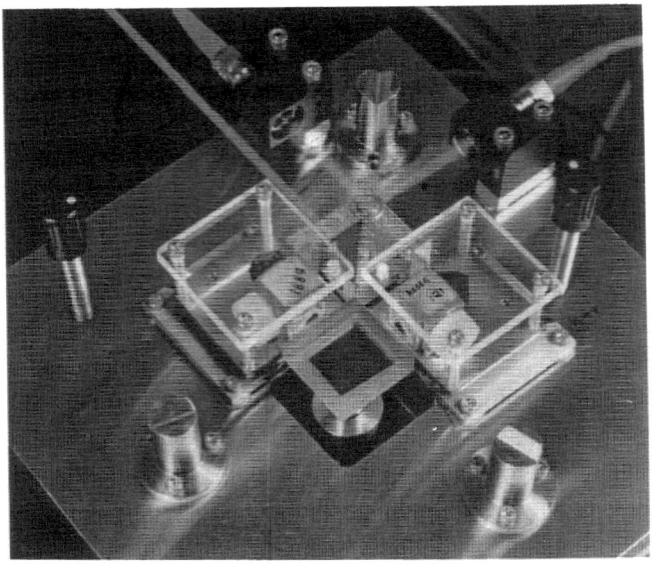

Figure 29. The assembled C-AFM but with the AFM force sensor removed, revealing interferometric optics and sample mount. Three pillars form the kinematic mount for the AFM force sensor.[177]

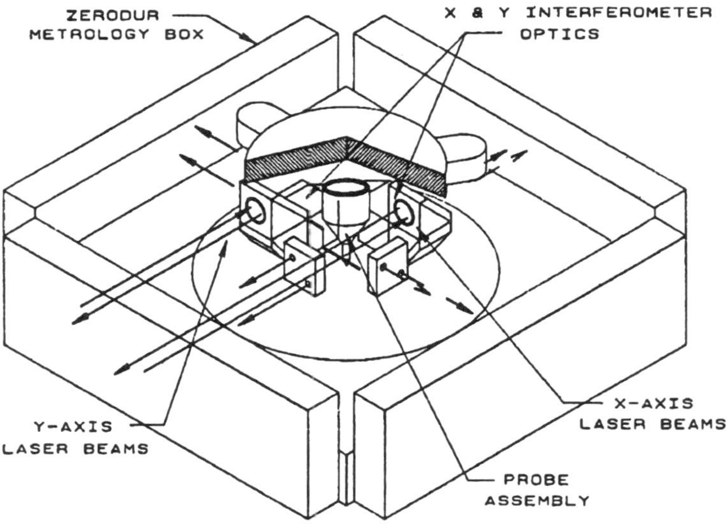

Figure 30. Schematic of the MMM showing the metrology reference system and the X and Y interferometer arrangement. The interferometer beams for X and Y displacement straddle the probe tip to minimize Abbe offset errors.[178]

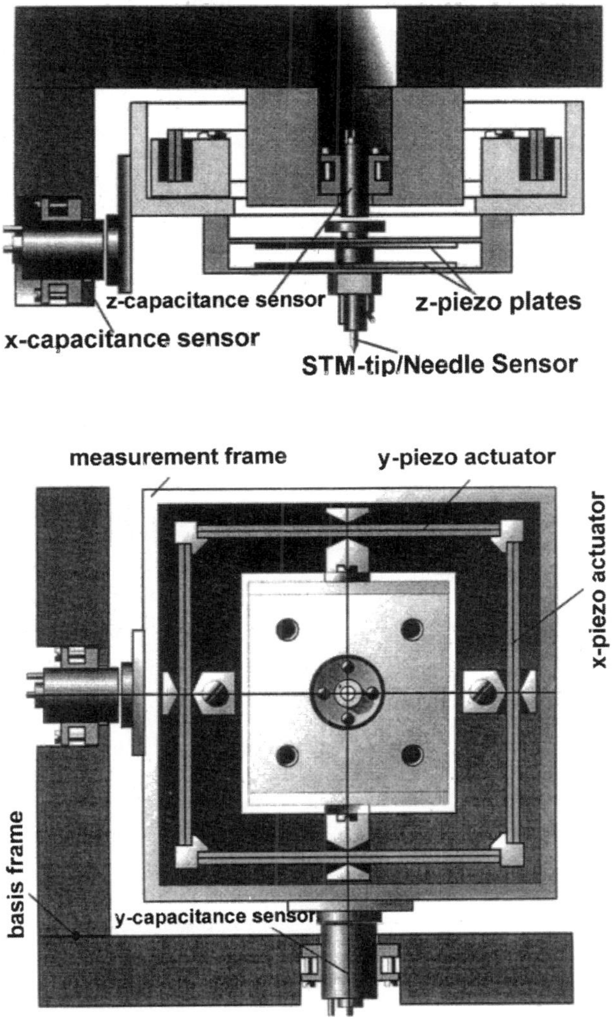

Figure 31. Probe-scan metrology SPM head, featuring capacitance gauges for measuring displacement in x, y, and z.[179]

Second, instrument makers and users have generally avoided the development of these elaborate tools in favor of more economical techniques to provide displacement calibration. Their primary considerations are low cost, compactness, ease of integration into the SPM, and ease of use. Several commercial instruments feature subsystems to correct scan motion errors in the x and y directions.[180–182] Important research and development for motion control is also being done at universities.

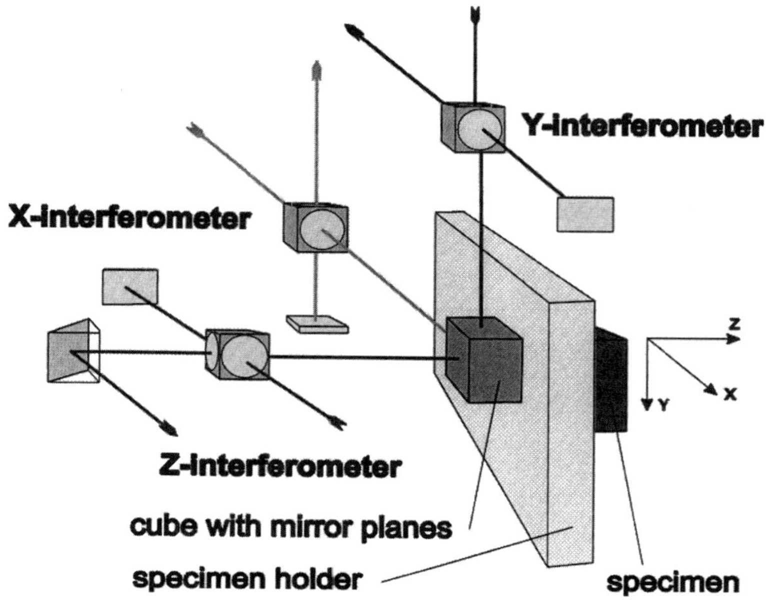

Figure 32. Schematic detail of the three-dimensional laser interferometric AFM, featuring an integrated plane-mirror laser interferometer system (Adapted from Hasche *et al.*[180]).

Holmes *et al.*,[183] for example, developed a magnetically suspended motion stage with a range of 100 μm and positioning noise of 0.2 nm.

5.2.2. Calibration Specimens for Displacement

Numerous pitch and height standards are available commercially for calibration of the *x*, *y*, and *z* axes of SPM instruments. In some cases the specimens contain calibrated pitch–height areas for calibrating all three axes at once. Figure 33 shows the waffle-like specimen,[95,184] discussed in connection with Fig. 2, that serves as a pitch–height standard. Pitch values to approximately 200 nm[185] are available commercially; the smallest available height is 7 nm.[95] As measurable features on semiconductors and data storage devices become smaller, there will be an increasing need to reduce the pitch length and the calibrated step heights to below these values. Specimens with rms periodic roughness as small as 0.5 nm[95] are also available commercially. Research is ongoing to develop atom-based step-height standards using the single-atom steps of height 0.314 nm on the (111) surface of silicon (Fig. 34).[184,186–188] An industrial standard has been written for calibration measurements using atom-based Si step-height standards.[189] Integration of a graphite, surface crystalline lattice into an instrument as a pitch standard for *x*,*y*

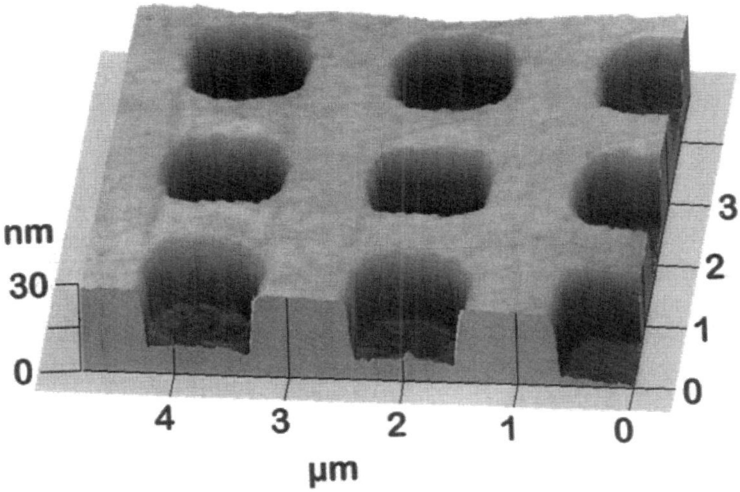

Figure 33. Topography map of a pitch and height calibration specimen as measured by the calibrated atomic force microscope (C-AFM).[96,186]

displacement has also been accomplished for STM work,[190,191] but has not yet been attempted for calibration of AFM measurements.

5.2.3. Instruments for Critical Dimensions and High Resolution

Although atomic resolution can be reached with both the AFM and STM when used on certain types of smooth surfaces, the lateral resolution for measurement of steep features depends on the overall size of the probe tip and its points of contact with the surface feature. These details determine how fine a feature can be imaged

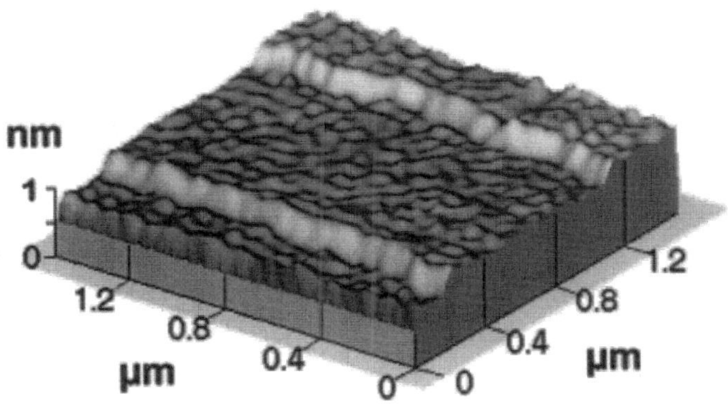

Figure 34. Single atomic steps on an oxidized (111) crystal face of Si as measured by the C-AFM.[186]

with sufficient contrast to distinguish it from noise and whether a probe can fully penetrate a hole to provide a measurement of its depth. They also determine the additive errors that accrue in measurements of critical dimensions on semiconductors. Instruments for high resolution generally require probe tips with very small radii and apex angles. Perhaps a carbon nanotube attached to the end of an ordinary probe tip[192] is the ultimate configuration (Fig. 35). A carbon nanotube has been used in research to probe the structure of trenches 400 nm wide and 800 nm deep,[192] but to our knowledge has not yet been used in manufacturing applications.

The balance-beam force microscope, developed by Griffith *et al.*,[193,194] uses sharp tips to probe the trenches and sidewalls of semiconductor structures. Another instrument that has become important to industry is the critical-dimension SPM first developed by Martin *et al.*[195] This instrument uses special probes with flared ends[195,196] (Fig. 36) and vertical and lateral force sensing to allow probing perpendicular to the surface to profile the steep sidewalls of a circuit element. These capabilities permit measuring the true linewidth provided that the width of the probe itself can be accurately determined. The measurement of critical dimension with

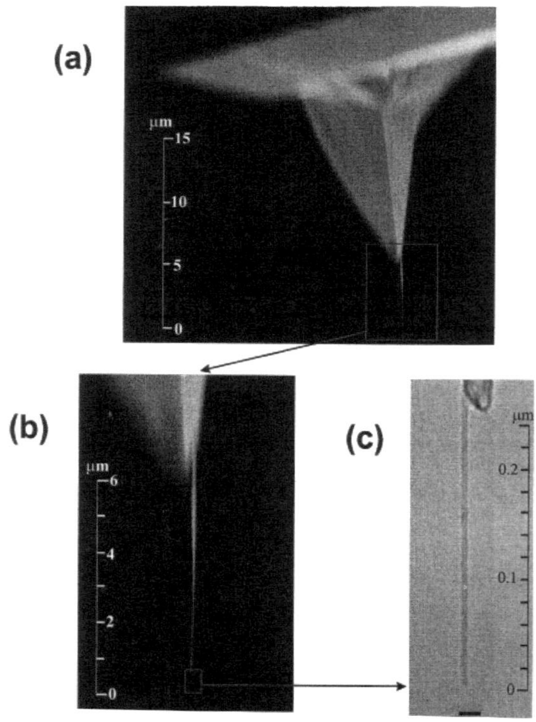

Figure 35. Carbon nanotube attached to an AFM tip on the end of an AFM cantilever(a). The nanotube in (c) is pulled down from a bundle of nanotubes shown in (b) (Adapted from Dai *et al.*[194] with permission from *Nature*, copyright 1996, Macmillan Magazines Ltd.).

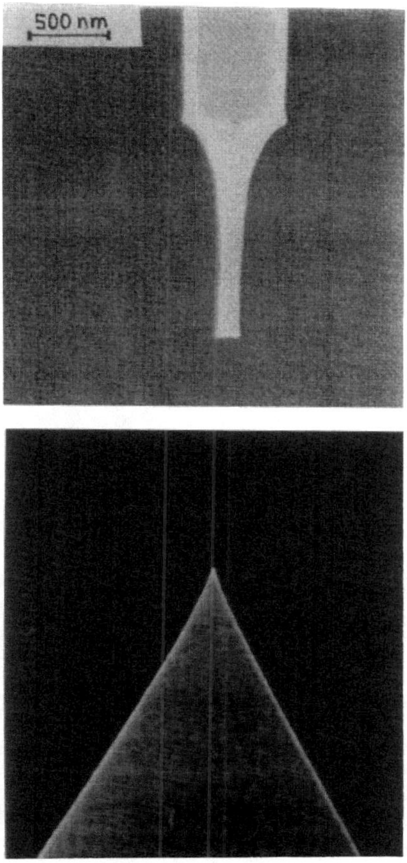

Figure 36. SEM of a high-aspect-ratio flared tip (top) and of a calibrating Si nanoedge tip characterizer (bottom). The narrow cylindrical tip is designed to profile narrow trenches with widths as small as 0.25 μm. The flared corners protrude by about 30 nm at the bottom (Adapted from Martin and Wickramasinghe,[197] copyright 1995, American Vacuum Society).

an AFM or STM is conceptually similar to the approaches developed previously for the semiconductor industry based on optical microscopy[197] or SEM.[174,198] These approaches include using procedures for modeling the specimen shape and the interactions of the probe with the specimen, and solving the inverse problem with those models as *a priori* knowledge.

The uncertainty in linewidth measurements resulting from the modeling is limited by the range of interaction of the probe with the specimen, which is about the wavelength of light for optical microscopy and the mean-free path of secondary electrons for SEM. Use of optical microscopy is no longer practical with critical dimensions of 0.35 μm or less in contemporary semiconductors. Successful SEM use will probably decrease as the design rules and critical dimensions shrink further. The range of interaction of an AFM or STM probe with the surface is subnanometer. Therefore, the accuracy of AFM or STM for measuring critical dimensions depends on having a capability for measuring the probe shape[(199)] as long as the probe and the feature are not distorted by the contact forces. It is possible to measure the probe shape by a high-resolution technique such as TEM,[(200)] but this requires removal of the probe from the instrument. Changes in the probe shape between TEM runs result in large uncertainties. Alternatively, surface standards known as *tip characterizers* may be used for *in situ* measurement of the probe during SPM operation. These are discussed next.

5.2.4. Specimens for Critical Dimensions

In the measurement of critical dimensions, the probe shape must be characterized accurately, otherwise it is folded into the width measurement. To charac-

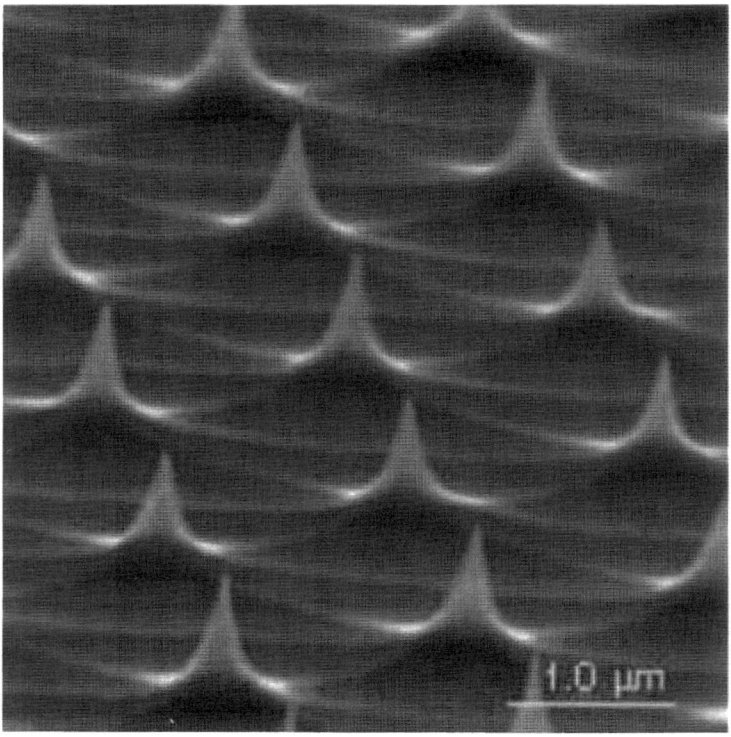

Figure 37. SEM micrograph of silicon tip array for characterizing AFM probe tips. The specified apex angle and tip radius are 20°–22° and <10 nm.[(203)]

terize the probe, we need a specimen feature whose shape is well known. When the probe is scanned over the feature, the measured profile, which is a convolution of the probe shape and the feature shape, can be unfolded to produce the shape of the probe. To accomplish this procedure with the needed accuracy, damage or distortion should be negligible to either the probe or the feature during the measurement. Also, the feature itself should be narrow or have sharp edges in comparison with the probe. The ideal specimen would resemble an array of rigid, indestructible Dirac delta functions.

Various specimens have been used to characterize the shapes of SPM tips. Recently, a promising specimen that approximates that ideal has been developed.[(201)] An SEM micrograph of the specimen surface is shown in Fig. 37. It contains a two-dimensional array of small spikes fabricated by etching a Si wafer. The spikes are widely spaced so that an individual spike can be measured at a time and the background level of the surrounding valley can be accurately sensed by the probe. If the individual spikes are sufficiently stable under ordinary conditions of SPM measurement, then a three-dimensional map over one of these spikes can be used to determine the topography of the scanned probe tip. The development of these spike arrays for probe characterization is recent, so it is not yet clear whether they are strong and stable enough to allow accurate measurements of SPM probes.

Another type of probe characterizer, known as the Si nanoedge, was developed by Bayer *et al.* (as discussed by Martin and Wickramasinghe[195]). This specimen consists of a row of sharp knife edges (bottom of Fig. 36) with nominal tip radii of 5 nm. Because of its two-dimensional character, the nanoedge is likely to be stronger and more stable than the spike array. However, the two-dimensional structure can only be used to characterize the probe in one direction. It cannot be used to characterize the full three-dimensional structure of the probe with a single map of the specimen. In addition, the apex angle is about 25°. Therefore, highly sloped areas on the shank of the probe tip cannot be profiled with this specimen. Nevertheless, Martin and Wickramasinghe[(195)] estimated that the specimen allows for measurements of width to an uncertainty of approximately 3 nm.

5.3. Other Types of Scanned Probe Microscopes

AFMs and STMs are currently the primary SPM instruments[(202)] manufactured and used. However, an amazing number of instruments have been developed with different sensing techniques that permit detection of a wide range of surface properties in addition to topography. Several of these have seen significant usage in industry. These are briefly summarized in Table 1, and their actual or potential industrial uses are listed. The following sections describe a few highlights. The techniques are classified by the types of quantities that are directly measured.

Table 1. Types of Scanned Probe Microscopes

Class of technique	Name[a]	Type of probe–surface interaction (if not clear from the name)	Typical applications
Force based (i.e., causes a change in cantilever deflection or oscillation)	Atomic force (AFM)	Normal mechanical contact or long-range noncontact interaction	Semiconductors, optics, data storage
	Frictional force (FFM)	Lateral contact	Polymers, contrast enhancement on surfaces with low AFM contrast
	Magnetic force (MFM)		Magnetic storage, materials science
	Electrostatic force (EFM)		Locally induced charges on insulators, triboelectrification, domain structures of ferroelectrics
	Kelvin probe force (KPFM)	Senses work function variations	Dopant profiles, localized charge defects, band bending
	Eddy current	Force between induced eddy current and ferromagnetic element	Soft magnetic materials
	Near-field acoustic (SNAM)		Surface roughness profiling
	Tunneling acoustic (TAM)	Detects probe-induced acoustic waves in surfaces	Subsurface defects
Optical	Near-field optical (SNOM or NSOM)		Biology, spectroscopy in materials science, semiconductors
Thermal	Thermal profiler (SThP)		Thermography in microelectronics
Electrical	Tunneling (STM)	Electronic tunneling	Semiconductors, optics
	Capacitance (SCM)		Lateral dopant profiling for semiconductors
	Electrochemical (SECM)	Electrolytic current	Noncontact profiling, chemistry of surfaces
	Nanoscilloscope	RF field coupling	*In situ* testing of semiconductors

[a]For most entries, the words "microscope" or "scanning" or both under the name are implied by the abbreviations M or S, respectively.

5.3.1. Force-Based Methods (Mechanical)

With conventional AFM repulsive or attractive contact forces normal to the surface of the sample are measured by detecting the induced bending of the cantilever or induced changes in its oscillatory motion. However, many other types of tip–surface interactions cause mechanical deflections or oscillations as well. For example, frictional (lateral) forces between the tip and surface induce torsional-mode deflections of the cantilever during scanning. In most cases the detection is accomplished by using the beam-deflection method, with which it is possible to detect normal and lateral forces simultaneously.[203] Theoretical studies have shown that the friction coefficient depends strongly on the electrical potential between the two materials in contact and on the intrinsic spring constant of the cantilever. Also, friction occurs only in a certain range of spring constants.[204] Therefore, lateral-force measurement currently serves as a qualitative technique for surface characterization.

Other force microscopes are based on interactions that arise from an additional force at the tip, sensed as additional bending or a resonance frequency shift of the cantilever. However, in the contact mode, the ion–ion ("atomic") force dominates other forces that in most cases cannot be separately detected; these other interactions can be sensed only in the noncontact mode. Noncontact sensing of long-range forces is best carried out in a vibrating mode because quasi-static cantilever deflections in most cases are beyond detection limits. Typical oscillation amplitudes are 1 to 10 nm. However, alternating current (AC) methods are sensitive to force gradients rather than to the interaction forces themselves, and a correct interpretation of the resulting image requires separation of the topographic and interaction signals.

Perhaps the magnetic force microscope (MFM) is the most important of the force-based methods. If a magnetic tip approaches a magnetic sample within a distance of typically 10 to 500 nm, the magnetic interaction between the tip and the stray magnetic field of the sample becomes measurable, and the resulting cantilever bending can be monitored. In most cases an AC sensing mode is used, and the magnetic force gradients rather than force magnitudes are measured.[205,206]

Because the interaction forces can be attractive or repulsive, problems with feedback loop stability may occur and impede the interpretation of the deflection signal. Therefore, an additional attractive force is required that can serve as a topographic signal. The van der Waals force, which is always present, serves this purpose. Also, an additional controllable attractive electrostatic force due to a bias voltage between tip and sample is often used. The Coulomb interaction signal is equivalent to a topographic signal and, thus, can serve as a distance control.[207] In most cases the separation of topographic and magnetic images is crucial for their interpretation. Sometimes the variation of the strength of the magnetic interaction can be measured, and a differential measurement enables the complete separation of topographic and magnetic contrast.[208] An example of the information concerning surface magnetic structures that MFM provides is shown in Fig. 38.

To maximize the lateral resolution, the effective magnetic tip volume and the stray magnetic field have to be as small as possible, and the effective total magnetic moment must be as large as possible. In addition, the coercivity of the tip should be large to reduce the influence of the sample stray field on the tip, and the magnetic behavior of the tip should be well defined. These requirements cannot be fulfilled simultaneously. There is a trade-off between achieving a small effective magnetic tip volume and a high magnetic moment (i.e., force sensitivity). The tip radius must be at least 6 nm for a minimum detectable force of approximately 10^{-12} N at room temperature. The theoretical lateral-resolution limit for MFM operation in air is about 5 to 10 nm; the typical resolution is about 50 to 100 nm.[207] For quantitative investigations the MFM tips may be characterized by calibration measurements on well-defined magnetic test structures, such as magnetic bits written on a recording medium.[209]

The eddy current microscope (ECM)[210] responds to an induced mechanical force. An oscillating ferromagnetic tip, such as those for MFM, induces eddy currents in a conducting material, which responds by producing an electrodynamic interaction between the tip and sample (Lorentz force). The forces resulting from the induced currents are much smaller than in MFM measurements, but they are detectable with elaborate AC methods. Alternatively, eddy currents are induced in nonmagnetic but conducting tips that oscillate above a magnetic sample. Again, a damping process occurs due to the resulting Lorentz force. Because the interaction

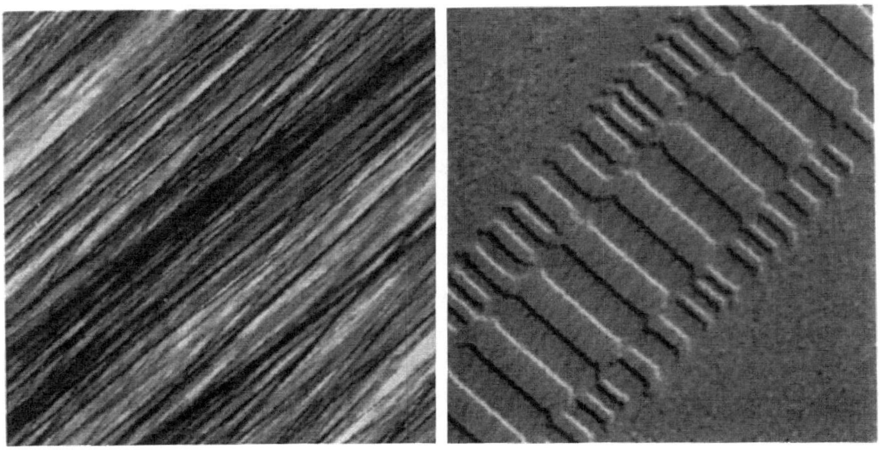

Figure 38. Force microscope images of overwritten tracks on a textured hard disk. Topography (left), magnetic image (right). The field of view is about 25 μm × 25 μm (Adapted from Babcock[207] with permission of Digital Instruments).

forces are very weak, the retroreaction on the magnetization structure of the sample is negligible.[210]

The scanning near-field acoustic microscope (SNAM) uses a high-Q-factor oscillator, such as a 32-kHz quartz tuning fork, driven at resonance. An edge of one leg of the tuning fork can serve as a probe tip, or an AFM tip may be glued to the edge. If this tip approaches the sample, the resonance frequency and the amplitude begin to decrease at distances typically smaller than 200 μm.[173] These decreases are due to hydrodynamic damping forces in the air gap; the forces strongly depend on the gas pressure and are absent in vacuum. The competing van der Waals forces are weaker at these separations. A lateral resolution of about 1 μm and a vertical resolution of about 10 nm have been demonstrated. Other force-based techniques include the electrostatic force microscope (EFM),[211–213] the Kelvin probe force microscope (KPFM),[214] the tunneling acoustic microscope,[215] and a technique relying on low-temperature magnetic resonance force detection.[216]

5.3.2. *Methods Based on Other Probe–Sample Interactions*

Many types of SPMs use probe–sample interactions that are not sensed as forces on the cantilever. Some of them use these interactions mainly to image the topography of a sample. Others use the interactions to monitor properties of the surface that are not related to topography. In the latter cases the topographic data should generally be measured separately to distinguish completely the topographic features and the nontopographic properties. We have classified these as optical, thermal, and electrical methods, and describe them further in the sequel.

5.3.2a. Optical Probe Microscopies. The spatial resolution achieved in classical optical microscopy is limited by diffraction to about $\lambda/2$. Classical optics is concerned with the far-field regime where only propagating waves with low spatial frequencies ($< 2/\lambda$) survive, whereas evanescent waves with high spatial frequencies ($> 2/\lambda$) are confined to subwavelength distances from the object interacting with the electromagnetic wave.[217] Near-field optical microscopies overcome this limitation by using optical probes with subwavelength dimensions positioned at subwavelength distances from the surface to detect the evanescent field with subwavelength spatial resolution.

For example, with the scanning near-field optical microscope (SNOM), the optical probes are formed by a sharpened glass fiber tip coated with a thin layer of metal, and a submicrometer aperture at the apex of the tip.[218] These fibers can be used as light emitters or detectors. In the first case, light is guided to the tip via the glass fiber. Because of the cutoff condition, only a small fraction of the intensity reaches the aperture. In the detection mode the aperture probes the evanescent waves of the surface under investigation. The light from the sample is guided to a photodetector via the fiber.

The tip–surface distance has to be kept constant by using long-range force interactions. Shear force ("dithering") methods[219] are often used for this purpose. SNOM can be performed in reflection and transmission and can be combined with other techniques in classical optical microscopy, such as spectroscopy and luminescence.

Attempts are being made to use metal-coated AFM cantilevers as SNOM tips.[220] In this case the tip–surface distance is controlled in the usual AFM way, either in the contact or noncontact modes. An alternative way to combine AFM and SNOM is to bend the tip of an optical fiber so that it serves both as an AFM cantilever and SNOM detector.[221]

The lateral resolution of the scanning near-field approach has been improved further in the scanning interferometric apertureless microscope (SIAM).[222] Here a tip is vibrated in the evanescent field emerging from a sample that is irradiated from the back by a laser. The backscattered laser light is detected with an interferometer. The smallest resolvable features are approximately 1 nm. Related near-field optical methods include the photon scanning tunneling microscope (PSTM)[223] and the scanning plasmon near-field microscope (SPNM).[224]

5.3.2b. Thermal Microscopies. The scanning thermal profiler (SThP) is a noncontacting, near-field probe with a miniaturized thermocouple[225] or resistance thermometer[226] sensor at the end of a sharp tip. The thermal transport between tip and sample in ambient air is due to conventional heat conduction for distances larger than the mean-free path of the air molecules (~60 nm). For other conditions, coupling to phonon or plasmon fields govern the process.

For the thermoelectric potential microscope, also known as the tunneling thermometer, an STM tip and a sample form a "tunneling thermocouple" that probes the product of the local gradient of the chemical potential times the temperature gradient. In the scanning optical absorption microscope (SOAM), which is based on this approach, the tunnel junction is heated by laser radiation to produce the necessary temperature gradient.[227] Alternatively, for the scanning chemical potential microscope (SCPM), the sample is heated from the rear to produce the necessary temperature gradient.[228]

5.3.3c. Electronic Microscopies. Besides the scanning tunneling microscope, many types of scanned probe microscopes rely on sensing of electrical quantities. With the scanning capacitance microscope (SCAM),[229] the tip–sample configuration serves as a miniature capacitor, and AC methods are used for the capacitance measurement. In recent developments the AFM and SCAM are combined by using AFM probes and AC detection.[230]

Other microscopes operate in electrolytic solutions. The scanning electrochemical microscope (SECM),[231] for example, uses an insulated conducting tip as an electrode probe. When this tip is brought close to a conducting or semiconducting sample in an electrolytic solution, the redox process between tip and sample yields topographic and "chemical images." For the scanning ion conductance microscope (SICM),[232] an electrolyte-filled micropipette is used as the probe in the electrolytic solution. The ion flow through the opening of the micropipette is blocked by the sample surface. The ion current is therefore a measure of the distance (topography). With this technique microsources (pores) can be detected.

For the nanoscilloscope[233] an AFM cantilever contains an integrated RF waveguide and coaxial tip. The RF is fed into integrated circuits under test. This instrument can be used *in situ* to measure the topography (via AFM) and the RF response simultaneously of electronic chips.

5.4. Applications of SPM Measurements

Scanned probe microscopes have been used in industry in a great variety of ways.[234–237] Table 2 is a summary indicating the various types of applications and the types of technical areas or industries that are applying STM and AFM techniques. The levels of application include applied research studies with potential industrial usage, studies by industry in research, manufacturing off-line usage, and manufacturing on-line usage.

5.4.1. Data Storage Industries

Data storage industries are probably the most advanced in applying the techniques of AFM and STM to product inspection. These industries manufacture

Table 2. Assessment of Industrial Applications of STM and AFM by Industrial Area and Level of Usage

Industry	Applied research	Industrial research	Manufacturing off-line	Manufacturing on-line
Data storage	X	X	X	X
Microelectronics	X	X	X	X
Polymers and coatings	X	X	?[a]	?
Optical elements	X	X	?	?
Mechanical parts	X	X	?	?
Electrochemistry	X	X	?	?
Biology and biomedicine	X	?	?	?

[a]The question marks indicate that we are not aware of an application in that area.

devices that write and sense surface features at the submicrometer level and require tolerances near the nanometer level. The important storage technologies are magnetic disk storage and optical disk storage; for magnetic data storage technology the AFM has been widely used.

Magnetic disk storage currently[205] involves disk media with bit densities of about 1.5 per μm^2. Recording heads scan the disks on a cushion of air with a separation of about 50 nm to read and write the information. AFM is used on-line in manufacturing to assess the slider roughness, which must be smooth enough to allow for a smooth flight over the disk, but rough enough to minimize sliding friction upon take-off and landing. In addition, the recession of the pole tip from the slider surface of the device is an important specification that needs to be measured to a high accuracy. Optical interferometry has been the traditional technique used to measure pole-tip recession (PTR). However, because of their limited lateral resolution, the use of optical devices becomes problematical as the dimensions of the read and write heads become smaller. Alternatively, because of its superior lateral resolution, AFM is being developed to perform in-line PTR measurement in manufacturing.[238] Magnetic force sensing for AFM (Sec. 5.3) is another critical measurement technology for the magnetic disk industry.

Atomic force microscopes are currently being introduced into the manufacturing line to measure laser-textured bumps on the head-landing zones of magnetic disks.[93] To accomplish this in a manufacturing environment, the speed, ease of use, and probe-tip lifetime of the production–inspection AFMs were significantly enhanced without compromising the profiling accuracy.

Optical compact disks contain a pattern of tiny valleys in a plastic disk. The medium is read from the back side by an infrared laser sensor as the spinning disk travels above the sensor at high speed. The spacing of these features is about 1.5 μm for compact disks and even smaller on high-density digital video disks. The quality of the pits is of great importance in optical disk technology, and AFM has been a key technique for assessing such properties as the variation in pit depth on a disk or the presence of ridges on the edges of the pits, which seem to cause visual blemishes.[239]

5.4.2. *Microelectronics Industries*

Because of its high lateral resolution, the AFM has been used extensively for performing dimensional measurements in the semiconductor industry where the sizes of the circuit elements are becoming smaller.[240] The measurements[241–248] include surface roughness, linewidths, via depths, step heights, edge roughness, edge slopes, and pitch dimensions between features. In some cases the AFMs are located in the clean areas of the semiconductor plants (also called fabs). Factors that moderate the continued integration of AFMs into the fabs are the need for a skilled operator and the slow speed at which the instrument operates to provide the needed lateral resolution.[247] The AFMs for semiconductor fabs must also be compatible with a clean-room environment. This requirement means AFMs cannot be a significant source of particle contamination and must be insensitive to the high-acoustical-vibration levels in fab facilities.[247]

5.4.3. *Polymers and Coatings Industries*

Great diversity in the types of products and processes can be addressed by SPM techniques in the chemical industry.[236,237,249–253] Some examples are the evaluation of automotive surface finishes, particle adhesion during printing, and analysis of component mixing during injection-molding and film-forming processes. Key performance criteria for coatings and polymer additives are adhesion, wettability, friction, wear, stiffness, scratch damage, and appearance. Studies of polyethylene, polyimide, latex emulsions, and other commercially important materials have been reported.

5.4.4. *Optical Element Industries*

In research and development AFM and STM have been used extensively to measure the surface roughness of smooth optical surfaces and optical gratings.[254–261] Surface roughness and defects on optical elements produce scattering and stray light in optical systems, with a resulting loss of contrast and intensity in the optical images and measurements. The AFM has been used extensively in research to study the roughness and defects of optical surfaces at nearly atomic levels of lateral resolution. For electrically conducting optical surfaces the STM may be used as well. For visible and infrared optical elements, AFM or STM measurements are not critical for assessing the optical surface quality because the finest surface imperfections are considerably smaller than a wavelength of visible or infrared light and thus cause very little diffraction or degradation in the performance of the optical element. However, tiny imperfections revealed by the AFM can yield information concerning the quality of the surface fabrication process and how to improve it.

For UV optical elements AFM measurements of the smallest topographical features on the surface become more important. Extreme ultraviolet (EUV) lithography at a wavelength of 13.6 nm is a strong candidate technology for producing future generations of semiconductor chips with critical dimensions smaller than about 100 nm. Windt *et al.*[254] formulated surface finish requirements for EUV optical lithography elements in terms of AFM measurements. Table 3, taken from their work, shows a comparison of rms roughness measurements by AFM over a spatial wavelength range between 5 and 1000 nm with the X-ray reflectance measured for a set of four mirror prototypes. Noting that the percent reflectance decreases as the rms roughness increases, Windt *et al.*[254] established 0.1-nm rms as a surface finish requirement for prototype EUV lithography optics operating at

Table 3. Surface Roughness (rms) Calculated from AFM Measurements of MoSi Multilayer Samples Compared with Measured Specular Reflectance at a Soft X-ray Wavelength of 14 nm

Sample	Rms roughness (nm)[a]	Soft X-ray reflectance
A	0.11	62.5%
B	0.10	60.8%
C	0.29	56.0%

[a]Calculated from power spectral density data over a 5- to 1000-nm band of spatial wavelengths. These results were taken from Windt *et al.*[(256)]

a wavelength of 13.6 nm. These research results clearly show the need for AFM in this potentially important industrial application.

In other work, Spiller *et al.*[(255)] described the quality of the reflecting surfaces of X-ray telescopes in terms of the power spectral density of the surface profile. For assessment of roughness at surface spatial wavelengths smaller than 0.5 μm, an STM was used as the measuring instrument. Other research applications include the measurement by AFM of optical fiber quality.[(256)]

Bennett *et al.*[(34)] reviewed the field of atomic force microscopy for studying optical surfaces and performed AFM measurements of a wide range of optical materials with three types of AFM instruments, as well as with other types of profiling instruments. These and other studies show that extremely smooth surfaces are produced with current industrial capabilities. For example, Bennett *et al.*[(34)] measured an rms roughness of 0.069 nm on a polished Zerodur[*] surface (Fig. 39) over an area of 50 μm × 50 μm. This is an example of the high quality of optical surfaces that can be made and the high quality of the instruments now used to verify surface quality. However, an important concern is the variability in these roughness measurements due to differences in the probe tips, instrument feedback, and electronic filters. Standardization of these properties is needed to produce consistent results for comparison with surface specifications.

5.4.5. Mechanical Parts Industries and Materials Science

The term "mechanical parts" is used to represent the automotive, aerospace, and related industries where important surface functions are sliding components and the tribological interactions between components. One interesting set of

*Certain commercial equipment or materials are identified in this paper to specify adequately an experimental procedure. Such identification does not imply recommendation or endorsement by NIST, nor does it imply that the equipment or materials are necessarily the best available for the purpose.

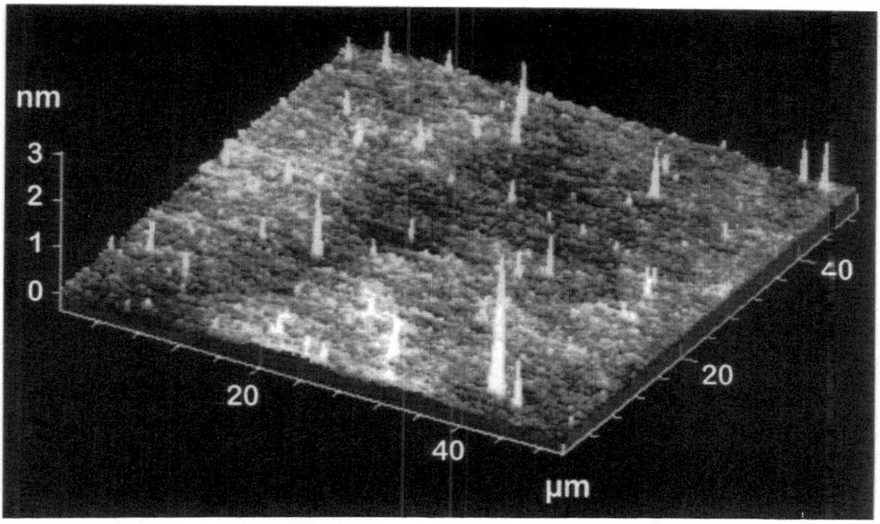

Figure 39. AFM image of a bare Zerodur surface with particles as high as 3 nm. The surface roughness including particles is 0.069 nm rms, which is close to the noise level of the instrument (Adapted from Bennett *et al.*[34]).

applications is the use of AFM by Lucca *et al.*[262] and Brinksmeier *et al.*[263] to measure the edge quality of tools used for machining. Figure 40 shows an AFM map of a diamond tool edge measured by Brinksmeier *et al.* The radius of curvature of the edge was estimated as between 40 and 50 nm after correction for the AFM tip radius. The AFM has also been used to study the fine structure of machined and polished surfaces.[263–266] In addition, STM and AFM have been used extensively in materials and mechanical engineering research to study the basic structures of industrially important materials and the tribological interactions between them.[264–273]

5.4.6. *Electrochemical Science*

The ability of STM and AFM to provide surface images with atomic resolution while functioning in solution makes them ideal techniques for *in situ* electrochemistry studies, and a wealth of interesting surface science research[274–280] has been performed. For example, Soriaga *et al.*[280] used STM to observe the structure of a Pd(111) surface before and after a catalyzed dissolution. The top section of Fig. 41 shows a section of the surface near a scratch (marker) that enables relocation of the area in the STM. The middle section shows the same area after 10 min of an electrochemical dissolution process in H_2SO_4, and the bottom section shows the same area after 20 min of dissolution. The figures reveal the progressive formation of wide-area smooth terraces as the surface dissolution progresses along the

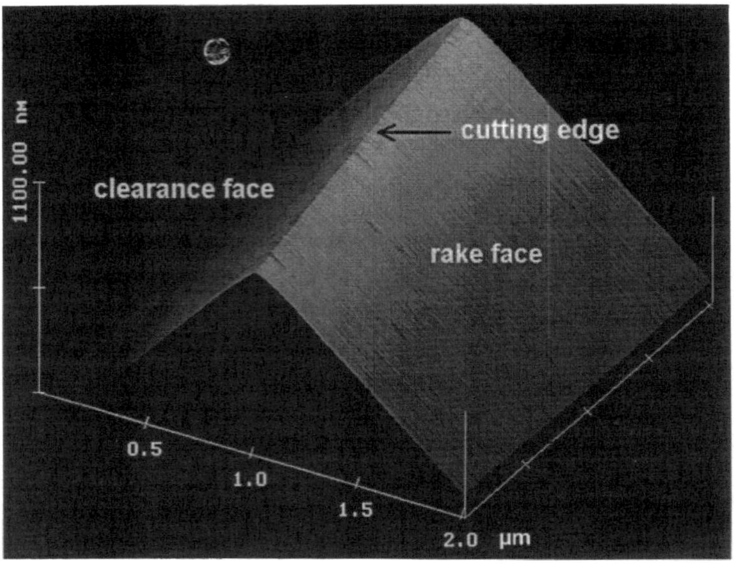

Figure 40. AFM image of a diamond tool edge.[265]

monatomic and diatomic steps. In addition, rapid etching at the scratch edge is also apparent.

5.5. Future Directions: Techniques and Instrumentation

As SPMs become more widely accepted as industrial tools, based largely on the applications reviewed in Sec. 5.4, it is important to consider how these applications will affect their further development. In this section we consider developments that are likely to play a critical role in new applications of the proximal probe concept.

Consideration of these future applications requires considering SPM not just as a purely imaging device, i.e., a microscope, but rather as an integrated set of active components that may be embedded in positioning sensors for nanoscale fabrication, active displacement controllers, and *in situ* process monitors for advanced manufacturing. This approach will enable SPMs to be used routinely in prototyping functional, critically dimensioned structures for nanoelectronics, and other applications. In addition, SPM-based components are being considered for use as fast and sensitive biomedical, environmental, and industrial sensors.

At the foundation of these advances are microfabrication techniques that provide a route to making SPM-based sensors more rugged, reliable, and compact and that provide opportunities for integrating several types of measurement sensors onto the SPM tip. Efforts are underway to provide electrical and optical waveguides

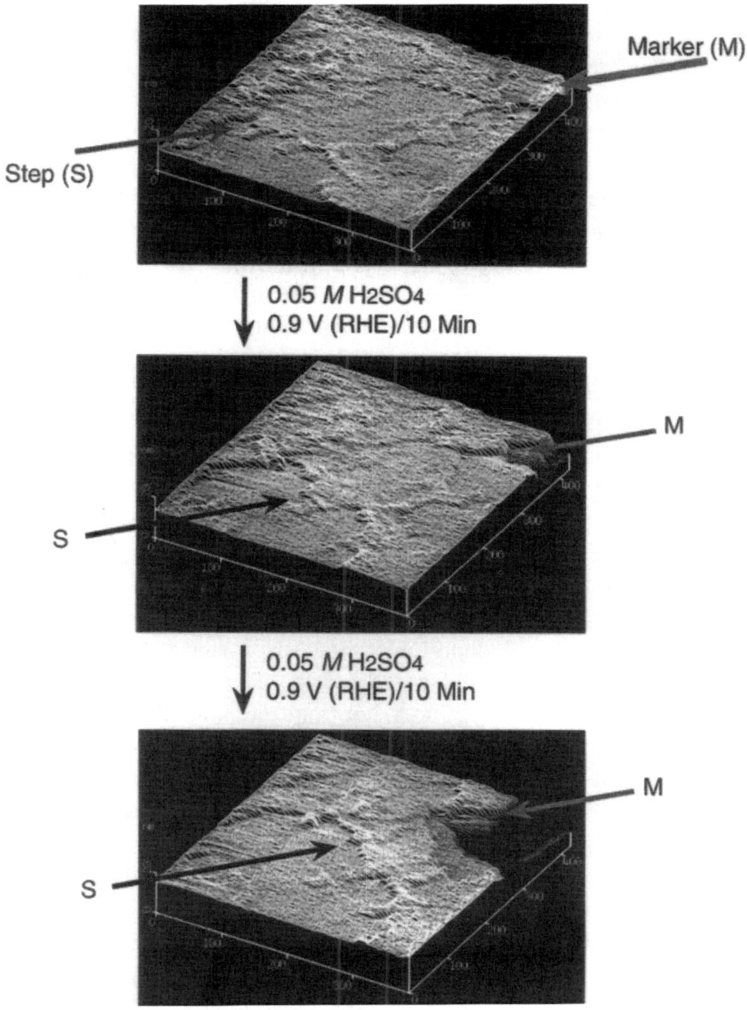

Figure 41. Scanning tunneling microscopy images of a facet on a single-crystal Pd bead at various stages of adsorbate-catalyzed dissolution. M: marker scratch; S: step (reprinted from Soriaga *et al.*[282] with kind permission of Elsevier Science-NL, Sara Burgerhartstraat 25, 1055 KV Amsterdam The Netherlands).

as well as shielded measurement circuits on the force-sensing unit. Because SPMs are capable of not only revealing nanoscale topographical features but also of acting as force sensors, measurement devices are being implemented for different sensors as outlined in Sec. 5.3.

Nanofabrication is one area where the SPM will play a role in the characterization of surface topography. SPM is widely perceived as an enabling technology for true control of nanoscale materials and the preparation of products with critical functional-device features on the nanometer-scale size. A complete set of tools that would routinely allow researchers to prepare prototype devices and test them in a range of microenvironments is the ultimate goal of these efforts.[281,282] Microenvironments include using a simple gas phase, self-assembling materials, electrolytes, biomaterials, and even living cells.

There have been several substantially different approaches to the fabrication of electronic and data storage devices of nanometric dimensions.[283–285] One example is SPM oxidation. The basic idea behind the SPM oxidation technique is straightforward: a conducting SPM tip, held in contact with or within a nanometer of an electrically biased, stable, homogeneous substrate in the presence of O_2 or H_2O, induces a highly localized oxidation of the substrate material.[286] Reactive oxygen species produced by the field near the tip–surface junction are swept into a field-defined reaction zone. Oxide linewidths of 10 to 20 nm are produced by using writing voltages of 3 to 10 V at speeds of up to 1 mm/s on several types of metallic, semiconductor, and insulating substrates. Because the field-enhanced oxide is typically more stable than the native oxide, it may be used as a nanoscale etch mask, as a template, or to form electrically isolated regions. Integration of SPM-based methods into practical device-processing schemes has resulted in the successful demonstration of functional, field-effect, and single-electron tunneling transistors.[287,288] SPM oxidation is illustrated in Fig. 42, as presented by Snow and Campbell.[287] This example illustrates not only the potential for using SPM in nanoelectronics technology to provide fine-scale patterning (e-beam lithography can produce sub-10-nm linewidths too), but also its unique potential to provide device engineers with simultaneous information about properties of the nanostructures, including topography, as they develop during fabrication. For example, if a 10-nm-thick oxide tunnel barrier is to be formed, then interfacial dimensions such as the average barrier thickness and the topography of each interface must be controlled to less than 1 nm to ensure satisfactory electrical properties. At present, SPM is the only characterization technique with sufficient spatial resolution and sensitivity to provide this information. The integration of SPM oxidation and electrical characterization techniques is a logical first step toward deducing predictive physical models of the oxidation mechanism; such models are needed to optimize SPM oxidation parameters. A model of this type would need to account for the role of breakdown voltage during the initial stages of oxidation, the electrochemical formation of oxyanions in the ambient surrounding the tip–sample junction, and the nonequilibrium distribution of holes set up by the highly nonuniform electric field at the evolving oxide–silicon interface. To return to the example of a 10-nm-thick oxide tunnel barrier, it must be possible, within the established SPM-based nanoelectronics methodology, to confirm experimentally the three-dimensional electrical characteristics of a 1-nm interfacial region of such a nanostructure.

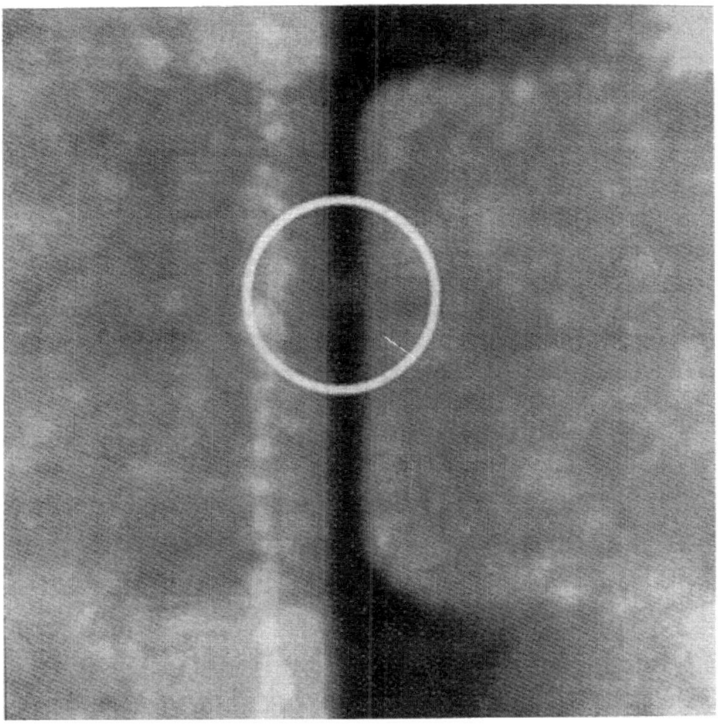

Figure 42. 900 nm × 900 nm AFM image of a 30-nm constriction in a Ti film manufactured by scanning the film with an electrically biased AFM. The dark vertical wire region is the 30 nm, electrically conducting Ti film; the lighter regions, left and right, are insulating oxide regions manufactured by scanning with the biased AFM; and the light bridge (circled) is a controlled resistance in the wire manufactured by this same approach, but also by slowly increasing the bias voltage (reprinted with permission from Snow and Campbell,[289] copyright 1995, American Association for the Advancement of Science).

As industrial applications begin to drive the development of SPM-based systems and components, an increasing divergence between dedicated production-level systems and general-purpose R&D level instrumentation will occur. Requirements for the former include throughput, reliability, and reproducibility, whereas multiple-mode imaging and measurement capabilities are valuable features for the latter. Since the first NIST Workshop in 1994 and subsequent workshops on industrial applications of scanned probe microscopy,[234–237] the response of SPM suppliers to issues such as large sample handling, automation, fast scan capabilities, closed-loop scanning, and pattern recognition serves as a signal that these features are in increasing demand on production-level systems. At the same time general-purpose SPM systems now typically include modes for sampling magnetic, electrical, optical, and mechanical properties in addition to topography.

Perhaps the most clear-cut example of an SPM-based manufacturing application is critical-dimension measurements in the semiconductor industry, as discussed in Sec. 5.3. The industry consensus that has emerged is that SPMs appear more likely to support the calibration of dedicated scanning electron microscopes than to perform routine full-wafer inspection.

6. Intercomparisons

Salient questions arising from the preceding discussions are: When measurements made with two types of profiling instruments on the same surface are compared, are the same results obtained? If not, why not? Church *et al.*[(22)] and McWaid *et al.*[(94)] are among those who have addressed these questions.

Church *et al.*[(22)] performed a careful intercomparison of the phase-shifting interferometric (PSI) profiler and a high-resolution stylus instrument by measuring profiles of diamond-turned Ge and Si with both instruments. The results with Si are particularly noteworthy because the diamond-turning process left a highly periodic surface with fundamental and harmonic components. Figure 43 shows the optical and stylus profiles of the Si surface. The stylus profile reveals a clear harmonic structure that is not as well resolved in the optical profile. The possibility that this harmonic structure might result from bouncing or vibration in the stylus instrument was checked by measuring the same profiles at widely varying stylus speeds. The difference in the lateral resolution of the two instruments of Fig. 43 is likely due to the optical point-spread function of the interferometer, to averaging of signals from neighboring detector elements of the interferometer, and to the use of a very fine stylus with a tip width of approximately 0.5 μm in the stylus instrument. The difference is apparent again in the PSD functions derived from the profiles and plotted in Fig. 44. The first harmonic of the periodic structure of the optical PSD, located at a spatial frequency of approximately 0.6 μm^{-1}, is significantly smaller than the fundamental component, whereas for the stylus PSD the first harmonic is slightly larger than the fundamental. However, the stylus PSD contains a surprising result that exposes a drawback of the instrument. The periodic components for this instrument are significantly broader than those of the optical PSD. The increased width was caused no doubt by velocity fluctuations in the stylus drive mechanism that led to frequency modulation and broadening of the discrete frequency components in the measured profiles of the stylus instrument. This problem was avoided in the optical instrument since lateral positions in the

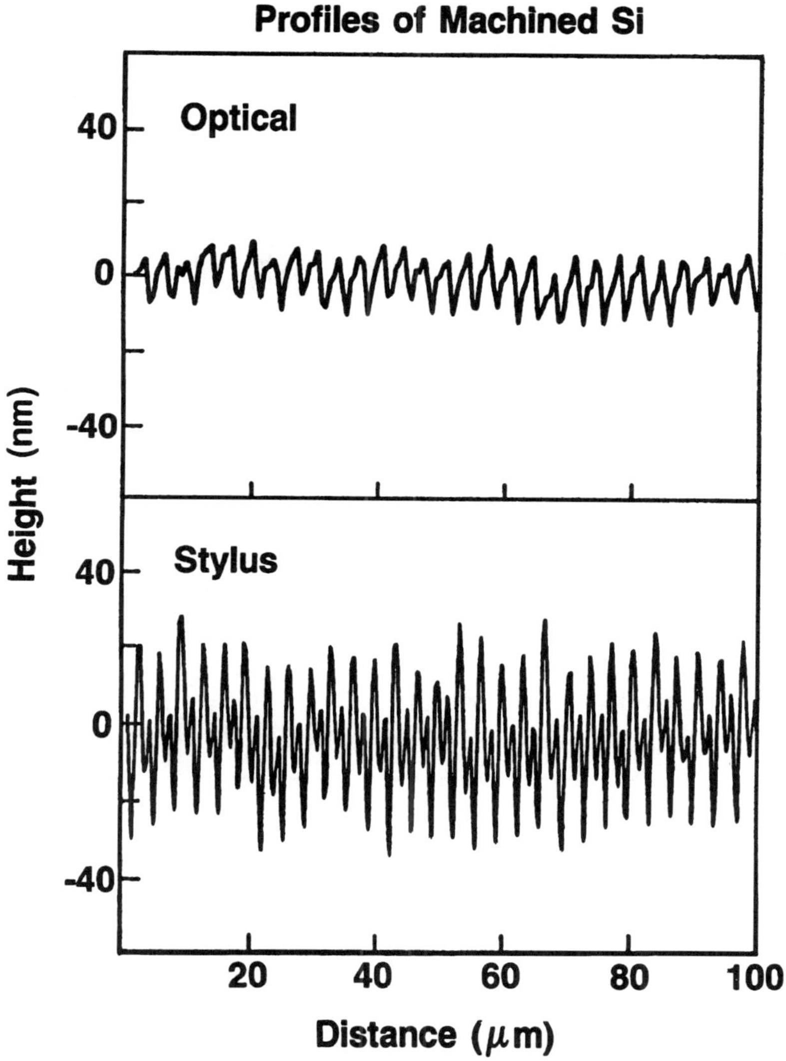

Figure 43. Optical: profile of a diamond turned Si surface measured with a phase-shifting interferometric profiler. Stylus: profile of the diamond-turned Si surface measured with a high-resolution stylus instrument (Redrawn from E. L. Church *et al.*[22]).

measured profile depend directly on the static positions of the elements in the linear detector array, and these were equally spaced with a small uncertainty.

McWaid *et al.*[94] compared STM, stylus, and phase-shifting interferometry on several different regions of a Si pitch and height specimen coated with Pt. In

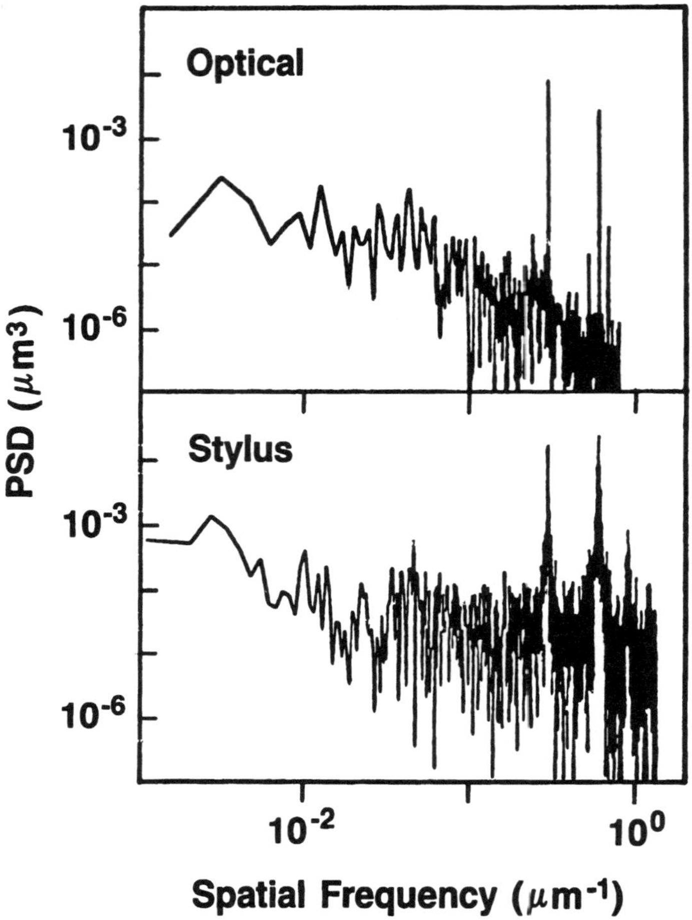

Figure 44. Optical: PSD for the optical profile of Fig. 43. Stylus: PSD for the stylus profile of Fig. 43 (Redrawn from E. L. Church *et al.*[22]).

particular, the same identical defect was profiled by STM and a PSI microscope. These results are shown in Figs. 45 and 46. Figure 45a shows the STM image of the defect, which is located near a step edge. This image was then digitally smoothed to model the smoothing effect of the point-spread function expected for the interferometer. The result is shown in Fig. 45b, and a single-line profile perpendicular to the step edge near the peak of the defect is shown in Fig. 45c. Figure 46 shows the comparable image and profile obtained with the PSI microscope. The smoothed STM image and the interferometric image are visually quite similar even though these two techniques are sensitive to completely different surface properties.

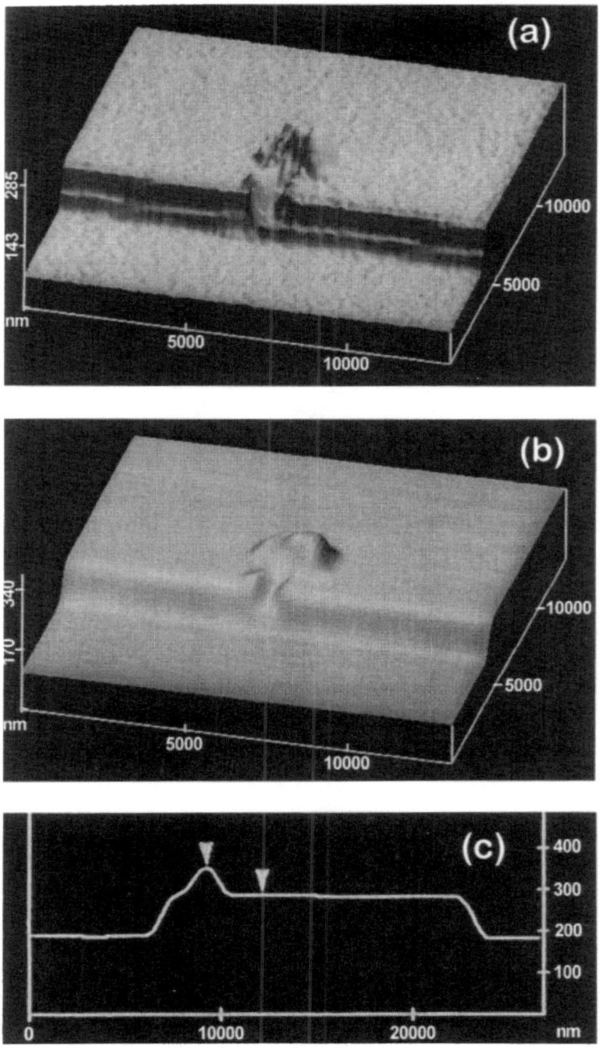

Figure 45. (a) Surface image of a defect on a Pt-coated pitch and height specimen measured with an STM by McWaid *et al.*[94] (b) The data from (a) after application of a smoothing filter to simulate the resolution of an optical area-profiling instrument. (c) A line profile through the defect obtained from the data in (b).

The height of the defect relative to the step height is about 60% for the STM image and only about 30% for the interferometric image. Such a difference could arise if there were a difference in the optical phase shift upon reflection from the defect and from the Pt coating. Such a phase-shift difference would then lead to error when

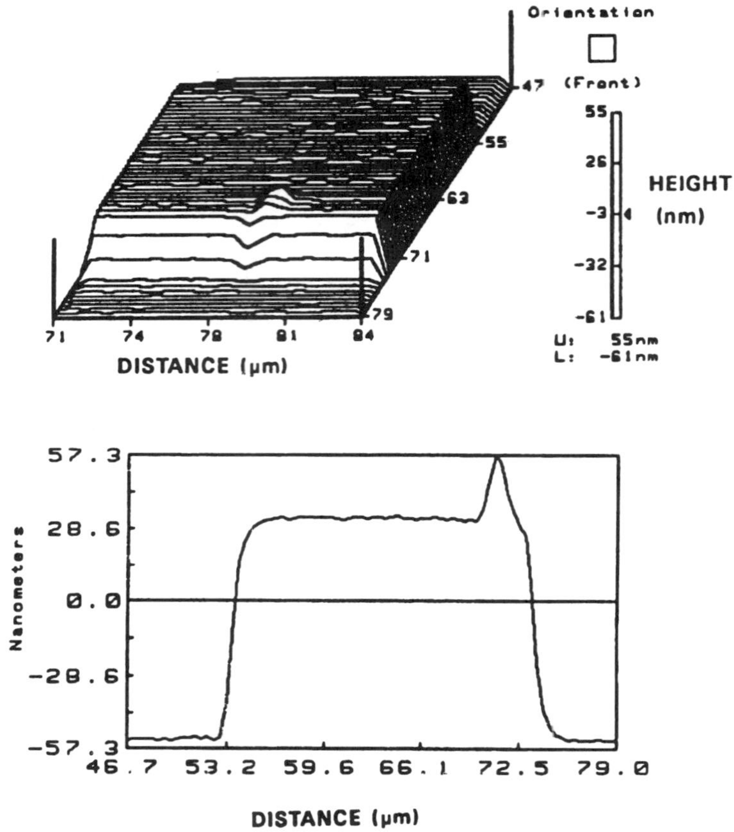

Figure 46. Surface image of the same defect shown in Fig. 45 measured with a PSI microscope by McWaid *et al.*[94]

measuring the defect height with the optical technique. This condition would likely occur if the defect were created after the Pt coating was applied.

Another key difference is that the data set taken with the PSI microscope is sparser then that taken with the STM. The PSI microscope data have a 400-nm point spacing fixed by the 100× magnification of the objective and the 40-µm center spacing of the pixels in the array camera. This spacing is appropriate considering that the nominal lateral resolution of the PSI microscope is about 600 nm.

Recently Hillman *et al.*[289] and Krüger and Krystek[290] have performed intercomparisons of stylus and phase-shifting interferometer measurements on different types of smooth roughness specimens and measured consistently smaller values of roughness with the interferometer. They ascribed the differences to a larger short-wavelength cutoff in the interferometer, which caused decreased sensitivity of the instrument to the finest spatial wavelengths on the surface.

These comparisons are examples where good agreement between different types of profiling techniques has been obtained under controlled measurement conditions and where differences in results have been explained through knowledge of the differences in the instruments. There are counterexamples where such good agreement was not obtained. However, we are convinced that the origins of differences between the results of profiling techniques can be uncovered with careful measurements and analysis.

7. Conclusions

We have provided an overview of topographic measurements with profiling and area-profiling techniques. We have discussed the strengths and limitations of the stylus, optical, and scanned probe methods and other considerations relevant to their use. In Fig. 47 we use a Church–Stedman diagram[291,292] to compare the approximate performance limits of three specific methods: the stylus instrument, interferometric microscopy (IM), including both the PSI and the vertical-scanning white-light interferometer, and the AFM. The Church–Stedman diagram plots the

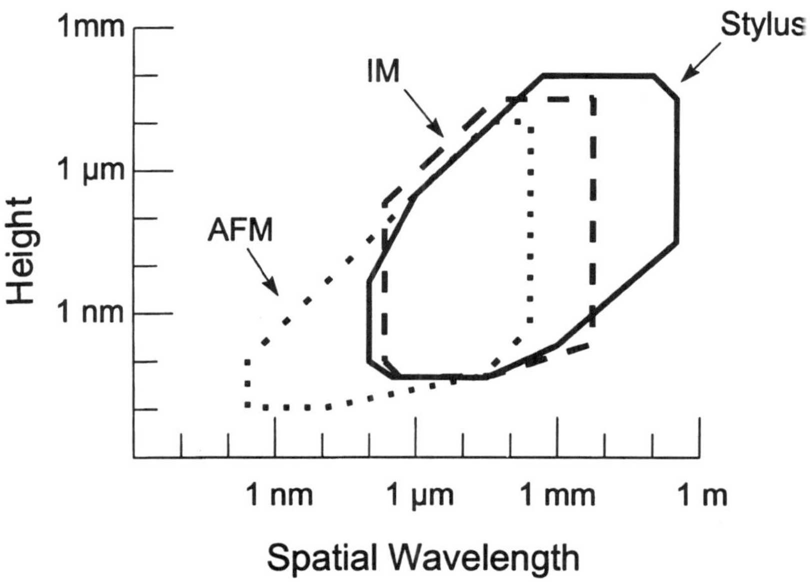

Figure 47. Height vs. spatial wavelength plots of the approximate performance limits of stylus instruments, interferometric microscopes (IM), and AFMs.

Table 4. Relevant Parameters for Three Surface Topographical Characterization Techniques

	Stylus	Interferometric Microscope	AFM
Requires vacuum?	No	No	No
Contacting?	Yes	No	Either
Requires a strongly reflecting surface?	No	Not necessarily	No
Requires a conducting surface?	No	No	No
Quantitative?	Yes	Yes	Yes
Commercial devices available?	Yes	Yes	Yes
Large-area/fast scan capability	$(200\ \mu m)^2/20$ s[37]	3 mm × 4 mm/10 s[295]	$(100\ \mu m)^2/6$ s[93]

measurable surface heights for each type of instrument along the vertical axis and measurable surface spatial wavelengths along the horizontal axis. The height resolution determines the bottom edge of each area, the spatial resolution determines the left-hand edge, the height range determines the top edge, and the lateral range determines the right-hand edge. These range and resolution limits are determined in some cases from published information of typical commercial instruments, such as stylus radius and apex angle, optical point-spread function, vertical range of interferometric microscopes,[293] and AFM tip size, and in some cases from measured results by ourselves or others.

Figure 47 shows considerable overlap in the measurable spatial wavelengths for the three profiling techniques. We have also reviewed, in Sec. 6, experiments that show that the profiling techniques compare well, if not perfectly, in the overlapping spatial frequency ranges. In Table 4 we compare requirements and factors relating to the use and selection of the stylus, interferometric microscope, and AFM methods.

Optical and SPM techniques are relatively new compared to stylus instruments, and an amazing number of developments have been made in these fields in the last 10 years. The extension of the stylus-profiling technique to perform area profiling rapidly has been an important development as well. In the future the rate of technology development should remain substantial, but perhaps not as explosive as during the past decade. There will likely be increasing use of SPMs in industry, and we expect that significant efforts will be made to standardize the procedures for use of area-profiling techniques for industrial purposes.

Acknowledgments

The authors are grateful to the following for providing material to use in this review or for generously sharing their time and expertise in discussion: S. Aoyama, K. Babcock, J. M. Bennett, J. Bielle, G. Binnig, E. Brinksmeier, E. L. Church, D. Cohen, D. Colbert, L. DeChiffre, R. A. Dragoset, W. Drews, C. J. Evans, M. Everson, R. Feenstra, J. Fu, Ch. Gerber, C. H. W. Giauque, D. E. Gilsinn, G. Goch, J. Griffith, K. Hasche, G. G. Hembree, P. Hopkins, C. Howard, K. Itaya, M. Kirk, R. Köning, H. Kunzmann, M. Lagally, D. Lucca, E. Mainsah, Y. Martin, Y. Namba, O. Riemer, F. E. Scire, P. Scott, E. Snow, J. F. Song, K. J. Stout, J. A. Stroscio, Y. Strausser, N. Sullivan, P. Sullivan, M. Suzuki, E. C. Teague, T. R. Thomas, V. Tsai, C. Van Haesendonck, V. Venkatesh, P. West, D. J. Whitehouse, J. C. Wyant, and R. D. Young. We also thank A. Czanderna, R. Dixson, A. W. Hartman, R. Krüger-Sehm, T. E. Madey, C. J. Powell, E. C. Teague, and E. Thwaite for reviewing the manuscript. V. Gagne and D. Stanfield expertly assisted in the preparation of the manuscript. The work was supported in part by the NIST Office of Microelectronics Programs and the National Advanced Manufacturing Testbed.

References

1. K. J. Stout and E. J. Davis, Surface Topography of Cylinder Bores—The Relationship between Manufacture, Characterization and Function, *Wear* **95**, 111–125 (1984).
2. J. E. Willn, Characterization of Cylinder Bore Surface Finish—A Review of Profile Analysis, *Wear* **19**, 143–162 (1972).
3. J. C. Campbell, Cylinder Bore Surface Roughness in Internal Combustion Engines: Its Appreciation and Control, *Wear* **19**, 163–168 (1972).
4. H. Lackenby, Resistance of Ships, with Special Reference to Skin Friction and Hull Surface Condition, *Proc. Inst. Mech. Eng.* **176**, 981–1013 (1962).
5. J. M. Bennett, D. K. Burge, J. P. Rahn, and H. E. Bennett, Standards for Optical Surface Quality Using Total Integrated Scattering, *Proc. SPIE* **181**, 124–132 (1979).
6. J. A. Detrio and S. M. Miner, Standardized Total Integrated Scatter Measurements of Optical Surfaces, *Opt. Eng.* **40**, 419–422 (1985).
7. T. E. Madey, The Role of Steps and Defects in Electron Stimulated Desorption: Oxygen on Stepped W(110) Surfaces, *Surf. Sci.* **94**, 483–506 (1980), see references therein.
8. T. E. Furtak, Current Understanding of the Mechanism of Surface Enhanced Raman Scattering, *J. Electroanal. Chem.* **150**, 375–388 (1983).
9. T. R. Thomas, ed., *Rough Surfaces*, Longman Group, London (1982).
10. J. M. Bennett and L. Mattsson, *Introduction to Surface Roughness and Scattering*, Optical Society of America, Washington, DC (1989).
11. D. J. Whitehouse, *Handbook of Surface Metrology*, Institute of Physics, Bristol, UK (1994).
12. ASME B46.1-1995, *Surface Texture*, American Society of Mechanical Engineers, New York (1995).
13. ANSI PH3.617-1980, *Definitions, Methods of Testing, and Specifications for Appearance Imperfections of Optical Elements and Assemblies*, American National Standards Institute, New York (1980).

14. Draft International Standard ISO/DIS 10110-7, *Optics and Optical Instruments—Preparation of Drawings for Optical Elements and Systems. Part 7: Surface Imperfection Tolerances*, International Organization for Standardization, Geneva (1992).
15. *Terminology for Particle Counting and Surface Microroughness of Silicon Wafers*, Sematech Tech. Transfer Report 95082941A-TR.
16. D. J. Whitehouse, in: *Characterization of Solid Surfaces* (P. E. Kane and G. R. Larrabee, eds.), Plenum Press, New York (1975), pp. 49–73.
17. J. M. Bennett and J. H. Dancy, Stylus Profiling Instrument for Measuring Statistical Properties of Smooth Optical Surfaces, *Appl. Opt.* **20**, 1785–1802 (1981).
18. G. Breitweiser, Surface Profile Measurements—A Survey of Applications in the Area of Vacuum Deposition, *J. Vac. Sci. Technol.* **11**, 101–105 (1974).
19. T. V. Vorburger, F. E. Scire, and E. C. Teague, Hydrodynamic Drag versus Roughness for Rotating Disks, *Wear* **83**, 339–349 (1982).
20. Y. Namba and H. Tsuwa, Mechanism and Some Applications of Ultra-Fine Finishing, *CIRP Ann.* **27/1**, 511–516 (1978).
21. J. C. Wyant, C. L. Koliopoulos, B. Bhushan, and O. E. George, An Optical Profilometer for Surface Characterization of Magnetic Media, *ASLE Trans.* **27**, 101–113 (1984).
22. E. L. Church, T. V. Vorburger, and J. C. Wyant, Direct Comparison of Mechanical and Optical Measurements of the Finish of Precision Machined Optical Surfaces, *Opt. Eng.* **24**, 388–395 (1985).
23. E. L. Church and P. Z. Takacs, Survey of the Finish Characteristics of Machined Optical Surfaces, *Opt. Eng.* **24**, 396–403 (1985).
24. J. M. Bennett, Comparison of Techniques for Measuring the Roughness of Optical Surfaces, *Opt. Eng.* **24**, 380–387 (1985).
25. S. Tolansky, *Multiple-Beam Interference Microscopy of Metals*, Academic Press, London (1970).
26. M. Sternheim, W. vanGelder, and A. W. Hartman, A Laser Interferometer System to Monitor Dry Etching of Patterned Silicon, *J. Electrochem. Soc.* **130**, 655–658 (1983).
27. R. Young, J. Ward, and F. Scire, The Topografiner: An Instrument for Measuring Surface Microtopography, *Rev. Sci. Instrum.* **43**, 999 (1972).
28. G. Binnig and H. Rohrer, Scanning Tunneling Microscopy, *Helv. Phys. Acta* **55**, 726–735 (1982).
29. J. A. Golovchenko, The Tunneling Microscope: A New Look at the Atomic World, *Science* **232**, 48–53 (1986).
30. R. A. Dragoset, R. D. Young, H. P. Layer, S. R. Mielczarek, E. C. Teague, and R. J. Celotta, Scanning Tunneling Microscopy Applied to Optical Surfaces, *Opt. Lett.* **11**, 560–562 (1986).
31. N. Garcia, A. M. Baro, R. Miranda, H. Rohrer, Ch. Gerber, R. Garcia Cantu, and J. L. Pena, Surface Roughness Standards, Obtained with the Scanning Tunneling Microscope Operated at Atmospheric Air Pressure, *Metrologia* **21**, 135–138 (1985).
32. G. Binnig, H. Rohrer, Ch. Gerber, and E. Weibel, 7/7 Reconstruction on Si (111) Resolved in Real Space, *Phys. Rev. Lett.* **50**, 120 (1983).
33. R. S. Becker, J. A. Golovchenko, and B. S. Swartzentruber, Tunneling Images of Germanium Surface Reconstructions and Phase Boundaries, *Phys. Rev. Lett.* **54**, 2678–2680 (1985).
34. J. M. Bennett, J. Jahanmir, J. C. Podlesny, T. L. Balter, and T. D. Hobbs, Scanning Force Microscope as a Tool for Studying Optical Surfaces, *Appl. Opt.* **34**, 213–230 (1995).
35. J. M. Bennett, M. M. Tehrani, J. Jahanmir, J. C. Podlesny, and T. L. Balter, Topographic Measurements of Supersmooth Dielectric Films Made with a Mechanical Profiler and a Scanning Force Microscope, *Appl. Opt.* **34**, 209–212 (1995).
36. K. Creath, Phase Measurement Interferometry Techniques, in: *Progress in Optics*, Vol. 261 (E. Wolf, ed.), Elsevier, New York (1988), p. 349.
37. E. Morrison, A Prototype Scanning Stylus Profilometer for Rapid Measurement of Small Surface Areas, *Int. J. Mach. Tools Manufact.* **35**, 325–331 (1995).

38. G. Rasigni, F. Varnier, M. Rasigni, J. P. Palmari, and A. Llebaria, Spectral-Density Function of the Surface Roughness for Polished Optical Surfaces, *J. Opt. Soc. Am.* **73**, 1235–1239 (1983); see references therein.
39. J. F. Nankivell, The Theory of Electron Stereo Microscopy, *Optik* **20**, 171–198 (1963).
40. D. W. Butler, A Stereo Electron Microscope Technique for Microtopographic Measurements, *Micron* **4**, 410–424 (1973).
41. W. Hillman, Research and Development in the Field of Roughness Measuring, Part 3: Measuring Methods by Means of Scanning Electron Microscope, *Technisches Messen* **47**, 273–283 (1980).
42. M. Schubert, A. Gleichmann, and M. Hemmleb, Determination of the Height of a Microstructure Sample by a SEM with a Conventional and a Digital Photogrammetric Method, *Ultramicroscopy* **63**, 57–64 (1996).
43. T. Kohno, N. Ozawa, K. Miyamoto, and T. Musha, High Presentation Optical Surface Sensor, *Appl. Opt.* **27**, 103–108 (1988).
44. A. M. Hamouda, A Precise Pneumatic Co-axial Jet Gauging System for Surface Roughness Measurement, *Prec. Eng.* **1**, 95–100 (1979).
45. J. N. Brecker, R. E. Fromson, and L. Y. Shum, A Capacitance-Based Surface Texture Measuring System, *CIRP Ann.* **25/1**, 375–377 (1977).
46. A. G. Lieberman, T. V. Vorburger, C. H. W. Giauque, D. G. Risko, and K. R. Rathbun, Comparison of Capacitance and Stylus Measurements of Surface Roughness, in: *Metrology and Properties of Engineering Surfaces 1988* (K. J. Stout and T. V. Vorburger, eds.), Kogan Page, London (1988), pp. 115–130.
47. T. V. Vorburger and E. C. Teague, Optical Techniques for On-Line Measurement of Surface Topography, *Opt. Eng.* **3**, 61–83 (1981).
48. T. V. Vorburger, E. Marx, and T. R. Lettieri, Regimes of Roughness for Measurements of Light Scattering, *Appl. Opt.* **32**, 3401 (1993).
49. J. M. Elson, H. E. Bennett, and J. M. Bennett, Scattering from Optical Surfaces, in: *Applied Optics and Optical Engineering*, Vol. 7 (R. R. Shannon and J. C. Wyant, eds.), Academic Press, New York (1979), pp. 191–244.
50. J. C. Stover, *Optical Scattering: Measurement and Analysis*, McGraw-Hill, New York (1990).
51. R. Brodmann, O. Gerstorfer, and G. Thurn, Optical Roughness Measuring Instrument for Fine-Machined Surfaces, *Opt. Eng.* **24**, 408–413 (1985).
52. T. A. Germer and C. C. Asmail, Proposed Methodology for Characterization of Silicon Wafer Haze Instrumentation, *Proc. SPIE* **2862**, 12–17 (1996).
53. T. A. Germer, Angular Dependence and Polarization of Out-of-plane Optical Scattering from Particulate Contamination, Subsurface Defects, and Surface Microroughness, *Appl. Opt.* **36**, 8798–8805 (1997).
54. G. A. Al-Jumaily, S. R. Wilson, and J. P. McNeil, Frequency Response Characteristics of an Optical Scatterometer and a Surface Profilometer, *Proc. SPIE* **675**, 14–22 (1986).
55. J. G. Valliant and J. M. Bennett, Instrument for On-line Monitoring of Surface Roughness of Machined Surfaces, in: *Proceedings of Advances in Surface Metrology*, American Society of Precision Engineers, Raleigh (1997), pp. 82–87.
56. ASTM F-1084-87, *Standard Test Method for Measuring the Effective Surface Roughness of Optical Components by Total Integrated Scattering*, American Society for Testing and Materials, Philadelphia (1987).
57. J. Lorincik, D. Marton, R. L. King, and J. Fine, Scanning Scattering Microscope for Surface Microtopography and Defect Imaging, *J. Vac. Sci. Technol.* B**14**, 2417–2423 (1996); D. Marton and J. Fine, Sputtering-Induced Surface Roughness of Metallic Thin Films, *Thin Solid Films* **185**, 79–90 (1990).
58. E. Spiller, Experience with the In Situ Monitor System for the Fabrication of X-ray Mirrors, *Proc. SPIE* **563**, 367–375 (1985).

59. J. W. Goodman, Statistical Properties of Laser Speckle Patterns, in: *Laser Speckle and Related Phenomena, Topics in Applied Physics*, Vol. 9 (J. C. Dainty, ed.), Springer-Verlag, Berlin (1975), pp. 9–75.
60. T. Asakura, Surface Roughness Measurement, in: *Speckle Metrology* (R. K. Erf, ed.), Academic Press, New York (1978), pp. 11–49.
61. J. Uozumi and T. Asakura, Granularity Growth of Laser Speckle and Its Application to Surface Inspection, *Opt. Laser Technol.* **9**, 177–182 (1977).
62. R. J. Ahlers, White-Light Method: A New Sensor for the Optical Evaluation of Rough Surfaces, *Opt. Eng.* **24**, 423–427 (1985).
63. F. Hottier, J. B. Theeten, A. Masson, and J. L. Domange, Comparative LEED and RHEED Examination of Stepped Surfaces; Application to Cu (111) and GaAs (001) Vicinal Surfaces, *Surf. Sci.* **65**, 563–577 (1977).
64. M. Henzler, Atomic Steps on Single Crystals: Experimental Methods and Properties, *Appl. Phys.* **9**, 11–17 (1976); M. Henzler and P. Marienhoff, High Resolution Measurement of the Step Distribution at the Si/SiO Interface, *J. Vac. Sci. Technol.* B **2**, 346–348 (1984).
65. T. E. Madey, J. T. Yates, Jr., D. R. Sandstrom, and R. J. H. Voorhoeve, in: *Treatise on Solid State Chemistry*, Vol. 6B (N. B. Hannay, ed.), Plenum Press, New York (1976), pp. 1–124.
66. S. J. Gregg and K. S. W. Sing, *Adsorption, Surface Area, and Porosity*, Academic Press, New York (1982); P. H. Emmett, in: *Catalysis*, Vol. I, Reinhold, New York (1965), pp. 63–67.
67. J. E. Benson and M. Boudart, Hydrogen-Oxygen Titration Method for the Measurement of Supported Platinum Surface Areas, *J. Catal.* **4**, 704–710 (1965).
68. J. R. Benford and H. E. Rosenbergen, in: *Applied Optics and Optical Engineering*, Vol. 4, Part 1 (R. Kingslake, ed.), Academic Press, New York (1967), Chap. 2.
69. J. H. Richardson, *Optical Microscopy for the Materials Sciences*, Marcel Dekker, New York (1971).
70. T. T. Saito and L. B. Simmons, Performance Characteristics of Single Point Diamond Machined Metal Mirrors for Infrared Laser Applications, *Appl. Opt.* **13**, 2647–2650 (1974).
71. E. Rabinowicz, *Friction and Wear of Materials*, Wiley, New York (1965).
72. G. N. Maracas, G. L. Harris, C. A. Lee, and R. A. McFarlane, On the Origin of Periodic Surface Structure of Laser-Annealed Semiconductors, *Appl. Phys. Lett.* **33**, 453–455 (1978).
73. C. P. Kirk and D. Nyyssonen, Measuring Linewidths with an Optical Microscope, *Test Measure. World* **6**, 68–79 (1986).
74. A. O. Christie, L. V. Evans, and M. E. Callow, A New Look at Marine Fouling, Part 4, *Shipping World Shipbuilder*, 121–124 (1976).
75. J. I. Goldstein, D. E. Newbury, P. Echlin, D. C. Joy, C. Fiori, and E. Lifshin, *Scanning Electron Microscopy and X-Ray Microanalysis*, Plenum Press, New York (1981).
76. R. C. Enger and S. K. Case, Optical Elements with Ultrahigh Spatial-Frequency Surface Corrugations, *Appl. Opt.* **22**, 3220–3228 (1983).
77. L. N. Gilbertson and R. D. Zipp, eds., *Fractography and Materials Science*, American Society for Testing and Materials, Philadelphia (1981).
78. W. F. Cordes and R. F. Leonard, Resist Materials for High Resolution Photolithography, *Proc. SPIE* **275**, 164–172 (1981).
79. C. W. Wilson, P. Roitman, and J. L. Blue, High-Accuracy Physical Modeling of Submicrometer MOSFET's, *IEEE Trans. Electron Devices* **ED-32**, 1246–1258 (1985).
80. For example, M. Kimizuka and K. Hirata, Pattern Profile Control of Polysilicon Plasma Etching, *J. Vac. Sci. Technol. B* **3**, 16–19 (1985), and other articles in the same issue.
81. T. Ichinokawa and Y. Ishikawa, Surface Analyses by Low Energy SEM in Ultra High Vacuum, *Ultramicroscopy* **15**, 193–204 (1984).
82. J. C. H. Spence, *Experimental High-Resolution Electron Microscopy*, Oxford University Press, Oxford (1981).

83. M. Field and J. F. Kahles, Review of Surface Integrity of Machined Components, *CIRP Ann.* **20/2**, 1–11 (1971).
84. L. D. Marks, Direct Imaging of Carbon-Covered and Clean Gold (110) Surfaces, *Phys. Rev. Lett.* **51**, 1000–1002 (1983).
85. D. G. Stearns, The Scattering of X-Rays from Nonideal Multilayer Structures, *J. Appl. Phys.* **65**, 491–506 (1989).
86. P. E. Hojlund-Nielsen and J. M. Cowley, Surface Imaging Using Diffracted Electrons, *Surf. Sci.* **56**, 340–354 (1976).
87. N. Osakabe, Y. Tanishiro, K. Yagi, and G. Honjo, Reflection Electron Microscopy of Clean and Gold Deposited (111) Silicon Surfaces, *Surf. Sci.* **97**, 393–408 (1980).
88. T. Jach, G. Hembree, and L. B. Holdeman, Observation of Gold Thin Film Growth with Reflection Electron Microscopy, *Thin Solid Films* **187**, 133–140 (1990).
89. N. Osakabe, Y. Tanishiro, K. Yagi, and G. Honjo, Image Contrast of Dislocations and Atomic Steps on (111) Silicon Surface in Reflection Electron Microscopy, *Surf. Sci.* **102**, 424–442 (1981).
90. N. Osakabe, Y. Tanishiro, K. Yagi, and G. Honjo, Direct Observation of the Phase Transition Between the (7/7) and (1/1) Structures of Clean (111) Silicon Surfaces, *Surf. Sci.* **109**, 353–366 (1981).
91. T. Wilson, ed., *Confocal Microscopy*, Academic Press, London (1990).
92. A. Bryant, D. P. E. Smith, and C. F. Quate, Imaging in Real Time with the Tunneling Microscope, *Appl. Phys. Lett.* **48**, 832 (1986).
93. M. Kirk and S. I. Park, Production Monitoring of Laser Texturing for Magnetic Disks by Ultra-fast AFM, in: *Technical Program, Fourth Workshop on IASPM*, Gaithersburg, MD, May 6–8 (1997), p. 25.
94. T. H. McWaid, T. V. Vorburger, J. Fu, J. F. Song, and E. Whitenton, Methods Divergence Between Measurements of Micrometer and Submicrometer Surface Features, *Nanotechnology* **5**, 33–43 (1994).
95. VLSI Standards, Inc., San Jose, CA.
96. J. F. Song and T. V. Vorburger, Stylus Profiling at High Resolution and Low Force, *Appl. Opt.* **30**, 42–50 (1991).
97. R. E. Reason, in: *Modern Workshop Technology, 2—Processes*, 3rd ed. (H. W. Baker, ed.), Macmillan, London (1970), Chap. 23.
98. B. Scheffer and C. Thurel, Donnees de base de la Realisation d'un Calculateur R et W, *Mecanique, Materiaux, Electricite*, **286**, 19–26 (1973).
99. J. Bielle, Functional Needs, Machining Conditions, and Economics of Surface Finishing (transl. T. V. Vorburger and V. B. Roy), *Prec. Eng.* **7**, 31–37 (1985).
100. ISO 13565-2, *Geometrical Product Specifications (GPS)—Surface Texture: Profile Method; Surfaces Having Stratified Functional Properties. Part 2: Height Characterization Using the Linear Material Ratio Curve*, International Organization for Standardization, Geneva (1996).
101. Draft International Standard, ISO/DIS 4287/1.2, *Geometrical Product Specifications (GPS)—Surface Texture—Profile Method—Part 1: Terms, Definitions, and Parameters of Surface Texture*, International Organization for Standardization, Geneva (1995).
102. R. D. Young, *The National Measurement System for Surface Finish*, NBSIR 75-927, US Department of Commerce, Washington, DC (1976).
103. D. J. Whitehouse and J. F. Archard, The Properties of Random Surfaces of Significance in Their Contact, *Proc. R. Soc. Lond.* **A316**, 97–121 (1970).
104. R. S. Sayles and T. R. Thomas, Thermal Conductance of a Rough Elastic Contact, *Appl. Energy* **2**, 249–267 (1976).
105. P. R. Nayak, Random Process Model of Rough Surfaces, *Trans. ASME, J. Lubric. Technol.* **93**, 398–407 (1971).

106. J. S. Bendat and A. G. Piersol, *Random Data: Analysis and Measurement Procedures*, Wiley-Interscience, New York (1971).
107. M. S. Longuet-Higgins, The Statistical Analysis of a Random, Moving Surface, *Trans. R. Soc. London* **249A**, 321–387 (1957).
108. R. B. Blackman and J. W. Tukey, *The Measurement of Power Spectra*, Dover, New York (1959).
109. R. S. Sayles, in: *Rough Surfaces* (T. R. Thomas, ed.), Longman Group, London (1982), pp. 91–118.
110. J. B. P. Williamson, in: *Rough Surfaces* (T. R. Thomas, ed.), Longman Group, London (1982), pp. 1–10.
111. J. M. Elson and J. M. Bennett, Relation Between the Angular Dependence of Scattering and the Statistical Properties of Optical Surfaces, *J. Opt. Soc. Am.* **69**, 31–47 (1979).
112. J. Peklenik, New Developments in Surface Characterization and Measurements by Means of Random Process Analysis, *Proc. Instn. Mech. Engrs.* **182**, Pt 3k, 108–126 (1967–68).
113. E. O. Brigham, *The Fast Fourier Transform*, Prentice-Hall, Englewood Cliffs, NJ (1974), pp. 147–171.
114. T. V. Vorburger, *FASTMENU: A Set of FORTRAN Programs for Analyzing Surface Texture*, NBSIR 83-2703, US Department of Commerce, Washington, DC (1983), pp. 102–119.
115. E. C. Teague, F. E. Scire, and T. V. Vorburger, Sinusoidal Profile Precision Roughness Specimens, *Wear* **83**, 61–73 (1982).
116. T. V. Vorburger, J. F. Song, C. H. W. Giauque, T. B. Renegar, E. P. Whitenton, and M. C. Croarkin, Stylus-laser Surface Calibration System, *Prec. Eng.* **19**, 157–163 (1996).
117. R. S. Sayles and T. R. Thomas, Surface Topography as a Nonstationary Random Process, *Nature* **271**, 431–434 (1978).
118. G. A. Somorjai, *Chemistry in Two Dimensions: Surfaces*, Cornell University Press, Ithaca, NY (1981), pp. 490–491; see references therein.
119. J. G. Endriz and W. E. Spicer, Study of Aluminum Films, I. Optical Studies of Reflectance Drops and Surface Oscillations on Controlled-Roughness Films, *Phys. Rev. B* **4**, 4144–4159 (1971).
120. E. L. Church, M. R. Howells, and T. V. Vorburger, Spectral Analysis of the Finish of Diamond-Turned Mirror Surfaces, *Proc. SPIE* **315**, 202–218 (1981). J. Schneir, R. Jobe, and V. W. Tsai, *Solid State Technology* **40**, 203 (June, 1997).
121. J. Fu, R. D. Young, and T. V. Vorburger, Long Range Scanning for Scanning Tunneling Microscopy, *Rev. Sci. Instrum.* **63**, 2200–2205 (1992).
122. E. J. Abbott and F. A. Firestone, Specifying Surface Quality, *J. Mech. Eng.* **55**, 569–572 (1933).
123. J. D. Garratt, New Stylus Instrument with a Wide Dynamic Range for Use in Surface Metrology, *Prec. Eng.* **4**, 145–151 (1982).
124. R. E. Reason, Surface Finish and Its Measurement, *J. Inst. Prodn. Engrs.* **23**, 347–372 (1944).
125. J. Bielle, private communication.
126. J. B. Bryan, The Abbe Principle Revisited—An Updated Interpretation, *Prec. Eng.* **1**, 129–132 (1979).
127. E. C. Teague, R. D. Young, F. E. Scire, and D. Gilsinn, Para-Flex Stage for Microtopographic Mapping, *Rev. Sci. Instrum.* **59**, 67–73 (1988).
128. D. J. Whithouse, Theoretical Analysis of Stylus Integration, *CIRP Ann.* **23/1**, 181–182 (1974).
129. T. V. Vorburger, E. C. Teague, F. E. Scire, and F. W. Rosberry, Measurements of Stylus Radii, *Wear* **57**, 39–49 (1979).
130. R. Krüger-Sehm, private communication, Taylor-Hobson Ltd., Leicester, UK.
131. J. S. Villarrubia, Morphological Estimation of Tip Geometry for Scanned Probe Microscopy, *Surf. Sci.* **321**, 287–300 (1994).
132. J. S. Villarrubia, Algorithms for Scanned Probe Microscope Image Simulation, Surface Reconstruction, and Tip Estimation, *J. Res. NIST* **102**, 425–454 (1997). Also located at http://nvl.nist.gov/pub/nistpubs/jres/jres.htm

133. T. R. Thomas, ed., *Rough Surfaces*, Longman Group, London (1982), pp. 24–25.
134. O. Krantz, Untersuchungen des Abtastvorganges bei der Rauheitsmessung (Investigation of the Probing Process in Roughness Measurement), PTB-Bericht ME 29 (1980).
135. R. C. Spragg, Accurate Calibration of Surface Texture and Roundness Measuring Instruments, *Proc. Instn. Mech. Engrs.* **182**, Pt 3k, 397–405 (1968).
136. E. C. Teague, *Evaluation, Revision, and Application of the NBS Stylus/Computer System for the Measurement of Surface Roughness*, NBS Tech. Note 902, US Department of Commerce, Washington, DC (1976); E. C. Teague, private communication.
137. M. N. H. Damir, Error in Measurement Due to Stylus Kinematics, *Wear* **26**, 219–227 (1973).
138. S. Ajioka, The Dynamic Response of Stylus, *Bull. Jpn. Soc. Prec. Eng.* **1**, 228–233 (1966).
139. J. I. McCool, Assessing the Effect of Stylus Tip Radius and Flight on Surface Topography Measurements, *J. Tribol. Trans. ASME* **106**, 202–210 (1984).
140. J. F. Song and T. V. Vorburger, Stylus Flight in Surface Profiling, *J. Manuf. Sci. Engr.* **118**, 188 (1996).
141. H. Bosse, R. Krüger, W. Löhr, and F. Lüdicke, Oberflächensimulator für die dynamische Kalibrierung von LängenmeBtastern (Surface Simulator for Dynamic Probe Calibration), *Technisches Messen* **61**, 82–88 (1994).
142. For example, Veeco Surface Metrology, Santa Barbara, CA, or KLA-Tencor, Milpitas, CA.
142. D. J. Whitehouse, Some Ultimate Limits on the Measurement of Surfaces Using Stylus Techniques, *Measurement and Control* **8**, 147–151 (1975).
143. J. F. Verrill, Use of the Talystep in Investigating Diffraction Grating Groove Profiles, *J. Phys. E.* **6**, 1199–1201 (1973).
144. J. B. P. Williamson, Microtopography of Surfaces, *Proc. Inst. Mech. Engrs.* **183**, Pt. 3k, 21–31 (1967–68).
145. R. S. Sayles and T. R. Thomas, Mapping a Small Area of a Surface, *J. Phys. E.* **9**, 855–861 (1976).
146. R. S. Sayles, T. R. Thomas, J. Anderson, I. Haslock, and A. Unsworth, Measurement of the Surface Microgeometry of Articular Cartilage, *J. Biomech.* **12**, 257–267 (1979).
147. E. C. Teague, F. E. Scire, S. M. Baker, and S. W. Jensen, Three-Dimensional Stylus Profilometry, *Wear* **83**, 1–12 (1982).
148. K. J. Stout, P. J. Sullivan, W. P. Dong, E. Mainsah, N. Luo, T. Mathia, and H. Zahouani, *The Development of Methods for the Characterization of Roughness in Three Dimensions*, ECSC-EEC-EAEC, Brussels (1993).
149. T. Kohno, N. Ozawa, K. Miyamoto, and T. Musha, Practical Non-Contact Surface Measuring Instrument with One Nanometre Resolution, *Prec. Eng.* **7**, 231–232 (1985).
150. L. Mattsson and P. Wågberg, Assessment of Surface Finish on Bulk Scattering Materials: A Comparison Between Optical Laser Stylus and Mechanical Stylus Profilometers, *Prec. Eng.* **15**, 142–149 (1993).
151. S. Inoue, Foundations of Confocal Scanned Imaging in Light Microscopy, in: *Handbook of Biological Confocal Microscopy* (J. B. Pawley, ed.), Plenum Press, New York (1995), pp. 1–17.
152. J. C. Wyant, Precision Optical Testing, *Science* **206**, 168–172 (1979).
153. J. F. Biegen and R. A. Smythe, High Resolution Phase-measuring Laser Interferometric Microscopy for Engineering Surface Metrology, in: *Metrology and Properties of Engineering Surfaces, 1988* (K. J. Stout and T. V. Vorburger, eds.), Kogan-Page, London (1988), pp. 287–299.
154. T. C. Bristow, Surface Roughness Measurements over Long Scan Lengths, in: *Metrology and Properties of Engineering Surfaces* (K. J. Stout and T. V. Vorburger, eds.), Kogan-Page, London (1988), pp. 281–285.
155. K. H. Guenther, P. G. Wierer, and J. M. Bennett, Surface Roughness Measurements of Low-Scatter Mirrors and Roughness Standards, *Appl. Opt.* **23**, 3820–3836 (1984).
156. P. Z. Takacs, S. K. Feng, E. L. Church, S. Qian, and W. Liu, Long Trace Profile Measurements on Cylindrical Aspheres, *Proc. SPIE* **966**, 354–364 (1988).

157. R. Smythe, L. Selberg, and L. Deck, Pole Tip Recession Measurement of Transducers on Thin Film Sliders for Rigid Disk Drives, presented at *International Disk Conference*, Tokyo, April (1992).
158. C. J. Evans, Cryogenic Diamond Turning of Stainless Steel, *CIRP Ann.* **40/1**, 571–575 (1991).
159. K. Creath and J. C. Wyant, Absolute Measurement of Surface Roughness, *Appl. Opt.* **29**, 3823–3827 (1990).
160. J. H. Richardson, *Optical Microscopy for the Materials Sciences*, Marcel Dekker, New York (1971), p. 136.
161. P. J. Caber, Interferometric Profiler for Rough Surfaces, *Appl. Opt.* **32**, 3438–3441 (1993).
162. L. Deck and P. deGroot, High-speed Noncontact Profiler Based on Scanning White-light Interferometry, *Appl. Opt.* **33**, 7334–7338 (1994).
163. J. C. Wyant and J. Schmit, Large Field of View, High Spatial Resolution, Surface Measurements, in: *Proceedings of the Seventh International Conference on Metrology and Properties of Engineering Surfaces* (B. G. Rosen and R. J. Crafoord, eds.), Chalmers University of Technology, Göteborg, Sweden (1997), pp. 294–301.
164. G. Binnig, C. F. Quate, and Ch. Gerber, Atomic Force Microscope, *Phys. Rev. Lett.* **56**, 930 (1986).
165. O. C. Wells, *Scanning Electron Microscopy*, McGraw-Hill, New York (1974).
166. C. J. Chen, *Introduction to Scanning Tunneling Microscopy*, Oxford University Press, New York (1993).
167. J. A. Stroscio and W. J. Kaiser, eds., *Scanning Tunneling Microscopy, Methods of Experimental Physics*, Vol. 27, Academic Press, Boston (1993).
168. B. S. Swartzentruber, Y. W. Mo, R. Kariotis, M. G. Lagally, and M. B. Webb, Direct Determination of Step and Kink Energies on Vicinal Si(001), *Phys. Rev. Lett.* **65**, 1913–1916 (1990).
169. R. M. Feenstra, J. A. Stroscio, J. Tersoff, and A. P. Fein, Atom-Selective Imaging of the GaAs (110) Surface, *Phys. Rev. Lett.* **58**, 1192–1195 (1987).
170. S. Alexander, L. Hellemans, O. Marti, J. Schneir, V. Elings, and P. K. Hansma, An Atomic-Resolution Atomic-Force Microsope Implemented Using An Optical Lever, *J. Appl. Phys.* **65**, 164 (1989).
171. R. Erlandsson, G. M. McClelland, C. M. Mate, and S. Chiang, Atomic Force Microscopy Using Optical Interferometry, *J. Vac. Sci. Technol. A* **6**, 266 (1988).
172. M. Tortonese, R. C. Barrett, and C. F. Quate, Atomic Resolution with an Atomic Force Microscopy Using Piezoresistive Detection, *Appl. Phys. Lett.* **62**, 834 (1993).
173. P. Günther, U. Ch. Fischer, and K. Dransfeld, Scanning Near-Field Acoustic Microscopy, *Appl. Phys.* **B48**, 89 (1989).
174. M. T. Postek, Critical Issues in Scanning Electron Microscope Metrology, *J. Res. NIST* **99**, 641 (1994).
175. J. Schneir, T. H. McWaid, J. Alexander, and B. P. Wilfley, Design of an Atomic Force Microscope With Interferometric Position Control, *J. Vac. Sci. Technol. B* **12**, 3561 (1994).
176. E. C. Teague, The National Institute of Standards and Technology Molecular Measuring Machine Project: Metrology and Precision Engineering Design, *J. Vac. Sci. Technol. B* **7**, 1898 (1989).
177. O. Jusko, X. Zhao, and G. Wilkening, Scanning Probe Microscopes for the Measurement of Micro- and Nano Structures, in: *Proceedings of the International Seminar on Quantitative Microscopy* (K. Hasche, W. Mirande, and G. Wilkening, eds.), PTB-F-21, Physikalisch-Technische Bundesanstalt, Braunschweig, Germany (1995), p. 32; O. Jusko, X. Zhao, H. Wolff, and G. Wilkening, Design and Three Dimensional Calibration of a Measuring Scanning Tunneling Microscope for Metrological Applications, *Rev. Sci. Instrum.* **65**, 2514 (1994).
178. M. Biennias, S. Gao, K. Hasche, R. Seemann, and K. Thiele, Scanning Probe Microscopes for Dimensional Length Measurements, *Surface and Interface Analysis* **25**, 606 (1997).

179. G. B. Picotto, S. Desogus, S. Lanyi, R. Nerino, and A. Sosso, Scanning Tunneling Microscopy Head Having Integrated Capacitive Sensors For Calibration of Scanner Displacements, *J. Vac. Sci. Technol. B* **14**, 897 (1996).
180. R. C. Barrett and C. F. Quate, Optical Scan-Correction System Applied to Atomic Force Microscopy, *Rev. Sci. Instrum.* **62**, 1393 (1991).
181. P. E. West, R. Jobe, and T. Van Slambrouck, Linearity and Calibration of Scanning Probe Microscope Images, in: *Atomic Force Microscopy/Scanning Tunneling Microscopy* (S. H. Cohen, M. T. Bray, and M. L. Lightbody, eds.), Plenum Press, New York (1994), p. 321.
182. D. R. Marshall, E. M. Fray, J. D. Mueller, L. M. Courtney, III, J. C. Podlesny, J. B. Hayes, T. L. Balter, and J. Jahanmir, A Closed-Loop Optical Scan Correction System for Scanning Probe, in: *Atomic Force Microscopy/Scanning Tunneling Microscopy* (S. H. Cohen, M. T. Bray, and M. L. Lightbody, eds.), Plenum Press, New York (1994), p. 437.
183. M. Holmes, D. Trumper, and R. Hocken, Atomic Scale Precision Motion Control Stage (the Angstrom Stage), *CIRP Ann.* **44/1**, 455 (1995).
184. T. V. Vorburger, C. J. Evans, V. Tsai, J. Fu, E. D. Williams, R. Dixson, P. J. Sullivan, and T. McWaid, Finish and Figure Metrology for Soft X-ray Optics, in: *JSPE Proceedings on Soft X-ray Optics: Technical Challenges* (T. Namioka, H. Kinoshita, and K. Ito, eds.), Japan Society for Precision Engineering, Tokyo (1997), p. 298.
185. MOXTEK, Inc., Orem, UT.
186. M. Suzuki, S. Aoyama, T. Futatsuki, A. J. Kelly, T. Osada, A. Nakano, Y. Sakakibara, Y. Suzuki, H. Takami, T. Takenobu, and M. Yasutake, Standardized Procedure for Calibrating Height Scales in Atomic Force Microscopy on the Order of 1 nm, *J. Vac. Sci. Technol. A* **14**, 1228 (1996).
187. V. Tsai, X. S. Wang, E. D. Williams, J. Schneir, and R. Dixson, Conformal Oxides on Si Surfaces, *Appl. Phys. Lett.* **71**, 1495 (1997).
188. A. J. Pidduck, A. B. J. Smout, P. Wagner, M. Suhren, D. C. Gupta, and S. Yang, Microsc. Semicond. Mater. Conf., Oxford, Int. Phys. Conf. Ser. No 157 (1997), p. 601.
189. UC Standard, *One-Nanometer Order Z-Axis Calibration of AFM*, Ultra Clean Society, Tokyo (1995).
190. M. Aketagawa, K. Takada, S. Sasaki, K. Kobayashi, S. Suzuki, K. Yamada, and Y. Nakayama, Direct Comparison of the SEM Standard Grating and Crystalline Lattices, in: *Progress in Precision Engineering and Nanotechnology* (H. Kunzmann, F. Wäldele, G. Wilkening, J. Corbett, P. McKeown, M. Weck, and J. Hummler, eds.), Physikalisch-Technische Bundesanstalt, Braunschweig, Germany (1997), p. 49.
191. H. Kawakatsu and H. Kougami, Automated Calibration of the Sample Image Using Crystalline Lattice for Scale Reference in Scanning Tunneling Microscopy, *J. Vac. Sci. Technol. B* **14**, 11 (1996).
192. H. Dai, J. H. Hafner, A. G. Rinzler, D. T. Colbert, and R. E. Smalley, Nanotubes as Nanoprobes in Scanning Probe Microscopy, *Nature* **384**, 147 (1996).
193. J. E. Griffith, G. L. Miller, C. A. Green, D. A. Grigg, and P. E. Russell, A Scanning Tunneling Microscope With a Capacitance-Based Position Monitor, *J. Vac. Sci. Technol. B* **8**, 2023 (1990).
194. J. E. Griffith, G. L. Miller, and C. E. Bryson, The Balance Beam Force Sensor, in: *Technical Program, Fourth Workshop on IASPM*, Gaithersburg, MD, May 6–8 (1997), p. 56.
195. Y. Martin and H. K. Wickramasinghe, Toward Accurate Metrology With Scanning Force Microscopes, *J. Vac. Sci. Technol. B* **13**, 2335 (1995).
196. D. Nyyssonen, L. Landstein, and E. Coombs, Two-Dimensional Atomic Force Microprobe Trench Metrology System, *J. Vac. Sci. Technol. B* **9**, 3612 (1991).
197. D. Nyyssonen and R. D. Larrabee, Submicrometer Linewidth Metrology In the Optical Microscope, *J. Res. Natl. Bur. Stand.* **92**, 187 (1987).
198. D. Nyyssonen and M. T. Postek, SEM-Based System For Calibration of Linewidth SRMs for the IC Industry, *Proc. SPIE* **565**, 180 (1985).

199. ASTM E1813-96, *Standard Practice for Measuring and Reporting Probe Tip Shape in Scanning Probe Microscopy*, American Society for Testing and Materials, Philadelphia (1996).
200. J. Schneir, J. A. Dagata, H. H. Harary, C. J. Evans, A. J. Melmed, and H. B. Elswijk, Scanning Tunneling Microscopy of Optical Surfaces, *Proc. SPIE* **1164**, 112 (1989).
201. NT-MDT Company, Zelenograd Research Institute of Physical Problems, Moscow, Russia. http://www.ntmdt.ru/grating.htm (6/17/97).
202. R. Wiesendanger, *Scanning Probe Microscopy and Spectroscopy*, Cambridge University Press, New York (1994).
203. G. Meyer and N. M. Amer, Simultaneous Measurement of Lateral and Normal Forces With an Optical-Beam-Deflection Atomic Force Microscope, *Appl. Phys. Lett.* **51**, 2089 (1990).
204. D. Tomanek, W. Zhong, and H. Thomas, Calculation of an Atomically Modulated Friction Force in Atomic-Force Microscopy, *Europhys. Lett.* **15**, 887 (1991).
205. K. Babcock, Magnetic Force Microscopy: High-Resolution Imaging for Data Storage, *Data Storage*, 43 (Sept. 1994).
206. Y. Martin, D. Rugar, and H. K. Wickramasinghe, High-Resolution Magnetic Imaging of Domains in TbFe by Force Microscopy, *Appl. Phys. Lett.* **52**, 244 (1988).
207. D. Rugar, H. J. Mamin, P. Guethner, S. E. Lambert, J. E. Stern, I. McFadyen, and T. Yogi, Magnetic Force Microscopy: General Principles and Application to Longitudinal Recording Media, *J. Appl. Phys.* **68**, 1169 (1990).
208. C. Schönenberger, S. F. Alvarado, S. E. Lambert, and I. L. Sanders, Separation of Magnetic and Topographic Effects in Force Microscopy, *J. Appl. Phys.* **67**, 7278 (1990).
209. P. Rice, S. E. Russek, and B. Haines, Magnetic Imaging Reference Sample, *IEEE Trans. Magn.* **32**, 4133 (1996).
210. B. Hoffmann, R. Houbertz, U. Hartmann, Eddy Current Microscopy, in *STM'97*, Hamburg, Germany (1997), p. 357.
211. Y. Martin, D. W. Abraham, and H. K. Wickramasinghe, High-resolution Capacitance Measurement and Potentiometry by Force Microscopy, *Appl. Phys. Lett.* **52**, 1103 (1988).
212. B. D. Terris, J. E. Stern, D. Rugar, and H. J. Mamin, Contact Electrification Using Force Microscopy, *Phys. Rev. Lett.* **63**, 2669 (1989).
213. C. Schönenberger and S. F. Alvarado, Observation of Single Charge Carriers by Force Microscopy, *Phys. Rev. Lett.* **65**, 3162 (1990).
214. M. Nonnenmacher, M. P. O'Boyle, and H. K. Wickramasinghe, Kelvin Probe Force Microscopy, *Appl. Phys. Lett.* **58**, 2951 (1991).
215. K. Takata, T. Kure, and T. Okawa, Observation of Deep Contact Holes and Conductive Components Underlying Insulator in a Memory Cell by Tunneling Acoustic Microscopy, *Appl. Phys. Lett.* **60**, 515 (1992).
216. K. Wago, O. Zuger, R. Kendrick, C. S. Yannoni, and D. Rugar, *J. Vac. Sci. Technol. B* **14**, 1197 (1996). D. Rugar, O. Zuger, S. Hoen, C. S. Yannoni, H. M. Vieth, and R. D. Kendrick, Force Detection of Nuclear Magnetic Resonance, *Science* **264**, 1560 (1994).
217. M. A. Paesler and P. J. Moyer, *Near Field Optics*, Wiley-Interscience, New York (1996).
218. E. Betzig and J. K. Trautman, Near-Field Optics: Microscopy, Spectroscopy, and Surface Modification Beyond the Diffraction Limit, *Science* **257**, 189 (1992).
219. E. Betzig, P. L. Finn, and J. S. Weiner, Combined Shear Force and Near-Field Scanning Optical Microscopy, *Appl. Phys. Lett.* **60**, 2484 (1992).
220. N. F. van Hulst, M. H. P. Moers, O. F. J. Noordman, T. Faulkner, F. B. Segerink, K. O. van der Werf, B. G. De Grooth, and B. T. Bölger, Operation of a Scanning Near-Field Optical Microscope in Reflection in Combination with a Scanning Force Microscope, *Proc. SPIE* **1639**, 36 (1992).
221. K. Nakajima, H. Muramatsu, N. Chiba, T. Ataka, and M. Fujihara, Optical Processing by Scanning Near-Field Optical Atomic Force Microscopy, *Thin Solid Films* **273**, 327 (1996).

222. F. Zenhausern, Y. Martin, and H. K. Wickramasinghe, Scanning Interferometric Apertureless Microscopy: Optical Imaging at 10 Angstrom Resolution, *Science* **269**, 1083 (1995).
223. D. Courjon, J. M. Vigoureux, M. Spajer, K. Sarayeddine, and S. Leblanc, External and Internal Reflection Near Field Microscopy: Experiments and Results, *Appl. Opt.* **29**, 3734 (1990).
224. M. Specht, J. D. Pedarnig, W. M. Heckl, and T. W. Hansch, Scanning Plasmon Near-Field Microscope, *Phys. Rev. Lett.* **68**, 476 (1992).
225. C. C. Williams and H. K. Wickramasinghe, Scanning Thermal Profiler, *Appl. Phys. Lett.* **49**, 1587 (1986).
226. Digital Instruments (Santa Barbara, CA).
227. J. M. R. Weaver, L. M. Walpita, and H. K. Wickramasinghe, Optical Absorption Microscopy and Spectroscopy With Nanometre Resolution, *Nature* **342**, 783 (1989).
228. C. C. Williams and H. K. Wickramasinghe, Scanning Chemical Potential Microscope: A New Technique For Atomic Force Scale Surface Investigation, *J. Vac. Sci. Technol. B* **9**, 537 (1991).
229. C. C. Williams, J. Slinkman, W. P. Hough, and H. K. Wickramasinghe, Lateral Dopant Profiling With 200 nm Resolution By Scanning Capacitance Microscopy, *Appl. Phys. Lett.* **55**, 1662 (1989).
230. R. C. Barrett and C. F. Quate, Charge Storage in a Nitride-Oxide-Silicon Medium by Scanning Capacitance Microscopy, *J. Appl. Phys.* **70**, 2725 (1991).
231. A. J. Bard, Fu-Ren Fan, J. Kwak, and O. Lev, Scanning Electrochemical Microscopy. Introduction and Principles, *Anal. Chem.* **61**, 132 (1989).
232. P. K. Hansma, B. Drake, O. Marti, A. C. Gould, and C. B. Prater, The Scanning Ion-Conductance Microscope, *Science* **243**, 641 (1989).
233. K. Keilmann, D. W. van der Weide, T. Eichelkamp, R. Merz, and D. Stöckle, Extreme Sub-Wavelength Resolution with a Scanning Radio-Frequency Transmission Microscope, *Opt. Comm.* **129**, 15 (1996).
234. J. A. Dagata, A. C. Diebold, C. K. Shih, and R. J. Colton, eds., *Workshop Summary Report: Industrial Applications of Scanned Probe Microscopy (IASPM)*, NISTIR 5550, National Institute of Standards and Technology, Gaithersburg, MD (1994).
235. J. A. Dagata, A. C. Diebold, C. K. Shih, and R. J. Colton, eds., *Summary Report: Second Workshop on IASPM*, NISTIR 5752, NIST, Gaithersburg, MD (1995).
236. *Technical Program, Third Workshop on IASPM*, Gaithersburg, MD, May 2–3 (1996).
237. *Technical Program, Fourth Workshop on IASPM*, Gaithersburg, MD, May 6–8 (1997).
238. P. Hopkins and K. Babcock, AFM Pole Tip Recession Measurements of Magnetic Recording Heads, in: *Technical Program, Fourth Workshop on IASPM*, Gaithersburg, MD, May 6–8 (1997), p. 21.
239. W. Morris, Scanning Probe Microscopy of Compact Disks, in: *Technical Program, Fourth Workshop on IASPM*, Gaithersburg, MD, May 6–8 (1997), p. 18.
240. *The National Technology Roadmap for Semiconductors*, Semiconductor Industry Association, San Jose, CA (1994).
241. H. M. Marchman, Nanometer-Scale Dimensional Metrology Using Non-Contact-AFM, in: *Technical Program, Third Workshop on IASPM*, Gaithersburg, MD, May 2–3 (1996).
242. H. M. Marchman, Scanned Probe Metrology and Its Application to CD SEM Matching, in: *Technical Program, Fourth Workshop on IASPM*, Gaithersburg, MD, May 6–8 (1997), p. 13.
243. N. T. Sullivan, Current and Future Applications of In-line Atomic Force Microscopy for Semiconductor Wafer Processing, in: *Summary Report: Second Workshop on IASPM*, NISTIR 5752 (J. A. Dagata, A. C. Diebold, C. K. Shih, and R. J. Colton, eds.), NIST, Gaithersburg, MD (1995).
244. J. E. Griffith and D. A. Grigg, Dimensional Metrology with Scanning Probe Microscopes, *J. Appl. Phys.* **74**, R83 (1993).
245. J. A. Dagata, SPM Based Metrology in the Semiconductor Industry, *Future Fab Int.* **1**(1), 263 (1996).

246. M. Niwa and H. Iwasaki, SiO_2/Si Interfaces Studied by STM, *Jpn. J. Appl. Phys.* **28**, L2320 (1986).
247. C. G. Masi, Scanning Probe Microscopy Earns Its Keep in Industry, *R&D Mag.*, p. 22 (March 1997).
248. R. C. Barrett and C. F. Quate, Imaging Polished Sapphire with Atomic Force Microscopy, *J. Vac. Sci. Technol. A* **8**, 400 (1990).
249. S. H. Cohen, M. T. Bray, and M. L. Lightbody, eds., *Atomic Force Microscopy/Scanning Tunneling Microscopy*, Plenum Press, New York (1994).
250. L. A. Bottomley, J. E. Coury, and P. N. First, Scanning Probe Microscopy, *Anal. Chem.* **68**, 185R (1996); see references therein.
251. H. Fuchs, ed., Proceedings of the First Workshop on Development and Industrial Applications of Scanning Probe Microscopes (SXM-1), *Thin Solid Films* **264**, 135 (1995).
252. B. Bhushan, ed., *Handbook of Micro/Nanotribology*, CRC Press, Boca Raton, FL (1995).
253. Papers presented at the SXM 2 Workshop, Vienna, Austria, *Surface and Interface Analysis* **25**, 481ff (1997).
254. D. L. Windt, W. K. Waskiewicz, and J. E. Griffith, Surface Finish Requirements for Soft X-Ray Mirrors, *Appl. Opt.* **33**, 2025 (1994).
255. E. Spiller, R. A. McCorkle, J. S. Wilczynski, L. Golub, G. Nystrom, P. Z. Takacs, and C. Welsch, Normal Incidence Soft X-Ray Telescopes, *Opt. Eng.* **30**, 1109 (1991).
256. C. R. Kurkjian, R. A. Frantz, and H. H. Yuce, Light Guide Fibers: Problems and Prospects, *Opt. Photon. News* **8** (7), 35 (1997).
257. Y. Namba, Chubu University Report, Kasagai, Japan (1996).
258. V. C. Venkatesh, F. Fang, and W. K. Chee, On Mirror Finishing Obtained With and Without Polishing, *CIRP Ann.* **46/1**, 505 (1997).
259. P. G. Burkhalter, D. B. Brown, J. V. Gilfrich, J. H. Konnert, P. D'Antonio, H. Rosenstock, L. M. Shirey, M. Thompson, and V. Elings, Surface Characterization of Multilayer X-Ray Diffraction Specimens, *J. Vac. Sci. Technol. B* **9**, 845 (1991).
260. O. Citterio, F. Mazzoleni, E. Monticone, and G. B. Picotto, Surface Characterisation of Electroformed Mirrors for an X-Ray Telescope, *Surf. Sci.* **337**, 98 (1997).
261. T. Oshio, Y. Sakai, S. Ehara, and K. Takashima, STM Observation of Grating Surfaces, *Modern Phys. Lett.* **B4**, 181 (1990).
262. D. A. Lucca, Y. W. Seo, and R. L. Rhorer, Aspects of Surface Generation in Orthogonal Ultraprecision Machining, *CIRP Ann.* **43/1**, 43 (1994).
263. E. Brinksmeier, R. Hoper, and O. Riemer, Characterisation of Micromachined Surfaces by Atomic Force Microscopy, *Ind. Diam. Rev.* **2**, 59 (1996).
264. Y. Ichida and K. Kishi, Nanotopography of Ultraprecise Ground Surface of Fine Ceramics Using Atomic Force Microscope, *CIRP Ann.* **42/1**, 647 (1993).
265. M. Touge and T. Matsuo, Removal Rate and Surface Roughness in High-Precision Lapping of Mn-Zn Ferrite, *CIRP Ann.* **45/1**, 307 (1990).
266. L. DeChiffre, H. N. Hansen, and A. Bronstein, Investigation on the Surface Topography in Polishing Using Atomic Force Microscopy, *CIRP Ann.* **45/1**, 523 (1996).
267. R. Miranda, N. Garcia, A. M. Baro, R. Garcia, J. L. Pena, and H. Rohrer, Technological Applications of Scanning Tunneling Microscopy at Atmospheric Pressure, *Appl. Phys. Lett.* **47**, 367 (1985).
268. I. Heyvaert, E. Osquiguil, C. Van Haesendonck, and Y. Bruynseraede, Etching of Screw Dislocations in $Yba_2Cu_2O_7$ Films With a Scanning Tunneling Microscope, *Appl. Phys. Lett.* **61**, 111 (1992).
269. J. Krim, I. Heyvaert, C. Van Haesendonck, and Y. Bruynseraede, Scanning Tunneling Microscopy Observation of Self-Affine Fractal Roughness in Ion-Bombarded Film Surfaces, *Phys. Rev. Lett.* **70**, 57 (1993).

270. C. Schönenberger, S. F. Alvarado, and C. Ortiz, Scanning Tunneling Microscopy as a Tool to Study Surface Roughness of Sputtered Thin Films, *J. Appl. Phys.* **66**, 4258 (1989).
271. D. R. Denley, Scanning Tunneling Microscopy of Rough Surfaces, *J. Vac. Sci. Technol.* **A8**, 603 (1990).
272. N. P. Suh and M. Mosleh, The Minimum Coefficient of Friction: What Is It?, *CIRP Ann.* **43/1**, 491 (1994).
273. T. L. Altshuler, Examination of Plain Carbon Steels Using An Atomic Force Microscope, in: *Atomic Force Microscopy/Scanning Tunneling Microscopy* (S. H. Cohen, M. T. Bray, and M. L. Lightbody, eds.), Plenum Press, New York (1994), p. 167
274. R. Sonnenfeld, J. Schneir, and P. K. Hansma, Atomic-Resolution Microscopy in Water, *Science* **232**, 211 (1986).
275. H. Y. Liu, F. R. Fan, C. W. Lin, and A. J. Bard, Scanning Electrochemical and Tunneling Ultramicroelectrode Microscope for High-Resolution Examination of Electrode Surfaces in Solution, *J. Am. Chem. Soc.* **108**, 3838 (1986).
276. R. Sonnenfeld, J. Schneir, and P. K. Hansma, Scanning Tunneling Microscopy: A Natural for Electrochemistry, in: *Modern Aspects of Electrochemistry*, Vol. 21 (R. E. White and J. O'M. Bockris, eds.), Plenum Press, Oxford (1990), p. 1.
277. S. Manne, H. J. Butt, S. A. C. Gould, and P. K. Hansma, Imaging Metal Atoms in Air and Water Using the Atomic Force Microscope, *Appl. Phys. Lett.* **56**, 1758 (1990).
278. S. Manne, P. K. Hansma, J. Massie, V. B. Elings, and A. A. Gewirth, Atomic-Resolution Electrochemistry With the Atomic Force Microscope—Copper Deposited on Gold, *Science* **251**, 183 (1991).
279. C. J. Chen, *Introduction to Scanning Tunneling Microscopy*, Oxford University Press, New York (1993), Chaps. 15, 16.
280. M. P. Soriaga, J. A. Schimpf, A. Carrasquillo Jr., J. B. Abreu, W. Temesghen, R. J. Barriga, J. J. Jeng, K. Sashikata, and K. Itaya, Electrochemistry of the I-on-Pd Single-Crystal Interface: Studies by UHV-EC and In Situ STM, *Surf. Sci.* **335**, 273 (1995).
281. J. Itoh, Y. Nazuka, S. Kanemaru, T. Inuoe, H. Yokoyama, and K. Shimizu, Microscopic Characterization of Field Emitter Array Structure and Work Function by Scanning Maxwell-stress Microscopy, *J. Vac. Sci. Technol. B* **14**, 2105 (1996).
282. J. A. Dagata, Evaluation of Scanning Maxwell-stress Microscopy for SPM-based Nanoelectronics, *Nanotechnology* **8** (1997).
283. H. J. Mamin, S. Chiang, H. Birk, P. H. Guethner, and D. Rugar, Gold Deposition from a Scanning Tunneling Microscope Tip, *J. Vac. Sci. Technol. B* **9**, 1398 (1991).
284. I. Fujiwara, S. Kojima, A. Machida, and J. Seto, High Density Charge Storage Memory with Semiconductor Nonvolatile Memory Media by Scanning Probe Microscopy, *SP-NANO: Tsukaba Workshop on Scanning Probe-based Nanoelectronics*, Tsukuba, Japan, December (1996), p. 21.
285. S. Hosaka, A. Kikukawa, H. Koyanagi, T. Shintani, K. Nakamura, and K. Etoh, SPM-based Data Storage for Ultrahigh Density Recording, *SP-NANO: Tsukaba Workshop on Scanning Probe-based Nanoelectronics*, Tsukuba, Japan, December (1996), p. 23.
286. J. A. Dagata, Device Fabrication by Scanned Probe Oxidation, *Science* **270**, 1625 (1995).
287. E. S. Snow and P. M. Campbell, AFM Fabrication of Sub-10-Nanometer Metal-Oxide Devices with in Situ Control of Electrical Properties, *Science* **270**, 1639 (1995).
288. K. Matsumoto, S. Takahashi, M. Ishii, M. Hoshi, A. Kurokawa, S. Ichimura, and A. Ando, Application of STM Nanometer-size Oxidation Process to Planar-type MIM Diode, *Jpn. J. Appl. Phys.* **34**, 1387 (1995).
289. W. Hillman, U. Brand, and M. Krystek, Capabilities and Limitations of Interference Microscopy for Two- and Three-Dimensional Surface Measuring Technology, *Measurement* **19**, 95–102 (1996).

290. R. Krüger and M. Krystek, Diamond Turned Surface Roughness Standards for the Calibration of Interference Microscopes, *Proceedings of the Seventh International Conference on Metrology and Properties of Engineering Surfaces* (B. G. Rosen and R. J. Crafoord, eds.), Chalmers University of Technology, Göteborg, Sweden (1997), addendum pp. 1–3.
291. E. L. Church, The Measurement of Surface Texture and Topography by Differential Light Scattering, *Wear* **57**, 93–105 (1979).
292. M. Stedman, Basis for Comparing the Performance of Surface-Measuring Machines, *Prec. Eng.* **9**, 149–152 (1987).
293. WYKO Corporation, Tucson, AZ.

5

Depth Profiling Using Sputtering Methods

H. W. Werner and P. R. Boudewijn

1. Introduction

The advanced performance of modern electronic systems (e.g., computers, telecommunication systems, electron optical systems), the stability and corrosion resistance of mechanical constructions (e.g., buildings, cars, mechanical tools), and many other applications are based on the respective properties of devices and materials used to fabricate these systems. The development of technologies to produce materials and devices (e.g., in the electronics industry, steel industry, and in the production of ceramics, glasses and synthetic materials) was only possible by the simultaneous development of instruments for the assessment of materials and devices emerging from such technologies.

For the assessment of a material or a device, a specimen is analyzed after the necessary treatment (see Volume 4 in this series) by an appropriate technique Two types of properties are determined from such an analysis: (a) physical properties, such as, surface and interface roughness, layer thickness, extension, and orientation of inclusions in a matrix, crystallographic properties and (b) chemical properties, i.e., the concentration of one or several elements in a given specimen. Specimen

H. W. Werner and P. R. Boudewijn • Department of Materials Analysis, Philips Research Laboratories, 5656 AA Eindhoven, The Netherlands. Present address (Werner): Technical University, Vienna, Austria. Correspondence address (Werner): Ansbalduslaan 17, 5581 CV Waalre, The Netherlands.

Beam Effects, Surface Topography, and Depth Profiling in Surface Analysis, edited by Czanderna *et al.* Plenum Press, New York, 1998

assessment may also involve (c) bulk analysis, by determining the average concentration of a given element across a given (probed) volume, and (d) structure analysis, by using techniques in which the distribution of a given element in the specimen is determined. The determination of elemental distribution can include (1) obtaining a point analysis or determining the composition in a monolayer,[1] (2) determining the concentration along a line in the surface or in depth (one-dimensional analysis), (3) determining the distribution of an element X along the surface (two-dimensional analysis), and (4) securing a combination of (2) and (3), i.e., three-dimensional analysis.

Table 1. List of Acronyms

AE	Auger electrons
AES	Auger electron spectroscopy, with three excitation modes: viz. by photons (P-AES), electrons (E-AES), and ions (I-AES)
BLE	Bombardment-induced light emission
E_{AE}	Energy of Auger electron
E_f	Final energy of bombarding ion beam in RBS
EPMA	Electron probe microanalysis, used in two modes, viz. energy dispersive X-ray analysis (EDX) and wavelength-dispersive X-ray analysis (WDX)
ESCA	Electron spectroscopy for chemical analysis = XPS
FWHM	Full width half maximum (of an electron, neutral, or ion beam)
GDMS	Glow discharge mass spectrometry
GDOS	Glow discharge optical spectrometry
HEIS	High-energy-ion scattering = RBS
IMFP	Inelastic mean-free path
ISS	Ion scattering spectroscopy = LEIS
LAMMA	Laser microprobe mass analysis = LMP
LEIS	Low-energy-ion scattering = ISS
LMP	Laser microprobe analysis = LAMMA
LOES	Laser optical emission spectrometry
m/e	mass-to-charge ratio
NRA	Nuclear reaction analysis
RBS	Rutherford backscattering spectrometry = HEIS
SAM	Scanning Auger microprobe
SEM	Scanning electron microscopy
SIMS	Secondary-ion mass spectrometry
SNMS	Sputtered or secondary neutral mass spectrometry
STEM	Scanning transmission electron microscope (microscopy)
TEM	Transmission electron microscope (microscopy)
XPS	X-ray photoelectron spectroscopy = ESCA
XRF	X-ray fluorescence spectrometry
λ	Wavelength of electromagnetic radiation
TEELS	Transmission electron energy loss spectrometry

This chapter deals with determining the distribution of an element X in a given specimen as a function of depth by using sputtering methods to make the bulk accessible to the methods of surface compositional analysis. Sputter depth profiles are necessary to characterize the effect of diffusion or the implantation of an element in a sample or across an interface (solid–solid interface, solid–vacuum interface = surface) in heterostructure layers. Such layers may be present in solid-state lasers, integrated circuits, coating on metals, etc. A list of acronyms used is given in Table 1.

1.1. Principle of Sputter Depth Profiling

For determining a depth profile, the sample is mounted in a suitable vacuum (or UHV) chamber of an instrument. The surface of the sample (also called the target) is bombarded with a beam I_b of energetic (0.5–5 keV) ions (e.g., Ar^+, O^+, Cs^+) or neutral particles (e.g., fast argon atoms; Fig. 1). Typical ion currents are between 10^{-9} A and 10^{-6} A and full-width-half-maximum (FWHM) beam diameters are between 0.1 μm and 1 mm. The total ion currents are not proportional to the beam cross section because the current density is different, depending on the type of ion source used. The resulting sputter process (Sec. 2) causes the ejection of atoms or clusters of atoms of the target material in a neutral, excited, or ionized state (charge state q = 0, *, +, –, ++). For simplicity we assume (a) that the bombarded target area A_b is eroded atomic layer by atomic layer (the ion beam here has the same function as a mechanically or chemically based force to erode the surface; it is the aim of every experimental procedure to find conditions in which the bombardment time (erosion time t) is proportional to the thickness of the removed target layer or the erosion depth, z) and (b) that the actual bombarded target surface (Fig. 1) recedes proportional to the bombardment time t.

In the methods of class I (Table 2), the target is bombarded with one and the same ion beam or atom beam $I_b \equiv I_e$ to achieve sputtering *and* excitational emission of a particle relevant for elemental detection in a given method. The emitted current here consists of (a) sputtered target atoms or cluster atoms in any charge state q and (b) bombardment-induced visible light and electrons. In the methods of class II (Table 3), the ion beam I_b is only used for sputtering the target surface A_b. An additional beam I_e here is used to excite a specific bombarding-beam–target interaction, which leads to the emission I_{em} of element specific particles. In AES[(2)] and XPS these additional beams are electrons (3–5 keV) and X-rays (e.g., Al Kα ≅ 1.5 keV), respectively; in ISS (LEIS) the conditions for the ion beam used for elemental identification (from the final energy of the backscattered ions) are chosen such that a low sputter yield is obtained. Moreover, the current density of the exciting beam in ISS is kept low. For obtaining a sputter profile, an additional ion beam is typically used that has a high sputter yield and high ion current density.

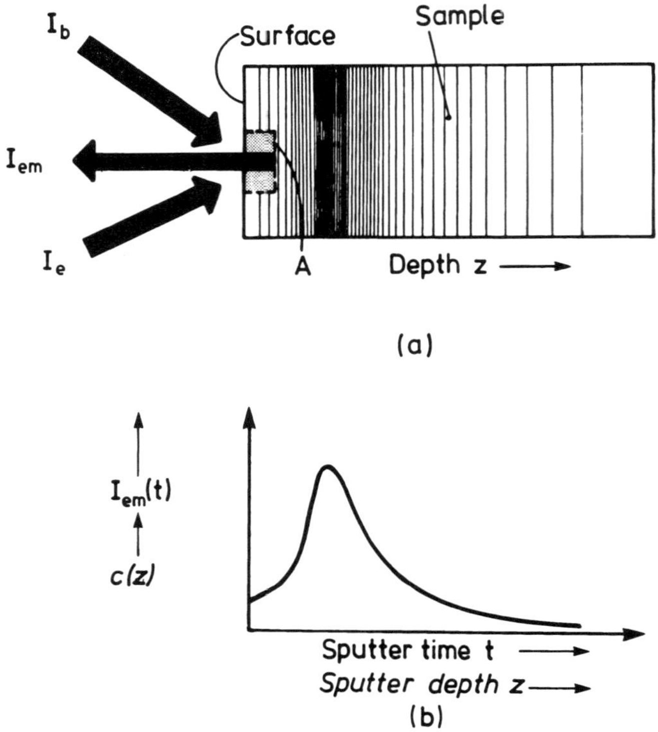

Figure 1. Principle of sputter depth profiling. (a) Experimental setup (schematic): I_b = bombarding beam for continuous sputter erosion of the sample; I_e = additional beam for excitation (emission) of an element-specific bombarding-beam target interaction; I_{em} = beam of specific emitted particles; A = area of bombardment (A_b), excitation (A_{ex}) and emission (A_{em}). In SIMS $I_e \equiv I_b$ and $A_b = A_{ex} = A_{em}$. The density of the vertical bars in the z direction roughly indicates $c(z)$. (b) Sputter depth profile $I_{em}(t)$ obtained where I_{em} is the intensity of the emitted beam as a function of sputter time t. This depth profile can be converted into a concentration profile $c(z)$ by using the relations between concentration c and I_{em}, and bombardment time t and z, respectively.

There are two advantages of class II methods compared with class I methods: (a) the sputtering can be stopped and the composition measured at a given erosion depth z for any reasonable length of time; (b) the beam diameter of I_b can be made much larger than that of I_e, which results in better depth resolution (Sec. 4). In class I, however, the analyzed area $A_{em} < A_b$ can be obtained by rastering techniques (Sec. 4). The emitted beam I_{em} contains information about the composition of the actual surface; it consists of contributions from all target elements X, Y, . . . that may vary as a function of depth.

Information about the content of one or several elements as a function of depth can only be obtained when the emitted beam I_{em} is additionally element analyzed

Table 2. Methods Used for Sputter Depth Profiling[a]

Acronym	Particles used for Sputtering–Excitation	Specific parameter and detected particle used for element identification[b]
SIMS	Ions	(m/e) of sputtered ions X^q
FAB	Fast atoms	(m/e) of sputtered ions X^q
SNMS	Ion beam	(m/e) of sputtered, post-ionized neutrals $(X^o)^q$
GDMS	Glow discharge	(m/e) of sputtered ions X^q
GDOES	Glow discharge	Wavelength of emitted (visible) light
BLE	Ions	Wavelength of emitted (visible) light
Ion–AES	Ions	Energy of ion-induced Auger electrons

[a]Class I: The same beam (I_b) is used for sputtering (bombarding) and excitation ($I_b \equiv I_e$).
[b]For SIMS, FAB, SNMS, and GDMS, q is the charge state of the sputtered ions or postionized neutrals.

in a suitable spectrometer positioned in front of the target (Fig. 2): the spectrometer is tuned to an element (X) in question, and the particle current, which can be related to the concentration of element X, is detected as a function of time. The resulting current (taken as $I_{em}f = I$, where f is the overall instrument transmission) $I(X,t)$ as a function of time is called the sputter (depth) profile; by suitable means this sputter depth profile can be converted into a concentration depth profile $c(Z)$.[3]

1.2. Methods for Sputter Depth Profiling

Tables 2 and 3 show different analytical techniques employing sputter depth profiling; Fig. 3 shows schematically the principle of these techniques together with

Table 3. Methods Used for Sputter Depth Profiling

Acronym	Particles used for Sputtering	Particles used for Excitation	Specific parameter and detected particle used for element identification
(E)-AES	Ions	Electrons	Energy of electron (E) or photon (P)
		Soft X-rays	Induced Auger electrons
XPS, (ESCA)[b]	Ions	Soft X-rays	Energy of photoelectrons
ISS (LEIS)	Ions (high sputter yield)	Ions (low sputter yield)	Final energy of backscattered bombarding ions

[a]Class II: Different beams are used for sputtering (I_b) and excitation of element-specific bombarding-beam target interaction ($I_b \neq I_e$).
[b]When compounds are destroyed by the sputtering process, XPS obviously can only be used to detect the *elemental* composition as a function of depth.

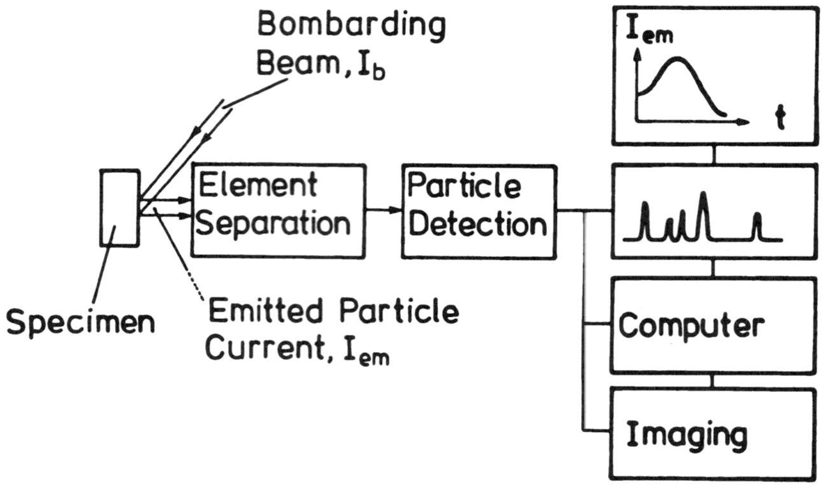

Figure 2. Principle of sputter depth profiling using SIMS for detecting the surface composition of one particular element.

other (nonconsumptive) techniques, which do not need to erode the sample for depth profiling (cf. Sec. 1.4).

Two classes of methods are evident: class I methods, in which only one beam I_b is used for sputter erosion and excitation of a particle (I_e) relevant to a given method ($I_b \equiv I_e$); and class II methods, in which separate beams are used for sputtering (I_b) and for element-specific bombarding-beam–target interaction ($I_b \neq I_e$). The two classes have the following differences: (a) In class II, sputtering and excitation are separated, the sputtering can be stopped, and the composition of the actual surface can be easily measured. (b) The intensity of the exciting beam, and therefore the measured signal in class II, can be chosen independently of the sputter rate (αI) (c) The methods of class II measure the composition of the remaining surface *after* sputtering, while methods of class I measure the particle current, i.e., the composition of the surface that has been sputtered away. The signal in class I is proportional to the sputter rate. These differences become particularly evident when preferential sputtering occurs (Sec. 2.1.4b).

1.3. Different Modes of Sputter Depth Profiling

1.3.1. Planar Sputter Depth Profiling

The method described thus far is called planar sputter depth profiling; it is practical as long as the sputter time does not exceed several hours. However, when the depth exceeds about 5 μm, the necessary sputter time for depth analysis

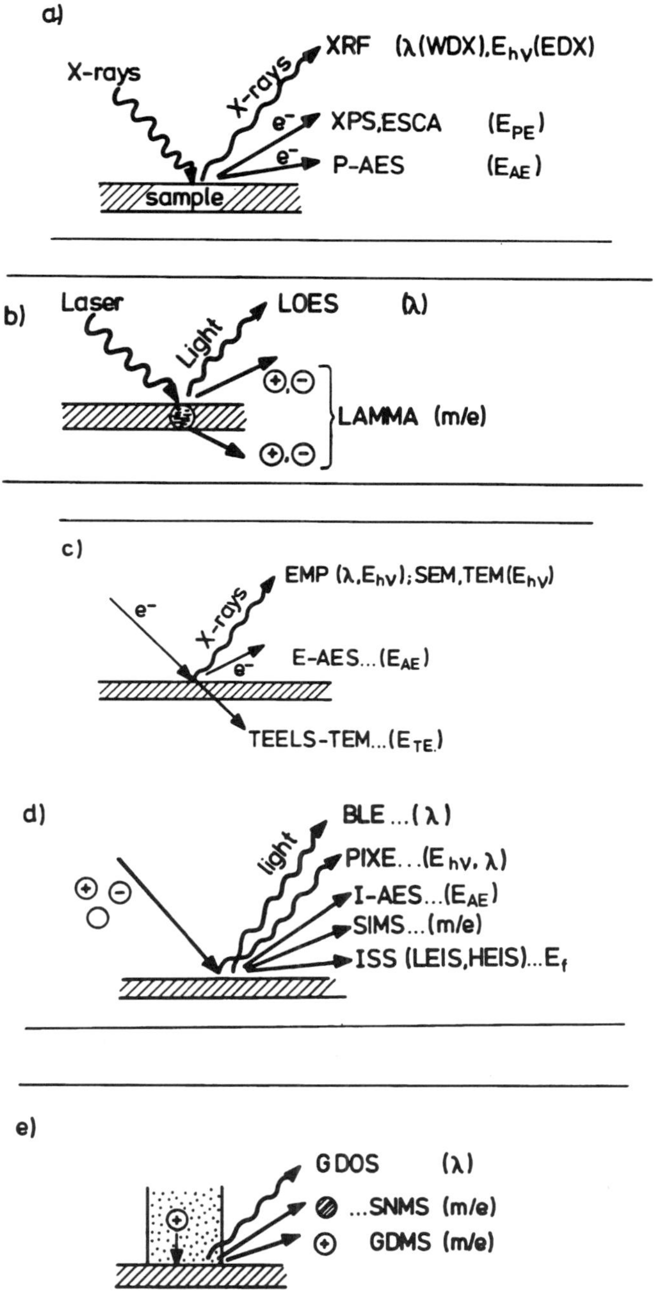

Figure 3. Excitation and emission in different methods for detecting surface composition. For symbols and acronyms, see Table 1 (From Werner[(1)]).

consequently exceeds 5 even with sputter rates of 1 μm/h. In this case, planar sputter depth profiling is no longer advisable, and one must use the method described in Sec. 1.3.2.

1.3.2. Crater-Wall, Tapered-Section, or Angular-Mapping Depth Profiling

This method is similar to one that has been used for decades in electron microprobe analysis. A tapered Sec. is obtained by mechanically abrading the sample with a rotating ball.[3,4] When scanning with a fine analyzing (exciting) beam across the slope of the curved crater, which has been obtained from the abrading treatment, one probes the composition along the crater wall; this composition, in turn, is a measure of the depth composition (see Fig. 4 from Ref. 5).

Instead of a rotating ball, an *inhomogeneous* ion beam can also be used to obtain a curved crater bottom. Here, in contrast to planar sputter depth profiling, the requirements for beam homogeneity, with respect to the current density distribution and, hence, ion energy distribution, are less stringent. Therefore, larger current densities and crosion rates can be obtained. Independently of how such a crater is obtained (mechanically or by an inhomogeneous ion beam), the following considerations hold: (a) the depth scale is strongly enlarged if the slope of the crater wall is very small; (b) the method requires the information depth to be smaller than the thickness of the layer to be analyzed, and the beam diameter must be sufficiently small (<1 μm). Depth resolution is then approximately independent of depth and can be as small as 2.5 nm in optimal cases. The tapered-section method has long been used in EMP, but depth resolution is much less than in, for example, scanning Auger microprobe analysis (SAM). However, the information depth (1 μm) is much larger than that in SAM. SAM has been used with the tapered-section method for thin-layer (typically 10 nm thick) depth profiling and selected interface analysis; (c) the advantage of the crater-wall method over the planar sputtering method is that artifacts due to atomic mixing (Sec. 2; see also Chap. 3, this volume) can be removed from the surface of the crater by gentle chemical etching before probing across the crater wall.

1.4. Comparison of Sputter Depth Profiling with Other Methods

1.4.1. Consumptive Methods for Depth Profiling

In sputter depth profiling the actual surface is continuously removed by sputtering with energetic ions or neutrals while a preselected element is detected as a function of bombardment time (proportional to depth). This method involves the removal of sample atoms; i.e., the sample is consumed (consumptive method) and obviously destroyed (destructive method). Any methods in Tables 2 and 3 is therefore consumptive (and *a fortiori* destructive) when applied to *depth profiling*. However, when used as such in class II methods *only*, i.e., for surface analysis, no consumption of material is necessary, whereas the methods of class I are intrinsically based on sample sputtering, i.e., are intrinsically consumptive even for surface analysis. However, the methods of class I carry with them the information about the composition of the surface before it was sputtered or destroyed.

1.4.2. Nonconsumptive Methods and Modes for Depth Profiling

Any consumptive method is *a fortiori* destructive. However, a nonconsumptive method may also be considered as destructive, although to a much lesser degree.

1.4.2a. Methods Based on Ion Bombardment. As described in Sec. 2, the sputter yield has a maximum at a few kiloelectron volts and depends on the atomic number of target and projectile. The sputter yield corresponds to the behavior of the cross section for nuclear stopping as a function of energy; at megaelectron-volt energies, the cross section rapidly decreases, and the interaction is mainly reduced to electronic stopping; destruction by atomic displacements become negligible. In this way, high-energy, MeV ions are used in RBS, PIXE, and NRA in order to obtain (Table 4 and Fig. 3) elemental depth profiles without sample consumption.(6–10) In RBS the energy spectrum (final energy of the backscattered projectiles, usually MeV–He+ particles) is measured.(10) The backscattered energy from collision at the surface gives the element information, while depth information about a particular element can be obtained from the known energy loss/cm (stopping power) In NRA, element-specific nuclear reactions are brought about by impact of heavy MeV particles.(10) The cross section of such a reaction is strongly energy dependent (resonance reaction). By variation of the projectile energy, the energy for a

Table 4. Exciting and Emitted Particles used in Different Methods

	Emission[a]		
Excitation	Photons	Electrons	Ions
Photons	XRF (λ_{XR}, E_{XR})	XPS (E_{PE})	LAMMA, LMP ($(m/e)^{\pm}$)
	LOES (λ_{opt})	(ESCA)	
Electrons	EMP (λ_{XR}, E_{XR})	AES (E_{AE})	
	SEM–EDX		
	TEM–EDX (E_{XR})	TEELS (E)	
	STEM–EDX		
Ions	PIXE		SNMS ($((m/e)^0)^{\pm}$)
	NRA (λ_{XR})		SIMS ($(m/e)^{\pm}$)
	BLE (λ_{opt})		ISS (LEIS)
			RBS (HEIS) (E_{final})

[a]The physical properties used for element identification are given in parentheses.
From Werner.[1]

specific reaction will therefore be available at different depths; hence a depth information for a given element is obtained.

1.4.2b. Methods and Modes Based on Electron Bombardment. In AES when an element has Auger electrons with at least two different energies and therefore different escape depths, a rough estimate of the depth profile can be obtained by comparing the intensities of the two or more corresponding Auger peaks. In EMP lowering the energy of the exciting-electron beam shifts the excited volume closer to the surface. Thin layers can be analyzed in this way. After determination of the layer thickness of such layers, a kind of rough depth profiling is possible, since for a given excitation energy the correction factor k depends on the layer thickness.(11) In SEM–EDX, depth profiles can be obtained by preparing a cross section (perpendicular to the surface) and scanning along this surface with an exciting-electron beam and detecting the generated characteristic X-rays. Scanning along such a cross section is equivalent to the tapered-section method (cf. Sec. 1.3.2) with a tapering angle of 90° (b in Fig. 4). The lateral resolution along this cross section (z resolution in the original sample) is approximately 1 mm. The poor resolution results from the exciting-electron beam(12) that produces an excitation

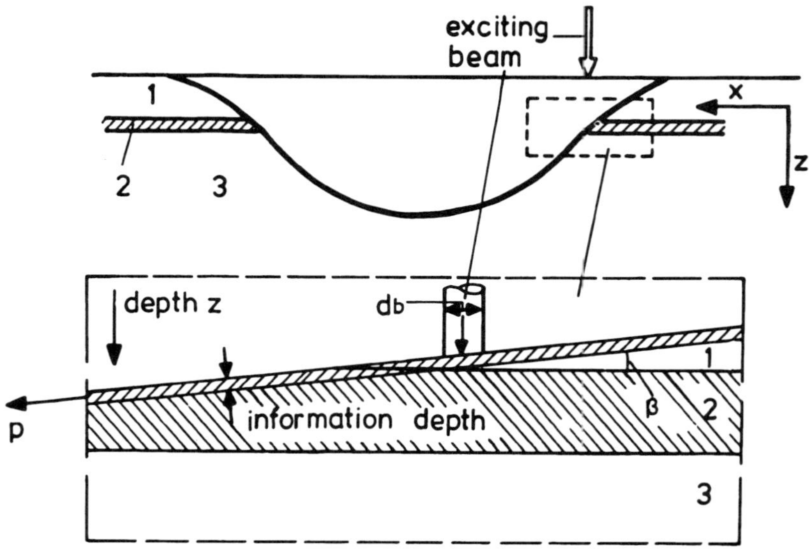

Figure 4. Principle of the tapered-section (crater-wall) method.[3] The curved sample surface obtained makes an angle β of about 1° or less with the interface. The depth scale z is thus magnified. On this magnified scale (p), with $\beta = 0.01°$ and a beam diameter d_p of 1 μm, a depth resolution of about 2 nm should be obtainable (From Werner[5]).

and information volume of about 1 mm in depth. In electron microprobe analysis this mode is sometimes also called the tapered-section method.

In a combination of AES and EDX signals,[13] use is made of the difference in escape depth for Auger electrons (typically, a few nm) and of the characteristic X-rays (typically μm) in EDX, generated by the same exciting-electron beam. Comparing the two signals (converted into element concentration) for one element lets a rough estimate be made about the presence of an element in the near surface or deeper-situated parts of the sample.

With TEM cross sections as in SEM, a section perpendicular to the original surface is made; in addition, however, the sample is thinned to about 100 nm. The element depth distribution is obtained, just as in the mode described for SEM, by means of characteristic X-rays. The lateral resolution (z resolution in the original sample) is considerably better (10 nm) since, unlike in SEM, the straggling is much smaller due to the smaller sample thickness. By using TEM high-resolution images (no element information), interfaces can be studied with z resolutions of about 1 atomic diameter.[14]

1.4.2c. Methods Based on Excitation with X-rays. In XPS, tilting the sample with respect to the analyzer axis changes the effective layer thickness that must be traversed by a photoelectron of a given energy.[15,16] An estimate of a given element is present in a layer of a given thickness (typically a few nanometers comparable to the information or escape depth) can then be made by comparing the intensities obtained from the two experimental conditions. As in AES, the intensity ratio of photoelectrons with different escape depths is a measure of the elemental composition in depth. In X-ray fluorescence, tilting the sample just as in XPS can give rough depth information. In XDC with *single crystals*, a narrow nearly parallel beam of X-rays is deflected according to Bragg's law. The peak width can be made sufficiently small[17] in a high-resolution diffractometer (typically 10 arc s). Note that in normal X-ray diffraction, the peak width is about 0.1°. Therefore, lattice constant changes of about 10^{-5} can be detected.[18] Incorporation of several percent of Al in GaAs, for example, results in enough change in the lattice constants to allow quantitative analysis of the multilayer layer composition by this method.[19]

1.4.2d. Acoustic Methods: Scanning Acoustic Microscopy.[20,21] The propagation, absorption, and scattering of ultrasonic waves depends on the acoustic density of the propagating medium and gives information on the composition, but with only very small mass dispersion. The exciting acoustic beam can be focused onto the surface or to an arbitrary depth of about 1 μm. In this way a rough impression of the composition in depth is possible.

2. Physical Basis of the Sputtering Process

2.1. Introduction

In this section a brief description of the physical processes involved in sputtering is presented, together with some results of the most widely applied theory of the sputtering process. A sputtering theory should be able to predict values of the measurable quantity in sputtering, i.e., the sputtering yield Y; it is defined as the mean number of atoms removed in any state per incident particle. Furthermore, it allows estimates to be made of the dependence of the sputtering yield on parameters such as the energy and the mass of the incoming projectiles and the target atoms, respectively, and the angle of incidence of the projectiles. The validity of the model is demonstrated by comparing theoretical values with experimental data on sputtering yields.

Apart from sputtering, several other effects are observed when a solid surface is bombarded with energetic particles. In this section only those processes are discussed that influence the results of surface analytical techniques employing sputtering as a depth-profiling method.

2.2. Theory of the Sputtering Process

2.2.1. *Binary Collision Theory*

The erosion of a solid surface by sputtering is observed for a variety of bombarding species, i.e., ions (atomic or molecular), (neutral) atoms, electrons, neutrons and energetic photons. Here the discussion is restricted to sputtering by "heavy" particle bombardment such as ions or neutral atoms. Details on sputtering by electrons, photons, or neutrons may be found in recent reviews.[22,23]

Energetic ions or atoms will penetrate the solid and transfer energy to target atoms by nuclear collisions. Bohr[24] showed that atomic collision processes can be described as elastic, which means that the conservation laws of energy and momentum apply. A further simplification follows from the assumption that atomic collisions are binary; the atomic motion of the projectile–target atom pair is completely determined by the energies and momenta of the colliding partners; interactions of the target atom with the surrounding lattice are not taken into account. This is justified because the interaction time during an atomic collision at low energies is at least 100 times smaller than the vibrational period of the lattice; the target atom and lattice are considered decoupled. The validity of the binary collision model for atomic motion at the lowest energies (<20 eV) is still open to discussion.[25]

The binary collision theory forms the basis for the description of sputtering and scattering phenomena. The probability of a given collision process is determined by the magnitude of the differential cross section. This cross section depends on the type of collision and the interaction potential chosen to describe the interaction between the colliding particles. At very high bombarding energies, a Coulomb potential may be used, but at low energies screening of the Coulomb interaction must be taken into account. The Thomas–Fermi (TF) model of atomic interactions is often used. With the power approximation of the TF potential, an expression for the differential cross section can be derived, which is valid over a rather broad energy range.[26]

2.2.2. *Classification of Sputtering Events*

We now distinguish sputtering events. A classification scheme[25] based on the energy of the recoiling target atoms after collision with the projectile and on the spatial density of the recoiling atoms is shown in Fig. 5. In the single knock-on regime, the amount of energy transferred between the projectile and the target atom is rather small, and at the surface sputtering of the target atoms takes place that, possibly after a small number of collisions, have sufficient energy to overcome the surface binding energy. Single-knock-on sputtering may be expected for ion bombardment at energies in the region well below 1 keV.

At bombarding energies exceeding a few hundred electron volts, secondary recoiling target atoms and higher-order recoils are generated. This situation is called a collision cascade. The difference between a linear collision cascade and a spike (cf. Fig. 5b, c) is as follows: only a small fraction of the atoms are in motion in a certain cascade volume in a linear cascade; essentially all atoms move in a certain spike volume. Experimentally this difference is illustrated by bombardment with diatomic molecular ions.[27,28] At the impact point the molecule will dissociate immediately and result in a superposition of two cascades (or spikes) at the same time. In the linear cascade regime the dilute character of the cascade will prevent a strict overlap; only the number of moving atoms is about twice as large as in the single cascade. The response of the system is linear, since the sputtering yield will be approximately twice that for bombardment with an atomic ion. When two spikes are superimposed, energy will be shared by atoms within a single spike volume, and consequently the available energy per atom approximately doubles. The system is nonlinear; a drastic increase in sputtering yield may be expected as the fraction of atoms able to overcome the surface barrier increases. For ion or atom bombardment at energies in the kiloelectron-volt range, the linear cascade approximation is appropriate, except for heavy projectiles on heavy substrates where spikes can be expected.

2.2.3. *Sputtering from Linear Collision Cascades*

The sputtering theory developed by Sigmund[29] describes sputtering from linear collision cascades. An important parameter in the theory is the nuclear

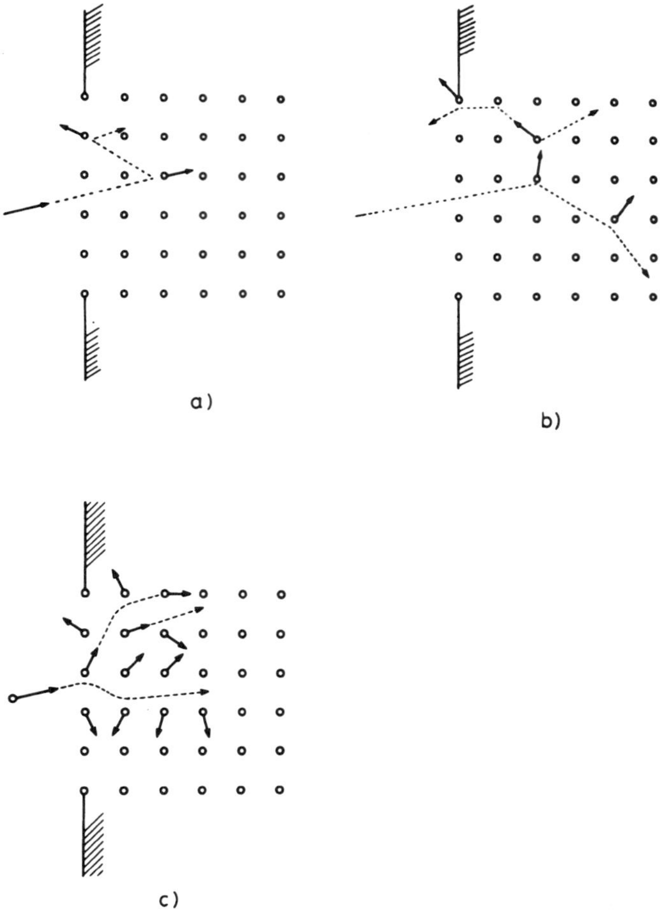

Figure 5. Three regimes of sputtering by elastic collisions. (a) The single-knock-on regime. Recoil atoms from ion–target collisions receive sufficiently high energy to be sputtered, but not enough to generate recoil cascades. (b) The linear cascade regime. Recoil atoms from ion–target collisions receive sufficient energy to generate a recoil cascade. The density of recoil atoms is low so that knock-on collisions dominate and collisions between moving atoms are infrequent. (c) The spike regime. The density of recoil atoms is so high that the majority of atoms within a certain volume (the spike volume) are in motion [Courtesy P. Sigmund (1984)].

stopping cross section $S_n(E) = dE/dx$, defined as the mean projectile energy spent in elastic, nuclear collisions per unit path length of the projectile. It $S_n(E)$ is a function of the projectile energy and further depends on the projectile–target atom mass ratio. A more universal expression, $s_n(\varepsilon)$, which depends only on the form of the screened Coulomb interaction,[26] is obtained when the reduced energy ε is used.[30]

Using the Boltzmann transport theory to describe the statistical motion of atoms in a cascade volume, Sigmund derived the following expression for the sputter yield Y:[29]

$$Y(E_i) = \Lambda F_D(E_i, \theta, z = 0) \tag{1}$$

where E_i is the energy of the projectile, θ is its angle of incidence with respect to the surface normal, F_D is the deposited-energy density at the surface ($z = 0$), and Λ is an averaged escape probability of target atoms and is inversely proportional to a surface escape barrier U_0. Generally the sublimation energy U_0 is taken for metals, the cohesive energy for covalent materials. Equation (1) holds for backsputtering, semi-infinite, random (amorphous) media.[29] The following trends may be qualitatively derived from Eq. (1): the sputtering yield is expected to increase (a) with increasing mass of the bombarding projectile at a fixed energy due to an increased nuclear stopping and (b) with increasing angle of incidence, since Y is proportional to the energy deposition at *the surface*. Furthermore Y will decrease with an increasing surface escape barrier.

By using the proportionality between F_D and $S_n(E_i)$, one may obtain numerical estimates of the sputtering yield from

$$Y(E_i) = 0.042\alpha S_n(E_i)/10^{-16}U_0 k \tag{2}$$

where α is a dimensionless parameter that depends on the ratio of projectile to target atom mass, the angle of incidence, and the energy of the projectile. The latter dependence results from the influence of electronic stopping.[31] If $S_n(E_i)$ is expressed in eV-cm^2 and U_0 in eV, a constant $k = 1$ cm^2 has to be introduced for dimensional reasons. The determination of Y is then obtained from a numerical calculation of α[29] or by extracting its value from figures in the literature[29,31] and inserting the appropriate values of the nuclear stopping cross section and surface escape barrier.

An interesting simplification of Eq. (2) was proposed by Zalm,[32] who used the form $Y(E_i) = C_{pt}S_n(E_i/E_{pt})$, where C_{pt} and E_{pt} are constants that depend on the projectile and target parameters (viz., atomic number, mass, and U_0), and E_{pt} [keV] $= (1/32.5)(1 + M_p/M_t)Z_pZ_t(Z_p^{2/3} + Z_t^{2/3})^{1/2}$, with M_p, M_t and Z_p, Z_t the mass and atomic numbers of the projectile and target, respectively. If α in Eq. (2) is replaced with a simple, approximate analytical expression and $S_n(E)$ as proposed by Wilson *et al.*,[33] C_{pt} can be calculated; a convenient formula for the sputtering yield at normal incidence is then obtained:[32]

$$Y(E_i) = (Z_tZ_p)^{5/6}S_n(E_i/E_{pt})/3U_0 \tag{3}$$

The agreement between Eqs. (3) and (2) is within 7% for $1/16 < Z_t/Z_p < 5$.[32]

2.2.3a. Deviations from Linear Cascade Theory. As discussed in the previous section, the assumption underlying Sigmund's sputtering theory(29) is no longer valid in the spike regime and the single-knock-on regime. In the latter situation we may distinguish between light-ion sputtering at low energies, where the concept of an isotropic recoil cascade is no longer valid due to insufficient energy transfer, and near-threshold sputtering at energies well below 1 keV. The existence of a threshold energy in sputtering follows because a minimum energy is needed to displace an atom in the solid.

Several corrections to Eq. (2) for near-threshold sputtering have been proposed.(34,35) These propositions were criticized by Zalm,(32) who showed that low-energy sputtering yields can be estimated from:(36)

$$Y(E_i) = 1.9/U_0 = Z_t^{1/2}[E_i^{1/2} - 0.09U_0^{1/2}]/F \tag{4}$$

with $F = [(Z_t/Z_p)^{2/3} + (Z_p/Z_t)^{2/3}]/2$. In Eq. (4), U_0 must be expressed in eV and the projectile energy E_i in keV. Since F varies slowly with Z_p/Z_t, the sputtering yield is nearly independent of projectile type (Z_p) but depends only on the projectile energy E_i.

Spike effects in sputtering were investigated by several authors.(27,28,37,38) Based on experimental systematics derived from such studies and theoretical considerations, a dedicated search for spike effects in Xe^+ sputtering of Zn was performed.(39) Surprisingly, excellent agreement with linear cascade theory was found at low-current-density bombardments. With high-flux 20-keV Ne^+ bombardment, the sputtering yield was found to depend on the primary-beam current density,(39) and yields up to 200 atoms/ion were observed. This was ascribed to beam-induced heating of the target, but the data could not be explained by a simple evaporation process. Therefore, it was tentatively concluded that temperatures might play an important role in invoking spike effects.

Deviations from Sigmund's predictions at high bombarding energies, especially for light ions (see Sec. 2.3), are partly explained by the neglect of electronic stopping in the linear cascade theory. Projectiles and energetic recoil atoms also suffer energy loss due to inelastic cross section $S_e(E)$, which is proportional to $E^{1/2}$ for values of E that are not too high.(40) The effect of electronic stopping can be incorporated into Eq. (2) by making the factor α slightly energy dependent,(31) but this is not adequate to explain the deviation, thus making the concept of a universal nuclear stopping questionable.(32)

2.2.3b. Sputtering of Monocrystals and Compounds. In the derivation of Eq. (2), it was assumed that the target atoms in the solid were distributed randomly. Strictly speaking, the results of Sec. 2.2.3 only apply to amorphous solids or to a first approximation to polycrystalline materials with randomly oriented crystallites. For monocrystalline materials the influence of the ordered crystal lattice on the development of the collision cascade must be taken into account. Channeling of incident projectiles or recoiling target atoms may occur, while further blocking or shadowing can take place. The selvedge region of the crystal may have a different structure, with possibly different binding energies, and the crystal surface may not be flat; all these factors can influence the collisions that result in sputtering.

Experimentally, the sputtering yield of monocrystals is lower than for a polycrystalline material if the bombarding beam is incident parallel to a low-index crystallographic direction thus "channeling" deep into material; higher yields were observed for beams along a high-index direction.(41) Using the channeling concept, Onderdelinden(42,43) pointed out that channeled particles generate collisions at such a depth in the material that they do not contribute to sputtering; only the nonchanneled fraction of the beam gives a contribution that can be calculated according to the theoretical scheme in Sec. 2.2.3. More detailed information on experimental data and theoretical aspects of monocrystal sputtering can be found in recent reviews.(44,45)

In multicomponent materials (e.g., alloys and compounds) the different components are not generally sputtered according to the stoichiometry of the material during the initial phases of particle bombardment. The process, called preferential sputtering (Chap. 3, this volume), results from a different sharing of energy in the collision cascade for target atoms of different mass, different binding energies of the target components or segregation, and diffusion effects. Preferential sputtering alters the composition of the target in the near-surface region and leads to the formation of a so-called altered layer. The low-yield component will accumulate on the sample surface, and its contribution to the sputtered particle flux will increase with increasing projectile fluency until the composition of the sputtered flux equals the bulk composition of the target (steady-state sputtering). Theoretical considerations for binary alloys(46) show that preferential sputtering occurs for the least-bound component for equal masses or the lighter component for different masses. A compilation of sputtering yield data for multicomponent materials is given in Ref. 47.

2.3. Sputtering Yields (Experimental)

The dependence of sputtering yield on the various parameters in Eq. (2) will now be briefly illustrated, and a comparison with experimental data will be given. For the projectile energy E_i the sputtering yield Y depends on E_i mainly through the nuclear stopping cross section $S_n(E_i)$. At low energies the latter increases with increasing energy E_i, and reaches a maximum before decreasing again at high incident energies. This behavior is illustrated in Fig. 6, taken from the original work of Sigmund.(29) For the projectile atomic number Z_p (also denoted Z_1) the sputtering yield increases with increasing mass of the bombarding ion due to an enhanced energy deposition at the surface, as shown in Fig. 7. For the target atomic number

Z_t (also dented Z_2) enhanced nuclear stopping causes Y to increase with increasing target atomic number, as shown by Eq. (3), and a comparison with experimental data is given in Fig. 8a. However, note that Y also depends on the magnitude of the surface escape barrier, which differs from element to element. This is illustrated in Fig. 8b, which that shows the reciprocal of the sublimation energy as a function of

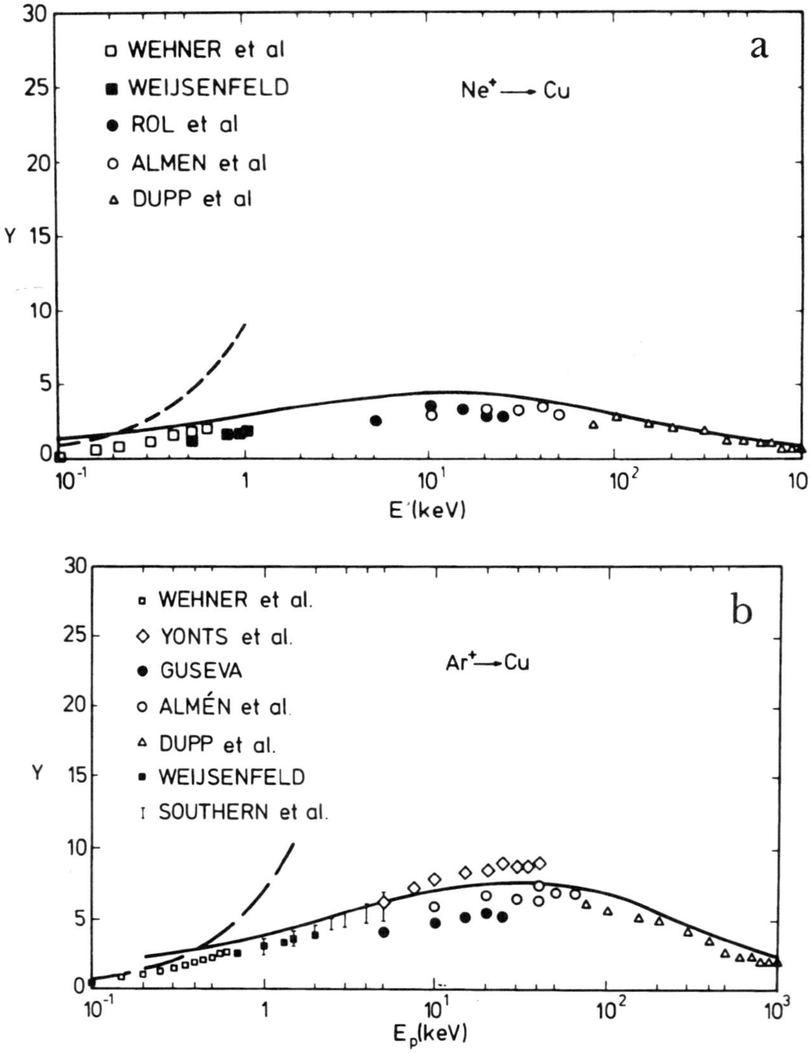

Figure 6. The sputtering yield (Y) of Cu as a function of the primary-ion energy for (a) Ne^+ and (b) Ar^+ bombardment (After Sigmund[29]).

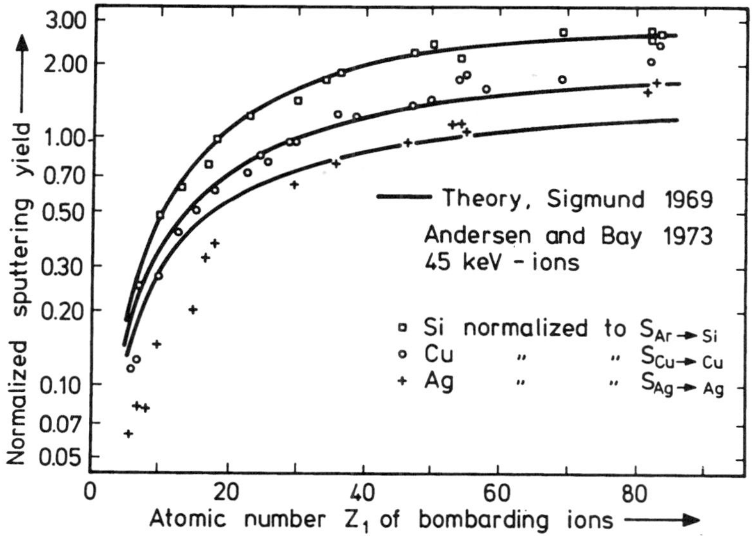

Figure 7. Sputtering yield as a function of the atomic number Z_1 of the bombarding ion [After H. Oechsner, *Appl. Phys.* **8**, 185 (1975)].

atomic number. For the projectile angle of incidence θ with respect to the surface normal, a comparison of the theoretical and experimental dependencies of Y on the projectile angle of incidence is shown in Fig. 9. Generally, the yield reaches a maximum at angles between 50° and 80° before decreasing rapidly at larger angles. This decrease is related to an increased reflection of projectiles at near grazing incidence.

The dependence of the sputtering yield of a given target on projectile energy and projectile atomic number is conveniently demonstrated by plotting the normalized sputtering yield as a function of the reduced energy. Yield data for noble-gas-ion bombardment of silicon are shown in Fig. 10, and it can be seen that the reduced energy dependence in Eq. (3) is obeyed at low and intermediate energies. Deviations at higher energies will be discussed later. Finally, Table 5 is a survey of experimental sputtering yields for various elemental targets obtained with Ar^+ bombardment in the kiloelectron-volt energy range; these data were extracted from a compilation by Andersen and Bay.[(47a)]

2.4. Information Depth

The information depth is an important parameter for evaluating the quality of a depth profile, since this parameter sets a lower limit on the best-obtainable depth resolution. The information depth depends on the analytical method used and is

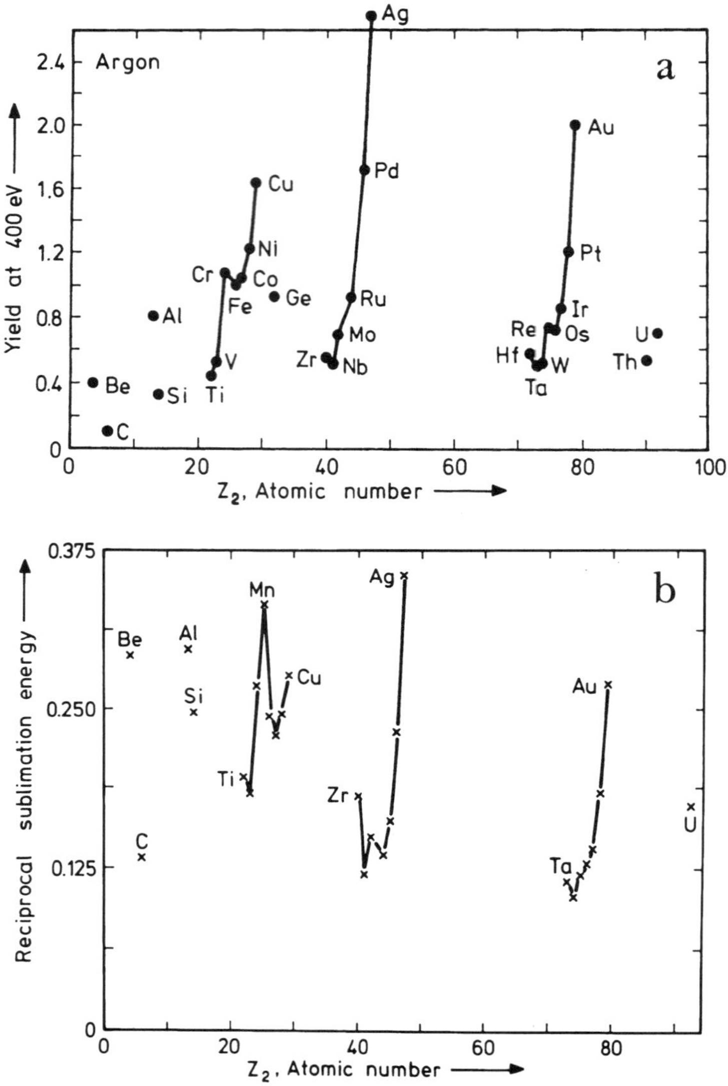

Figure 8. (a) Sputtering yield as a function of atomic number Z_2 of the target [After N. Laegreid and G. K. Wehner, *J. Appl. Phys.* **32**, 365 (1961)]. (b) Reciprocal of sublimation energy as a function of atomic number Z_2 of the target [After D. Rosenberg and G. K. Wehner, *J Appl. Phys.* **33**, 1842 (1962)].

related to the escape depth of the information-carrying particles. For the electron spectroscopies AES and XPS, the analytical information is contained in the kinetic energy of the emitted electrons; thus, the information depth depends on the inelastic mean-free path (IMFP) of these electrons. A compilation of IMFP data is given by

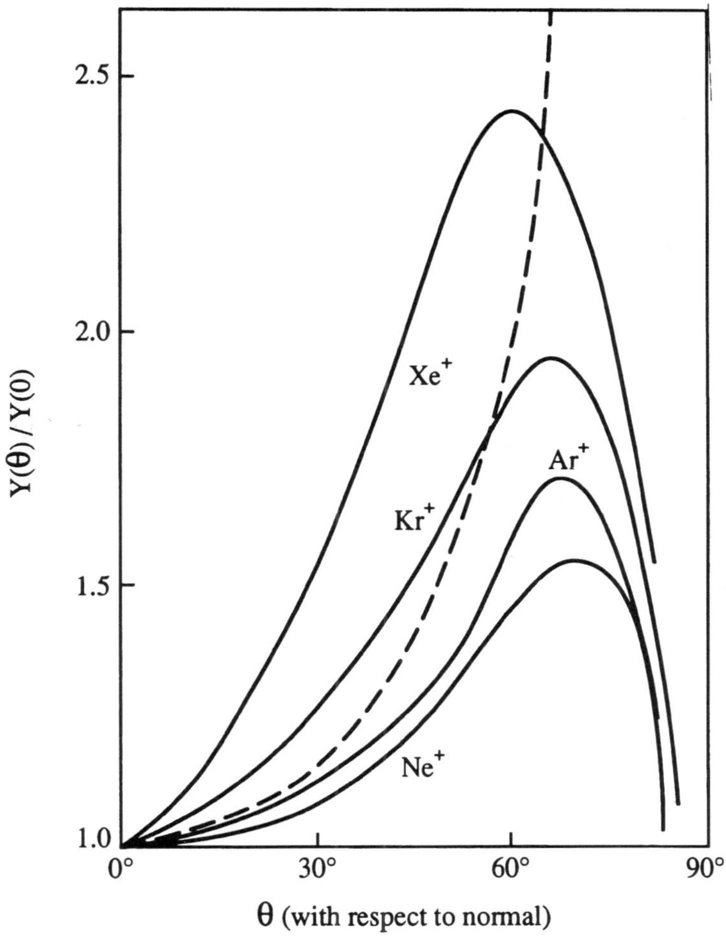

Figure 9. Normalized heavy-ion sputtering yields from polycrystalline Cu with bombarding energy of 1.05 keV as a function of incident angle θ from the surface normal dashed curve = 1/cos θ (After H. Oechsner[48]).

Seah and Dench,[49] from which an information depth d between $2 < d < 10$ atomic layers, which depends on the energy of the ejected electrons, can be derived.

In SIMS and SNMS the information depth depends on the mean escape depth of sputtered particles, which depends on the energy of the bombarding species.[50] It is limited to 1 to 2 atomic layers for bombarding energies below 1 keV and may increase to 10 atomic layers or more for higher energies.

In ISS the analytical information is carried by backscattered, low-energy (0.5–5 keV) ions. Since the neutralization probability of the primary ions is very large

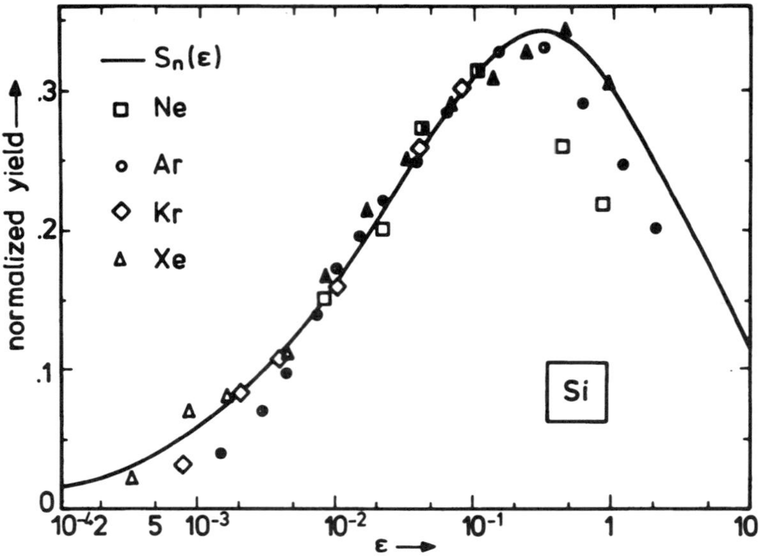

Figure 10. Sputtering yield data (symbols) for Ne-, Ar-, Kr-, and Xe-ion bombardment of Si as a function of reduced energy compared with nuclear stopping $s_n(\varepsilon)$ to prove the relation Y = coust. $s_n(\varepsilon)$ (After Zalm[32]).

beyond the first atomic layer, the information depth is effectively the surface monolayer, and consequently ISS is the most-surface-sensitive technique.

2.5. Processes Related to Sputtering

The collision processes following energetic particle bombardment of a solid result in displacement of target atoms from their original lattice positions. Apart from sputtering, other transport processes occur that deserve attention in discussing sputtering as a depth-profiling method. In general, these transport processes modify the composition profile of a multicomponent target in the near-surface region or tend to smear out the composition depth profile of an originally atomically sharp interface. Three processes are commonly distinguished. (i) Primary recoil mixing[51] is a process resulting from direct projectile–target (atom) collisions and results in transport of surface atoms into the bulk of the solid (recoil implantation). (ii) Cascade mixing[52–56] results after several nuclear collisions and at lower energies of the recoiling target atoms in which transport over smaller distances takes place. In this process the mixing is characterized by involving a larger number of particles and is in contrast to primary recoil mixing; the direction of the target atoms is no longer related to the direction of incidence of the projectile. By reducing the primary-ion energy in SIMS to 200 eV, Clegg[57] obtained a recoil-mixed layer only

Table 5. Sputtering Yield Data for Some Elemental Targets Bombarded with 5-keV Argon Ions at Normal Incidence[a]

Element	Sputtering yield
C	2.3
Al	2.0
Si	1.5
Ti	1.6
Fe	2.5
Ni	2.8
Cu	5.0
Ge	3.0
Mo	1.2–2.6
Ag	8.3
Cd	21
In	3–5
Ta	0.9
W	1.7
Pt	3.6
Au	7–8
Pb	10.5

[a]Compilation after Andersen and Bay.(47a)

1 nm thick. (iii) Radiation-enhanced diffusion can occur after the displacement of target atoms from their lattice positions forms vacancies and interstitials (Chap 3); this may influence the diffusional behavior of mobile ions in the solid. Generally, processes (i) through (iii) take place simultaneously during particle bombardment; the range over which they play a role is determined by the spatial extent of the collision cascade. The influence of these displacement processes on the quality of sputter depth profiles will be considered more explicitly in Sec. 4. Implantation of the projectile atoms in the surface layers of the target is a final point of attention. The incorporation of the bombarding ions as atoms may change the surface escape barrier or influence the development of the collision cascade, especially when the mass difference between projectile and target atoms is large. An equilibrium concentration of trapped projectiles generally results above a certain (steady state) fluency, typically (>2 × 10^{17} atoms/cm^2). The steady-state projectile concentration is determined by a balance between implantation and diffusion of the projectiles into the bulk and to the surface and removal by sputtering.(58) With reactive primary ions (e.g., O_2^+), formation of a compound surface layer has been observed.(59) Simple models of projectile retention(58,60) indicate that the fractional surface concentration

of implanted projectiles at steady-state bombardment conditions is proportional to θ/Y, where Y is the partial sputtering yield of the substrate atoms and θ is the retention coefficient.

The development of a compound surface layer was also observed for noble-gas-ion bombardment in a reactive-gas ambient.[59] In this case the compound layer results from recoil implantation and cascade mixing or from diffusion of atoms adsorbed on the surface.[61]

3. Experimental Aspects

3.1. Introduction

In many cases the capability of a surface analytical technique may be extended toward sputter depth profiling by simply adding an ion beam source to the instrument. This is certainly true for the electron spectroscopies XPS and AES, whereas other techniques (SIMS, ISS) require an ion beam source for generating the information-carrying particles. In this section the various types of ion beam sources will be briefly described, while further attention will be paid to aspects such as beam species, beam purity, maximum current, and beam diameter.

3.2. Ion Beam Sources

The beam of fast, heavy projectiles required for sputtering generally consists of energetic ions generated in an ion source. In most sputter-depth-profiling applications, the energy of the ions is a few hundred electron volts up to about 20 keV. In principle, any element can be used as the bombarding species as a positive or negative ion, or a neutral atom; molecular ions are also available. In practice, noble-gas-ion bombardment is preferably used for depth profiling, and AES or XPS for analysis, since disturbing chemical effects from implanted primary ions are undesirable. On the other hand, primary ions from reactive elements such as O_2^+, N_2^+, or Cs^+ are often used for depth profiling when SIMS is used for the analysis, since it is well known that the presence of these elements in the surface layers of the sample enhances the secondary-ion yield.

Various types of ion beam sources are available for sputtering. The choice of a particular type of source is determined by its performance characteristics, which involve parameters such as ion species, beam purity, total current output, energy spread, brightness, and obtainable beam diameter. Electron-impact and plasma-discharge ion sources are mainly used for noble- or reactive-gas ions. In plasma-discharge sources a magnetic field is often used to constrict the plasma; ions are extracted through a small axial bore in the anode. A particularly favorable type of source with respect to energy spread and brightness (i.e., high current density in a small angular aperture) is the duoplasmatron[62,63] (Fig. 11). Both the magnetic field and the shape of the intermediate electrode constrict the plasma to a narrow space on the beam axis. Primary-beam mass filtering is mandatory for gas-discharge ion sources because of the presence of polymeric gas ions, multiply charged ions, and impurity ions.[63,64] Cs^+-ion beams are generally obtained from a surface ionization source.[65] Cs atoms evaporate from a reservoir and diffuse toward a heated tungsten frit; there Cs^+ ions are produced by thermal ionization. Beam purification is not necessary because the beam contains only singly charged Cs ions.[64]

Liquid-metal ion sources (LMIS), now called focused ion beam (FIB) sources,[271] have received much attention because of their ability to produce extremely small ion beams at a relatively high current. In these sources a liquid metal (In, Ga, or Cs) wets the tip of a wire, which is then shaped into a cone by a high electrostatic field. Emission of ions is assumed to be the result of a combination of field desorption and thermal evaporation combined with field ionization.[66] Emission of polymer and impurity ions[67,68] necessitates primary-beam mass filtering. Care must be taken to prevent the emission of metal droplets, as reported for an indium LMIS.[69] Ion beams of submicron diameter at a relatively high current can be generated with a LMIS because of its high brightness; this parameter and aberrations of the ion optical system determine the ion current that can be focused onto a small spot.[70] In Table 6 are some data about the relationships among source brightness, beam diameter, and beam current. Note the advantage of the Ga LMIS

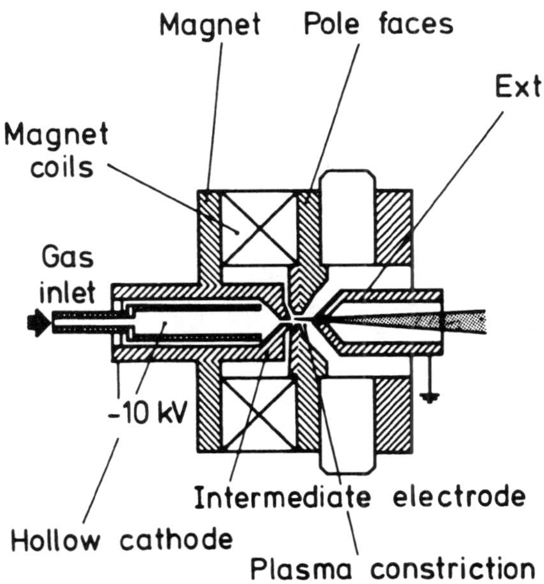

Figure 11. Schematic of the duoplasmatron ion source with hollow cathode.

Table 6. Characteristic Data for Optimized Microfocus Ion Guns

		Beam current (A)	
Source type	Brightness (A/cm^2 sr)	d^a = 100 nm	d^a = 10 nm
Duoplasmatron	2×10^2	4×10^{-14}	2×10^{-18}
Surface ionization (C_s^+)	1×10^3	1×10^{-11}	1×10^{-14}
LMIS (Ga^+)	1×10^6	1×10^{-10}	1×10^{-15}

[a] d = FWHM (full width at half-maximum) beam diameter.
[b] After Rüdenauer.[271]

in the 100-nm range. At much smaller beam diameters the Cs surface ionization source is preferable, mainly because of the smaller energy spread of the ion beam. The use of a LMIS for obtaining secondary-ion images of submicron lateral resolution has been demonstrated by several authors.[71,72]

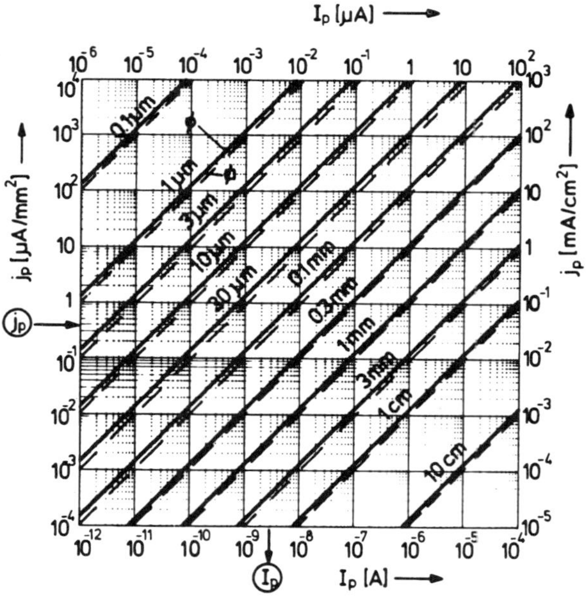

Figure 12. Nomogram for the determination of ion current density j_p ($= I_p/A$) from the primary-ion current I_p, and *vice versa*, for different beam dimensions (The circular beam diameter is indicated by solid lines, and a square beam diameter is indicated with the dashed lines.) [From H. W. Werner, in: *Electron and Ion Spectroscopies of Solids* (Fiermans *et al.*, eds.), with permission of Plenum Press (1978), p. 340.

For sputter depth profiling of insulating or poorly conducting materials, a beam of neutral particles is often preferred to reduce or eliminate charging effects (Sec. 4). A beam of neutral atoms (e.g., noble gases) can be obtained from an electron-impact or a plasma-discharge source by adding a neutralization chamber.[73] Interaction of the ion beam from the ion source with the gas at a relatively high pressure in the neutralization chamber (charge exchange processes) results in a beam of neutral particles.

3.3. Time Needed to Obtain a Depth Profile

The time necessary to sputter a given layer thickness is related to the sputter rate. The latter, in turn, can be described by the formula.[74]

$$\frac{dz}{dt}\,(\mu\text{m/h}) = \frac{3.6 \times 10^{-4}\, M\,(\text{amu})\, j_b\,(\mu\text{A/cm}^2)\, Y}{\rho\,(\text{g/cm}^3)} \tag{5}$$

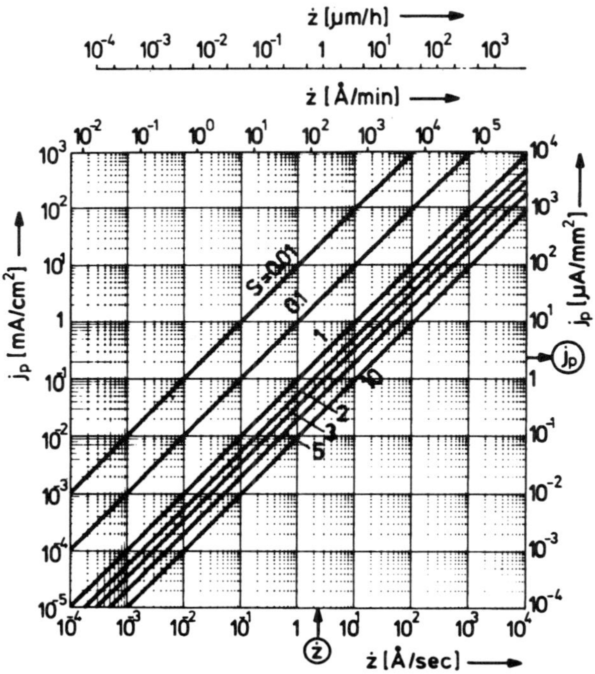

Figure 13. Erosion rate dz/dt for different sputtering yields (indicated as S in the figure) at various current densities.

As can be seen, the erosion rate depends on the mass number M and the density ρ of the sample, the current density j_b of the bombarding ion beam, and the sputter yield Y. For a given target, M and ρ are fixed; the sputter yield can be influenced by the choice of the bombarding ion and its energy (Sec. 2). The current density can vary by several orders of magnitude, depending on the ion source used (Sec. 3.2). Typical values for the erosion rate are between 1 μm/h for a duoplasmatron and 100 nm/h for some types of electron-impact ion sources. The relations among ion current, ion current density, beam diameter, and erosion rate can be rapidly estimated from Figs. 12 and 13.

4. Analysis of Sputter Depth Profiles

4.1. Introduction

The objective of a sputter-depth-profiling experiment is to obtain information about the variation of the elemental composition of a sample as a function of depth into the solid.[75] Generally the sputter depth profile is obtained in the form of an analytical signal (specific for the analytical method) that is measured as a function of sputtering time. This necessitates two corrections to be applied to the as-measured data: conversion of the sputtering time into a depth scale and conversion of the intensity of the analytical signal into an elemental concentration. Procedures for these two conversions are the subject of this section. We also address the important question: how accurately does a $c(z)$ profile represent the elemental in-depth distribution originally in the sample? Answering this question involves a discussion of the possible artifacts introduced into a measured depth profile by the sputtering process; artifacts may be related to sample properties, instrumental factors, or may result from the interaction process between the bombarding, energetic particles and the solid sample (Chap. 3). Furthermore the information depth of the respective analytical technique must be taken into account.

4.2. Conversion of the Sputter Depth Profile of an Element X, $I(X, t)$, into a Concentration Depth Profile $c(z)$

4.2.1. Determination of the Depth Scale

For targets of constant composition, the erosion rate dz/dt is expected to be constant at a constant primary-beam current density. The depth scale z then follows from $dz/dt = zt$, and dz/dt may be determined from the total sputtered depth Z and the total bombardment time T as $dz/dt = Z/T$.

Several methods are available for determining the total crater depth: in interference microscopy[76] the minimum depth and accuracy of this method depend on the detectable fraction of an integral fringe offset ($n \times$ wavelength of the light used) and amount to approximately 30 nm; with stylus techniques[77] crater depths can be determined with an accuracy of 5%, and the minimum detectable depth amounts to 10 nm (problems arise for strongly curved or very rough sample surfaces); with ellipsometry,[78,79] if an optically transparent overlayer with a refractive index different from that of the substrate (e.g., SiO_2 on Si) is partly sputter-removed, the remaining overlayer thickness can be determined; with X-ray methods[80] the intensity of characteristic X-rays from an overlayer is measured. If the penetration depth of the exciting electrons is much larger than the overlayer thickness, the intensity is proportional to the residual overlayer thickness. The erosion rate dz/dt can also be derived from the time t necessary to sputter through a layer of known thickness. This method requires that the experimental sputtering conditions (i.e., ion beam current density, beam species, etc.) be kept constant and that the composition of the sample to be investigated be identical to that of the calibration sample. Furthermore, the erosion rate can be calculated from the known sputtering yield Y and the atomic density n of the sample according to $dz/dt = Yj_p/n$. In this case, the primary-beam current density j_p must be accurately known.[81] For the latter two methods, certified calibration standards (Ta_2O_5/Ta) are available* and often used in sputter depth profiling with analysis by AES and XPS.

For depth-profiling with SIMS analysis, especially in semiconductor materials, implantation profiles can be used as depth calibration standards; from the known projected range R_p and straggling ΔR_p of the implantation profile, the erosion rate can be determined accurately.[82] An additional advantage is that implantation profiles can also be used for quantification of SIMS data (Sec. 4.2.2).

For multilayer samples consisting of several layers of different composition, z for each layer can be found from the time t of breaking through the layer; the thickness of the layers must be known or be determined by an independent method, such as cross-sectional SEM or TEM, RBS, or X-ray fluorescence.

For samples varying strongly in composition as a function of depth, the erosion rate is no longer constant but changes with bombardment time [$dz/dt = dz(t)/dt$]. The depth scale can be found from $z(t) = \int[dz(t)/dt]dt$, and $z(t)$ or $dz(t)/dt$ must be determined *in situ* by a suitable method. Illustrative examples are the methods of determining the loss of mass of the sample with the aid of a quartz crystal microbalance incorporated into the instrument[83] or by using laser interferometry.[84]

4.2.2. *Conversion of the Measured Signal I into an Elemental Concentration c*

The measured intensity I (I_{em}, Sec. 1.1) must be related to an elemental concentration. It can be a current of photons, electrons, or ions, depending on the

*Ta_2O_5/Ta certified reference material can be purchased from the National Physical Laboratory, Division of Materials Applications, Teddington, Middlesex TW11 OLW, England.

analytical method used (Tables 2 and 3). Generally the relation between I_A and c_A of element A can be written as $I_A = kfc_A(dN/dt)$, where dN/dt is the flux or flux density (depending on the method) of probing particles, f is the transmission of the instrument, and k is a yield factor that depends on the physical processes relevant for the specific analytical technique considered. The quantity c_A may be expressed as a fractional concentration (SIMS, SNMS, ISS) or as an atomic volume concentration (AES, XPS). Furthermore, it is assumed that the sample composition is homogeneous over a depth much larger than the escape depth of the analytical particles. Quantification is then achieved by using (a) a theoretical calculation of the yield factor k based on a knowledge of all physical parameters involved in the interaction process; (b) semiempirical methods, which use relations containing parameters that can be adjusted by the experimental results; and (c) purely empirical methods based on constructing working curves from calibration standards. In fact, approaches (a), (b), and (c) need well-characterized calibration samples to verify a theoretical prediction, to adjust or refine parameters, or to construct a working curve.

In the following we consider different analytical methods separately. In XPS the yield factor k depends on the photoabsorption cross section and the IMFP of the emitted photoelectrons.[(85)] Quantification schemes based on first-principle calculations[(86–89)] and models of the photoemission process have been developed.[(90–92)] Quantification is commonly performed by using elemental sensitivity factors.[(93)] The sensitivity factor s_A is defined relative to the fluorine 1s line (s_F = 1) by $I_A/I_F = N_A s_A/N_F$, with N_A and N_F the number of atoms of element A and F, respectively. A review of sensitivity factors is available.[(94)] The influence of matrix effects on s_A is often minimized by making reference to an internal standard line. With this method analytical results with a relative accuracy of ±20% have been obtained.[(94–98)] It is important to note that in XPS, I_A stands for the integrated intensity of an electron signal; the peak shape and peak position are strongly influenced by the chemical environment of element A. The corresponding change in peak position is referred to as a chemical shift[(98)] and provides information about the valence state or oxidation state of the element.

In AES the yield factor k depends on the ionization cross section, the Auger transition probability, the IMFP of the Auger electrons, and the backscattering factor of the primary electrons in the sample matrix.[(99)] Auger intensities are usually given as peak-to-peak heights (PPH) when Auger spectrometers are operated in the "derivative mode" of the spectrum, as is usually the case to increase the signal-to-noise ratio and to lower the detectability.[(100,101)] For quantitative analysis it is often recommended that the intensity be determined from the area under the Auger peak in a nondifferentiated spectrum;[(102–105)] the PPH is only a reliable quantity when the Auger peak shape does not change with varying matrix composition. In practical work this is often assumed, but less often verified.

Calculations of the Auger intensity from theory are generally of poor accuracy due to limited knowledge of the physical parameters involved.(106,107) The best quantitative results can be obtained by calibration with standards that closely match the composition of the sample to be investigated. In practical work the method based on relative elemental sensitivity factors s_A, derived from pure elemental standards, is often used. The fractional concentration c_A is found from $c_A = I_A s_A / I_n s_n$.

The elemental sensitivity factors s_A, s_n (normalized to the Ag Auger MNN reference line) can be found in a handbook.(108) The method is regarded as semiquantitative because it neglects variations in the backscattering factor and escape depth with a varying matrix. Improved accuracy was observed by including an empirically determined matrix correction term.(109–111)

In SIMS the yield factor k depends on the sputtering yield and the secondary-ion yield.(112) Sputtering yields can be calculated from the model in Sec. 2, but problems are encountered for bombardment with reactive (O_2^+, N_2^+) due to retention of implanted primary ions in the sample surface. The secondary-ion yield is much more difficult to calculate since it was found that it depends strongly on the elemental ionization potential (positive ions) or electronegativity (negative ions), the primary-ion species, and the composition of the matrix.(59,113)

Of all the surface analytical techniques used in combination with sputtering, SIMS shows the strongest matrix effects. Several authors proposed theoretical models for the secondary-ion emission process ranging from a kinetic model(114) to an autoionization model(115) and to several surface ionization models.(116–121) Although the accuracy of these models is reasonable for the cases considered, their applicability is limited to a few model systems. Of more general value is the semitheoretical approach based on the local thermal equilibrium (LTE) model.(122–124) Using two internal standards, one can calculate the parameters "ionization temperature"(125) and "electron density;" these parameters are necessary for deriving the concentration of unknown elements from the secondary-ion intensities. Simplified versions of the LTE model only require knowledge of the "ionization temperature" of a particular matrix for quantitative analysis; in some cases only one internal standard is needed.(126–128) With these LTE models quantitative SIMS analysis may yield error factors of up to 2.(129–131) More accurate quantitative analysis is possible by calibration with suitable standards. For relative accuracies to better than 10%, the matrices of the standard and unknown samples should be identical. Obviously, the applicability of this method is limited by the availability of standard samples, since for each element–matrix combination a standard is needed. Ion implantation is a powerful method for preparing such standards;(82,132) in principle, a well-defined quantity of almost any element can be implanted in nearly all matrices with good lateral homogeneity.(133) Implantation standards are extensively used in quantitative SIMS depth profiling of semiconductor materials.(134,135)

In SNMS the yield factor k depends on the sputtering yield and the postionization efficiency for sputtered neutral particles.(136) Compared to SIMS, SNMS shows a much less pronounced matrix effect because ion yields are determined by the well-characterized electron collision ionization process instead of the matrix-dependent emission of secondary ions. Using standard samples Müller *et al.*(137) obtained accurate quantitative results for some alloy and semiconductor materials.

In ISS the yield factor k depends on the differential scattering cross section and the probability that the scattered particle leaves the surface as an ion.(138) Quantitative analysis with ISS is difficult because (a) the differential cross section for scattering at low energies cannot be calculated easily because the precise form of the screened Coulomb potential appropriate for interaction in this energy region is not well known;(139) (b) neutralization effects are not easy to predict; and (c) the shadowing effect(139) is a further complicating factor in the quantitative interpretation of scattered ion intensities, in which signals of one element may be suppressed by other atoms above it; shadowing effects have been demonstrated by varying the angle of incidence of the primary ions. For (b) strong oscillations of the scattered ion yield for He^+ impinging on Pb have been observed as a function of the primary-ion energy.(140) A possible solution to this problem is to measure both scattered ions and neutrals in time-of-flight instruments.(141) Another alternative might be the use of alkali primary ions that tend to stay ionized.(142) It must be realized, however, that these methods degrade the extreme surface sensitivity of ISS.

In conclusion, the scattered ion yield in ISS is mainly determined by experimental conditions, such as the type and energy of the projectiles, angle of incidence, and scattering angle. Chemical (matrix) effects are thought to be small,(143) which means that pure elements can be used for calibrating elemental sensitivity.(143,144)

4.3. Artifacts in Sputter Depth Profiles

4.3.1. *Artifacts Related to the Interaction Process between Energetic Projectiles and the Solid*

Preferential sputtering(7,8,55,145–148) alters the near-surface composition of an alloy or compound sample (Sec. 2.2.4b). As a consequence, analytical results obtained with AES, XPS, or ISS may deviate from those obtained with SIMS or SNMS, since the former techniques probe the composition of the instantaneous, sputtered surface, whereas with SIMS and SNMS the flux of sputtered particles is determined. After the initial steady-state sputtering conditions are established, the altered surface composition for a compound, which may be different from that of the bulk, results in altered signal intensities I_A and I_B in AES, XPS, and ISS; i.e., $I_A/I_B \propto c_A Y_B / c_B Y_A$, where c_A, c_B and Y_A, Y_B are the bulk concentrations and partial sputtering yields of elements A and B, respectively. In SIMS and SNMS the intensities are proportional to the bulk concentration: $I_A/I_B \propto c_A/c_B$. Before steady-state sputtering conditions are reached, transients in signal intensities may be observed, which are related to compositional changes not originally in the solid. The extent of such changes depends on many parameters, such as alloy composition,[149,150] energy,[151] and mass of the bombarding ions.[152,153]

As discussed in Sec. 2, sputtering resulting from the collision cascade is accompanied by transport processes resulting in displacement of target atoms from their original lattice positions, i.e., primary recoil mixing and radiation-enhanced diffusion. Because of this intrinsic relation to the sputtering process itself, these processes form one of the main disadvantages of sputter depth profiling. They cannot be circumvented, and, at the most, their influence can be minimized by selecting the bombardment conditions properly. The principal effect of these mixing processes is to degrade the depth resolution by smearing out sharp concentration gradients and broadening interfaces. This is demonstrated in Fig. 14 by the SIMS depth profile of Al_x across $Al_xGa_{1-x}As$–GaAs and AlAs–GaAs interfaces;[154–160]

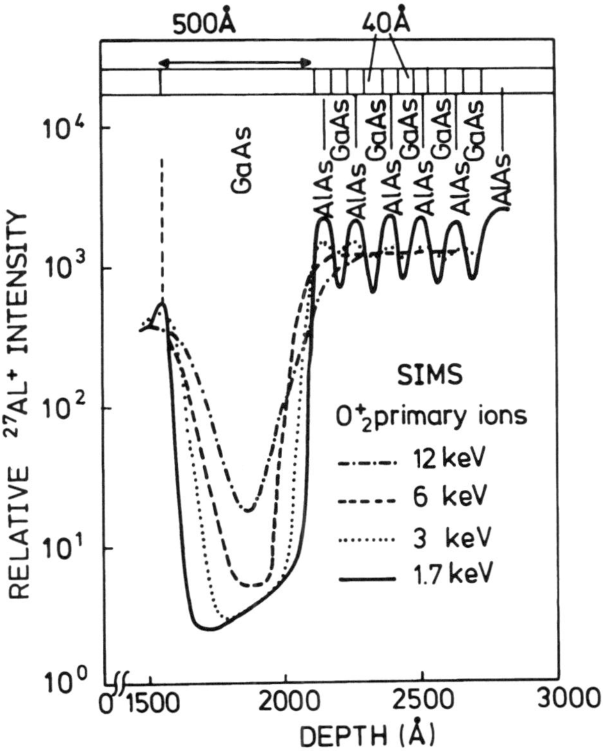

Figure 14. SIMS $^{(27)}Al^+$ depth profiles of a GaAs–AlAs multilayer structure as a function of the primary-ion O_2^+ energy. The $^{(27)}Al^+$ intensity is given relative to the $^{(27)}Al^+$ background signal obtained from a GaAs substrate [From P. R. Boudewijn, M. R. Leys, and F. Roozeboom, *Surf. Interface Anal.* **9**, 303 (1986)].

these interfaces were shown to be atomically sharp.[14] It can further be seen in Fig. 14 that a large improvement in depth resolution can be obtained by lowering the primary-ion energy, which decreases the spatial extent of the collision cascade and reduces the range over which mixing effects play a role. Optimal depth resolution is obviously reached when the extent of mixing is limited to a depth from which the information-carrying particles originate (information depth), i.e., 1–2 atomic layers (Sec. 2.4 and Chap. 3). This would require beam energies of about 100 eV, which are not only difficult to obtain but they also greatly reduce the erosion rate. Other measures to reduce projectile range include bombarding with primary ions of a large atomic mass and choosing a large angle of incidence relative to the surface normal. All these considerations have been confirmed by considerable experimental data (for recent reviews see Refs. 154 and 161); it has been shown that a depth resolution of 2 to 3 nm can be obtained by using sputtering.[153] Artifacts caused by the sputtering process can be delineated by comparing profiles with these obtained by nonsputtering profiles, e.g., RBS.[8,162–164]

Theoretical modeling of the mixing process has been performed by various approaches. Andersen[165] proposed treating the relocation of target atoms as a random-walk problem from which an effective diffusion constant describing the broadening of tracer profiles could be derived. Hofer and Littmark[166] used standard transport theory to describe the multiple collisions in the solid; they calculated that for 5-keV Ar^+ incident perpendicular onto silicon, tracer profiles show a broadening of 10 nm and a shift up to 4 nm, dependent on tracer position. It was further predicted that the trailing edge of the tracer profile decreases exponentially (see also Fig. 14). According to Wittmaack,[167–169] the characteristic decay length depends on the mixing function and the differential escape probability. Consequently, the decay length is expected to be element dependent (in a fixed matrix), as was observed experimentally.[170] In certain cases the exponential decay length strongly depends on the bombardment conditions; it has been suggested that segregation of impurity ions at the interface with the surface layer modified by reactive-gas ion (O_2^+) bombardment leads to delayed sputtering of the impurity ions.[171–173]

It has been argued that the statistical nature of the sputtering process is another intrinsic factor limiting the depth resolution. The sequential-layer sputtering (SLS) model, originally proposed by Benninghoven[174] and later extended by Shimizu[175] and Seah,[176] predicts a surface roughness on an atomic scale. Model calculations showed that, after sputter removal of six monolayer equivalents, the number of monolayers contributing to an area fraction of 95% of the actual microstructured surface is 5.[177] According to this model the contribution of microroughness to the depth resolution is about 1.5 nm. The influence of sample roughness on depth resolution has been considered by many authors.[178–180]

4.3.2. Artifacts Related to the Properties of the Sample

As discussed in Sec. 1, the measured $I(t)$ curve can only be interpreted as a depth profile if the element of interest is distributed homogeneously in planes parallel to the original sample surface. For samples with a laterally inhomogeneous composite, meaningful sputter depth profiling is only possible if the elemental composition in the x–y plane across the analyzed area is homogeneous. In some cases this can be realized by reducing the size of the analyzed area, but this may lead to sampling errors since the area or volume analyzed may no longer be representative for the average sample composition.(181) A method to determine the inhomogeneity of a sample is to use the lateral-imaging capability of the surface analytical technique (cf. SIMS ion microprobe or ion microscope or AES with a scanning capability). Lateral inhomogeneities are in the SIMS sputter profile in Fig. 15, which shows the Na^+-ion current as a function of time, obtained from an aluminum target.* Assuming a homogeneous lateral distribution, this curve would be interpreted as the Na concentration profile in Al. SIMS imaging has been used to check the validity of this assumption. In Fig. 16, which shows the Na distribution across the surface, it is evident that Na is present only on certain parts of the target surface. Figure 16, taken at $t = t_2$, explains the decrease in the Na^+ peak; the area covered by Na is diminished, but the intensity of the Na line on the remaining spots is the same as in the beginning. At $t = t_3$ (Fig. 16) only a faint Na^+ image is still visible. The decrease of the Na^+ peak in Fig. 15 must therefore be interpreted as a decrease in the Na-covered area—the remaining spots emitting the same secondary-ion current densities as in the beginning—and not as a diffusion of Na into the sample.

Distortion of depth profiles may result from the decomposition of compounds under heavy-particle bombardment. For example, ion bombardment results in reduction of oxide compounds,(182) and therefore sputter depth profiling combined with XPS cannot provide information about the original chemical state of atoms in deeper layers.(183)

Serious problems may arise when sputter-depth-profiling insulating materials or thin films, especially, when ion bombardment is used for sputtering. Severe charging of the sample surface can occur, resulting in primary-beam deflection or problems with secondary-particle extraction. The latter point mainly applies to SIMS and ISS and is of lesser importance for the electron spectroscopies AES and XPS; these techniques use a different particle beam for sputtering and excitation, and they can be applied alternately. Charging resulting from the photon or electron beam bombardment is discussed in Chaps. 1 and 2, respectively. A further conse-

*Ta_2O_5/Ta certified reference material can be purchased from the Nation Physical Laboratory, Division of Materials Applications, Teddington, Middlesex TW11 OLW, England.

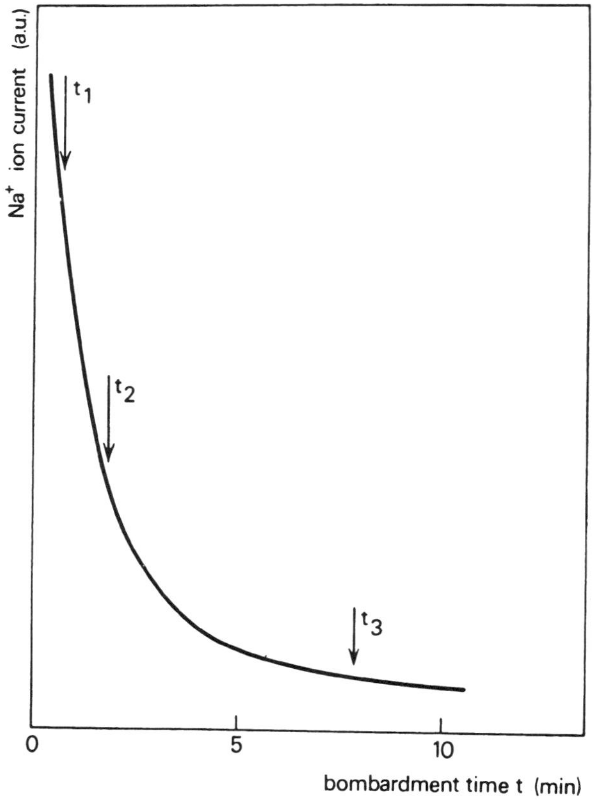

Figure 15. Na^+-ion current as a function of bombardment time obtained with SIMS.[83]

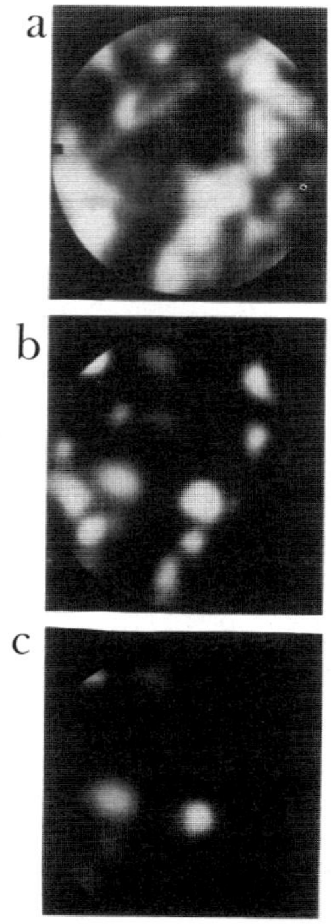

Figure 16. Na^+-ion image from the same sample as in Figure 15 at time t_1 (upper), t_2 (middle), and t_3 (lower) part.[83]

quence of the presence of an electric potential at the sample surface can be the field-induced migration of mobile ion species. This is illustrated by SIMS depth profiles of Na implanted in a SiO_2 layer on Si (Figs. 17 and 18). A dramatic migration of Na toward the SiO_2–Si interface was observed with O_2^+ primary ions, while O^- bombardment resulted in the expected profile.[184] The surface charging of insulators in SIMS profiling has been discussed extensively.[185] Several methods have been proposed to reduce the charging; a survey is available.[185]

Surface roughness is another sample property influencing the quality of a sputter depth profile; it may be initially present or induced by the bombarding ion beam. In the latter case, the crystalline quality and defect structure of the sample

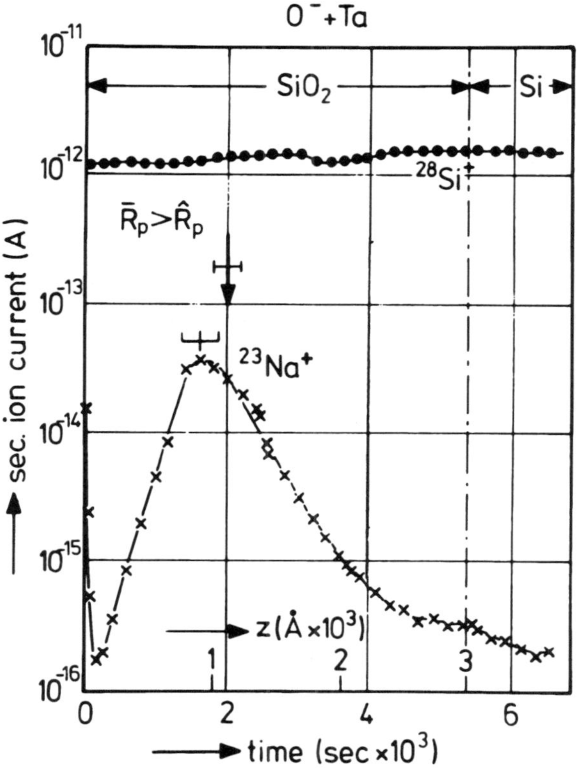

Figure 17. Profile of Na (80 keV, 10^{15} ions/cm^2) implanted in SiO_2 film (thickness 300 nm) on Si, rastered with an O_2^- beam of 0.7 μA and Ta diaphragm with orifice 600 μm in diameter. The calculated average projected range R_p is 110 nm (From Gossink *et al.*[185]).

are of major importance[186] because the sputtering yield depends on them (Sec. 2). Both initial and bombardment-induced surface roughness degrade the depth resolution. Further problems arise with quantitative XPS or AES analysis because of the reduced escape probability of the electrons.[187]

4.3.3. Artifacts Related to Instrumental Parameters

One of the major instrumental factors influencing the quality of a sputter depth profile is the homogeneity across the analyzed area of the bombarding-particle-beam current density. Generally, the current density profile of the bombarding beam will be more or less Gaussian, resulting in a nonflat crater bottom. Thus, the information-carrying particles in SIMS and ISS originate at different depths beneath the original sample surface, leading to a corresponding loss in depth resolu-

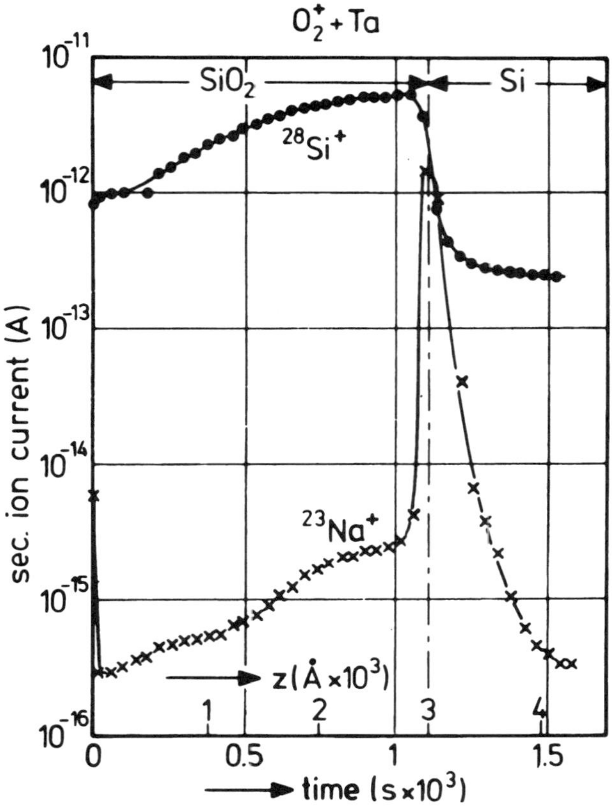

Figure 18. SIMS profile of the same sample as in Figure 17, rastered with an O_2^+ beam of 0.5 μA and using a Ta diaphragm with orifice 600 μm in diameter (From Gossink *et al.*[185]).

tion. The same situation applies to XPS since, generally, the large analyzed area is of the same order of magnitude as the sputtering beam diameter. In AES this problem can be reduced by choosing the diameter of the probing electron beam to be considerably smaller than that of the sputtering beam; over such a small diameter the sputtering beam current density is assumed to be constant. Note, however, that systematic errors in depth scale calibration might result from misalignment of the sputtering and electron beams. Furthermore, the combination of electron and ion beams produce in an enhanced sputtering yield in the area probed by the electron beam.

The nonuniform sputtering crater can be improved enormously by rastering the bombarding ion beam; generally this capability is standard in present surface analytical instrumentation. Wittmaack has shown that a flat-bottomed crater can be obtained by suitable choice of the ratio between raster width and beam diameter;[188] a ratio of 3 is generally sufficient to obtain a reasonable flat portion. An improvement in depth resolution is obtained by limiting the analyzed area to this central, flat portion of the crater. In AES and XPS this is realized by careful alignment of the sputtering and probing beams, while in SIMS and ISS electronic gating can be used.[189] In ion microscopes the analyzed area is defined by an aperture situated in an image plane of the ion optical system of the instrument.[190]

Sputtering beam inhomogeneity gives a depth-dependent contribution to the depth resolution; i.e., depth resolution decreases with increasing sputtered depth. Even for a rastered ion beam, this contribution may not be neglected at large sputtered depths because the tail of the ion beam profile often drops much more slowly than that of a Gaussian profile. The influence of the resulting curvature of the crater bottom on depth resolution can be minimized by decreasing the size of the analyzed area.[191] In practice, the minimum allowable size is determined by the required sensitivity.

The aforementioned methods not only serve to improve the depth resolution of the depth profile, they also lead to an increase in the dynamic range of the profile by reducing crater-wall effects; this is clearly demonstrated by the SIMS depth profiles in Fig. 19.

The dynamic range of a sputter depth profile generally depends on the analytical method and on the performance of the specific instrument used. The best values can be obtained with SIMS (10^6–10^7); for AES, ISS, and RBS the range is approximately 10^3, and for XPS it is about 10^2. In SIMS the dynamic range depends on three instrumental factors:

(1) Neutral component(s) in the primary-ion beam. Charge exchange processes in the high-pressure part of the ion source result in a neutral, nonfocusable contribution. These neutrals may hit the sample outside the crater (high concentration region) and thus give a dynamic-range-limiting background, especially in instruments using electronic gating. It has been shown that deflection of the primary beam suppresses this background, and 6 decades of dynamic range have been obtained for B implanted in Si.[192]

(2) Memory effects related to the deposition of sputtered material in the instrument, which may be resputtered by energetic secondary particles or scattered primary ions and subsequently detected. Memory effects may result from the analysis itself, when material deposited in the high concentration part is detected in the low-concentration region, or previous analyses.[113] In the latter case, the memory background can be suppressed by sputtering a solid not containing the element of interest prior to the actual analysis.[193]

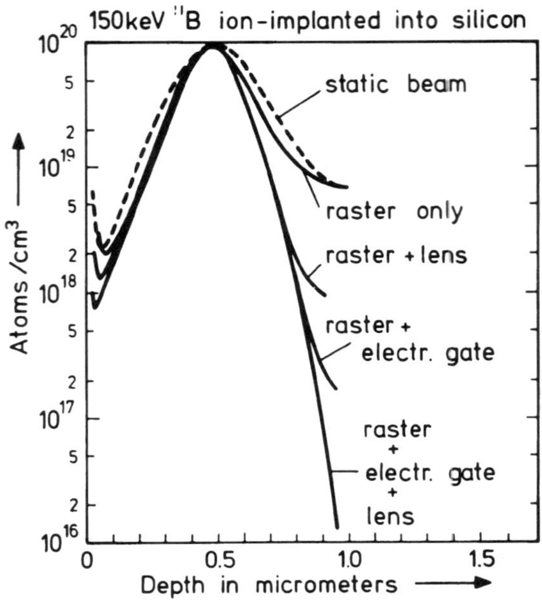

Figure 19. SIMS depth profile of ^{11}B implanted in Si, obtained using different experimental conditions, i.e., static beam, scanning beam, mechanical gating (ion lens), and electronic gating (From Magee *et al.*[190]).

(3) Adsorption of residual gas atoms and subsequent sputtering results in a variety of molecular-ion peaks in a SIMS mass spectrum. These peaks may interfere with the analytical ion at the same nominal mass, resulting in a dynamic-range-(and sensitivity; Sec. 4.4) limiting background.

Residual-gas effects can be minimized by choosing the erosion rate such that the removal rate is high compared to the adsorption rate from the ambient. In practice, this condition is met if the residual-gas pressure P (Pa) is kept below a certain value, according to $P < 10^{-4}\, j_p$, where P is the pressure in Pa and j_p is the primary-beam current density in $\mu A/mm^2$. In present SIMS instrumentation the low-pressure requirement is easily met because residual-gas pressures are well below 10^{-6} Pa. Ultrahigh vacuum conditions are equally important for the electron spectroscopies and ISS; the sputtered surface to be analyzed must not be contaminated by adsorption of gases from the ambient.

4.4. Evaluation of a Measured Depth Profile

4.4.1. Depth Resolution

The shape of a measured depth profile $m(z)$ [for sputter profiles of one element X (I(X,t)) in which the time t has been converted into depth z, we use the symbol

$m(z)$ for the remainder of this chapter] depends on both instrumental performance (instrumental depth resolution or instrumental response function) and the state of the sample,[1] i.e., the element distribution as a function of depth, called the true profile $p(z)$. To separate the influence of instrument and sample, one must use depth-profile calibration standards; in such samples the concentration $p(z)$ (Fig. 20) of element X changes abruptly from 0% to 100% (or *vice versa* in a step-down). Figure 20 also shows the measured profile $m(z)$, which can be assumed to be a Gaussian error function. The depth resolution of such a sample is then defined[125,154,194] as the depth interval Δz across which the intensity changes from 16% to 84%, i.e., the 2-sigma definition of a Gaussian error function. It can be shown[191] that for such a sample the measured depth resolution is equal to the instrument response function $a(z)$. The measured profile of any depth profile in such an instrument can then be described as the convolution of the instrument response function $a(z)$ with the true depth profile $p(z)$: $m = a * z$,[194] where * is the convolution operator. (Apart from the sign of the depth scale, convolution and cross-correlation are identical for symmetric functions.)[194] Typical values obtained for $a(z)$ are about 2–2.5 nm. The values obtained depend on beam inhomogeneity and other artifacts (Sec. 4.3). Applications of the convolution operator concept and its limitations have been discussed.[195–198]

Depth resolution can also be defined by the decay length, which is the depth interval across which the signal changes by a factor of 10.[20,21,57,169,195] The best reported experimental results have been obtained by Clegg *et al.*[57a] and Iltgen *et*

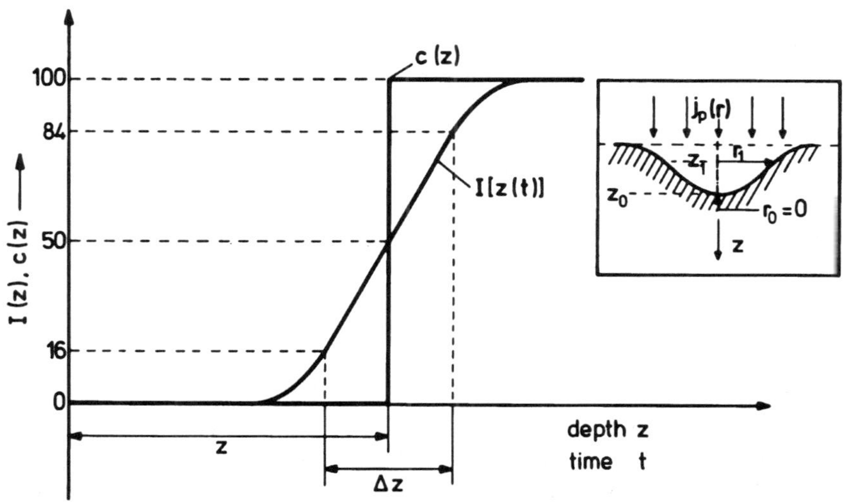

Figure 20. Definition of the depth resolution from a step-function concentration profile (From Werner[1]).

al.[57b] i.e., a leading-edge Δz value of 0.2 nm (~0.5 nm per decade of concentration for 200-eV ions) and Iltgen *et al.*[57b] (a leading edge λ_u of 0.4 nm with 1-keV Ar ions). Excellent reviews on the limits and possibilities of SIMS as an analytical tool for investigating ultrathin, atomically sharp dopant distributions (delta layers) have been given by Zalm,[199] Dowsett,[200] and Iltgen *et al.*[57b] (See also Sec. 6 in Chap. 3.)

Depth-resolution calibration standards with extremely abrupt compositional transitions (< 2 nm) have been identified, e.g., Ta_2O_5–Ta layers.[80,201] Viegers[14] and Bulle-Lieuwma *et al.*[202] have shown by means of cross-sectional TEM that interfaces between GaAs–AlAs and Si–GaP[203] and Si–GaAs[202] can be made with a sharpness of one monolayer.

For the interpretation of depth profiles, the following assumptions must be fulfilled, or else an additional broadening of the measured depth profile occurs:

1. The sample surface at the beginning of the measurement is flat and smooth,
2. The sample surface remains flat and smooth during the measurement; i.e., it must be further assumed that
   a. The sputter yield is, and remains, constant across the bombarded area during the measurement.
   b. A continuous "smooth" sputtering takes place (the more likely sputtering phenomena such as sequential-layer sputtering, etc., are discussed in Sec. 2.
3. The sample composition is homogeneous across the whole sample, so we can consider the element distribution $c(x,y,z)$ as a one-dimensional distribution $c(z)$.

When the sample surface at a given time is curved (Fig. 20, insert), the actual surface penetrates the sample at different depths (from $z = 0$ until $z = z_0$), depending on the radial extension of the emitting area A_e. The sputter depth z in this case is not unambiguously defined and must be related, for example, to the leading edge ($z = z_0$ in Fig. 20) of the profile. Similar considerations hold for rough surfaces. The influence of these depth variations on the depth resolution is discussed in Sec. 4.3.

4.4.2. Detection Limit for a Given Element

An element is said to be detected with a 97.73% single-sided confidence level when its signal is twice the average background.[131,204] For the evaluation of the detection limit, one must take into account which value of k has been used and whether the background has been subtracted (see the foregoing). Typical values for the detection limit (expressed in fractional concentration) are 10^{-3} for AES, XPS, ISS and 10^{-6} or better for SIMS. The value for the (absolute) detection limit in SIMS is about 10^{13} atoms/cm^3 in the best case; more typical values are 10^{15}–10^{17}

atoms/cm^3 (1 ppm in Si = 5×10^{16} atoms/cm^3).[205] The detection limit within a given range of parameters becomes lower (i.e., better) when the analyzed area A_e is larger. On the other hand, the depth resolution deteriorates with increasing A_e (insert of Fig. 20 and Sec. 4.3). In practice, one would therefore have to compromise.

5. Application of Sputter Depth Profiling to Various Thin-Film Materials

Most applications of all thin-film analysis techniques deal with sputter depth profiling. The applications range from stable isotope profiling to semiconductors and high-tech applications. Selected examples of applications in these fields are given in this section.

5.1. Stable Isotope Tracers

Magee, Harrington, and Honig designed a secondary-ion quadruple mass spectrometer for depth profiling.[189] Using this instrument, Magee and Botnick[206] detected traces of H and ^{2}D. They encountered problems such as contamination, stray secondary ions, the measurement of mass 1, and matrix-related artifacts. They found that negative SIMS, under Cs^+ bombardment, should be employed when profiling for deuterium in the presence of large amounts of hydrogen. This was confirmed by Magee in a paper on the analysis of H and ^{2}D by SIMS as applied to fusion technology.[207] He showed fusion-technology-related examples, such as flux and energy measurements of H and ^{2}D in the etch plasma, H and ^{2}D isotope saturation and exchange studies, and species-mix determination for H and ^{2}D in neutral beams from high-power injectors. Depth profiles of H and ^{2}D implanted into Si and C are also shown.

Czanderna *et al.*[208] used ISS to analyze copper films oxidized in ^{18}O. Thirty-six Cu films 20 nm thick were deposited in vacuum onto Pt, Au, and glass substrates and oxidized to $CuO_{0.67}$ in $^{16}O_2$ or in 98% pure $^{18}O_2$. The ISS depth profiles of the oxide films were obtained with a 1500-eV $^3He^+$ beam at 200 nA in 8×10^{-3} Pa of helium. The high mass resolution permitted measurement of different ^{18}O-to-^{16}O ratios of oxides. The sum of the peak heights of ^{18}O and ^{16}O is practically equal to that for pure Cu $^{16}O_{0.67}$. In 36 films a peak height ratio of about 16.4 was measured for Cu–O. The latter two facts suggest that isotopic labeling could be used to study these oxidation mechanisms. Depth profiles of polypropylene deposited onto Cu $^{18}O_{0.67}$ before and after oxidation of the polymer in $^{16}O_2$ are presented. The profile data agree with a model of reduction of the oxide by the polymer and partial reoxidation of the copper.[208]

5.2. Thin-Film Interdiffusion and Multilayer Analysis

Czanderna and Summermatter[209] used ISS to obtain depth profiles of gold overlayers deposited onto silver films. The four overlayers, 5, 20, 40, and 80 nm thick, were deposited onto freshly deposited Ag films (60 nm), which were not completely covered by the 5- and 20-nm-thick Au layers. Silver could still be detected under the 40-nm-thick Au overlayer. After aging for about six months at room temperature, the depth profiles showed some bulk diffusion at the interface and considerable accumulation of Ag over the Au. This is attributed to preferred diffusion paths for Ag. A sputtering yield of 2.0 Au atoms per Ne ion was calculated for 1800-eV Ne^+ ions incident at 45°.

Wildman, Howard, and Ho[210] explored the use of AES for solute grain-boundary diffusion measurements in thin films. Diffusion couples of a sandwiched Al–Cu–Al structure were prepared. The composition profile was determined before and after annealing with the AES system, which employed argon sputtering for thin-film sectioning. Both the lattice and grain-boundary diffusivities could be extracted from the solute profile. The agreement between the measured lattice diffusivity and radioactive tracer data is satisfactory. Sputter depth profiling with AES includes the analysis of Ta–Si multilayers,[211] Fe–Ti multilayers,[212] Cr–Ni multilayers,[180,211,213] Ni–Ag multilayers,[214] AgNi bilayers,[215] and nitrided oxide layers.[216] XPS sputter depth profiling of Si–SiO_2 interfaces[217] and of Ta_2O_5 [146] have been reported.

5.3. Corrosion and Oxidation

Frankenthal and Malm[218] used low-energy ISS to obtain compositional depth profiles for the oxides on a series of Fe–Cr alloys in the composition range from 0 to 25 at% of Cr. Both $^{20}Ne^+$ and $^{3}He^+$ spectra were recorded. Relative sensitivities of the oxide components were determined with the aid of standards; they appeared to be independent of alloy or oxide composition and phase. No preferential sputtering of any component was observed. The profiles (Fig. 21) show that no region of constant concentration of any component exists within the oxide. Figure 22 shows that the average Cr content of the oxide varies linearly with alloy composition and exceeds the Cr content of the alloy. The kinetics of oxidizing Cr(100) has been studied by Arlow with AES, SIMS, and sputter depth profiling.[219]

Nagasaka *et al.*[220] used ISS techniques to make compositional versus depth profiles of Au (80 nm) films on a Cu substrate. The spectra were obtained with a CMA spectrometer that used a 1500-eV He^+ beam of 7 $\mu A/cm^2$. The profiles were determined before and after treatment, which consisted of heating in both vacuum and 10.6 kPa of oxygen at 163, 202, and 250°C. Heating times were 10, 100, and 1000 min. Heating the double-layer film in vacuum at 250°C resulted in a homogeneous Au–Cu film. The same procedure in oxygen resulted in an inverted

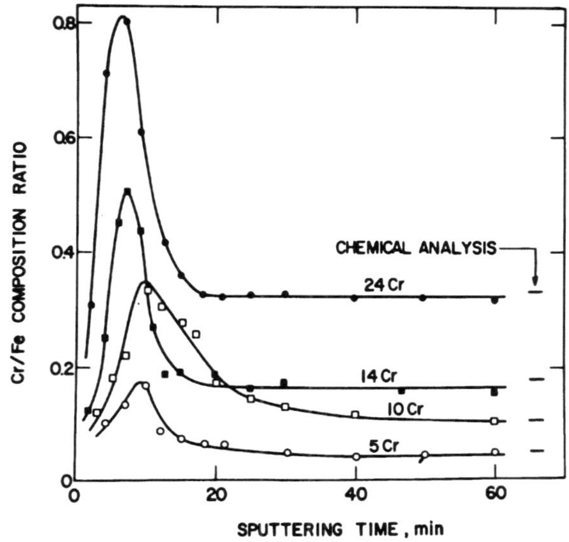

Figure 21. ISS Cr/Fe atom composition ratio as function of sputtering time for four iron–chromium alloys. The curves were obtained by correcting the experimental data for relative sensitivities of Fe and Cr in the oxides and in the alloys. The ratio for each alloy, as determined by chemical analysis, is shown to the right of each curve (From Frankenthal and Malm[219]).

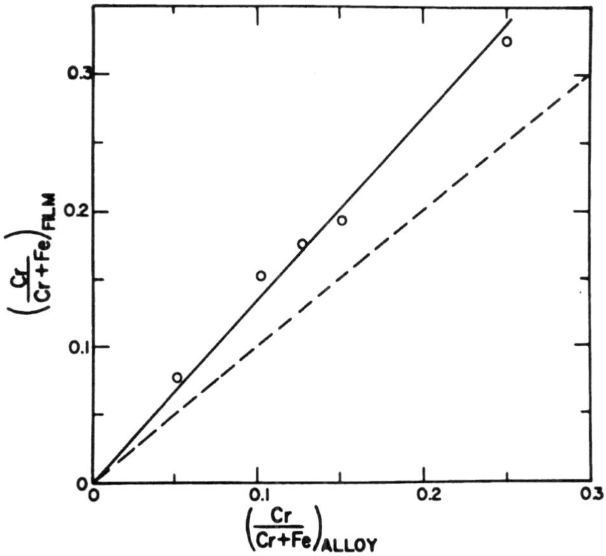

Figure 22. Average oxide film Cr content vs. alloy Cr content, showing a linear relation from 0% to 25% Cr in the alloy. The dashed line represents equal concentrations in the oxide and alloy (From Frankenthal and Malm[219]).

structure wherein the Au was found to lie beneath a gold-free copper-oxide overlayer (32.4 nm).

Miller and Czanderna[221] used ISS to study the oxidation and interfacial behavior of a vacuum-deposited thin copper film–gold foil substrate system. An ion beam of 1500-eV $^3He^+$ was used to obtain composition in-depth profiles of the oxide layer and the overlayer–substrate interface. An analytic cratering analysis technique was applied to correct the raw data. It was found that significant diffusion had occurred at the Cu–Au interface after 30 min of heating. After continued oxidation the interface became increasingly sharper and more distinct. In the later stages copper diffuses out of the Cu–Au alloy and migrates through the oxide layer. Thus, a final two-layer system of $CuO_{0.67}$ and Au is formed with a sharp interface.

Betz *et al.*[222] combined AES with simultaneous sputter removal of material to determine compositional depth profiles of surface layers (about 100 nm) on stainless steels that were air-oxidized at various temperatures. Oxidation of stainless steel 304 at temperatures up to 600°C results in the formation of an Fe-oxide layer with Cr and Ni appearing at the oxide–bulk interface. At 700°C, Cr-oxide formation near the surface begins. At 900°C, the Fe-oxide layer is replaced by a Cr-oxide film.

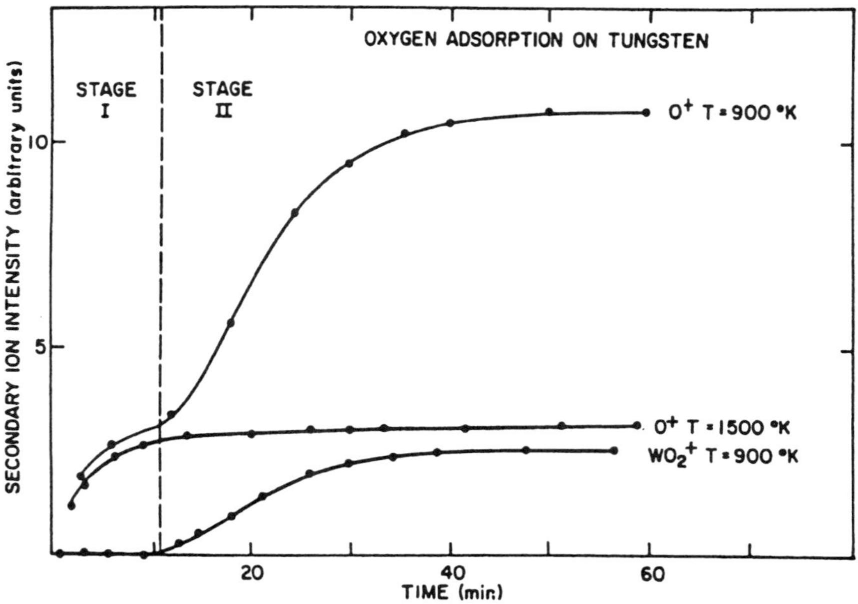

Figure 23. The variation with time of the secondary-ion intensities of O^+ and WO_2^+ from a W surface at 900 K and 1500 K. At $t = 0$, the surface of the W sample was free of adsorbed gases. The stage II process or compound formation is evident at the lower temperature (900 K) O^+ and WO_2^+ curves (From Rybalko *et al.*[224]).

Profiles for air-oxidized (900°C) stainless steel 321 and Inconel are also reported. All profiles were determined with a commercial thin-film Auger analyzer and an Ar-ion gun for simultaneous sputter etching.

Rybalko *et al.*[223] used SIMS to investigate the dynamic surface process of oxygen adsorption on W. They confirmed the two-stage interaction mechanism observed by McCarroll with other methods.[224] Figure 23 in stage I shows O_2 dissociating into atomic oxygen when adsorbed on the surface; in stage II, WO_2^+ indicates the presence of W compounds. The saturation level in the 900 K stage II region corresponds to equilibrium conditions between gas-phase O_2 and the surface oxide layer.

McIntyre[225] claims that a detailed analysis of the relationship between oxide and metal phases is only possible in XPS, where oxide–metal contributions to the

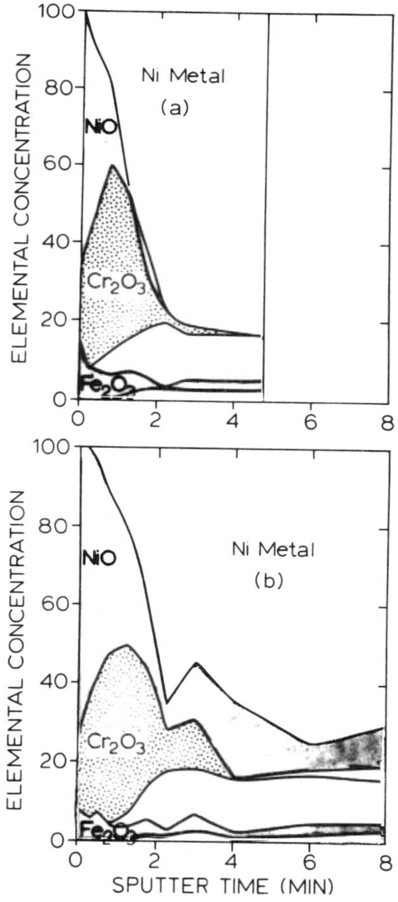

Figure 24. XPS depth profiles of surface films on Alloy 600 grown at 300°C under two oxygen pressures: (a) 0.01% O_2, 5 min; and (b) 5% O_2, 5 min (From McIntyre[226]).

spectrum are well separated. Figure 24 shows cumulative presentations of oxide and metal phases on NiCr 600 surfaces, oxidized at 300°C under two oxygen pressures: 0.01 and 5% O_2, both for 5 min. The sample oxidized at higher oxygen partial pressure shows a continuation of a nickel-oxide phase deep into the substrate. Only AES could probably detect such effects.

5.4. Catalysts

Chin and Hercules[226] used XPS, ISS, and SIMS to investigate the surface properties of a series of impregnated Co–Al_2O_3 catalysts. Those surface characteristics are affected by metal loading as well as calcination. This is probably due to metal-support interactions arising from Co-ion diffusion into lattice sites of the γ-Al_2O_3 support during calcination. The fraction of Co ions that interact with the support increases with decreasing metal loading or rising calcination temperature. It is also shown that the catalyst reducibility is directly related to the degree of metal-support interaction. The same authors[227] studied the influence of Zn on the surface properties of the catalysts mentioned above. It can be seen from the XPS and ISS results that adding a small quantity of Zn to the γ-Al_2O_3 support can enhance formation of a Co "surface spinel" if the Co concentrations are <8%. This is due to Co-ion diffusion into tetrahedral sites of the support. At higher Co concentrations, Zn induces a drastic increase in the XPS and ISS metal-support intensity ratios. X-ray diffraction data indicate that the latter increase results from an increase in dispersion. This means that smaller Co-oxide crystallites are generated in catalysts doped with Zn.

Zingg *et al.*[228] used XPS, ISS, and laser Raman spectroscopy to characterize a series of MoO_3–γ-Al_2O_3 catalysts. Raman spectra showed three different Mo species (see also Ref. 229). Reaction of molybdic acid with surface hydroxyl groups probably produced an interaction species, found at low concentrations. XPS and ISS have identified the presence of both tetrahedrally and octahedrally coordinated Mo, the latter being present above 4 wt% MoO_3. After monolayer coverage, $Al_2(MoO_4)$ is formed, and at higher concentrations bulk MoO_3 is formed. Before monolayer coverage, Mo forms no discrete compounds; geometrical configurations were determined by sites of the γ-Al_2O_3 lattice. It was found that both increased calcination temperature and time favor the formation of $Al_2(MoO_4)_3$.

5.5. Polymer–Metal Interfaces

Miller *et al.*[230] used ISS profiles to determine the role of the polymer–copper oxide interface in the catalyzed oxidative degradation of polypropylene. Copper films were deposited onto glass substrates and oxidized partly or completely to $CuO_{0.67}$ in ^{16}O or ^{18}O. The 44-nm-thick oxide films were then overlaid with 40 to

115 nm of isotactic polypropylene (PP). The PP–$CuO_{0.67}$–glass sandwiches were heated in ^{16}O at 90, 100, 110, and 120°C with and without getters for the product gases. ISS depth profiles were obtained for Cu films, partially and completely oxidized films (labeled and unlabeled), and for undegraded and oxidatively degraded PP-oxide sandwiches. With getters present, partial reduction to Cu of the oxide film is induced by PP, and the $^{16}O/^{18}O$ ratio in the oxide increased considerably after degradation. Extensive oxidation of the PP is evident from the ^{16}O signal, and again Cu is found throughout it. The results show that reduction of the Cu oxide by the PP is part of the mechanism by which Cu catalyzes PP degradation. Because partial reoxidation of Cu also occurs without getters, a reduction–reoxidation cycle at the Cu–polymer interface is suggested. See also Ref. 208.

Sparrow and Mishmash[(231)] used ISS with a cylindrical mirror analyzer and SIMS to study glass surfaces and the first 3 nm of treated polymers to understand its chemistry and properties. The information for the polymers contributed to the explanation of properties related to oxidation, degradation, reflectance, hardness, and particularly adhesion to polymers. The surface is subject to ion migration, which is disconcerting to the analyst. A commercial polyamide, Kapton, cleaned by acid, base, solvents, and various drying techniques was surface-analyzed. Charge neutralization was necessary. Surface information was correlated with original and residual contaminants and thin-film bondability. The concentrations of C, N, and O of the polyamide were substantially altered by simple cleaning to about 3 nm thick. The surface with the greatest bondability of coated Al showed minimum contaminations due to Na, Mg, Al, Si, K, Ca, and F.

5.6. Insulating Materials

Nonconducting samples, particularly when bombarded with ions to achieve sputter depth profiling, will charge electrically at the surface. This charging generally degrades the performance of the analytical technique used. Several methods have been developed to reduce surface charging,[(185,232)] and, generally, reliable sputter depth profiling of insulating materials can be achieved.

Aqueous corrosion profiles in soda–lime glass were studied by Gossink.[(184)] Sodium depletion of the surface as a function of time is illustrated in Fig. 25. Continuing this study, Smets[(233)] measured Na profiles as a function of leaching time, using SIMS. The measured S-shaped profile shows a Na depletion at the surface; the thickness of the depletion layer increases with the square root of the leaching time. This indicates that the depletion is governed by a diffusion process.

Compositional changes in yttrium–iron–garnet thin films were studied with SIMS depth profiles.[(234,235)] Undulations in the Pb content were found to be related to the rotation rate of the sample during growth, as illustrated in Fig. 26.

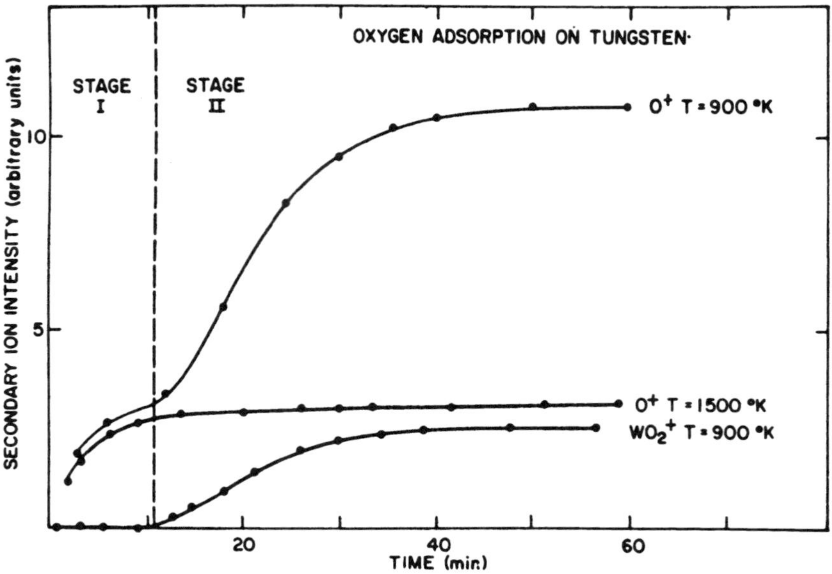

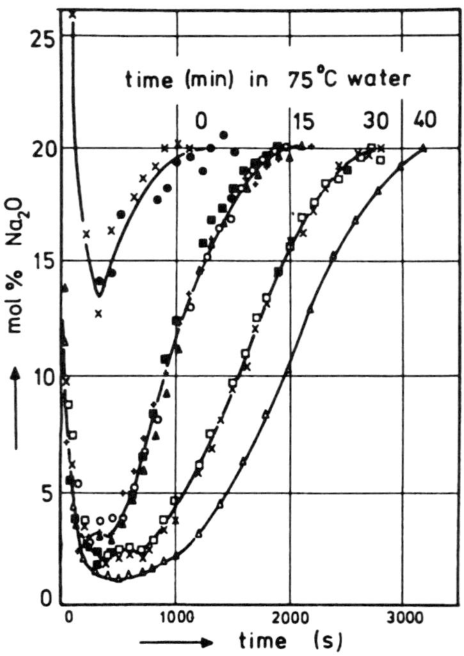

Figure 25. Na_2O concentration profiles in a 20% Na_2O·10% CaO·70% SiO_2 glass treated in water at 75°C for the times as indicated using SIMS with a rastered O_2^- beam of 0.7 μA and a Ta diaphragm. The approximate erosion rate is 0.4 nm/s (From Gossink *et al.*[185]).

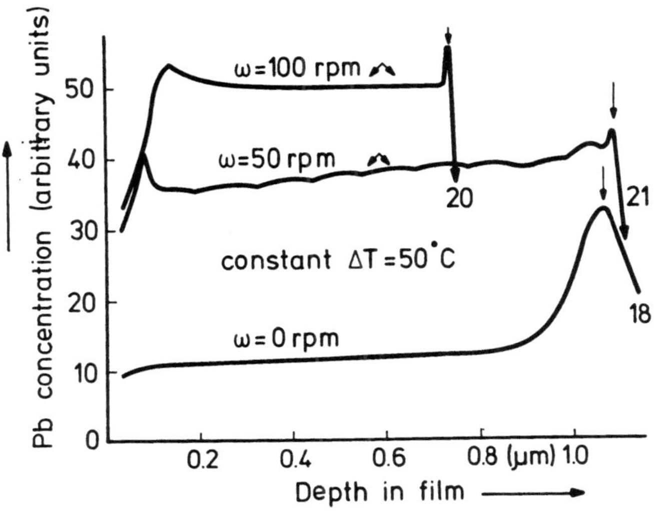

Figure 26. Concentration depth profiles of Pb in films grown at different rotation rates with constant supercooling ($T = 50°C$) (From Jonker *et al.*[235]).

5.7. Semiconductor Materials

Sputter depth profiling is widely used for the characterization of thin films of semiconductors and other materials used in the electronics industry. In the past two decades, numerous examples of the application of various analytical techniques using sputtering in this field have appeared in the literature. A survey may be found in recent review articles and books.[134,236–238] We will attempt to illustrate this broad application field by presenting some suitable examples.

Because of its high sensitivity, SIMS has been used extensively to monitor the incorporation of metallic impurities during growth of GaAs[239] and InP[240] epitaxial layers. Furthermore, it was shown by SIMS depth-profiling studies that information on the redistribution of intentional and nonintentional impurity elements at a very low concentration level is indispensable to understand the electrical properties of devices based on III–V compound semiconductors.[239] The experimental requirements for such ultratrace SIMS analyses in these materials have been reviewed by Clegg.[241]

The applications of SIMS in silicon technology are in materials research, process or device development, and failure analysis. A typical example of the application of SIMS to process development is the study of polycrystalline–monocrystalline silicon contact structures used, e.g., in SRAM chips.[242] Contacts of *n*-type can be obtained by doping the poly-Si layer with phosphorus from a PH_3 or $POCl_3$ source or by implantation of P^+ ions. The SIMS depth profiles of phosphorus after annealing for these three cases are shown in Fig. 27. After PH_3 or $POCl_3$ doping, the profile is characterized by a pronounced tail penetrating deeply into the monocrystalline Si, whereas such a tail is absent for the P^+-implanted sample. The extended diffusion tail is thought to be responsible for the low punch-through voltage to neighboring diffusions in contact structures obtained with PH_3 or $POCl_3$ doping; neighboring diffusions, especially in a dense layout, are insufficiently separated due to the lateral component of the diffusion. Much improved punch-through characteristics were obtained with P^+-implanted structures, and based on the understanding resulting from the SIMS depth profiles[(242)] this implantation step was implemented in the fabrication process of SRAM chips.

Another important application area of SIMS in present semiconductor research is the calibration and optimization of process simulation programs such as SUPREM.[(243)] Accurate and precise data on solubility,[(244)] segregation coefficient,[(147,245,246)] diffusivity,[(244,246)] and range distribution of implanted ions[(247,248)] for various doping elements can be derived from SIMS depth profiles; such a type of SIMS analysis requires well-defined quantification procedures.[(135)] Figure 28 shows the SUPREM3-simulated and SIMS depth profiles of the dopant elements in a bipolar transistor from a merged bipolar MOS process.[(135)]

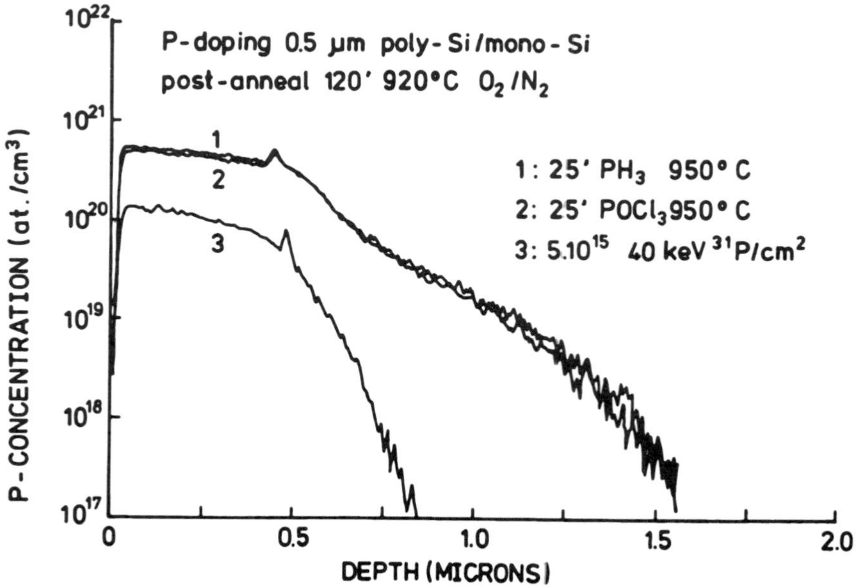

Figure 27. SIMS depth profiles of phosphorus in 0.5-μm polycrystalline Si on monocrystalline Si for different doping procedures [From P. R. Boudewijn and K. T. F. Janssen, *Fresenius Z. Anal. Chem.* **329**, 215 (1987)].

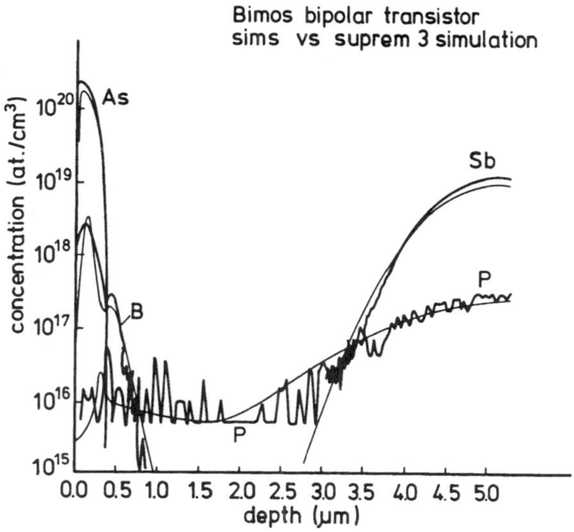

Figure 28. Comparison of B, P, As, and Sb depth profiles in a bipolar MOS transistor obtained with SIMS (thick lines) and with the SUPREM3 simulation program (thin lines).

AES in combination with sputter depth profiling is routinely used for the analysis of thin films and interfaces of materials used in the semiconductor industry.(211,236,237) AES is especially attractive in the scanning mode (SAM) for application to problems related to integrated circuit (IC) manufacturing due to the high lateral resolution (20–300 nm). As an example, the nature of hillocks (0.5 μm diameter) on a gold layer present on an IC for bonding purposes was investigated with SAM.(249) Using a 30-nm electron beam, the authors used a secondary-electron image to localize the hillocks (Fig. 29), and a 300-nm FWHM beam was then used for the actual Auger analysis. The larger beam is required to give sufficient Auger intensity, and the hillocks could be identified as gold precipitates. In another sample from the same batch, it was found that bonding problems occurred on the gold hillock areas. A depth analysis using sputtering showed that in this case the hillocks were covered with 50 nm of SiO_2 (Fig. 30); this explained the bonding failure.(249)

In the semiconductor industry, analytical techniques using sputter depth profiling such as SIMS have become indispensable tools for assessing materials and devices, particularly when focused-ion-beam (FIB) sources are used.(271) However, continuous research efforts into the basic principles of the analytical techniques and into the field of instrumentation are still required to ensure successful analytical assistance for future generations of materials and devices resulting from submicron technologies. XPS combined with sputter depth profiling has also been used in failure analysis.(250)

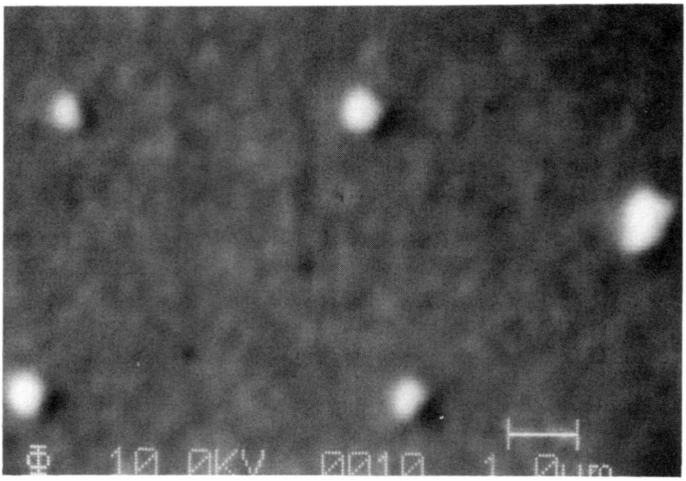

Figure 29. Micrograph (topographic contrast) of Au hillocks on a 0.1-μm Au layer on Si obtained using a SAM in a secondary-electron mode. (Courtesy B. Otterloo[250]).

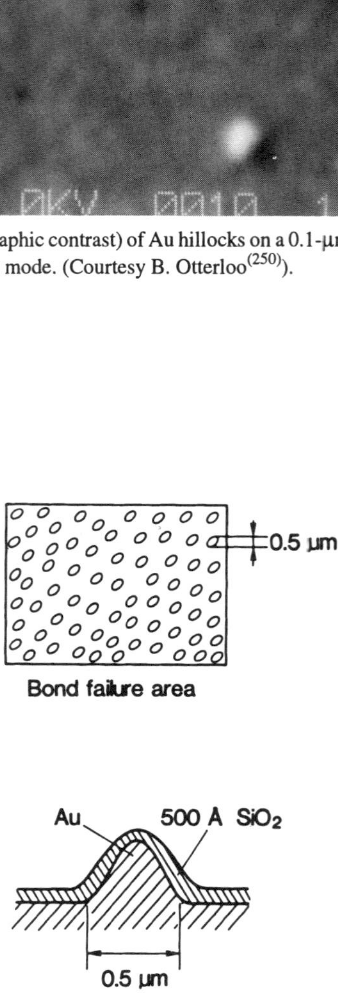

Figure 30. Cross section (schematic) of a Au hillock covered with SiO_2 in the bond failure area (Courtesy B. Otterloo[250]).

5.8. Miscellaneous Applications

Sputtering of monolayers in SIMS has been studied by Benninghoven,[251] who showed that the secondary-ion current I of species X_i as a function of time can be described by $I(X_i, t) = I(X_i, 0)\exp(-t/\tau)$, where X_i is contained in a monolayer, the average lifetime τ of the original monolayer $= 1/\nu\sigma$, where σ is the disappearance cross section of a surface molecule M from which the selected secondary ion X_i has emerged and ν is the arrival rate/cm^2 of primary ions = particle current density/cm^2-s. The given relation can be used (a) to detect if a layer is present as a monolayer and (b) to determine the disappearance cross section of molecule M.

For the study of Ag catalysts, Fig. 31[252] shows the decrease of some secondary ions during sputtering from the topmost layer of a poisoned Ag catalyst. The exponential decrease of some anion complexes indicates their presence in only the uppermost monolayer. For other negative secondary ions (e.g., OH^-, O^-, Cl^-), the decrease is much slower, indicating a multilayer presence of the corresponding parent elements or compounds.

The oxidation behavior of polycrystalline chromium has been studied with static SIMS by Benninghoven.[252] Different oxide structures, which can be described by characteristic fingerprint spectra,[253] are formed successively. The following model for surface oxidation of Cr (and other metals) has been derived from these measurements: during the first step of oxidation (oxide phase I, Fig. 32[252] CrO_2^- and, with increasing thickness of this phase, CrO_3^- ions are emitted; at the end of the oxidation, an oxide with a high oxygen abundance (phase II; Fig. 32) is formed on top of phase I, characterized by the emission of CrO^+ ions. This model can be confirmed by observing the behavior of different secondary ions during sputtering of the oxide layer. Figure 32b[252] gives the results for an oxide layer formed by an oxygen dose of a 1200 L. At the beginning of the sputtering process, a sharp decrease of CrO^+ is observed, when the parent structure of CrO^+ (oxide phase II) is sputtered away with a high sputtering rate. As CrO^+ decreases, oxide phase I becomes the uppermost layer of the bombarded surface. Thus, a simultaneous increase in CrO_2^- is observed. The CrO_2^- emission is constant for a relatively long period (Fig. 32b, upper curve of CrO_2^-), which means that oxide phase I has a thickness of more than one monolayer. At a primary-ion dose density of 10^{-4} A·s/cm^2, the CrO_2^- intensity decreases. Oxide phase I now vanishes, and the pure metal surface appears at the surface. This results in a corresponding increase in Cr_3^+ and is typical of the clean metal surface. If a cleaned Cr surface is oxidized by an oxygen dose of only 100 L, oxide phase I is formed on the surface with a thickness below one monolayer. This idea is supported by CrO_2^- decrease during sputtering of such a layer. Figure 32b (lower CrO_2^- curve) shows an exponential time dependence, indicating that oxide phase I is now free of a covering structure (no increase in CrO^- at the beginning of the sputtering process) and is not more than a monolayer thick.

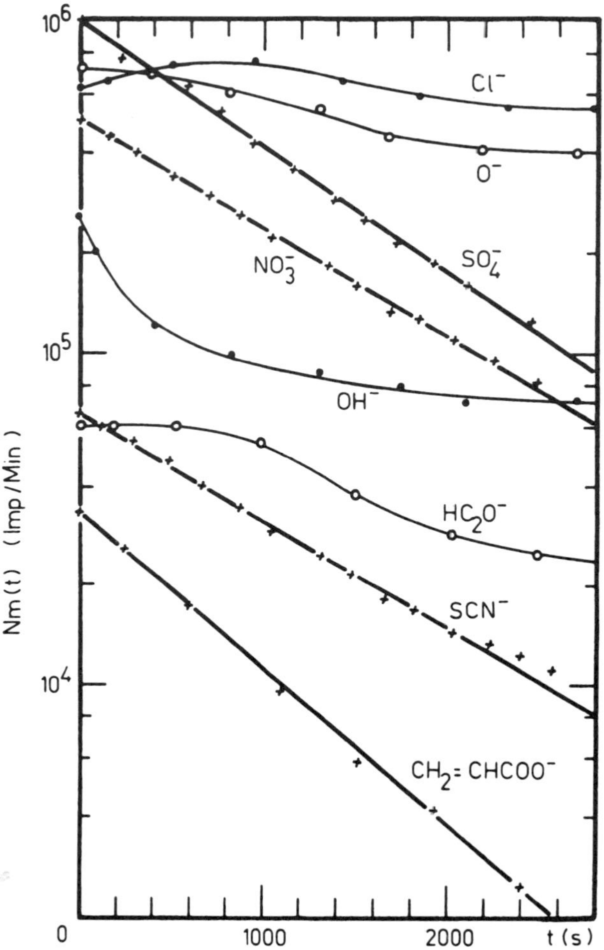

Figure 31. Decrease of some secondary-ion intensities N_m during sputtering of the uppermost monolayer of a Ag catalyst. The exponential decrease of some anion complexes indicates their presence in only the uppermost monolayer (From Benninghoven[253]).

Benninghoven[254] has studied the ion bombardment behavior of amino acid overlayers on metals by means of static SIMS. This resulted in a new model of amino acid–substrate interaction. In addition, HgCdTe (MCT) has been studied with AES.[255] Superconducting materials have been studied by AES[256] and laser ionization (SALI).[257] The formation of "magnetically dead layers" in ferromagnets by sputtering has also been studied.[258]

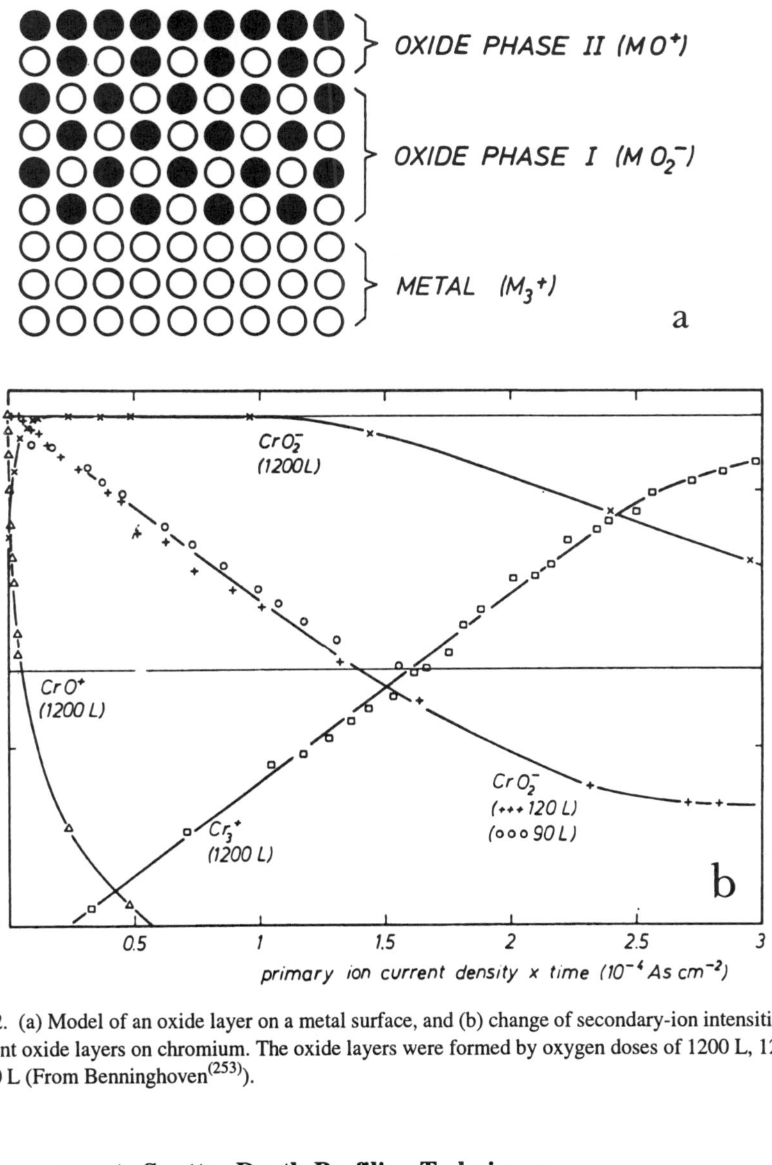

Figure 32. (a) Model of an oxide layer on a metal surface, and (b) change of secondary-ion intensities of different oxide layers on chromium. The oxide layers were formed by oxygen doses of 1200 L, 120 L, and 90 L (From Benninghoven[253]).

5.9. Newcomers to Sputter-Depth-Profiling Techniques

Sputter depth profiling using SNMS to detect the surface composition is being used with increasing frequency.[257,259,260] This has resulted from the pioneering and launching of SNMS by Oechsner,[261,262,263] of photoionization of sputtered neutrals by Becker,[257,264,265] and of electron-gas SNMS by Jede.[266] Even Raman spectroscopy has been combined with sputter depth profiling.[229]

6. Summary and Future Prospects

Sputter-depth-profiling techniques have become a mature tool, in particular for studies in the microelectronic industry; the latter can be derived from the relatively large number of SIMS instruments (many tens per year) purchased by leading companies in the field. The fundamental processes and the basis for reliable applications are being, and will be better, understood in comparison with sputter profiles obtained by nonsputtering ("nondestructive" techniques), e.g., Rutherford backscattering[6,10,163,164,267] and scanning acoustic microscopy.[268,269]

Acknowledgment

The authors wish to thank Mr. A. H. E. Barten for his help in updating the references.

References

1. H. W. Werner, in *Thin Film and Depth Profile Analysis, Topics in Current Physics,* Vol. 37 (H. Oechsner, ed.), Springer, Berlin (1984), pp. 5–38.
2. M. D. Hill, *Surf. Interface Anal.* **12**, 21 (1988).
3. J. M. Walls, D. D. Hall, and D. E. Sykes, *Surf. Interface Anal.* **1**, 204 (1979).
4. R. Voigtmann and W. Moldenhauer, *Surf. Interface Anal.* **13**, 167 (1988).
5. H. W. Werner, *Electron Microsc.* **3**, 200 (1980).
6. A. W. Czanderna, in *Ion Spectroscopies for Surface Analysis,* (A. W. Czanderna and D. M. Hercules, eds.), Plenum Press, New York (1991), pp. 1–44.
7. J. P. Zhang, P. L. F. Hemment, N. Hatzopoulos, S. M. Newstead, A. R. Powell, T. E. Whall, and E. H. C. Parker, *Mater. Sci. Eng.* B**12**, 21 (1992).
8. J. P. Zhang, P. L. F. Hemment, N. Hatzopoulos, S. M. Newstead, A. R. Powell, T. E. Whall, and E. H. C. Parker, *Nucl. Instr. Meth. Phys. Res.* B**62**, 325 (1992).
9. A. V. Drigo, *Mikrochim. Acta,* **114/115**, 89 (1994).
10. L. Feldman, in: *Ion Spectroscopies for Surface Analysis*, (A. W. Czanderna and D. M. Hercules, eds.), Plenum Press, New York (1991), pp. 311–361.
11. P. Willich and D. Obertop, Paper presented at 12th Colloquium on Materials Analysis, Vienna, May 13–15 (1985), in: *Proceedings of Progress in Materials Analysis, Mikrochim. Acta, 11*, (M. Grasserbauer and M. K. Zacherl, eds.), Springer, Wien (1985), p. 299.
12. H. W. Werner, *Mat. Sci. Eng.* **42**, 1 (1980).
13. W. v. Ooij and T. Visser, *Spectrochimica Acta B* **39**, 1541 (1984).
14. M. P. A. Viegers, A. F. de Jong, and M. R. Leys, *Spectrochim. Acta B* **40**, 835 (1985).
15. T .D. Bussing, P. H. Holloway. Y. X. Wang, J. F. Moulder, and J. S. Hammond, *J. Vac. Sci. Technol. B* **6**, 1514 (1988).
16. M. Wolff and J. W. Schultz, *Surf. Interface Anal.* **12**, 93 (1980).
17. W. J. Bartels, *J. Vac. Sci. Technol. B* **1**, 338 (1983).
18. W. J. Bartels and W. Nijman, *J. Cryst. Growth* **44**, 518 (1978).
19. H. W. Werner, *Fresenius Z. Anal. Chem.* **314**, 274 (1983).
20. P. C. Zalm, *Philips J. Res.* **47**, 287 (1993).

21. K. Wittmaack, *Surf. Interface Anal.* **21**, 323 (1994).
22. P. D. Townsend, in: *Sputtering by Particle Bombardment II*, (R. Behrisch, ed.), Springer, Berlin (1983), p. 147.
23. R. Behrisch, in: *Sputtering by Particle Bombardment II*, (R. Behrisch, ed.), Springer, Berlin (1983), p. 179.
24. N. Bohr, *K. Dan. Vidensk. Selsk. Mat. Fys. Medd.* **18**(8), 1 (1948).
25. P. Sigmund, in: *Sputtering by Particle Bombardment I*, (R. Behrisch, ed.), Springer, Berlin (1981), p. 9.
26. J. Lindhard, V. Nielsen, and M. Scharff, *K. Dan. Vidensk. Selsk. Mat. Fys. Medd.* **36**(10), 32 (1968).
27. H. H. Andersen and H. Bay, *J. Appl. Phys.* **45**, 953 (1975).
28. H. H. Andersen and H. Bay, *J. Appl. Phys.* **46**, 2416 (1975).
29. P. Sigmund, *Phys. Rev.* **184**, 383 (1969).
30. J. Lindhard, M. Scharff, and H. E. Schiott, *K. Dan. Vidensk. Selsk. Mat. Fys. Medd.* **33**(10), 1 (1963).
31. P. Sigmund, M. T. Matthies, and D. L. Phillips, *Radiat. Eff.* **11**, 34 (1971).
32. P. C. Zalm, *J. Appl. Phys.* **54**, 2660 (1983).
33. W. D. Wilson, L. G. Haggmark, and J. P. Biersack, *Phys. Rev.* B**15**, 2458 (1977).
34. J. Bodansky, J. Roth, and H. L. Bay, *J. Appl. Phys.* **51**, 2861 (1980).
35. N. Matsunami, Y. Yamamura, Y. Itihawa, N. Itah, Y. Katumata, S. Myyagawa, K. Morita, and R. Shimizu, *Radiat. Eff. Lett.* **57**, 15 (1980).
36. P. C. Zalm, *J. Vac. Sci. Technol. B* **2**, 151 (1984).
37. D. A. Thompson and S. S. Johar, *Appl. Phys. Lett.* **34**, 342 (1979).
38. D. A. Thompson, *J. Appl. Phys.* **52**, 982 (1981).
39. P. C. Zalm, *Rep. Prog. Phys.* **58** (10), 1321–74 (1995).
40. J. Lindhard and M. Scharff, *Phys. Rev.* **124**, 128 (1961).
41. P. K. Rol, J. M. Fluit, F. P. Viehböck, and M. de Jong, in: *Proceeding of the Fourth International Conference on Phen. Ion Gases*, (N. R. Nilsson, ed.), North-Holland, Amsterdam (1959), p. 257
42. D. Onderdelinden, *Appl. Phys. Lett.* **8**, 189 (1966), **46**, 739 (1968).
43. D. Onderdelinden, *Can. J. Phys.* **46**, 739 (1968).
44. M. T. Robinson, in: *Sputtering by Particle Bombardment I*, (R. Behrisch, ed.), Springer, Berlin (1981), p. 73.
45. H. E. Roosendaal, in: *Sputtering by Particle Bombardment I*, (R. Behrisch, ed.), Springer, Berlin (1981), p. 219.
46. N. Andersen and P. Sigmund, *K. Dan. Vidensk. Selsk Fys. Medd.* **39** (3), 1–45 (1974).
47. G. Betz and G. K. Wehner, in: *Sputtering by Particle Bombardment II*, (R. Behrisch, ed.), Springer, Berlin (1983), p. 11.
47a. H. H. Andersen and H. L. Bay, in: *Sputtering by Particle Bombardment I* (R. Behrisch, ed.), Springer, Berlin (1987), pp. 145–209.
48. H. Oechsner, *Z. Phys.* **261**, 37–58 (1973).
49. M. P. Seah and W. A. Dench, *Surf. Interface Anal.* **1**, pp. 1–11 (1979).
50. T. Ishitani and R. Shimizu, *Appl. Phys.* **6**, pp. 2–41 (1975).
51. J. A. Peinador, I. Abril, J. J. Jimenez-Rodriguez, and A. Gras-Marti, *Surf. Interface Anal.* **15**, 463 (1990).
52. Yang-Tse Cheng, A. A. Dow, B. M. Clemens, and Eun-Hee Cirlin, *J. Vac. Sci. Technol. A* **7**, 1641 (1989).
53. Yang-Tse Cheng, A. A. Dow, and B. M. Clemens, *Appl. Phys. Lett.* **53**, 1346 (1988).
54. P. A. Ronsheim and M. Tejawani, *J. Vac. Sci. Technol. B* **12**, 254 (1994).
55. H. S. Lee, R. T. Lareau, S. N. Schauer, R. P. Moerkirk, K. A. Jones, S. Elagoz, W. Vavra, and R. Clark, in: *Advanced III-V Compound Semiconductor Growth Processing and Devices Symposium*, (S. J. Pearton, D. K. Sadana, and J. M. Zavada, eds.), Materials Research Society, Pittsburgh (1992), p. 473.

56. S. G. Puranik and B. V. King, *Nuc. Instr. Meth. Phys. Res. B* **34**, 102 (1988).
57. (a). J. B. Clegg, N. S. Smith, M. G. Dowsett, M. J. J. Theunissen, and W. B. de Boer, *J. Vac. Sci. Technol.* **14**, 2645 (1996); (b). K. Iltgen, A. Benninghoven, and E. Niehuis, Wiley, Chichester, UK (1998).
58. F. Schulz and K. Wittmaack, *Radiat. Eff.* **29**, 31 (1976).
59. A. E. Morgan, H. A. M. de Grefte, N. and H. J. Tolle, *Appl. Surf. Sci.* **7**, pp. 372–392 (1981).
60. N. Warmoltz, H. W. Werner, and A. E. Morgan, *Surf. Interface Anal.* **2**, 46 (1980).
61. W. Wach and K. Wittmaack, *J. Appl. Phys.* **52**, 3341 (1981).
62. H. Liebl and R. F. K. Herzog, *J. Appl. Phys.* **34**, 2893 (1963).
63. H. Liebl, *J. Phys. E: Sci. Instrum.* **8**, 797 (1975).
64. J. J. Le Goux and H. N. Migeon, in: *Secondary Ion Mass Spectrometry SIMS III*, (A. Benninghoven, J. Giber, J. Laszlo, M. Riedel, and H. W. Werner, eds.), Springer, Berlin (1982), p. 52.
65. P. Williams, R. K. Lewis, C. A. Evans, Jr., and P. R. Hanley, *Anal. Chem.* **49**, 1399 (1977).
66. R. Gomer, *Appl. Phys.* **19**, 365 (1979).
67. R. Clampitt and D. K. Jefferies, *Nucl. Instrum. Meth.* **149**, 739 (1978).
68. M. J. Higatsberger, P. Pollinger, H. Studricka, and F. G. Rüdenauer, in: *Secondary Ion Mass Spectrometry SIMS III*, (A. Benninghoven, J. Giber, J. Laszlo, M. Riedel, and H. W. Werner, eds.), Springer, Berlin (1982), p. 38.
69. F. G. Rüdenauer, in: *Secondary Ion Mass Spectrometry SIMS IV*, (A. Benninghoven, J. Okano, R. Shimizu, and H. W. Werner, eds.), Springer, Berlin (1984), p. 133.
70. H. Liebl, in: *Secondary Ion Mass Spectrometry SIMS II*, (A. Benninghoven, C. A. Evans, Jr., R. A. Powell, R. Shimizu, and H. A. Storms, eds.), Springer, Berlin (1979), p. 176.
71. P. D. Prewett and D. K. Jefferies, *Inst. Phys. Conf. Ser.* **54**, 316 (1980).
72. A. R. Waugh, A.R. Bayly, and K. Anderson, in: *Secondary Ion Mass Spectrometry SIMS IV*, (A. Benninghoven, J. Okano, R. Shimizu, and H. W. Werner, eds.), Springer, Berlin (1984), p. 138.
73. F. Degreve and J. M. Lang, *Surf. Interface Anal.* **1**, pp. 177–187 (1985).
74. H. W. Werner and H. A. M. de Grefte, *Vakuum-Technik* **17**, 37 (1967).
75. S. Hofmann, *Prog. Surf. Sci.* **36**, 35 (1991).
76. H. W. Werner, in: *Applied Spectroscopy*, Vol. 7A (E. L. Grove, ed.), Plenum Press, New York (1969), p. 239.
77. C. J. Pellerin, J. Christensen, R. C. Jerner, and J. H. Peary, *J. Vac. Sci. Technol.* **12**, 496 (1975).
78. F. Meyer and C. J. Loyer, *Acta Electron.* **18**, 33 (1975).
79. W. Vandervorst, H. E. Mees, and R. F. de Keersmaecker, *J. Appl. Phys.* **56**, 1425 (1984).
80. J. Kirschner and H. W. Etzkorn, *Proceedings of the 7th International Vacuum Congress and 3rd International Conference on Solid Surfaces*, Vol. 3 (F. Rüdenauer, F. P. Viehböck, and R. Dobrozemsky, eds.), Technical Universität, Vienna, IAP (1977), p. 2213.
81. F. Lopez, M. V. Garcia-Cuenca, J. M. Asensi, and J. L. Morenza, *Appl. Surf. Sci.* **70–71**, 68–72 (1993).
82. H. W. Werner, *Surf. Sci.* **35**, 458 (1973).
83. P. Blank and K. Wittmaack, *J. Appl. Phys.* **50**, 1519 (1979).
84. J. E. Kempf and H. H. Wagner, in: *Thin Solid Films and Depth Profile Analysis, Topics in Current Physics*, Vol. 37 (H. Oechsner, ed.), Springer, Berlin (1984), p. 87.
85. W. M. Riggs and M. J. Parker, in: *Methods of Surface Analysis* (A. W. Czanderna, ed.), Elsevier, Amsterdam (1975), p. 103.
86. J. H. Ebel and M. F. Ebel, *X-ray Spectrosc.* **2**, 19 (1973).
87. R. D. Giauque, F. S. Goulding, J. M. Jaklovic, and R. H. Pehl, *Anal.Chem.* **45**, 671 (1973).
88. J. H. Scofield, *J. Electr. Spectrosc. Relat. Phenom.* **8**, 129 (1976).
89. S. T. Manson and D. Dill, in: *Electron Spectroscopy: Theory, Techniques and Applications*, Vol. 2 (C. R. Brundle and A. D. Baker, eds.), Academic, New York (1978), p. 158.
90. C. J. Powell and P. E. Larson, *Appl. Surf. Sci.* **1**, 186 (1978).

91. M. F. Ebel, *Surf. Interface Anal.* **2**, 173 (1980).
92. M. F. Ebel, H. Ebel, and K. Hirokawa, *Spectrochim. Acta.* **37B**, 461 (1982).
93. C. D. Wagner, *Anal. Chem.* **44**, 1050 (1972).
94. C. D. Wagner, L. E. Davis, M. V. Zeller, J. A. Taylor, R. H. Raymond, and L. H. Gale, *Surf. Interface Anal.* **3**, 211 (1981).
95. K. Hirokawa and Y. Danzaki, *Surf. Interface Anal.* **2**, 240 (1980).
96. H. E. Bishop, *Surf. Interface Anal.* **3**, 272 (1981).
97. W. D. Mross and J. Kronenbitter, *Chem.-Ing.-Techn.* **54**, 33 (1982).
98. K. Siegbahn, C. Nordling, A. Fahlmann, R. Nordberg, K. Hamrin, J. Hedman, G. Johansson, T. Bergmark, S. E. Karlsoon, J. Lindgren, and B. Lindberg, in: *ESCA—Atomic, Molecular and Solid State Structure Studied by Means of Electron Spectroscopy*, Nova Acta Regiae Societatis Sc en-turtum Apsaliensis Ser. IV, Vol. 20, Almqvist and Wichsells, Uppsala (1967).
99. P. W. Palmberg, *Anal. Chem.* **45**, 549A (1973).
100. C. C. Chang, in: *Characterization of Solid Surfaces*, (P. F. Kane and G. B. Larrabee, eds.), Plenum Press, New York (1974), p. 509.
101. M. P. Seah, *Surf. Interface Anal.* **1**, 91 (1979).
102. E. N. Sickafus, *Surf. Sci.* **100**, 529 (1980).
103. D. P. Griffis and R. W. Linton, *Surf. Interface Anal.* **4**, 197 (1982).
104. H. J. Mathieu and D. Landolt, *Appl. Surf. Sci.* **10**, 100 (1982).
105. H. H. Madden and G. C. Nelson, *Appl. Surf. Sci.* **11/12**, 408 (1982).
106. H. E. Bishop and J. C. Riviere, *J. Appl. Phys.* **40**, 1740 (1969).
107. C. J. Powell, in: *Quantitative Surface Analysis of Materials,* ASTM STP 643, (N.S. McIntyre, ed.), ASTM, Philadelphia (1974), p. 5.
108. L. E. Davis, N. C. MacDonald, P. W. Palmberg, G. E. Riach, and R. E. Weber, *Handbook of Auger Electron Spectroscopy*; 2nd ed., Perkin Elmer Corporation, Physical Electronics, Eden Prairie, MN, (1976).
109. P. M. Hall, J. M. Morabito, and D. R. Conley, *CRC Crit. Rev. Solid State Mater. Sci.* **8**, 53 (1977).
110. M. Keenlyside, F. H. Stott, and G. C. Wood, *Vacuum* **31**, 631 (1981).
111. H. Bubert, *Fresenius Z. Anal. Chem.* **314**, 237 (1983).
112. H. W. Werner, *Surf. Sci.* **47**, 301 (1975).
113. H. W. Werner, in: *Electron and Ion Spectroscopics. of Solids*, (L. Fiermans, J. Vennik, and W. Dekeyser, eds.), Plenum Press, New York (1978), p. 324.
114. P. Joyes, *J. Phys. (Paris)* **30**, 365 (1969).
115. G. Blaise and G. Slodzian, *J. Phys. (Paris)* **31**, 93 (1970).
116. Z. Sroubek, *Surf. Sci.* **44**, 47 (1974).
117. J. M. Schroeer, T. N. Rhodin, and R. C. Bradley, *Surf. Sci.* **34**, 571 (1973).
118. W. H. Griess and F. G. Rüdenauer, *Int. J. Mass Spectrom. Ion Phys.* **18**, 111 (1975).
119. M. Cini, *Surf. Sci.* **54**, 71 (1976).
120. J. Antal, *Phys. Lett.* **A55**, 493 (1976).
121. J. K. Norskov and B. I. Lundquist, *Phys. Rev. B* **19**, 5661 (1979).
122. C. A. Andersen and J. R. Hinthorne, *Anal. Chem.* **45**, 1421 (1973).
123. D. S. Simons, J. E. Baker, and C. A. Evans, Jr., *Anal. Chem.* **48**, 1341 (1976).
124. F. G. Rüdenauer and W. Steiger, *Vacuum* **26**, 537 (1976).
125. H. W. Werner, *Vacuum* **24**, 493 (1974).
126. F. G. Rüdenauer, W. Steiger, and H. W. Werner, *Surf. Sci.* **54**, 553 (1976).
127. A. E. Morgan and H. W. Werner, *Anal. Chem.* **48**, 669 (1976).
128. A. E. Morgan and H. W. Werner, *Microchim. Acta* **2**, 31 (1978).
129. M. A. Rudat and G. H. Morrison, *Anal. Chem.* **51**, 1179 (1978).
130. H. W. Werner, *Surf. Interface Anal.* **2**, 56 (1980).
131. H. W. Werner, *Mikrochim. Acta* **114/115**, 107 (1994).

132. D. P. Leta and G. H. Morrison, *Anal. Chem.* **52**, 514 (1980).
133. W. H. Griess, *Microchim. Acta* **1**, 335 (1981).
134. E. Zinner, *J. Electrochem. Soc.* **130**, 199L (1983).
135. P. R. Boudewijn and H. W. Werner, in: *Secondary Ion Mass Spectrometry SIMS V* (A. Benninghoven, R. J. Colton, D. S. Simons, and H. W. Werner, eds.), Springer, Berlin (1986), pp. 270–278.
136. H. Oechsner, in: *Thin Film and Depth Profile Analysis, Topics in Current Physics,* Vol. 37 (H. Oechsner, ed.), Springer Verlag, Berlin (1984), p. 63.
137. K. H. Müller, H. Oechsner, and T. Halden, *Fresenius Z. Anal. Chem.* **314**, 242 (1983).
138. H. H. Brongersma and T. M. Buck, *Surf. Sci.* **53**, 649 (1975).
139. J. A. van den Berg and D. G. Armour, *Vacuum* **31**, 259 (1981).
140. R. L. Erichson and D. P. Smith, *Phys. Rev. Lett.* **34**, 297 (1973).
141. S. B. Luitjens, Th. R. Verbeek, A. K. Algra, and A. L. Boers, *Surf. Sci.* **76**, L609 (1978).
142. J. Terzic, D. Cidric, and B. Perovic, *Surf. Sci.* **85**, 149 (1979).
143. H. H. Brongersma and T. M. Buck, *Nucl. Instrum. Meth.* **132**, 559 (1976).
144. E. Taglauer and W. Heiland, *Surf. Sci.* **47**, 234 (1975).
145. J. B. Malherbe, *Crit. Rev. Solid State Mater. Sci.* **19** (3), 129–195 (1994).
146. Dae Won Moon and Kyung Joong Kim, *Appl. Phys. Lett.* **62**, 3094 (1993).
147. E. Darque-Ceretti, M. Aucouturier, and A. Boutry-Forveille, *Surf. Interface Anal.* **18**, 229 (1992).
148. V. I. Zaporozchenko and M. G. Stepanova, *Prog. Surf. Sci.* **49**, 155–196 (1995).
149. G. Betz, *Surf. Sci.* **92**, 283 (1980).
150. G. Betz, J. Marton, and P. Braun, *Nucl. Instrum. Meth.* **168**, 541 (1980).
151. M. Opitz, G. Betz, and P. Braun, *Acta Phys. Acad. Sci. Hung.* **49**, 119 (1980).
152. Z. L. Liau, W. L. Brown, R. Homer, and J. M. Paate, *Appl. Phys. Lett.* **30**, 626 (1977).
153. P. R. Boudewijn, M. R. Leys, and F. Roozeboom, *Surf. Interface Anal.* **9**, 303–308 (1986).
154. S. Hofman, *Surf. Interface Anal.* **2**, 148 (1980).
155. J. F. Bresse, *Appl. Surf. Sci.* **66**, 825 (1993).
156. Y. Higashi, T. Marou, T. Tanaka, and Y. Homma, *Jpn. J. Appl. Phys. Part 1* **32**, 982 (1993).
157. I. L. Schwarzmann, M. G. Goldiner, and O. I. Plakhotnikov, *Int. J. Electron.* **73**, 977 (1992).
158. J. Morais, A. M. Machado, M. A. Sacilotti, and R. Landers, *Appl. Surf. Sci.* **44**, 161 (1990).
159. F. Matsunaga, H. Kakibayashi, T. Mishima, and S. Kawase, *Jpn. J. Appl. Phys. Part 1* **27**, 149 (1988).
160. T. Chasse, W. Heichler, J. Langhammer, and W. Zwanzig, *Crystal Res. Technol.* **22**, 1425 (1987).
161. K. Wittmaack, *Vacuum* **34**, 119–137 (1984).
162. E. A. Maydell in: *Quantitative Microbeams Analysis, Proceedings of the 40th Scottish Universities Summer School in Physics*, (A. E. Fitzgerald, B. E. Storey, and D. Fabian, eds.), Institute of Physics Publishing, Bristol and Philadelphia (1992), p. 329.
163. L. J. van Ijzendoorn, *Anal. Chim. Acta.* **297**, 55 (1994).
164. L. C. Feldman and J. W. Mayer, *Fundamentals of Surface and Thin Film Analysis*, North-Holland, Amsterdam (1986).
165. H. H. Andersen, *Appl. Phys.* **18**, 131 (1979).
166. U. Littmark and W. O. Hofer, *Nucl. Instrum. Meth.* **168**, 329 (1980).
167. K. Wittmaack, *J. Appl. Phys.* **53**, 4817–4820 (1982).
168. K. Wittmaack and D. B. Poker, *Nucl. Instrum. Meth. Phys. Res. B* **47**, 224 (1990).
169. K. Wittmaack, *Surf. Interface Anal.* **21**, 323 (1994).
170. W. Wach and K. Wittmaack, *Surf. Interface Anal.* **4**, 230 (1982).
171. P. R. Boudewijn, H. W. P. Akerboom, and M. N. C. Kempeners, *Spectrochimica Acta* **39B**, 1567 (1984).
172. S. M. Hues and P. Williams, *Nucl. Instrum. Meth. B* **15**, 206–209 (1986).
173. N. Trouche, E. Olivier-Rulliere, and J. P. Yonnet, *Conference Record of the 1991 IEEE Ind. Appl. Soc. Annual Meeting* **1**, 138 (1991).

174. A. Benninghoven, *Z. Phys.* **230**, 403–407 (1970).
175. R. Shimizu, *Appl. Phys.* **18**, 425–426 (1979).
176. M. P. Seah, J. M. Sanz, and S. Hofmann, *Thin Solid Films* **81**, 239–246 (1981).
177. H. Oechsner, in: *Festkörperprobleme*, Vol. 24, (P. Grosse, ed.), Vieweg, Braunschweig (1984), pp. 269–288.
178. A. Zalar, *Thin Solid Films* **193–194**, 258 (1990).
179. E.-H. Cirlin, Y.-T. Cheng, P. Ireland, and B. Clemens, *Surf. Interface Anal.* **15**, 337 (1990).
180. A. Zalar and S. Hofmann, *Surf. Interface Anal.* **12**, 83 (1988).
181. D. M. Drummer and G.H. Morrison, *Anal. Chem.* **52**, 2305–2310 (1980).
182. S. Hofman and J. M. Sanz, *Fresenius Z. Anal. Chem.* **314**, 1215–1219 (1983).
183. S. Storp and R. Halm, *J. Electron Spectrosc. Relat. Phenom.* **16**, 183–193 (1979).
184. R. G. Gossink, H. A. M. de Grefte, and H. W. Werner, *J. Am. Ceram. Soc.* **62**, 4 (1979).
185. H. W. Werner and N. Warmoltz, *J. Vac. Sci. Technol. A* **2**, 726–721 (1984).
186. G. Carter, B. Navinsek, and J. L. Whitton, in: *Sputtering by Particle Bombardment II*, (R. Behrisch, ed.), Springer, Berlin, Heidelberg (1983), pp. 231–269.
187. C. C. Chang, in: *Characterization of Solid Surfaces*, (P. F. Kane and G. B. Larrabee, eds.), Plenum Press, New York (1974), pp. 509–545.
188. K. Wittmaack, *Appl. Phys.* **12**, 149 (1977).
189. C. W. Magee, W. L. Harrington, and R. E. Honig, *Rev. Sci. Instr.* **49**, 477–485 (1978).
190. M. Lepareur, *Rev. Tech. Thomson-CSF* **12**, 225 (1980).
191. H. W. Werner, *Surf. Interface Anal.* **4**, 1 (1982).
192. K. Wittmaack and J. B. Clegg, *Appl. Phys. Lett.* **37**, 285 (1980).
193. V. R. Deline, *Nucl. Instrum. Meth.* **218**, 316–318 (1983).
194. A. Benninghoven, F. G. Rüdenauer, and H. W. Werner, *Secondary Ion Mass Spectrometry*, Wiley New York (1987), pp. 900, 1122.
195. G. A. Cooke, M. G. Dowsett, and P. Phillips, *J. Vac. Sci. Technol. B* **14**, 283 (1996).
196. G. A. Cooke, M. G. Dowsett, P. N. Allen, R. Collins, and K. Miethe, *J. Vac. Sci. Technol. B* **14**, 132 (1996).
197. V. V. Makarov and M. G. Dowsett, *Surf. Interface Anal.* **23**, 899 (1995).
198. P. C. Zalm and R. C. M. de Druif, *Appl. Surf. Sci.* **70/71**, 73 (1993).
199. P. C. Zalm, *Rep. Prog. Phys.* **58**, 1321 (1995).
200. M. G. Dowsett, R.D. Barlow, and P.N. Allen, *J. Vac. Sci. Technol. B* **12**, 186 (1994).
201. M. P. Seah and C. P. Hunt, *Surf. Interface Anal.* **5**, 33 (1983).
202. C. W. T. Bulle-Lieuwma, P. C. Zalm, and M. P. A. Viegers, in: *Proceedings of the Fourth Oxford Conference on Microscopy of Semiconducting Materials*, Institute of Physics, Conference Series, Vol. 75 (A. P. Cullis, and D. B. Holt, eds.), Adam and Hilger, Bristol, UK (1985), p. 123.
203. M. P. A. Viegers, C. W. T. Bulle-Lieuwma, P. C. Zalm, and P. M. J. Maree, in: *Mat. Res. Soc. Symp. Proc.,* Vol. 37, (J. M. Gibson and L. R. Dawson, eds.), Materials Research Society, Pittsburgh (1985), p. 331.
204. H. Kaiser and H. Specker, *Z. Anal. Chem.* **149**, 46 (1955).
205. F. Faupel, P. W. Huppe, K. Ratzke, R. Willecke, and T. Hehenkamp, *J. Vac. Sci. Technol. A* **10**, 92 (1992).
206. C. W. Magee and E. M. Botnick, *J. Vac. Sci. Technol.* **19**, 47, (1981).
207. C. W. Magee, *J. Vac. Sci. Technol. A* **1**, 901 (1983).
208. A. W. Czanderna, A. C. Miller, H. H. G. Jellinek, and H. Kachi, *J. Vac. Sci. Technol.* **14**, 227 (1977).
209. A. W. Czanderna and R. Summermatter, *J. Vac. Sci. Technol.* **13**, 384 (1976).
210. H. S. Wildman, J. K. Howard, and P. S. Ho, *J. Vac. Sci. Technol.* **12**, 75 (1975).
211. N. Tanovic, L. Tanovic, and J. Fine, *Nucl. Instrum. Meth. Phys. Res. B* **67**, 491 (1992).
212. J. Eymery, C. Senillou, J. C. Joud, and A. Chamberod, *Appl. Surf. Sci.* **47**, 127 (1991).

213. A. Zalar, E. W. Seibt, and P. Panjan, *Vacuum* **40**, 71 (1990).
214. N. Tanovic, L. Tanovic, and J. Fine, *Vacuum* **43**, 1177 (1992).
215. S. J. Simko, Yang-Tse Cheng, and M.C. Militello, *J. Vac. Sci. Technol. A* **9**, 1477 (1991).
216. K. Barla, D. Nicolas, R. Pantel, B. Vuillermoz, A. Straboni, and Y. Caratini, *J. Appl. Phys.* **68**, 3635 (1990).
217. O. Benkherourou and J. P. Deville, *Diffusion and Defect Data—Solid State Data, Part B (Solid State Phenomena)*, **19–20**, 593 (1991).
218. R. P. Frankenthal and D. L. Malm, *J. Electrochem. Soc.* **123**, 186 (1976).
219. J. S. Arlow, D. F. Mitchell, and M. J. Graham, *J. Vac. Sci. Technol. A* **5**, 572 (1987).
220. M. Nagasaki, R. Vasofsky, and H. F. Helbig, *J. Vac. Sci. Technol.* **16**, 151 (1979).
221. A. C. Miller and A. W. Czanderna, *Appl. Surf. Sci.* **4**, 481 (1980).
222. G. Betz, G. K. Wehner, L. Toth, and A. Joshi, *J. Appl. Phys.* **45**, 5312 (1974).
223. V. F. Rybalko, B. Ya Kolot, and Ya. Ma Fogel, *Sov. Phys. Tech. Phys.* **14**, 1290 (1970).
224. B. McCarroll, *J. Chem. Phys.* **46**, 863 (1967).
225. N. S. McIntyre, in: *Practical Surface Analysis*, (D. Briggs and M. Seah, eds.), Wiley, New York (1983), p. 407.
226. R. L. Chin and D. M. Hercules, *J. Catal.* **74**, 121 (1982).
227. R. L. Chin and D. M. Hercules, *J. Phys. Chem.* **86** (1982).
228. D. S. Zingg, L. E. Makovsky, R. E. Tischer, F. R. Brown, and D. M. Hercules, *J. Phys. Chem.* **84**, 2898 (1980).
229. M. Ramsteiner, J. Wagner, and P. Koidl, *Appl. Phys. Lett.* **59**, 3616 (1991).
230. A. C. Miller, A. W. Czanderna, H. H. G. Jellinek, and H. Kachi, *J. Colloid Interface Sci.* **85**, 244 (1982).
231. G. R. Sparrow and H. E. Mishmash, in: *Quantitative Surface Analysis of Materials*, ASTM STP 643 (N.S. McIntyre, ed.), American Society for Testing and Materials, Philadelphia (1978), pp. 164–181.
232. H. W. Werner and A. E. Morgan, *J. Appl. Phys.* **47**, 1232–1242 (1976).
233. B. M. J. Smets and T. P. A. Lommen, *Verres Refract.* **35**, 84 (1981).
234. H. D. Jonker, A. E. Morgan, and H. W. Werner, *J. Crystal Growth* **31**, 387–391 (1975).
235. A. E. Morgan, H. W. Werner, and J. M. Gourgout, *Appl. Phys.* **12**, 283–286 (1977).
236. P. A. Lindfors, R. W. Kee, and D. L. Jones, in: *Microelectronics Processing: Inorganic Materials Characterization*, (L. A. Casper, ed.), ACS Symposium Series 295, American Chemical Society, Washington (1986), pp. 119–143.
237. R. R. Olson, P. W. Palmberg, C. T. Holland, and T. E. Brady, in: *Practical Surface Analysis by Auger and X-ray Photoelectron Spectroscopy* (D. Briggs and M. P. Seah, eds.), Wiley , New York (1983), pp. 217–246.
238. A. Benninghoven, F. G. Rüdenauer, and H. W. Werner, *Secondary Ion Mass Spectrometry*, Wiley, New York (1987), pp. 995–1005.
239. A. M. Huber, G. Morillot, and A. Friederick, in: *Secondary Ion Mass Spectrometry SIMS IV* (A. Benninghoven, J. Okano, R. Shimizu, and H. W. Werner, eds.), Springer, Berlin (1984), p. 278.
240. A. M. Huber and G. Morrillot, in: *Secondary Ion Mass Spectrometry SIMS V* (A. Benninghoven, R. J. Colton, D. S. Simons, and H. W. Werner, eds.), Springer, Berlin (1986), p. 353.
241. J. B. Clegg and I. L. Gale, *Surf. Interface Anal.* **17**, 190 (1991).
242. P. R. Boudewijn and K. T. F. Janssen, *Fresenius Z. Anal. Chem.* **329**, 215–219 (1987).
243. SUPREM3. One-Dimensional Process Analysis Program, Technology Modeling Associates Inc.
244. M. Grasserbauer, G. Stingeder, E. Guerrero, H. Pötzl, R. Tielert, and H. Rijssel, in: *Secondary Ion Mass Spectrometry SIMS III* (A. Benninghoven, J. Giber, J. Laszlo, M. Riedel, and H. W. Werner, eds.), Springer, Berlin (1982), p. 321.
245. A. E. Morgan, T. Y. Chen, D. A. Reed, and J. E. Baker, *J. Vac. Sci. Technol. A* **1**, 1266 (1984).

246. C. W. Magee and K. G. Amberiadis, in: *Secondary Ion Mass Spectrometry SIMS V* (A. Benninghoven, R. J. Colton, D. S. Simons, and H. W. Werner, eds.), Springer, Berlin (1986), p. 279.
247. W. K. Hofker, *Philips Res. Rep. Suppl.* **8**, 1–121 (1975).
248. H. Rijssel, G. Prinke, K. Haberger, K. Hoffmann, K. Müller, and R. Hentelmann, *Appl. Phys.* **24**, 39 (1981).
249. B. Otterloo, private communication.
250. G. D. Davis, H. M. Clearfield, W. C. Moshier, and G. O. Cote, *Surf. Interface Anal.* **11**, 359 (1988).
251. A. Benninghoven, F. G. Rüdenauer, and H. W. Werner, *Secondary Ion Mass Spectrometry*, Wiley, New York (1987), p. 675.
252. A. Benninghoven, *Surf. Sci.* **35**, 427 (1973).
253. H. W. Werner, H. A. M. de Grefte, and J. van den Berg, in: *Advances in Mass Spectrometry*, Vol. 6 (A. R. West, ed.), Applied Science, Barking, Essex, UK (1974), p. 673.
254. W. Lange, M. Jirihowsky, and A. Benninghoven, *Surf. Sci.* **136**, 419 (1984).
255. E.-H. Cirlin, P. Ireland, S. Buckingham, and O. Wu, *J. Vac. Sci. Technol. A* **6**, 2631 (1988).
256. J. Colino, J. L. Sacedon, and J. L. Vincent, in: *Physics and Materials Science of High Temperature Superconductors II* (R. Kossowsky, B. Raveau, D. Wohlleben, and S. K. Patapis, eds.), Kluwer, Dordrecht, Netherlands (1992), p. 529.
257. J. B. Pallix, C. H. Becker, N. Missert, K. Char, and R. H. Hammond, *AIP Conf. Proc.*, Vol. 165, American Institute of Physics, Woodbury, New York (1988), p. 413.
258. J. Kirschner, K. Koike, and H. P. Oepen, *Phys. Rev. Lett.* **59**, 2099 (1987).
259. S. W. MacLaren, C. M. Loxton, E. Sammann, and C. J. Kiely, *J. Vac. Sci. Technol. A* **7**, 17 (1989).
260. U. Kaiser, G. P. Meeker, and J. C. Huneke, *J. Vac. Sci. Technol.* **6**, 1082 (1988).
261. H. Oechsner, in: *Analysis of Microelectronic Materials and Devices* (M. Grasserbauer and H. W. Werner, eds.), Wiley, Chichester, UK (1991), p. 493.
262. H. Oechsner, in: *Proc. VIII Int. Conf. Secondary Ion Mass Spectrometry* (A. Benninghoven, K. T. F. Janssen, J. Tuempner, and H. W. Werner, eds.), Wiley, Chichester, UK (1992), p. 95.
263. H. Oechsner, in: *Encyclopedia of Analytical Science*, Vol. 8 (A. Townsend, ed.), Academic Press, London (1995), p. 5014.
264. J. B. Pallix, C.H. Becker, and K.T. Gillen, *Appl. Surf. Sci.* **32**, 1 (1988).
265. C. H. Becker, S. G. MacKay, and D. G. Welkie, *J. Vac. Sci. Technol. B* **10**, 380 (1992).
266. R. Jede, H. Peters, G. Dunnebier, O. Ganschow, U. Kaiser, and K. Seifert, *J. Vac. Sci. Technol. A* **6**, 2271 (1988).
267. N. Menzel and K. Wittmaack, *Mat. Sci. Engr. B* **12**, 91 (1992).
268. J. van den Berg, J. van Oijen, and H. W. Werner, *Anal. Chim. Acta* **297**, 73 (1994).
269. A. Briggs, *An Introduction to Scanning Acoustic Microscopy*, Oxford University Press, Royal Microscopy Society, London (1985).
270. F. G. Rüdenauer, in: *Secondary Ion Mass Spectrometry SIMS III* (A. Benninghoven, J. Siber, J. Laszlo, M. Riedel, and H. W. Werner, eds.), Springer, Berlin (1982), p. 2.
271. R. J. Young, *Vacuum* **44**, 353 (1993).

Index